THE
ENCYCLOPEDIA OF
# Natural Insect &
# Disease Control

THE
ENCYCLOPEDIA OF

# Natural Insect & Disease Control

**The most comprehensive guide to protecting plants—vegetables, fruit, flowers, trees, and lawns—without toxic chemicals**

Edited by Roger B. Yepsen, Jr.

Rodale Press, Emmaus, Pennsylvania

Printed in the United States of America

Book and jacket design by Anita Groller.
Jacket photo by T. L. Gettings.

**Library of Congress Cataloging in Publication Data**
Main entry under title:

The Encyclopedia of natural insect and disease control.

    Rev. ed. of: Organic plant protection. c1976.
    Includes index.
    1. Plants, Protection of. 2. Organic gardening.
3. Garden pests. I. Yepsen, Roger B. II. Organic plant protection.
SB974.E53 1984        635.'0494        83-24643
ISBN   0-87857-488-3   hardcover

**Distributed in the book trade by St. Martin's Press**

     10            hardcover

This book is a revised and reordered edition of *Organic Plant Protection*, published in 1976; the earlier book borrowed in turn from *The Organic Way to Plant Protection* (1966), edited by Glenn F. Johns and the staff of *Organic Gardening* magazine.

Contributors to *The Encyclopedia of Natural Insect and Disease Control* include Jeff Cox, Tony Decrosta, Kim Foreman, Mark Kane, Mike Lafavore, Sara Price, and David Riggle of *Organic Gardening*, Diane L. Matthews of the Rodale Research Center, and consulting entomologists Tim White (East Coast) and Wayne S. Moore (West Coast). Barbara Yoder compiled the chart of insect emergence times. William Olkowski, Ph.D., helped with the chart of controls and strategies.

Contributors of line drawings include James R. Baker and the North Carolina Agricultural Extension; Roy Rings and the Ohio Agricultural Research and Development Center; Dr. Francisco Pacheco of Centro de Investi-gaciones Agricolas del Noroeste, Mexico; Metro Pest Management Consultants; J. W. Caslick and D. J. Decker's "Control of Wildlife Damage in Orchards and Vineyards," from Cornell University; Tom Quirk, from *Organic Gardening* magazine; and the U. S. Department of Agriculture.

# Introduction

This book is a guide, arranged in encyclopedic form, to protecting plants from bugs, diseases, and the environment at large. It is addressed to gardeners, small-scale orchardists, and homeowners who wish to avoid using chemical toxins.

These pages spend few words on the hazards and expense of chemical controls. The disadvantages are well known. Increasingly, people grow their own food to avoid chemical-laced produce in the supermarket; they filter their water or buy it bottled from certified sources to avoid agricultural contaminants; they see the futility in chemical agriculture in the face of insects' astounding adaptability (while in 1950 less than twenty pest species were known to be resistant, *most* pests in California today are considered resistant to at least one chemical).

Since an earlier version of this book was published (as *Organic Plant Protection,* in 1976), the organic grower's arsenal of methods and materials has grown considerably. Research here and abroad continues to identify new agents for biological control—insects, mites, molluscs, nematodes and pathogens—and many firms have sprung up to produce and market them. Some two dozen beneficial organisms have become available in the past decade, and more need only government approval. Although these organisms were studied with large-scale applications in mind, suppliers find that backyard growers make up an ever-increasing share of their buyers.

You don't have to spend money to benefit from biological control, of course. The biology around you is already operating under complex, interlocking systems of checks and balances. You, as gardener or orchardist or landscaper, interfere with these systems, but you cannot help yourself. For example, to grow a row of high-yield tomatoes is to invite trouble: these plants are native to the tropics, not your backyard;

they were developed with fat, early-ripening tomatoes in mind, rather than hardiness; and the tomatoes are planted in close proximity, so that a pest or disease may sweep from one to the rest. Nevertheless, you can encourage naturally occurring control.

Nonchemical sprays and dusts continue to be important to organic growers. This area has been less dynamic than biological control, because less research is devoted to it. And as growers gain more sophistication with balancing life in the garden, they realize even the tamest of sprays—water—can upset a happy status quo.

A less invasive means of managing bugs is use of repellent plants to ward them off. Marigolds, for example, have long been sown as a barrier to nematodes. Or weeds and spare crops can be grown as trap crops—appealing foods that divert would-be pests from valuable crops. Gardeners grow dill near tomatoes to keep the tomato hornworm on the herb and out of trouble. Most of these methods of interplanting are based on informal observation over many years, but one grower's success may be another's waste of time; ideally, repellent and attractive plantings should be undertaken in a spirit of experimentation.

Biological controls aren't ever likely to supplant cultural controls of hoeing, mulching, pruning, weeding, and so on. Again, experimentation is in order. Clean cultivation and mulching work wonders in some situations, but bring out the bugs in others.

Aside from the venus fly trap and a few others, plants seem pretty passive. But in fact they have elaborate defense mechanisms that are not yet fully understood. Well-nourished plants are better able to defend themselves than can plants suffering from a lack or a glut of nutrients.

Some methods of pest and disease control exact no extra labor, just planning. As you browse the seed and nursery stock catalogs in

## CONTROLS AND STRATEGIES

A graduated list, from those that interfere least with garden or yard or orchard ecology to those that interfere most. As a general rule, go no lower on the list than you must. Capitalized words are entries in this book.

---

### CULTURAL CONTROLS

NUTRITION, including soil building and watering
RESISTANT VARIETIES
INTERPLANTING with repellent plants, among or surrounding valued plants
Enhancement of naturally occurring BIO-LOGICAL CONTROL, by providing food, alternate hosts, and shelter
Crop rotation
TIMED PLANTING
MULCH and ground covers
TRAP CROPS of plants that appeal to pests
CLEAN CULTIVATION, pruning

### PHYSICAL AND MECHANICAL CONTROLS

Handpicking of pests
Physical barriers
TRAPS: light traps, PHEROMONE traps, and sticky traps

### BIOLOGICAL CONTROL

Augmenting BIOLOGICAL CONTROL with introduced PREDATORY INSECTS, PARASITIC INSECTS, and pathogens (MICROBIAL CONTROL)

### SPRAYS AND DUSTS

Repellent SPRAYS AND DUSTS (sprayed water, homemade preparations such as those based on garlic)
Strongly or broadly toxic SPRAYS AND DUSTS, such as insecticidal SOAP, dormant oil, DIATOMACEOUS EARTH, PYRETHRUM, ROTENONE, RYANIA, and SULFUR

---

winter, look for resistant and tolerant varieties; older ones may lose their special qualities over the years, but new ones are being developed all the time. Each growing season, bugs appear on time to intercept their favorite crops; by planting earlier or later than usual, you may be able to side-step them. Another way to disappoint bugs and diseases is to rotate crops from year to year.

The accompanying table of controls and strategies can help you to choose appropriate measures—those that will do the job without unduly upsetting the local ecology.

# How to use this book

Keep this list of controls in mind as you flip through the book to the pest or disease that troubles you. Most problems will be found under the entry for the endangered plant; insects are discussed first, and diseases second, both arranged in alphabetical order. If a problem is of only occasional or regional importance to a plant, it may be listed and cross-referenced at the end of the pest or disease section. Some of the more notorious pests and diseases are treated in their own entries. And general entries cover FLOWERS, FRUIT TREES, LAWNS, and TREES.

To aid you in identifying a plant's problem, the text is illustrated with many line drawings. In addition, more than one hundred pests and diseases are pictured in a color section, and are so noted by the word "color" in brackets [color]; these color photos are arranged alphabetically.

For further help in determining what ails your plant, read the following pages on insect and disease identification.

An important part of this book is the index. If you want to know all you can about the use of garlic as an insecticide, for instance, the index will lead you to the pertinent references scattered through the book. Index words set in capitals (such as JAPANESE BEETLE) are encyclopedia entries.

The appendix lists plants known to have insecticidal powers that might make them useful as sprays or dusts. An up-to-date list of natural sprays and dusts and biological controls can be had from the Organic Gardening Readers Service, 33 East Minor Street, Emmaus, PA 18049. Please include a business-size envelope, stamped and self-addressed.

# Insect identification

Insects may be identified either by their appearance or by their feeding habits. See entries for those in capital letters.

## APPEARANCE

### Winged insects with complete metamorphosis

**BEETLE,** curculio, weevil. Two pairs of wings; hard forewings cover membranous hindwings. Chewing mouthparts. Larvae are grubs or wireworms, with three pairs of legs. Chewing mouthparts, curculio and weevil with long snout.

**LACEWING,** antlion. Two pairs of clear, similar wings, long thin antennae. Larvae unlike adults. Predators of insects. Butterfly, moth. Broad wings. Larvae are caterpillars. Adults are not plant pests.

**WASP,** bee, sawfly, ant. Two pairs of clear wings, forewings larger. Larvae are maggots or caterpillars. Usually chewing mouthparts, adults may be predators or parasites of insects. Pollinators.

**FLY,** midge. One pair of membranous wings. Looks like housefly. Larvae are maggots. Sucking mouthparts. Some pollinators.

### Winged insects with simple metamorphosis (young resemble adults)

**BUG.** Two pairs of wings; forewings cover membranous hindwings. Triangle on back. Sucking mouthparts.

**SCALE, MEALYBUG, WHITEFLY, APHID,** leafhopper, psyllid. Two pairs of membranous wings (on some). Soft bodies. Some scales immobile. May produce honeydew, with sooty mold to follow. Sucking mouthparts.

**THRIPS.** Two pairs of slender, fringed wings. Sucking mouthparts.

**EARWIG.** Pincers on rear end. Chewing mouthparts.

**GRASSHOPPER,** cricket, **PRAYING MANTIS.** Chewing mouthparts.

### Wingless insects

**Springtail.** Flealike. Chewing mouthparts. Nymphs resemble adults. See SCALE, MEALYBUG, APHID, SPIDER MITE.

### Noninsects

**SPIDER MITE, SPIDER.** Four pairs of legs (three on some mite nymphs). Sucking mouthparts.

**MILLIPEDE,** centipede, **SYMPHYLAN.** Manylegged. May concentrate on decaying plant material.

**SNAIL AND SLUG.**

**NEMATODE.** Tiny, nearly invisible roundworms in and on belowground plant parts.

**SOWBUG,** pillbug. Humped, overlapping scales. Pillbugs roll up into ball when disturbed. Rarely a problem.

## FEEDING HABITS

### Chewing aboveground

| | | |
|---|---|---|
| armyworm | cricket | leaf skele- |
| bagworm | curculio | tonizer |
| bee | cutworm | leaftier |
| billbug | earwig | sawfly |
| blister beetle | flea beetle | snail and slug |
| bollworm | fruitworm | springtail |
| budworm | grasshopper | webworm |
| bug | hornworm | weevil |
| cankerworm | leafminer | |
| caterpillar | leafroller | |

### Sucking aboveground

| | | |
|---|---|---|
| aphid | mite | tarnished plant |
| chinch bug | psyllid | bug |
| harlequin bug | scale | thrips |
| lace bug | spittlebug | treehopper |
| leafhopper | squash bug | whitefly |
| mealybug | | |

### Feeding belowground

| | | |
|---|---|---|
| billbug | root aphid | symphylan |
| corn root aphid | root borer | white grub |
| corn rootworm | root maggot | wireworm |
| onion maggot | rootworm | woolly apple |
| nematode | root weevil | aphid |

### Borers

Various borers by name, plus:

| | | |
|---|---|---|
| apple maggot | European | spruce bud- |
| bark beetle | apple sawfly | worm |
| carpenterworm | melonworm | tomato fruit- |
| cherry fruit- | oriental fruit | worm |
| worm | moth | white pine |
| codling moth | pickleworm | weevil |
| corn earworm | plum curculio | |

# Disease identification

An ailing plant may be the victim of two or more diseases, complicating your diagnosis. Be certain that symptoms aren't caused by insects; for example, a stippled pattern on foliage may be caused by leafhoppers; aphids can cause leaves to curl; wilted foliage may be the work of squash bugs.

Diseases are classified as either parasitic (caused by organisms) or nonparasitic (caused by environmental problems, including pollution and nutritional imbalance). A 10× magnifying glass will help you to identify symptoms.

Once you diagnose the disease, consider how it can be prevented next year. Diseases tend to recur, their virulence depending on weather and other factors, unless plants are relocated. Perhaps a resistant variety will get you past a disease.

See DISEASES.

# ACEROLA

**Root-knot nematodes** can cause acerola considerable damage, especially when grown in soil lacking in organic matter. A deep, year-round mulch encourages the growth of a fungus that preys on the pests. If problems become severe, it may be necessary to graft plants onto resistant roots. See NEMATODE.

**Scale** may attack this fruit in some areas; a dormant spray can be used in serious cases. See SCALE.

# AIR POLLUTION

Several billion dollars worth of crops may be lost each year to air pollution—primarily ozone, sulfur dioxide, and nitrogen dioxide. Gardens, ornamentals, and wild plants are also harmed, although this damage has not been assessed. Only recently have experimental data been collected, and Environmental Protection Agency cutbacks mean that the full impact of air pollution on plant life won't be known for some time.

Leaves may be bronzed by ozone in the air. Ozone is a gas; much of it may be formed by the effect of ultraviolet light on exhaust emissions. Lower, older leaves are the first to show symptoms: look for small dead patches on the top surface. This condition is aggravated by periods of heat and much sun. New, resistant varieties of bean may appear on the market in response to ozone trouble.

# ALLELOPATHY

The effect on one plant of a toxin produced by another plant. Some plants are thought to suppress the growth of weeds with chemicals released into the soil.

# APHID

A great number of aphid species [color] are of economic significance to growers. There are violet aphids, green peach aphids, mint aphids, blackberry aphids, and dozens of other species named for a food preference.

The aphid is a small, soft-bodied insect distinguished by its pearlike shape, long antennae, and pair of cornicles—tubelike appendages that project from the back end. A defensive secretion is given off by these tubes. A winged aphid holds its wings vertically when at rest.

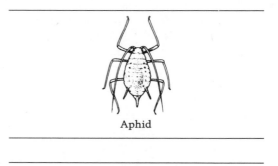

Aphid

Winged aphid

Aphids suck plant sap, and cause withering of foliage and a general loss of vigor. Excess sugars and sap emitted from the anus are known as honeydew. This substance may make leaf surfaces sticky, and can support a black mold that blocks light from leaves.

Aphids may carry diseases from plant to plant. For example, aphids transmit beet western yellows virus, and mosaic diseases of tomato.

Several parthenogenetic generations (produced by unfertilized females) of female aphids may be born alive, rather than as eggs, in a season. After one or two generations, winged forms are born and fly off to other plants. Toward the end of the season, males and females mate to produce eggs that overwinter. This method of reproduction can generate great numbers of aphids within a short time.

## CONVENTIONAL CONTROLS.

Aphids respond positively to elements in plants, especially nitrogen. In other words, the more nitrogen fertilizer you put on the soil, the happier the aphids. In particular, nitrogen has been found to favor bean, cabbage, and cotton aphids. Cultural controls include planting as far as possible from infested areas, clearing the garden neighborhood of aphid host plants (especially plantain, or bindweed and lamb's-quarters around orchards bothered by the green peach aphid), and seeing to it that plants are properly nourished and watered or irrigated as necessary.

Growers have come up with a great number of home remedies for aphid trouble. The simplest of all is to gently rub leaves between thumb and forefinger. These pests can be repelled by growing nasturtiums between vegetable rows and even around fruit trees. Other companion plants are garlic, chives, and other alliums, and coriander, anise, and petunias. Their effectiveness can be amplified by crushing a leaf now and then. As for sprays, aphids can be washed from the plants with a forceful jet of water. A tobacco water spray can be made by soaking tobacco stems, available from florists and seedsmen, in warm water for 24 hours. Dilute the brew to the color of weak tea and syringe it on the foliage, being careful to hit the undersides of the leaves. Or you can just buy a tobacco extract and follow the directions on the package. Try mincing shallots or green onions in a food chopper for a spray to keep aphids off roses. A spray recipe from England calls for three pounds of rhubarb leaves, boiled a half hour in three quarts of water and mixed when cool with one quart of water in which an ounce of soap flakes has been dissolved. Elder leaves can be substituted for the rhubarb leaves, with the added benefit of discouraging mildew on roses. Safer Agro-Chem's insecticidal soap works well against aphids and other soft-bodied insects. An exotic spray is made by boiling 3¼ ounces of quassia chips and 5 teaspoons of larkspur in 7 pints of water until the mixture is reduced to 5 pints. A spray of strong limewater will also take care of aphids. The pests are vulnerable to neem extract and to dusting with diatomaceous earth.

On apple and pear trees, aphids can be discouraged by repruning a quarter of the length of leading shoots in July, and a like length of the side shoots after the second leaf (not counting the little leaves around the base). Pruning shouldn't be necessary every year.

Aphids are attracted to the color yellow, and this can be used against them. Bright yellow plastic dishpans make a good trap when filled with slightly soapy water and set near infested or vulnerable plants. Or paint a sheet of plywood or Masonite bright yellow and coat it with a commercially formulated goo such as Tanglefoot. Traps can forewarn of insect troubles to come: winged aphids will usually appear in traps before they hit nearby crops.

Winged aphids won't land on low plants if sheets of aluminum foil are placed around the base as a mulch. Apparently the reflection of the sky confuses them—they don't know which way is up. This simple method works against other small flying pests, too.

## THE ROLE OF ANTS.

Ants feed on honeydew, and tend aphids like cows. Control is made much more difficult when ants distribute aphids from plant to plant. Ants stroke and tap the belly of aphids to milk them of the honeydew.

(This must be a very delicate process: Darwin tried in vain to coax drops from aphids by stroking them with fine hairs.) In fall, ants carry aphid eggs into their nests, to be carried back out in spring and set on plants. To keep their charges from straying, ants may chew off aphids' wings. Thorough control of aphids often requires doing something about the ants as well. These aphid-tenders are best kept from trouble by setting up barriers. Ants are loath to cross lines of bone meal or powdered charcoal. A band of cotton, made sticky with a commercial goo, can be wrapped around the bases of some plants and trees. Ants are sometimes drawn to trees that are leaking sap, and the trees can be made less attractive by cleaning out the wound and coating it with a dressing. If ants are carrying aphids into your house, track them to find where they are getting in and squeeze the juice of a lemon into the hole or crack. Then, slice up the lemon and put the peeling all around the entrance.

## IN THE GREENHOUSE.
Greenhouse growers can harvest their own aphid predators and parasites. Look for these beneficial bugs whenever you find large congregations of aphids, by using a magnifying glass and checking both sides of leaves for eggs, larvae, and adults. The ladybug [color] is the best-known aphid enemy. Its eggs are bright orange; the larva is shaped like an alligator and bears short bristles. Parasitic wasps may have laid their eggs in aphids; collect dead, discolored pests and allow them to incubate as described below. The eggs of the beneficial syrphid fly (called flower or hover fly as well) [color] occur singly and are white and sausage shaped; the eggs hatch into predatory maggots.

You can rear beneficial insects simply by placing aphid-infested foliage in a plastic bag, aphids included. Beneficial larvae likely will emerge in a few days, and can be picked up with the tip of a small paintbrush as they crawl about the inside of the bag.

## ROOT APHIDS.
Aphids and mites damage bulbs by sucking out nutrients and causing wounds through which fungal and bacterial decay organisms can enter. Badly infested bulbs produce weak growth, and leaves are stunted and yellowish with brown tips. Those flowers that mature are usually small, streaked, and off-color. For control methods, see ASTER.

## BIOLOGICAL CONTROL.
Naturally occurring controls, if permitted to operate, will often keep aphids from causing intolerable damage. Find a crowd of aphids, and you'll likely spot one or more examples of biological control. Ladybug adults and larvae increase in number to meet infestations. Ladybugs can be purchased from a number of suppliers. Other important predators include soldier bugs, damsel bugs, bigeyed bugs, pirate bugs, spiders, assassin bugs, and syrphid flies. Predatory lacewings are commercially available; see LACEWING. Gall midges are important natural enemies of aphids. Larvae are small maggots, bright orange in color. A maggot bites an aphid in the leg, paralyzes the pest, then drains it. The shriveled victims can be seen on plants. The midges may not appear early enough in the season to control aphids at first. The pale larvae of syrphids (or flower flies) hold aphids aloft as they eat.

The University of California imported the ladybug *Exochomus quadripustulatus* from Europe in 1921, and it has since caught on as an important predator of the woolly apple aphid. A parasite, *Aphelinus mali*, was introduced to West Coast orchards in 1929, and with the above ladybug, reduced populations of the woolly apple aphid to manageable levels. Perennial apple canker, spread by the aphid, was eradicated. The widespread use of DDT and other hydrocarbons in the late 1940s set these beneficials back so seriously that both aphid and canker returned. But *A. mali* has persisted and is now established in countries all over the world.

Wasp laying egg in aphid

In California the walnut aphid is kept in check by the parasite *Trioxys pallidus*, brought to North America from Iran in the 1960s. A fungus, *Entomorphthora exitialis*, has been distributed in California to help control the spotted alfalfa aphid.

Aphids are kept in check by naturally occurring diseases, and you can help spread a fungal disease among healthy, heartily eating aphids. Look for darkened or fuzzy aphids and collect as many as you can. Mash them (or mix them in a blender that will not be used for food preparation) with a little water, then strain through a coffee filter, add a bit of liquid soap to improve the consistency, and spray. Diseases flourish among aphids in cool and moist weather. Britons can buy the aphid disease *Verticillium lacanii*, but it is not available yet in North America.

You might also benefit by gathering parasitized aphids from wild plants and then distributing them among garden or greenhouse plants. These pests are called aphid mummies [color] for their enlarged, rigid appearance. Another naturally occurring control is rain—hard, driving rains can crush large numbers of aphids. See BUG JUICE. Damp weather favors development of aphid diseases.

# APPLE

Producing a good harvest of apples isn't easy. Even orchardists who blitz their trees with carefully timed sprays have trouble: a pamphlet on chemical disease control in the orchard put out by the Indiana Cooperative Extension Service lists four important tips to remember, of which number four is, "If all else fails, fruit trees make excellent firewood."

But such pessimism is unnecessary if you follow a program of cultural and biological control, along with judicious dormant-oil spraying. You should be able to bring an excellent crop of apples through the season, without poisonous herbicides, pesticides, or fungicides. Chemically sprayed apples may be flawlessly complected, but then so was that famous prop in *Snow White.*

In winter or first thing in spring, scrape off all loose bark with a paint scraper, old hoe, piece of saw blade, or part of a mower blade. This will destroy overwintering eggs and pupae, and permit more thorough coverage with dormant spray. Scrape carefully so that the blade doesn't lay open the living wood. Put a sheet or piece of canvas under the scene of activity to catch all the scrapings, which should be either burned or carted away to the compost pile.

Old decaying wounds, resulting from broken limbs, poor pruning cuts, and woodpeckers, are a favorite hiding place for codling moth larvae and other destructive pests. Dig out as much wet, dark wood as possible. Then paint the newly exposed, healthy wood inside with tree wound dressing.

The "cloverleaf method" is used to open up a tree so that air can circulate through the foliage; dry foliage is less prone to fungi that get their start on wet surfaces. Prune off all dead, split, or broken branches and stubs cleanly at the point of origin, leaving a slight stub (less than ⅛ inch long) to promote healing. Any running cavities and stubs that have rotted back into the tree should be cleaned out with a chisel to solid, live wood. This can be carried out in almost any season, except in wet weather. Trim the edges of cavities back to live bark and bring the cavity to a point at top and bottom, in a vertical ellipse; a sharp knife or straight chisel is the best tool for this. All cuts over two inches in diameter should be painted over with tree wound dressing. It's usually not a good idea to fill cavities with cement—the tree may look neater, but the filling will hide any further decay.

The danger of bacterial diseases such as fire blight can be minimized by cutting out cankers and diseased limbs during winter, and rubbing off any suckers and water sprouts while they are still very small. Bitter rot, blister canker, black rot, and blotch fungi also live in old cankers, and attack the tree through new open wounds. The bark on old cankers should be removed to healthy tissue, and the wound must be painted over promptly.

Many organic fruit growers rely on dormant-oil sprays to suffocate various pests; see FRUIT TREES for a description of a comprehensive spray program.

## Insects

A number of **aphid** species [color] may damage various parts of the tree. The yellow green apple aphid, trimmed with black, specializes in sucking the life from terminals all season. Leaves may become curled. Black sooty mold tends to grow on the honeydew secretions of this pest. The shiny black eggs spend the winter on tree bark, and a thorough going-over with the scraper in early spring should take care

Aphid

of many of them. Don't scrape so hard as to remove anything but loose bark. Spray with dormant oil to catch most of those you miss. It also helps to remove suckers, and some gardeners claim excellent results when winter-hardy onions are planted around trees.

Another important species, the rose-colored rosy apple aphid, feeds on buds and opening leaves. If sufficient numbers are present, the leaves will curl around and protect the pests, and fruit will be stunted. Much honeydew is secreted. A number of beneficials—syrphid flies, ladybugs, lacewings, and parasitic wasps —are of much help in keeping down populations of this pest, but aphids go relatively unbothered in spells of cold and wet weather. Spray with dormant oil to kill overwintering eggs. In spring, before the leaves curl severely, blast the tree with water or soapy water.

The woolly apple aphids [color] appear as white, cottony masses on the tree trunk and limbs. Large warty growths occur where aphids have been feeding several years. This aphid also occurs on the roots, from which it migrates to the trunk and branches. Trees decline in health under heavy, prolonged infestations. Use a dormant-oil spray, or a thorough soapy water spray on trunks and limbs in summer as needed. Aphids on roots are difficult to control; try soaking the base of the trunk with a soapy spray. Several tiny parasitic wasps are important natural control agents.

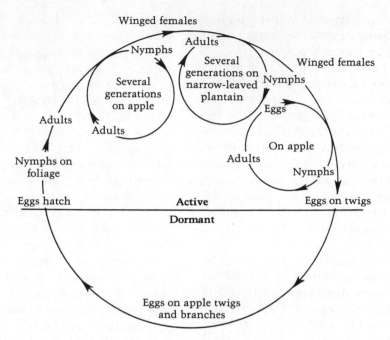

Life history of the rosy apple aphid

Reprinted, by permission, from Pyenson, *Fundamentals of Entomology and Plant Pathology*
(fig. 3.10), AVI Publishing Co., PO Box 831, Westport, CT 06881.

East of the Mississippi, the **apple curculio** injures young fruit by puncturing it. Unlike the marks left by the plum curculio, these punctures are not crescent-shaped. This is a brown snout beetle, smaller than the plum species and distinguished by four humps on its back. There is one generation a year. Eggs are deposited in young apples, and larvae mature within drops falling in June and in mummified fruit on the trees. Control measures are similar to those for the plum curculio (see PLUM).

The **apple maggot** [color], also known as apple fruit fly or railroad worm, makes brown tunnels or burrows inside apples. The unsuspecting grower may discover the presence of this pest the hard way—by biting into an infested apple. Maggots cause slight depressions on the surface and tiny holes where they emerge,

but these external symptoms are often so slight as to go unnoticed. After an infested apple falls or is picked, the flesh may break down into a brown pulpy mess.

The adult apple maggot is a fly, similar to the common housefly. In orchards throughout the Northeast and in northern parts of the Midwest, most serious infestations usually occur in July. Flies lay their eggs in the flesh of apples (usually sweet or subacid, thin-skinned varieties that ripen during summer or early fall), and legless white maggots develop in the flesh of the fruit. The insect passes winter in the soil in its pupal stage.

Collect dropped apples to reduce the population of apple maggots. Dropped fruit of summer varieties should be collected twice a week, and that of later varieties, once a week. These apples can be fed to livestock, dumped into water to kill the maggots, or even made into cider if not too badly damaged.

You can catch this pest in the fly stage, before it has a chance to cause trouble, with a simple baited trap. Mix up one part blackstrap molasses or malt extract to nine parts water, add yeast to encourage fermentation, and pour the beery liquid into wide-mouthed jars. Once fermentation subsides, hang the jars up in the trees. To monitor apple maggot fly populations, use a bait made of 2 teaspoons household ammonia and ¼ teaspoon soap powder mixed in a quart of water. Fill quart fruit jars with the bait and hang them shoulder-high on the outside of the sunny sides of the trees. Ten or so jars distributed throughout the orchard should give a good picture of maggot fly activity. Examine the traps every two or three days to count flies, and renew the bait if necessary; bait should be changed once a week or when diluted by rain.

One of the most effective homemade traps is nothing more than a piece of plastic fruit made sticky with Tanglefoot or Stikem, commercially available compounds that foul up the feet of insects.

If you have problems with the apple fruit fly or its relatives—the blueberry maggot fly, cherry fruit fly, and black cherry fruit fly— hang a sticky plastic orange or two in each tree, on the south side. The crop should be cleaner, although the results might not be so spectacular the first season because of a high initial infestation. To help reduce such an infestation, dropped fruit should be picked twice a week and burned; burn infested fruit or feed it to animals.

If apple maggots have consistently been a big problem, the answer may lie in growing tough-skinned later varieties, such as Winesap, Jonathan, and York Imperial.

The **apple red bug** [color] (along with the false apple red bug) is an active red insect about ¼ inch in length, found in the north central and northeastern United States. The eggs spend winter in bark crevices. Hatching occurs before blossoming, and the bugs mature in June. When the bugs feed on terminal leaves, distortion results. Feeding on fruit produces dimples or a series of small russet scars. Foliage feeding doesn't cause significant damage. A delayed spray of superior dormant oil should take care of these pests.

The **apple seed chalcid** is a tiny wasp active in the northeastern United States. It lays eggs in apples and related fruit. The tiny larvae feed on seeds, giving the apple a dimpled appearance similar to that caused by plant bugs. Destroy infested fruit to keep this minor pest at bay.

Green **buffalo treehoppers** [color] and other treehoppers have a distinctive triangular shape with short, stout horns at each shoulder. As with the periodical cicada, damage is caused when these insects puncture the bark to deposit their eggs, leaving double rows of crescent-shaped slits. Young trees are more susceptible than old trees to severe infestations, and may be stunted. The slits made when the treehopper lays the eggs can give entry to a variety of diseases. There is one generation a year. Egg laying occurs in September and October. Eggs overwinter in twigs. Nymphs hatch in spring and drop to the ground to feed on plants there.

If treehoppers are a problem, do not plant alfalfa, sweet clover, or other legumes as cover crops. Keep bindweed out of young orchards. Treehoppers are more of a problem when these favored plants are available to them. A thorough dormant spray will kill many of the overwintering eggs.

**Cankerworms** [color] are small, striped measuring worms, or inchworms. They are either black or green and appear in one generation a year. Cankerworms hatch from eggs laid on the tree in late fall or early spring by wingless moths. The male moths are gray and have a wingspan of about one inch.

After fall frosts, the fall cankerworms leave their pupal stage in the ground and the wingless females crawl up tree trunks to deposit their gray eggs on the trunk, branches, or twigs. The worms appear just in time to feed on unfolding foliage in spring; they are colored brown on top and green below, and are marked with stripes along the length of their bodies. You will likely see them hanging from trees on fine silk threads if you jar the limb.

These threads serve to complicate control of the worm. When winds catch a thread, the attached cankerworm may be lofted to a neighboring tree. This thwarts growers who band tree trunks with a sticky material to keep the females from making their egg-laying missions each fall, as one infested tree may spread trouble to those near it. Still, a sticky band around the trunk is the most effective means of control. Before applying the sticky compound, whether commercial or homemade, you should first lay down a layer of cotton batting or heavy paper to protect the tree bark. To catch this species of cankerworm, the bands should be applied in mid-October. Tanglefoot is a commercial preparation made for snaring pests. Or you can stir up your own goo from pine tar and molasses, or resin and oil. In any case, the sticky band should be renewed from time to time to ensure its effectiveness.

You can also spray, safely and sanely, for cankerworms; see *BACILLUS THURINGIENSIS.* This biological control is sprayed in April or May. Jar the limb. If the worms drop down on silk, apply BT. Repeat when live worms are present.

As for the spring cankerworm, the moths appear early in spring but are otherwise similar in habit to the fall species. The worms are green, brown, or black, and may show a yellow stripe. They have two pairs of prolegs, while the fall

cankerworm has three. Control as for the fall species, except that sticky bands should be applied in February to catch the wingless females.

The larva of the **codling moth** [color] is distinguished from other caterpillars by its color (white tinged with pink), its habit of tunneling directly to the core of the apple, and its appetite. Found in all apple-growing areas, this worm is considered one of the most serious apple pests. Fortunately, there are a number of effective ways of saving the crop without resorting to the poisons that many growers have accepted as a necessary part of fruit growing.

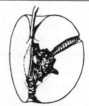

Codling moth larva in fruit

A hundred years ago, orchards were the scene of huge bonfires in June evenings. Growers theorized that because moths were attracted to light, they would be drawn to their deaths. Today, one of the best organic controls involves banding tree trunks in spring with strips of corrugated cardboard to draw larvae looking for a place to spin their cocoons. The bands should be wrapped in several thicknesses, and can be wired or tied on. The exposed ridges on corrugated cardboard must be at least 3/16 inch wide, and should face toward the tree; otherwise, the larvae will not form cocoons. Remove the band once a week in warm weather (once every two weeks in cooler weather when larvae develop more slowly), and kill the larvae. Continue until just after you have harvested all the fruit. The codling moth pupae, like many other apple

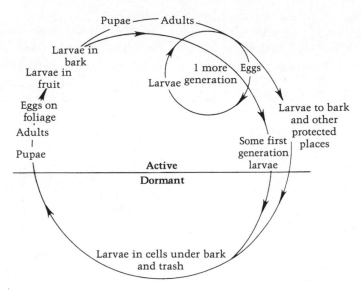

Life history of the codling moth

Reprinted, by permission, from Pyenson, *Fundamentals of Entomology and Plant Pathology* (fig. 3.20), AVI Publishing Co., PO Box 831, Westport, CT 06881.

pests, are reduced in number by a careful scraping of rough bark each spring. Scrape off all loose bark with a paint scraper or hoe. Place a sheet of some kind under the tree to catch the scrapings, and destroy them. Seal pruning wounds. Greet the first generation of codling moths in the orchard with a spray of two heaping tablespoons of ryania per gallon of water at petal fall. Since there are three broods in most areas, plan on repeating the ryania spray in five or eight weeks, and a third time five to eight weeks after that.

Codling moths have been found to be vulnerable to several sprays. Soapy-water and fish-oil sprays kill exposed insects, and can be applied weekly. Nasturtiums can be planted around the trunks as a repellent barrier.

BT sprays are only moderately effective, since the larva spends so little time feeding on the fruit surface. A virus is being developed for control. The tiny trichogramma wasp, present naturally and available through the mail as eggs, parasitizes codling moth eggs. Making successful use of them depends on timing—the eggs should hatch just about the time the moths are laying their eggs. If the number of wasp eggs seems paltry in comparison with the task at hand, keep in mind that trichogramma can have 25 or more generations in a favorable season.

You can aid naturally occurring parasitic insects in the orchard by growing a flowering cover crop, such as clover, mustard, sweet alyssum, buckwheat, or a member of the daisy family. Many adult parasites benefit from a good nectar source. It may well be worth your while to put out suet in winter and spring to attract woodpeckers to the orchard area, as these birds pick out a good number of pest larvae from tree bark.

The silken nests of the **eastern tent caterpillar** [color] signal trouble for orchards. Also known as the appletree tent caterpillar, the larvae ravage foliage by day and spend nights and rainy weather in the nests. There is one generation a year. Overwintering eggs hatch

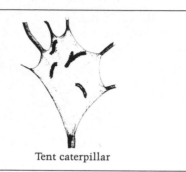

Tent caterpillar

very early in spring, and reddish brown moths appear three weeks after the dirty-white cocoons are spun. Pest infestations usually occur in cycles of seven to ten years (the result of the effects of predation, parasitism, and disease), and the extent of damage ranges from none at all to complete defoliation. For controls, see TENT CATERPILLAR.

The **European apple sawfly** [color] is brown and yellow in color, marked with numerous transverse lines, and a little larger than a housefly. They are active in the northeastern United States. Larvae feed in the fruit just below the

Sawfly damage

skin until they are about one-third grown, at which time they start boring directly through the fruit. They sometimes pass from one small apple to another, leaving a chocolate-colored sawdust on the surface. The winter is passed in the soil as mature larvae. Pupation occurs late in spring and the sawflies appear at about the time the trees blossom. Eggs are laid in the flesh of the calyx cup, and the young larvae feed there before beginning to tunnel.

If necessary to protect a crop, spray trees at pink and petal fall with ryania or rotenone. Fruit that falls prematurely often contains larvae which will continue to develop. Traditionally, growers have allowed hogs to clean up the drops. If you don't have a hog at your disposal, pick up fallen fruit every day or two. Destroy any larvae within before adding fruit to the compost pile; you can do this easily by placing it in a sealed plastic bag and leaving it in a sunny place for two weeks.

The pinkish white larva of the **oriental fruit moth** behaves much like the larva of the codling moth, but enters apples through the stem, seldom tunnels to the core, and thus gives little external sign of damage. See PEACH for control methods.

The **roundheaded appletree borer** [color] tunnels deep into the trunk of apple trees, usually near the ground line. The adult is a slender, long-horned beetle, colored grayish brown and marked with two conspicuous longitudinal white stripes. Infestations can weaken or girdle a tree; young trees may be killed. Adults may be seen eating leaves and fruit,

Roundheaded appletree borer

though this damage is usually insignificant, and lay their eggs in slits in the bark near the ground. Rusty brown shavings show at the borer's hole.

Look for the telltale brown castings just above or below ground level. You can destroy some borers by snaking a wire into their holes; this should be done each fall and spring. A thick wash of soap can be applied to the lower trunk in order to discourage beetles from laying eggs there. Also see peachtree borer under PEACH.

Healthy vigorous trees are rarely bothered by the **shothole borer,** known also as the fruit tree bark beetle, so an organic fertilizing program is the best way to prevent trouble. You may see shot holes in twigs of healthy trees and in the branches of trunks of weak ones. Black beetles exit the holes early in summer. For control, see APRICOT.

Small, active **white apple leafhoppers** are only ⅛ inch long, but their feeding can remove significant amounts of chlorophyll, leaving foliage spotted or white. Nymphs are pale yellow in color and feed on the undersides of leaves, leaving behind dark dots of excrement. Strong water sprays may wash leafhoppers from leaves, and sprays of rotenone or ryania should prevent serious infestations. In heavy infestations, honeydew will drip onto the fruit, producing small dark spots that readily wash off. Begin control measures at petal fall against the first generation. If this proves successful, the second generation in August will be much smaller.

Some pests of apple do not limit themselves to one fruit, and their life histories and control are included under FRUIT TREES: see European red mite, fall webworm, flatheaded appletree borer, leafrollers, periodical cicada, plant bug, and San Jose scale.

## Diseases

**Apple blotch** can be a severe problem in warm, rainy seasons, infecting leaves, twigs, and fruit. Leaf infections appear on the veins, midribs, and petiole as pale, elongated, sunken areas with numerous black dots. If the infection is severe enough, heavy leaf drop may occur. Twig infections enlarge each season to produce a canker that encircles the branch and is much larger in diameter than the branch. Lesions on fruit are shiny black blotches of varying sizes, the larger ones having irregular margins which often lead to cracking. This fungus is active from petal fall to harvest.

Fortunately, most apple varieties are not susceptible to blotch, though Rhode Island Greening, Duchess, and Transparent can be severely infected. Prune out cankers found on branches of any size and burn to remove the overwintering source of spores.

**Baldwin spot** (or bitter pit) is characterized by brown corky flecks in the flesh of the fruit. The flecks are most frequently found just under the skin of the apple, but may be scattered through the flesh as far as the core. Baldwin spot differs from drought spot in that the spots are sharply defined, whereas drought spot is a solid browning of the affected tissue. The name bitter pit derives from the bitter flavor of the corky tissue. These troubles are not caused by an organism, but by a calcium deficiency coupled with optimum temperature and humidity. The more susceptible varieties include Baldwin, Greening, Red Delicious, York, Stayman, and Northern Spy. The McIntosh group is seldom affected. Pruning and fruit thinning are keys to avoiding trouble. Heavy pruning of vigorous trees both stimulates growth and increases the chances of bitter pit. Encourage heavy cropping and delay thinning as long as possible. Cut down on fertilizers rich in nitrogen and potassium, and maintain a uniform level of soil moisture throughout the growing season. When planting, apply lime as called for by a soil test; lime lightly (five pounds per year per tree), unless more is called for based on soil tests. Fruit from overly vigorous or light-cropped trees should be harvested early, segregated, cooled rapidly, and used (or marketed) as soon as possible, especially if the fruit is large.

Once Baldwin spot develops, it can be countered with two sprays of calcium chloride, using 1 to 1½ pounds per 100 gallons of water six weeks and three weeks before harvest. Older trees will benefit from ½ pound of borax, spread within the drip line; use less for young trees.

Especially in warmer apple-growing regions, **bitter rot** [color] can infect the unbroken skin of an apple and cause terrific damage, especially if temperatures hover near 85°F. and the humidity is between 80 and 100 percent. The fungus over-

winters on mummified fruit, in bark crevices, and especially on jagged ends of broken limbs; the first symptoms usually don't appear before July. These circular, slightly sunken, brown lesions will appear four to five days following a rainy or moist period and rapidly expand to ½ or ¾ inch, with concentric rings. During subsequent wet periods, a mass of pink, creamy spores will appear. These spread the infection. Careful attention to sanitation will be rewarded with low levels of this potentially serious fungus; remove any mummies in fall, and keep pruning cuts clean and without stubs. As limbs break, promptly remove them with a saw and leave the cut area smooth.

**Black pox** is a fungal disease that produces many (more than a hundred have been noted) small, black, circular, sunken spots on the fruit and a scaling of the bark on twigs similar to

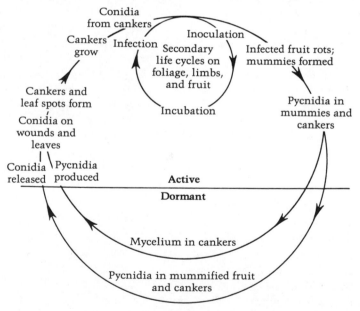

Life history of the black rot fungus on apple

Reprinted, by permission, from Pyenson, *Fundamentals of Entomology and Plant Pathology* (fig. 9.17), AVI Publishing Co., PO Box 831, Westport, CT 06881.

measles. Little is known of the life cycle, but you can reduce infections with wettable sulfur. Prune out twigs that show a flaking of the smooth bark. In the Midwest, Grimes Golden is considered susceptible, while infections have developed on Golden Delicious, Red Delicious, Maiden Blush, Rome Beauty, and Duchess. York Imperial, Gano, and Transparent are thought to be resistant.

**Black rot** [color] is a warm-weather fungus that overwinters in mummies, dead wood, and unlike bitter rot, cankers. Fruit symptoms, however, are similar to those caused by bitter rot. As the canker grows, spores are formed in spring and appear as small black pimples on the bark surface. Cankers become obvious when the bark loosens, turns a reddish brown, and takes on a slightly sunken appearance. Since this fungus cannot attack living wood, the simplest control method is to remove all deadwood from the orchard, burn all prunings rather than let them lie around the trees, keep pruning stubs short, and remove mummified fruit.

**Blossom-end rot** is caused by any of several fungi that are abundant in orchards. It is worse on varieties (like Red Delicious) whose calyx does not close completely after bloom and allows the fungus to enter the seed cavity. Losses are usually minimal and there are no effective control measures.

**Botryosphaeria rot** (or white rot) has fruit symptoms like those of black and bitter rot; it produces cankers like those of black rot but is more virulent and can kill entire limbs of susceptible varieties, such as Golden Delicious, Rome, Gallia Beauty, and Yellow Transparent. Fruit infections are common on Golden Delicious. As with black rot, a healthy tree is a resistant tree, so try to maintain proper nutrition, reduce winter injury, and water when drought conditions occur. Further, prune lightly on an annual basis rather than severely every few years; this minimizes the heavy cuts which open the tree to sunscald injury.

**Cedar-apple rust** [color] is a curious disease, living one stage of its life on apple and the other on red cedar. During spring rains, spores are blown from infected cedars to apple trees, where they cause bright orange spots on the foliage and to a lesser degree on fruit. The infections on apple foliage produce another spore stage, borne in delicate cuplike structures that reinfect the cedar and thereby keep the two-year cycle going. A related fungus, quince rust, infects the fruit only, causing a hard green lesion on the calyx end.

If cedar rust presents a real problem, remove red cedars within a radius of 300 yards of the orchard—assuming this can be done legally—and cut down nearby susceptible varieties, such as flowering crabapple. Resistant apple varieties are available; see RESISTANT VARIETIES.

**Fire blight** can be resisted to some extent with the apple varieties Delicious, Dolgo Crab, Duchess, Florence Crab, Gano, Haralson, Hibernal, Jonathan, McIntosh, and Northern Spy. The disease is apt to trouble Hopa Crab, Hylsop Crab, Ida Red, Transcendent Crab, Wealthy, Whitney Crab, and Yellow Transparent. The degree to which any one apple variety is resistant will vary from year to year, depending on local weather conditions during bloom, the proximity of pear trees with holdover cankers, and the number of insect vectors.

Among the many rootstocks available to apple growers, M26 is the only one that is highly susceptible to blight; it should be neither planted in areas where blight is severe nor grafted to blight-susceptible varieties. See PEAR for further controls.

**Manganese toxicity,** also known as measles and most severe on newly planted Red Delicious, is caused by the uptake of excess manganese from very acid soils. Water blisters in the outer bark of the trunk do not penetrate to the cambium. Usually these blisters burst and the brown liquid contents streak down the bark. Manganese toxicity in itself is not injurious to the tree, but the low pH it indicates will lead to reduced growth and yield.

An obvious control method is to raise the soil pH. This can be done by adding crushed limestone or dolomite limestone when the tree is planted. Wood ashes, marl, and ground oyster shells are all helpful. Any of these may be added to the compost pile. This problem is most common when an old orchard site is replanted.

**Nectria twig blight** is a minor disease and causes little damage unless ignored, but it may be confused with fire blight, a very serious disease that must be dealt with immediately. A rapid wilting of the leaves on one shoot is usually the first symptom (as with fire blight); but unlike blight, there are no blighted blossoms adhering to the twigs, and the leaves at the shoot tip are not discolored and appear to die from the base rather than the tip. For certain diagnosis, look for tiny pinkish bumps appearing in mid to late summer at the shoot base. These fruiting structures produce the spores that will spread the disease. Nectria and several other diseases will be only occasional headaches if wilting shoots are pruned out and burned as they appear. Varieties with large cluster-bud bases (such as Rome Beauty, Gallia Beauty, Ben Davis, and Northern Spy) seem to be particularly susceptible.

**Powdery mildew** fungus is a troublesome disease that overwinters in the apple buds. When the buds open in spring, the mildew fungus sends threadlike growth over the surface of the unfolding leaves and soon covers them with a mass of fungus filaments. The fungus growth robs the trees of nutrient material that the leaves normally provide for development of fruit. Spores are released by the fungus filaments, and these spread the infection to leaves, twigs, and fruit. Infected leaves are smaller and have a white or gray powdery growth on both the upper and lower surfaces. In addition, the leaves may be curled, wrinkled, or folded. Leaves at the tips of twigs are small, distorted, and covered with gray mildew. Twig growth is stunted and the tip or terminal part may be killed. On one-year-old twigs the fungus causes the same powdery white appearance, but by midsummer the powdery condition begins to disappear and the whitish growth turns into a brown feltlike covering in which numerous fruiting bodies are embedded, giving the twigs a speckled appearance. On fruit the injury appears as a network of fine lines of russet. In addition, the apples may be stunted, shriveled, or cracked. Warm days and cool, moist nights favor this disease.

For control, prune out distorted, badly mildewed twigs when you notice them, whether during the dormant season or the growing season. Wettable sulfur (95 percent) is very effective if applied before bloom and every other week after bloom until June 15, using 4 pounds per 100 gallons of water.

**Rosette,** known also as little leaf, is most often due to a deficiency of zinc. This condition affects many plants, including pecan, apricot, peach, plum, grapes, and numerous ornamentals. Both homegrown trees and commercial orchards may sustain serious damage. A deficiency of zinc usually is found in alkaline, sandy soils low in organic matter.

Symptoms are usually noted in pecan and fruit trees during either the second or third year after planting. When first purchased, most young trees have enough zinc stored in reserve to last for about one year. In its early stages, or on lightly affected trees, the condition appears as a yellowish mottling of the leaves, particularly in the treetops. However, this symptom is sometimes absent in apple trees. In advanced stages, leaves are small, narrow, and crinkled. Foliage is often sparse. Reddish brown areas or perforations appear between the veins of pecan leaflets. New shoot growth ceases, and the internodes (distance between leaves) are shortened, giving the foliage a bunched appearance known as rosetting.

In the final stages of this malady the shoots die back from the tips. Usually the dieback is confined to the current year's growth, but it sometimes extends to older branches of considerable size. When viewed at a distance in late summer, severely affected trees may have a bronze cast to them. Many branches will be seen to have only a small clump of leaves at the tips, and severely affected branches have no foliage and eventually may die.

Prevent trouble by laying down plenty of organic material. If only a few trees are involved, they can be protected by driving galvanized metal strips or nails into the trunk and limbs. For fruit trees, use four nails for each inch of trunk diameter.

**Scab** [color] is one of the most universally destructive of all apple diseases. The leaves, fruit, and occasionally the twigs of apple trees are affected. Scab generally appears first on the undersides of the leaves as pale yellow spots that gradually darken until they are nearly black. Both surfaces of the leaves may show numerous scab spots, and in severe cases curling, cracking, and distortion of the leaves result in defoliation. Symptoms on the fruit are quite similar. As the spot on the fruit enlarges and becomes older, the velvety appearance in the center becomes brown and corky. Heavy infection may cause the dropping of young fruit and distortion and cracking of growing fruit. Infection on twigs will blister and rupture the bark to produce a scurfy appearance.

The fungus overwinters on the fallen infected leaves, producing a spore stage in spring. The spores are wind-borne and infect the young leaves and fruit during periods of rain. No infection will occur below 36°F. As the temperature increases, the number of hours that the tree must remain wet for an infection to take place decreases. Warm, rainy weather is ideal for scab. During summer a different spore form is produced on the leaves and fruit, and these spores continue to produce new infections when washed onto leaves and fruit by rain. Infection may also occur during moist periods in fall. Summer spores are responsible for late fruit infection that will develop in storage either as typical scab spots or as small black lesions.

Damage to crops may come about in several ways: the premature dropping of young apples during or immediately following the blossoming period; the dropping of scabby apples before harvest; impairment of quality with a consequent reduction in value; the formation of so-called pin-point scab at harvest or storage scab, caused by late infection and responsible for lowering the quality of the fruit; and reduced vigor and health of the tree as the result of damaged leaves.

Because the fungus spends the winter on fallen infected leaves, these should either be raked up carefully or plowed under in fall or early spring. One inch of rototilling under the trees in early winter will help the leaves with overwintering infections decay. Varieties resistant to scab are available: Prima, Priscilla, Liberty, Sir Prize, Nova Easy Gro, Priam, MacFree, Nova Mac, and Florina. If your plants are in serious trouble, use a sulfur spray in spring. Ninety-five percent sulfur is mixed up at a ratio of 5 pounds to 100 gallons of water; do not apply within two weeks of an oil spray.

The first symptom of **silverleaf** disease is the appearance of a metallic, silvery discoloration on the leaves of some branches, caused by an air pocket between the leaf tissues. The causal parasite is a wound fungus—that is, the spores can infect living trees only through fresh wounds or injuries. The fungus does not spread through the soil, unless roots are grafted from diseased trees, but is airborne. Infected wood may show a brown stain. Infected branches gradually die. Vigorously growing trees are less susceptible to infection and are sometimes able to recover from infection. The disease is most prevalent in cool, humid climates and is encouraged by winter injury.

As branches begin to die, they should be pruned off, and seriously ill trees should be removed and burned. To make sure that the pathogen cannot enter trees, promptly cover pruning wounds with grafting wax or a tree dressing; if possible, avoid large wounds by pruning often. Do not water trees excessively.

**Storage scald** is not the doing of a disease organism, but results from a breakdown of fruit tissue caused by absorption of volatile compounds. These compounds are given off by the fruit itself in the normal respiration processes incidental to ripening. This trouble may manifest itself in either of two ways: a browning of the fruit skin in varying patterns, or a complete breakdown of the flesh so that the apple looks baked. These two symptoms are sometimes known as hard and soft scale, respectively. Oilpaper wraps and controlled air storage have proven effective. However, if fruit is picked when mature and not green, and proper attention is given to temperature and humidity control in storage, scald shouldn't be a problem.

**Water core** is a common problem in apples allowed to hang too long past their usual harvest time. It shows as translucent areas in the flesh of the apple; the affected areas may be confined to the flesh around the core, or encompass the entire fruit to give it the characteristic glassy look. Fruit affected with water core is heavier and harder than normal, and has a sweet, syrupy flavor. King, Fall Pippin, and Red Delicious are particularly prone to this trouble, though most varieties left too long on the tree will suffer some water core.

# Environmental problems

A variety of environmental conditions lead to symptoms that might appear at first to be the work of a disease. Pits in the flesh of fruit may result from ozone in the atmosphere. Leaf scorch sometimes follows a deficiency of magnesium. Chlorosis is usually a sign of too little iron in the soil. Dry weather may bring on a condition known as Jonathan spot. Also see manganese toxicity, rosette, storage scald, and water core.

If severe enough, hail injury is immediately apparent, but slight bruises or nicks may not be evident until picking time. Harm to the twigs will be readily apparent in serious cases, but slight injury won't show until the following year, when a series of sunken eye-shaped spots will appear on the uppersides of the twigs. Within a few years, these spots should be covered completely by new growth.

Prolonged winter temperatures below $-25°F$. can kill the bark and cambium on the lowersides of those apple branches nearest the ground. Trees that have borne a heavy crop of fruit the preceding summer are most likely to be injured by low temperatures. Such trees are low in stored carbohydrates, and it is this that renders them vulnerable. Late frost may cause both fruit russeting and foliage injury. On young leaves, the lowersides stop growing, and this results in a downward cupping and a loosening of the lower epidermis, which will be stretched across the cupped leaf like a drumhead. Russeting on fruit may appear as a random pattern, often in the form of equatorial bands or as a random pattern of arcs, circles, or blotches.

Sunscorch hits fruit on the south side of the tree or on the side exposed to the sun. The requisite conditions are air temperatures above 90°F. coupled with low humidity and crystal-clear skies. The injury shows as circular areas varying in color from pale yellow through buff to brown. In extreme cases the areas are flattened and black with a light halo. (High temperatures do not often directly bring about foliage scorch, but cause irregular light brown areas.) Heavy deposits of sulfur on fruit or foliage may increase the severity of damage on sulfur-sensitive varieties such as Stayman and Red Delicious.

When old trees are pruned, exposing formerly shaded areas to the sun, the bark on large branches may be sunburned; this occurs as anything from roughening of the bark to death of large areas. The loss of entire branches may follow if wood-rotting fungi enter through these dead areas. The solution is to leave some shoots in the treetop to shade limbs.

# APRICOT

When pruning apricot trees, be sure to remove all broken branches and diseased areas, and then rinse the trees with a naphtha soap solution to discourage bugs and fungus.

## Insects

A small greenish insect covered with a powdery substance is the mealy plum **aphid**, occurring on apricot, plum, prune, and other fruit trees. Should it strike, the foliage will curl from loss of vital plant juices, the tree becomes weak and stunted, and the fruit is either split or covered with sooty mold that grows on excreted honeydew. See PLUM.

Branch-and-twig **borers** are sometimes a problem to apricot growers. These brown or black beetles are about ½ inch long and cylindrical. They can be found burrowing into fruit buds or limb jointures, killing branches and weakening trees.

Prune off infected twigs from smaller fruit trees and remove all prunings from the vicinity of the tree as soon as possible. If infested wood is held for fuel, dip it for a moment in stove oil to kill the larvae under the bark.

With sharp appendages at the back end, the **earwig** [color] looks full of trouble. However, there's no need to be alarmed. It is more beneficial than harmful, especially if you ensure a good mix of insect life for it to prey on. It also feeds on decaying plant material and is active at night.

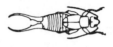

Earwig

You can keep earwigs out of the trees by luring them away with an organic mulch. Thorough plowing or discing in spring may drive them to hide at the base of tree trunks. They can be trapped by rolled up newspapers or jute sacking placed in the main crotch of the tree. The traps must then be collected and burned or placed in soapy water. Also, sticky barriers such as Tanglefoot and Stikem may be useful. See EARWIG.

The **shothole borer** is a little dark brown beetle that breeds under the bark of peach trees, among others, and emerges through small circular holes resembling buckshot holes. The larval galleries nearly girdle the stems and branches, and on stone fruits you will notice gum exuding from the exit holes. The adult beetle is about $\frac{1}{10}$ inch long and appears in

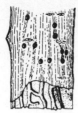

Shothole borer damage

early spring. There are usually two generations a year in the northern states. This pest seeks out injured and weakened trees; keep trees as vigorous as possible. Prune and burn during the winter any limbs with shot holes.

Other insects apt to affect apricots are oriental fruit moth, peachtree borer, lesser peachtree borer, and peach twig borer (see PEACH); leafrollers and plant bugs (FRUIT TREES); and plum curculio (PLUM).

## Diseases

Small purple spots on apricot leaves and black lesions on fruit are signs of **bacterial canker.** Resistant varieties have yet to be developed. Arrest the spread of the disease by pruning infected twigs and branches.

**Black heart** is a fungal disease that also causes verticillium wilt on the first hot days of summer, and the wood may show black streaks an inch below the surface. Trees may die of black heart, but often grow new foliage and recover.

No true cure is known, but proper feeding and watering should help to enable trees to resist disease. Do not plant tomatoes, cotton, potatoes, strawberries, or Persian melons near infected trees, as these crops are also susceptible to the verticillium fungus. This disease may be arrested by pruning affected branches.

**Crown gall** is a bacterial disease that girdles roots and crowns. It causes swellings where bacteria enter through wounds in bark. The swellings often occur at the soil line where cultivating tools have been carelessly used, and they may appear below the ground or on branches damaged by ladders.

Do not purchase trees that show such protuberances. Set apricots in soil that has not previously hosted this disease. Remove the gall with a sharp, clean knife and treat the wound with tree surgeon's paint to prevent further infection. Disinfect tools in alcohol to prevent the spread of contamination. Trees will fare better if this is not done in the hot part of the summer; to prevent damage from sunburn after treatment, cover the affected parts with soil. If the gall encompasses most of the trunk, remove the tree. Japanese apricots are resistant. Grafting can allow the bacterium access to trees; budding is a safer means of propagation.

A nonpathogenic strain of the bacterium is registered for control of crown gall on almond, apple, apricot, cherry, grape (in California), nectarine, peach, pear, plum, prune, berries, and walnut. It works by occupying the same ecological niche as the crown gall bacterium. New plants are dipped in a solution of the bacterium before planting. The product is sold under the brand name Gall-trol.

Apricot trees bloom very early and are especially susceptible to **frost** and freezing weather. The flower buds freeze and the entire fruit spur may be killed. If only the bud is affected, it turns brown and the flower never opens. Sometimes only the pistil (the central organ which ripens into the fruit) may be killed. You might try inspecting those flowers that will open to see if the pistil is still green and lifelike rather than dark and shriveled after a freeze. To reduce the likelihood of frost damage, plant the tree on a north slope where it will warm slowly and bloom later.

Deficiency of zinc in the soil causes the leaves to appear in spring as rosettes of small, narrow, stiff leaves mottled with yellow, a condition known as **little leaf.** In severe cases, twigs die and the tree may be lost. The disease is also responsible for fruit dropping, low yields, and delayed maturity.

One of the safest and easiest ways to supply zinc is sprinkling raw phosphate rock around the base of the tree. The use of plenty of manure, a natural source of zinc, will also help. A few zinc-plated (galvanized) nails can be driven into the trunk for a quick injection of the mineral.

Like peaches, apricots are subject to **pit burn.** When the fruit ripens in very hot weather, the area around the pit turns soft and brown, as if the pit were heated. A thick light-colored mulch under the tree tends to keep the tree cooler; if the heat is very bad, it may be helpful to sprinkle the mulch with water during the warmest part of the day.

Another possible cause of cracking fruit is **sunburn.** This affects the skins most often in hot, humid weather. When the fruit is thinned, particular attention should be given to those growing on the exposed tips of the branches. If the leaves are lacking at the tips of the branches, it is better to remove the fruit. This will send the tree's nourishment to other fruit that is protected and likely to survive.

Other diseases affecting apricots are bacterial spot, brown rot, cytospora canker, scab, and stem pitting (see PEACH), and verticillium wilt (CHERRY).

# ARACHNID

Spiders and spider mites are members of the Arachnida class of phylum Arthropoda. They are not true insects and usually are distinguished by four pairs of legs rather than an insect's three pairs (although immature mites may be a pair short). See SPIDER and SPIDER MITE.

# ARBORVITAE

## Insects

The **arborvitae leafminer** may cause browning at the tips of leaves. The small greenish larva is a caterpillar that becomes a light gray moth as an adult. Eggs are laid on the leaves in early June, and later that month the larvae begin mining the leaves. Shaded plants are more heavily infested.

Burn cocoons and leaves that appear to be infested. Prune back branches until healthy growth remains. Larvae can sometimes be repelled by spraying trees with a soap solution in late June or early July. See EVERGREEN.

The **arborvitae weevil** is a small black insect; body and elytra are covered with metallic green scales and fine short hairs. It emerges from the soil sometime around early May and is present usually until late July. The larvae are legless and white with a light brown head; they pass through seven instars. During this time they feed on the roots of host plants—arborvitae and red or white cedar. The larval period extends from June or July to midwinter or the following spring. Adults attack the foliage of host plants, feeding lightly here and there so that no large area of plant foliage is consumed at one time. Foliage feeding scars may be observed for months after the adults have disappeared.

A good control method makes use of the weevil's natural habits. When disturbed, it drops readily to the ground and remains immobile for a short period. Place a drop cloth under the tree and violently shake the main limbs until the bugs play possum and fall. Weevils may be drawn in large numbers to light traps. The adult is not highly migratory and will spend most of its time on the foliage of the host plant. Do not plant red or white cedars near arborvitae, as these trees may harbor the pest. Carefully check balled nursery stock before planting to make sure it is not infested.

Small bags hanging from your trees are likely evidence of the **bagworm.** This caterpillar lives in a silken cocoon that resembles a bag, often with bits of leaves attached to the outside. It carries this bag with it as it feeds. A fully developed bag is about two inches long. Eggs are laid in fall and hatch in May or June of the following year. During the winter, handpick the bags from your trees and burn them. Burpee sells a pheromone trap with bagworm bait.

**Juniper scales** may infest arborvitae; see JUNIPER.

Arborvitae may occasionally be troubled by **spruce spider mites,** but these can usually be kept off the trees by spraying leaves with water. See SPIDER MITE.

## Diseases

The fungus that causes **twig blight** of juniper may also attack arborvitae, causing tips of infected twigs to brown and die back gradually. Old as well as new leaves and twigs are involved. Black pinpoint fruiting bodies of the fungus may be found on affected portions. During wet weather, disease-producing spores that ooze out of the fruiting bodies in long threads are scattered by wind, rain, and insects. This disease may be confused with a fall browning of the inner leaves that often follows unfavorable growth conditions.

Control by pruning out and destroying those twigs and branches that appear affected. The Dwarf Greenspike arborvitae is fairly resistant.

**Twig browning** and shedding (cladoptosis) cause whole twigs throughout the tree to turn brown and later drop. This is caused by dry soil, and trees located on lawns or along streets probably will require watering during summer. Soak the soil to a depth of about two feet by slow application of water through a porous hose or with sprinklers. Do not water more often than at two-week intervals, to allow necessary aeration of the soil between waterings.

**Water injury** causes a late winter or early spring browning of the previous season's growth arborvitae, pine, and juniper. Drying winds and hot sun are the linked causes; trees in exposed locations and transplants are most severely affected. The discoloration is due to evaporation of moisture from the leaves or needles at a rate faster than the roots can pick up water. Minimize damage by mulching and thoroughly soaking the ground around trees before the ground freezes. Evergreens will fare better if planted in protected places. The reddish brown discoloration of the tips of branches should not be confused with the browning and dropping of leaves or needles on the inside of the tree nearest the trunk. This leaf fall may take place in spring or fall and is a natural shedding: it may take place every year, or every second or third year.

# ASH

## Insects

Larval **ash borers** tunnel into the trunk of trees at or below ground level, often damaging trees so badly that they can be easily pushed over. The adult moth is black with yellow bands and blackish brown wings.

The adult moth is drawn to light traps. Cut out and burn parts of infested trees, or even whole trees if necessary, to reduce infestation.

The borer's damage renders trees vulnerable to disease. Keep the ground perfectly clear of grass and weeds, especially between mid-May and mid-August, so that birds can easily spot the eggs and worms. Other applicable controls may be found under peachtree borer (PEACH).

The injurious stage of the **mountain-ash sawfly** is a small green caterpillar, ½ inch long, that is covered with numerous spines. These creatures strip the leaves by skeletonizing them into shreds, baring the ribs and veins. They particularly damage the leaves at the top of the trees. The adult is a ¼-inch-long fly with a wingspread of ½ inch. Body and wings are black, except for some body segments that are rusty or yellowish on the underside. The male fly is smaller than the female. The sawflies appear about the middle of May in the northeastern United States and the female soon deposits her eggs under the skin of the leaf, forming white spots on the top surface.

If you can reach high enough, collect the egg clusters on the lower side of the leaves and destroy them. If not, sticky traps may catch flies. See TRAPS.

The **obliquebanded leafroller** is troublesome to many fruits, vegetables, and flowers, as well as trees and shrubs. The greenish larvae roll the leaves and tie them together, feeding in the resulting protective cocoon. Eggs are often laid in branches and on rose leaves.

Where only a few plants are infested, the affected parts can be pinched off and destroyed. If infestation is heavy, dust plants with a mixture of equal parts of tobacco dust and pyrethrum powder. Or spray with a mixture of pyrethrum or rotenone. Dust or spray in two applications 30 minutes apart. The first drives the caterpillars from hiding, the second kills them.

Several insect pests of lilacs are also pests of ash trees: the lilac borer, the lilac leafminer,

and oystershell scale. For control, see LILAC. For carpenterworms, see TREES.

## Diseases

Spring's wet weather is apt to encourage the spread of **ash anthracnose,** a fungal disease also known as blight. Spores spread readily in times of rain and fog, infecting new foliage and turning it brown; older leaves may show whitened, dead margins. To block the disease, prune away branches that show signs of damage. General pruning helps to prevent problems by permitting air to circulate. You can avoid anthracnose by planting the resistant Moraine ash.

A small mite is responsible for **ash flower gall,** a distortion of the staminate flowers of white ash that forms bunches or masses from ¼ to ¾ inch in diameter. These masses finally dry and remain on the tree over winter. To keep the mites at bay, spray trees with a mild dormant-oil emulsion in late winter.

There are three types of **leaf blotch** to which ash trees are susceptible. *Gloeosporium aridum* causes irregular brown blotches to appear on leaves in the midsummer of wet seasons, and leaves may fall. *Phyllosticta fraxinicola* makes small yellow spots in late summer, and some years *Piggotia fraxini* causes an epidemic of small purple spots with yellow borders. Burn fallen leaves to remove much of the source of inoculum for the following year.

**Rust** describes raised crescent-shaped or irregular swellings on leaves or petioles that burst open to release an orange powder. The infection then travels to a marsh grass which produces spores the following spring to reinfect ash. Control is not necessary, since the rust seems to do little real harm.

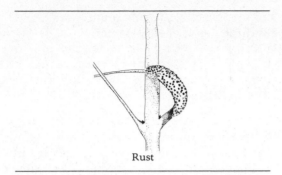

Rust

# ASPARAGUS

## Insects

The **asparagus aphid** is tiny ($1/16$ inch) and blue green. The young plant is damaged, and the bush may be severely rosetted. Protection is important in newly seeded plantings and young cutting beds. For control measures, see APHID.

The **asparagus beetle** is one of the most destructive pests of asparagus, attacking both garden variety and wild plants throughout the United States. The adult is metallic blue black, with three yellow to orange squares along each wing cover, and is $1/4$ inch long. The larva is $1/3$ inch long, colored olive green to dark gray with black legs and head, and is soft and wrinkled. The beetle lays dark shiny eggs, no bigger than specks, that are attached at one end to spears. Both adult and larva eat foliage and disfigure shoots. The beetle hibernates in trash around the garden and emerges in spring to eat the tender asparagus shoots and lay eggs.

Asparagus beetle

It follows that an important means of control is garden cleanliness. If trash is removed, the beetle may have to leave the garden to find housing. Turn over the soil around the plants in fall to disturb overwintering pests. The beetle has a dislike for tomato plants, and interplanting serves a second purpose—asparagus is a natural nematocide, and will keep nematodes from nearby tomatoes. The beetles are also turned off by nasturtiums and calendula (pot marigolds). If beetles have already become established in your patch, you might turn to the old trick of letting chickens, ducks, guinea hens, or other fowl run in the asparagus planting. These birds invariably do an efficient job of wiping out the beetles and their larvae, but keep in mind that they may also take a swipe at nearby tomatoes, ripe or not.

Naturally occurring chalcid wasps and ladybugs give some control. Release purchased ladybugs just as stalks nose out of the ground. Bone meal serves as both a repellent and a fertilizer. Dust plants with rock phosphate. Cloth or gauze netting helps to give young asparagus a good start; support the fabric with stakes between the plants and secure the ends so that no bugs can enter. As the plants grow larger, they become less susceptible to the beetle's attacks, and the protection can be removed. See TIMED PLANTING. See also the discussion of the spotted asparagus beetle, below.

The **asparagus miner** is an insect which tunnels near the base of the stem and just beneath the epidermis. Some of the miners start a foot above the soil and work downward, often to below the ground. As a result, foliage may turn yellow and die prematurely. The adult is a two-winged fly. The miner passes the winter in a tunnel in the form of a pupa, resembling a flax seed. The miner afflicts asparagus in the Northeast and in California, and is considered a minor pest.

Control is not usually necessary. Pull up and burn old stocks in late fall to destroy the overwintering pupae. Rust-resistant strains are also resistant to the miner (see asparagus rust, below).

The **spotted asparagus beetle** [color] is slender, a bit larger than the common asparagus beetle, and reddish brown or orange with six black spots on each wing cover. It occurs east of the Mississippi. The greenish eggs are glued singly by their sides to asparagus leaves just before the berries form. In a week or two, orange larvae appear and bore into the developing berries. Larvae pupate in the soil, and the mature beetles emerge in July to lay eggs for the over-wintering generation, which appears in September. You may find this pest on a variety of plants, from asters to zinnias, and it is especially common on all cucurbits. Since the spotted asparagus beetle is unable to fly in the cool of morning, it may be controlled by handpicking. Inspect shoots and berries daily, and pick and destroy affected berries. For further control measures, see asparagus beetle (above).

The **spotted cucumber beetle** is another pest of asparagus, having similar habits to the asparagus beetle. The adult is slender and about ¼ inch long. Its wing covers are greenish yellow, marked with 12 spots varying in size. For habits and control, see CUCUMBER.

Asparagus is also bothered by the bulb mite (see GLADIOLIS and SNAPDRAGON), cutworm (TOMATO), harlequin bug (CABBAGE), JAPANESE BEETLE, several types of scale (including black thread, dictyospermum, latania, and lesser snow), stink bug (SNAPDRAGON), and yellow woollybear caterpillar (imported cabbageworm under CABBAGE). The asparagus fern may host the green peach APHID, beet armyworm (BEET), onion thrips (ONION), and twospotted SPIDER MITE.

The **symphylan** [color] is a fragile white pest that grows up to ⅜ inch long. Adults have 12 pairs of legs, the young have fewer. It travels rapidly through tiny cracks in the soil, constantly waving its antennae, and is rarely seen aboveground. It eats numerous tiny holes, or pits, into the underground parts of plants. Roots of affected plants have a blunted appearance. The pest's presence may first be made known by stunted growth. Besides asparagus, the symphylan damages flowers and a number of vegetables: beans, peas, sweet corn, beets, carrots, celery, spinach, lettuce, potatoes, and radishes. It's found throughout the humid areas of the United States and is particularly injurious to asparagus crops in California. All told, it is considered an important soil pest in 25 out of the 31 states in which it has been reported.

Control is difficult. You can take advantage of this pest's aversion to light by planting asparagus in a sunny area and by clearing out rocks or trash that might serve as hiding places. Take these precautions early in the season, as it is most troublesome in spring. Dig up the soil around plants to disturb its habitat and cause it to look elsewhere for a home. On a large scale the symphylan is controlled by flooding asparagus fields.

## Diseases

**Asparagus rust** is the most serious disease affecting asparagus. It is a fungal disease that appears as elongated, dusty orange red or reddish brown spots (blisters) on leaves and stems. Later these blisters darken, become firmer, and burst to release a fine, rust-colored cloud of spores. Plant tops turn yellow and die early. Depending on local conditions, an asparagus planting may become rapidly infected and die. The spores require dampness for germination, and therefore areas subject to heavy dews and damp mists are poor locations for asparagus; the disease is worst in moist seasons, and is found throughout the United States.

Asparagus rust has been eliminated in many areas by the use of rust-resistant varieties. Martha Washington and Mary Washington were once very popular varieties, but new fungal biotypes have overcome them. Newer varieties that should work are Waltham Washington, Seneca Washington, and California 500. The New Roberts variety may be suited for the Midwest.

Since rust lives on the diseased plant tops from year to year, cut affected tops close to the ground and burn them in fall. Plant rows in the direction of prevailing winds to lessen the possibility of spores alighting on uncontaminated shoots.

# ASTER

## Insects

Certain species of **aphid** [color], especially the corn root aphid, attack the roots of asters and a number of other plants, including brownallia, calendula, primrose, and sweet pea. An aphid attack results in little or no plant growth, and leaves turn yellow and wilt under bright sunlight. To check for this pest, examine the roots; the aphids are small and either white or pale blue green in color.

Ants are the agents of root aphid damage. They carry the young aphids through their tunnels to plant roots. Once asters or another susceptible plant has been infested, rotate it. Thorough spading in fall will disturb the ant nests that often house caches of aphid eggs.

Aster flowers and foliage are often the target of the **black blister beetle,** also known as the Yankee bug and just plain blister beetle, is the best-known species. It is a fairly long (up to ¾

inch) and slender beetle, with soft, flexible wing covers. The entire body is black or dark gray, and the covers may be marked with white stripes or margins. Another species, the margined blister beetle, is distinguished by a narrow gray or yellow margin on the covers. Blister beetles are very active and frequently appear in large numbers in the latter part of June and through July.

Handpicking is effective in controlling this pest, but you should protect your hands with gloves, as the beetles discharge a caustic fluid that is harmful to the skin. Some growers achieve control by dusting with equal parts of lime and flour. This should be done at the warmest time of the day. For other controls, see BEET.

Brown areas on lower leaves signal the underground feeding of the **foliar nematode.** The discolored areas are wedge-shaped and occur between the veins, and the symptom progresses from the bottom of the plant upward. This wirelike pest also frequently affects garden chrysanthemums.

Unless the soil has been sterilized, don't plant susceptible flowers in succeeding seasons; these flowers include chrysanthemum, calendula, dahlia, and scabiosa. Interplanting with marigolds, on the other hand, may drive nematodes from flower beds. For other control measures, see NEMATODE.

**Grubs** of the Japanese beetle and Asiatic garden beetle feed on the roots of asters to cause an overall weakening of the plants. Milky spore disease is used to control both of these larvae; see JAPANESE BEETLE.

The **sixspotted leafhopper** damages asters by sucking plant juices from leaves, which may turn brown and die. It is greenish yellow with six black spots, and grows to ⅛ inch long. The young, or nymphs, are grayish. Look for leafhoppers on the undersides of leaves, especially on the lower foliage. The pest is fond enough of asters to have picked up the alternate name of aster leafhopper, and it is the vector of a disease known as aster yellows (see below). The insects pick up the disease in the spring while feeding on infected wild plants.

Because this leafhopper prefers wide open spaces, asters grown near the house or next to walls will stand a better chance. If possible, do not plant the flowers near carrots or lettuce, two vulnerable crops. Sixspotted leafhoppers spend the winter on weeds, and growers with many asters occasionally burn over all nearby weed patches in early spring, before the pests can reach the garden. The only way to ensure that asters will make it through the season unscathed is to grow them under a shelter of cheesecloth, muslin, or fine-meshed netting.

The **stalk borer** is a long, slender caterpillar that frequents flower gardens. It first makes its presence known by the effects of its boring: stems break over and leaves wilt. A close examination of the plants will reveal a small round hole in the stem from which the borer expels its castings. By splitting the stalk with a fingernail, you may find the caterpillar at work. It is brownish and is marked with a dark brown or purple band around the middle, with several conspicuous stripes running the length of the body.

Sprays are of little use in controlling the stalk borer, as it spends most of its life within plant stems. An effective, if time-consuming, remedy is to slit each affected stem lengthwise, remove the borer, and then bind the stem together. You can get quicker results by removing and destroying the stems. The best remedy of all is to discourage local borers before they get a chance to cause trouble. This involves clean cultivation and the burning of all stems and plant remains that are likely to harbor overwintering eggs. Adult borers (grayish brown moths) lay their eggs in weeds, especially ragweed, and a preventive program might include burning nearby weeds before the caterpillars can migrate to garden plants.

The **tarnished plant bug** [color] is an active, green to brown insect that sucks the life from young shoots and especially buds, leading to deformed or dwarfed flowers. The ¼-inch-long bug is mottled with markings of yellow, brown, and black. Look for a black-tipped yellow triangle

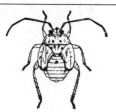

Tarnished plant bug

on each side. Nymphs are smaller, greenish yellow, and bear black dots on the thorax and abdomen. While feeding, the bugs apparently inject a poisonous substance into the plant that kills the surrounding tissue. The adults hibernate under leaf mold, stones, or tree bark, or in clover, alfalfa, and whitetop. They appear in early spring, and are most numerous toward the end of summer.

Because these bugs are so active, sprays are of limited use unless used early in the morning; at this time of day they are usually a bit stiff with cold. Prevention through clean culture is the best control. Sabadilla dust will take care of serious infestations.

Other insects that dine on asters include the CUTWORM, flower thrips (PEONY and ROSE), fourlined plant bug (similar to the tarnished plant bug; see CHRYSANTHEMUM), potato flea beetle (POTATO), SPIDER MITE, spotted cucumber beetle (see CUCUMBER), and WIREWORM.

# Diseases

**Aster wilt** is the most troublesome disease of the china aster (callistephus). The first noticeable symptom is a greatly reduced and blackened root system, which precedes wilting of the plant above. Brown streaks then appear on stems; this discoloration is most evident when a stem is cut.

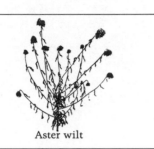

Aster wilt

In time, entire stems become darkened, and a grayish pink fungal growth sometimes may form. Plants may be attacked at any time: young plants dry up suddenly, and older plants first show a pale green color and wilting of lower leaves on one side of the plant, followed by death of the entire plant. Wilting is most pronounced in heat of day, and plants seem to recover somewhat at night.

Prevention of aster wilt is the best medicine. The plants should not be grown continuously in the same soil, and crop rotation and changing seedbeds will also help. The disease is aggravated by the use of nitrogenous fertilizers, such as fresh manure that has not had the benefit of the compost heap. The disease hits hardest in warm soil, and you can counteract this by growing alternate rows of corn or other tall crops that provide shade. Similarly, greenhouse asters that receive some shade are less severely attacked than those that get direct sunlight. Resistant aster strains are available, but do not work well in all areas; consult with the local agricultural agent before selecting a variety.

Next to aster wilt, **aster yellows** is the most destructive disease. Young infected plants show a slight yellowing along leaf veins—a symptom known as clearing of the veins. As the disease becomes more severe, the entire plant is stunted and yellowed, and the unusual number of shoots that develop cause the plant to become very bushy and upright. Secondary shoots are usually

Aster yellows

spindly, and flowers are curiously deformed. Regardless of the normal color of the variety, flowers are often a sickly yellow green. This disease is carried by the sixspotted leafhopper (see above).

**Damping-off** hits aster seedlings and causes stems to collapse and rot off at the soil surface. The disease usually occurs in seedbeds and flats in which the soil is wet, the temperature high, and the plants crowded.

To control damping-off, water the seedbed only as needed. Plant seedlings far enough apart so that air can circulate between them and carry away excess moisture.

**Gray mold** is a fungus that usually picks on white or light-colored flowers, showing first as a petal spot and then quickly rotting the entire flower. Leaves and stems are sometimes covered with a dirty-gray mold. Cool, wet weather favors the development of and spread of this disease, and a change to sunny weather can save the flowers. Plant asters in a location that affords plenty of ventilation and sunlight. Unless the

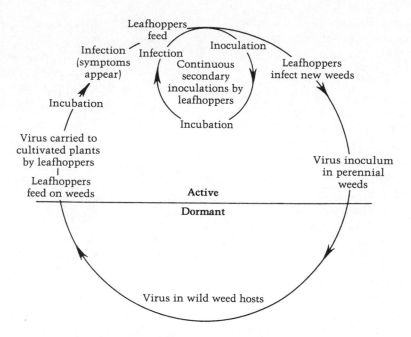

Life history of the virus causing aster yellows

Reprinted, by permission, from Pyenson, *Fundamentals of Entomology and Plant Pathology*
(fig. 10.5), AVI Publishing Co., PO Box 831, Westport, CT 06881.

season is unusually humid, this precaution should save you from any trouble with gray mold.

**Powdery mildew** appears as a whitish dust or bloom on leaves, and flowers become dull-colored. This fungal disease is encouraged by hot, dry weather. While there isn't much you can do to change the weather, you can allow ample space between flowers and burn all aboveground refuse in fall. If asters are really in trouble, dust with sulfur in the afternoon.

**Rust** causes bright orange pustules on the undersides of aster leaves. Common alternate hosts are the two- and three-needle ornamental pines, wild aster, and goldenrod.

Rust spores are easily carried about by the wind, so try to avoid planting asters on the leeward side of a possibly infected area. Rogue out infected plants, and consider long rotations if the asters are in trouble.

# AVOCADO

## Insects

Avocados, particularly in California, may be bothered by the **avocado caterpillar,** a leaf-roller moth with brownish red forewings. The larvae are yellow green and skeletonize leaves or web them together. They often damage the fruit also.

Rotenone or pyrethrum will take care of serious infestations. The moth stage of the insect can be lured to light traps. A trichogramma wasp is available commercially as eggs: *Trichogramma platerni* is marketed especially for avocado groves. The wasps should be hatched before the pest larvae appear, as they are egg parasites.

## Diseases

**Anthracnose,** or black spot, is a fungal disease that may not be noticed until the fruit starts to ripen. It causes brown or tan-colored spots on green-colored fruit, a lighter-than-normal spot on dark fruit. Depending on the amount of moisture present, spore masses may spread to form a pink layer over the surface. It does not take long for the rot to penetrate the entire fruit. Use resistant varieties; locate plantings in relatively dry areas.

There are several notable environmental diseases of avocado. One of the most common is zinc deficiency, causing a condition known as **little leaf.** Fruit is often deformed, and if the deficiency is not corrected, branches may die back. Zinc can be supplied in the form of raw phosphate rock around the base of the tree. Zinc is also found in manure, an excellent natural source.

A chlorotic yellow pattern on leaves, with veins remaining green, is characteristic of **iron deficiency.** Fruit size and shape and leaf size are little affected. Give less water to trees in problem soils.

If avocados are irrigated with chlorinated water, the result may be **tipburn** of the leaves—a considerable reduction of green leaf area. This leads to a general weakening of the tree, and the dead areas may be invaded by fungi which then spread to the fruit.

# AZALEA

## Insects

A pest of many ornamentals is the **azalea bark scale** (azalea mealybug). Plants growing outdoors show white, cottony, or woolly masses fastened to twigs, usually in the forks of branches or close to buds. Both sexes are enclosed in a feltlike sac. The honeydew secreted by the scale may give rise to black, sooty mold on foliage and branches. Affected azaleas become weakened and do not produce healthy flowers, and may even die.

Cut out and burn weakened branches. Use dormant-oil sprays of a weak white-oil emulsion; spray in late winter or early spring, when the nymphs are on the plants. For early-blooming varieties, spray on a warm day in late fall to avoid injuring the early spring blossoms.

Certain varieties of azalea are highly susceptible to the little **azalea leafminer** (azalea leafroller). This is a garden pest in the South,

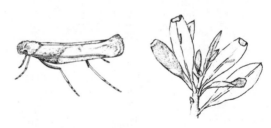

Azalea leafminer adult and damage

and more of a greenhouse pest in northern states. The adult moths are about ⅜ inch in length, are colored golden yellow with purplish markings, and hide under leaves when disturbed. The mining stages of the larvae are flat, while the leafroller stage has the characteristic caterpillar shape. The eggs are very small, white, and disc-shaped, and are found along the midrib or veins on the undersides of leaves. The leafminers typically winter in the leafroller stage, sheltered by rolled leaves. Yellow-and-purple moths emerge from the latter part of March through April. The larvae make mines that are visible only on the bottom side, but keep an eye out for the topside blisters that result from their feeding. All stages of the pest can be found during summer months, which suggests that there are a number of generations each season. In the greenhouse, the leafminer breeds continuously. Severe infestations may result in the complete destruction of foliage.

Handpick and destroy any blotchy and anemic-looking leaves. Burn them or place them in the depths of the compost pile, where the heat will kill the insects.

Azaleas held for a long while in the greenhouse may be severely damaged by the **cyclamen mite.** Although the mite is too tiny to observe in action, its feeding on new growth aborts or deforms both leaves and flower buds. Greenhouse temperatures between 60° and 80°F. favor this pest; growers have much less trouble with plants grown under lath or outdoors. Spray plants with water from time to time. Don't place newly potted azaleas with others until you're sure they're free of mites. Most nurseries are able to control mites, and you can buy mite-free plants from them.

The azalea **lace bug** [color] (as well as the rhododendron lace bug) injures azaleas by sucking plant juices from the undersides of leaves. The adult bugs have squarish lacy wings, are about ⅛ inch long, and have black markings. The nymphs are nearly colorless at first, and darken with time. Should the undersides of leaves show a reddish orange discoloration, suspect this pest.

Soap sprays have been used to good effect on the azalea lace bug, but if not many bugs are about, it may suffice to crush the bugs by drawing leaves between the thumb and forefinger.

For other insect pests of azalea, see RHODODENDRON.

## Diseases

The first sign of **azalea bud and twig blight** is dwarfing of flower buds, followed by their browning and death in late summer or early fall. The fruiting bodies resemble tiny pins stuck into the buds. Picking and burning infected plant parts either in late fall or early spring will help to reduce infection. Seedpods can be removed as soon as flowers have withered.

Perhaps the most serious disease attacking azaleas is **azalea petal blight** (also known as azalea flower spot and Ovulinia flower blight). The disease is found primarily in the southeastern United States, where it is particularly severe on azaleas of the Kurume and Indian types. The fungus causing this disease attacks only the flowers of the azaleas and, infrequently, the flowers of closely related plants, such as rhododendron and mountain laurel.

Azalea petal blight appears as small pale spots on the inner surfaces of the petals of colored flowers and as brown spots on white flowers. These spots rapidly enlarge until the whole flower collapses. In the diseased flowers, small dark resting bodies of the fungus serve to carry the disease through seasons when azaleas are not in flower. The resting bodies fall to the ground with the flower and remain there until favorable weather conditions return, at which time they germinate to produce little cups from which spores are shot. These spores, once on a moist azalea petal, germinate to produce an infection.

Azalea petal blight is controlled by avoiding overhead watering while the plants are in flower. Remove and destroy diseased blossoms and seed pods to keep the resting stage from perpetuating the disease. It also helps to cover beds with several inches of mulch. The disease is discouraged indoors by lowering the humidity.

**Chlorosis** is usually the result of environmental conditions rather than the attack of a disease-producing organism. Too much lime in the soil prevents plants from taking up certain elements, particularly iron. This is shown when leaves, particularly those at the tip of new growth, turn pale green or yellow. Leaf margins and the tissue between veins turn a lighter color, while the veins themselves stay dark green. The soil imbalance can be righted by adding plenty of organic matter to neutralize the basic soil.

Another soil-borne disease is **cutting and graft decay,** which results from infection by a soil fungus. In cutting and grafting frames, this fungus primarily attacks the tops of plants, growing over the leaves and stems as a cobwebby growth. Under moist conditions, it will cause the decay of the infected parts; otherwise, it may cause a rot at the base of the cuttings. It is controlled by using sterilized soil or sand in the cutting beds and by avoiding overwatering.

**Root injury** may result from hoeing or cultivating the soil around azaleas. These plants are surface feeders, meaning that the roots which absorb water and nutrients are found very near the surface. Disturb the soil around plants as little as possible.

**Salt injury** affects azaleas grown in containers. The symptoms include browning and death of the tissues around the margins of the leaves. Leaves may turn yellow, and small, dark or warty growths often appear on the undersides of leaves. Give an occasional heavy watering to leach away excess salts that have accumulated in the soil. If the plants are in small containers and can be handled easily, soak them overnight in a tub.

**Shoestring root rot** (known also as armillaria root rot) causes discoloration in both azalea and rhododendron. The disease is most easily diagnosed by cutting the bark of the crown and larger roots: if the fungus is present, there are large white fans or plaques between the bark and the hardwood and frequently small, dark, shoestringlike strands on the outside of the bark of the infected roots. Mushrooms in late fall or early winter are another indication of the fungus.

Once the root rot fungus becomes established, it is very difficult to control. Expose the crowns of infected plants to air: the fungus cannot exist under dry conditions. Because the fungus thrives in wet, heavy soils, improve drainage and avoid overwatering. Hybrid groups have been ranked for resistance to the disease. From least to most susceptible, they are: Indian Rutherford, Pericat, Glenn Dale, Whitewater, Satsuki, Black Acres, Gable, Kurume, and NCSU.

Under extremely wet soil conditions, a group of soil fungi known as **water molds** will attack the roots and crown area of young plants to cause wilt (also known as root rot or water mold root rot). Plants wilt and the leaves turn a dull green, followed by leaf drop and a permanent wilting. The main stem turns brown at the soil line, and a cut into this area reveals only dead bark and wood.

This disease only affects young plants growing in excessively wet soil, and it is easily prevented by providing good soil drainage before the plants are set out. In a heavy soil, avoid digging holes, even though they are filled with the proper soil mixture for the plant. Such holes act as a pocket to catch water, and thus provide the conditions that favor disease development. Planting too deeply also encourages trouble. See to it that the soil pH is between 4.5 and 5.5.

For other diseases of azalea, see RHODODENDRON.

# *BACILLUS POPILLIAE*

This bacterial control agent is used on Japanese beetles. It is marketed under several brand names, and infects grubs. See JAPANESE BEETLE.

# *BACILLUS THURINGIENSIS*

This bacterial pathogen is widely used as a biological control of pest larvae. It is safe to humans, other vertebrates, and plants.

*Bacillus thuringiensis* (or simply BT) was named in 1911 for the German province Thuringia, in which the pathogen was first encountered by a European researcher (the Japanese had discovered it a decade earlier). Today it is marketed under several trade names, including Dipel, Thuricide, and Biotrol. It is easily mass-produced, and is sold as endospores, in the form of a wettable dust or emulsion. BT is stable during storage, but the pathogen loses strength under sunlight. Additives are used to hold BT on plant surfaces; this is vital, because the bacterium must be ingested by insects as they feed.

BT was not used extensively in North America until the 1960s, but it is readily available today by mail order and through garden centers and nurseries. A pelleted bait containing BT is now marketed as Soilserv Bacillus Bait and has been registered for use in California for the control of some lepidopterous larvae on alfalfa, artichokes, broccoli, cucumbers, cabbage, celery, lettuce, melons, potatoes, spinach, and strawberries. The accompanying list shows which pests are successfully controlled with BT, arranged by the crops they are most apt to damage.

## INSECTS CONTROLLED BY BT

| Crop | Pest | Remarks |
|------|------|---------|
| Almond | peachtree borer | |
| Alfalfa | alfalfa caterpillar | As effective as chemicals. |
| Apple | codling moth worm | Because worms feed on the surface only a short time before tunneling, a good number of them won't ingest the pathogen and spraying will not result in flawless apples. |
| | tent caterpillar fall webworm eyespotted bud moth winter moth tentiform leafminer apple rust mite redbanded leafroller | Populations of predatory mites also dropped. |
| Artichoke | artichoke plume moth | |
| Castor bean | castor semi-looper | |
| Celery | celery looper, cabbage looper | |

*(continued on next page)*

| Crop | Pest | Remarks |
|------|------|---------|
| Corn | corn earworm | Variable results; works best in encapsulated form and when mixed with standard clay formulations, as earworms crawl down into the whorl and are hard to get at. |
|  | European corn borer |  |
| Cotton | cotton leafworm cotton leaf perforator bollworm | Good control. |
| Crucifers | cabbageworm diamondback moth cabbage looper | Control offered is good to excellent. Good to excellent control. Variable results (poor to excellent); especially useful on pests that have become resistant to traditional chemical sprays. |
| Grape | grape leaffolder |  |
| Orange | fruit tree leafroller orangedog |  |
| Peach | peachtree borer |  |
| Stored crops | various insects | BT has proven to be effective in controlling a number of insect pests of stored crops, but because the bacterium harms almost nothing but lepidoptera, its use has not yet proven practical. |
| Tomato | tomato hornworm, tobacco hornworm, cabbage looper |  |
| Trees, various | blackheaded budworm, fall cankerworm, spring cankerworm, California oakworm, elm spanworm, fruit tree leafroller, Great Basin tent caterpillar, gypsy moth, linden looper, saltmarsh caterpillar, spruce budworm, fall webworm western tussock moth, winter moth |  |

Adapted from *Microbial Control of Insects and Mites*, H. D. Burges and N. W. Hussey, editors. Academic Press, 1971.

# BACTERIA

For bacterial diseases of plants, see DISEASES.

Of the various insect diseases, those caused by beneficial bacteria are most numerous. Bacteria are simple one-celled animals, little more than plants. They usually enter insects through the mouth and then multiply in the bloodstream. Much like humans with the flu, insects become sluggish, lose their appetite, and get diarrhea. Unlike humans with the flu, infected insects dry up and may darken to brown or black.

Three bacterial insecticides are on the market. *Bacillus popilliae* is used against the Japanese beetle and *B. thuringiensis* against many pest larvae. These bacteria produce spores that, when eaten by a larva, blossom into rapidly growing plants that feed on the insect and occupy more and more of its body. After the insect's death, its decomposing body releases spores into the soil.

Fortunately, larvae cease to eat long before dying because the bacteria produce toxic crystals that paralyze the victim's gut. While sick larvae will hang on from two to four days, their stomachs don't work and they are no longer a threat to crops. See *BACILLUS THURINGIENSIS* and JAPANESE BEETLE. A third bacterial pathogen, sold by the name Gall-trol, is used as a root dip to combat crown gall disease.

Plants can be immunized against disease much as humans are. In his work at the University of Kentucky, Joseph Kuć protected cucumbers against at least ten diseases with a single immunization. Plants can be inoculated against fungal diseases, viruses, and bacteria.

The inoculant can be applied to the seed, soil, or leaf surface. It works systematically, spreading throughout the plant. Kuć believes the inoculant triggers a chemical response in the plant. The effect may last the life of the plant, or a second, booster inoculation may be necessary. Extracts of immunized plants can confer immunization on other plants, suggesting to Dr. Kuć that fields of crops could be immunized by spraying or treating the seed.

# BANANA

## Insects

**Nematodes** may attack the roots of banana plants to cause root knot. See NEMATODE.

## Diseases

Dwarf Chinese bananas are subject to **fingertip rot.** This disease starts in the flowers and advances along the peel more rapidly than in the fleshy pulp within. The malady must be prevented by inspecting the plants at periodic intervals and removing diseased fruit or withered sterile flowers. The infected material should be burned or buried.

**Freckle disease** occurs commonly on the fruit and leaves of Chinese bananas, causing black spots to appear. The fruit can be protected by wrapping it with paper before it matures. This will prevent the spores on the leaves above from spreading to the bananas. Also, if fruit becomes ripe before it is harvested, wrap it to prevent fruit fly damage.

**Panama wilt** is a soil-borne disease that can be brought into a new planting area on diseased planting material. This disease causes the plants to decay internally and eventually die. Leaves gradually die, often breaking in the middle of the leaf stalk. Frequently only one central leaf remains on the stalk, and it too will die in time.

It is extremely difficult to control this disease once it gains a foothold in the growing area. Be certain not to use diseased plants, and not plant bananas in soil that is known to have been recently infected with Panama wilt.

# BARBERRY

## Insects

The **barberry webworm** webs together leaves and twigs, forming a nest within which it feeds. This webbing usually starts after midsummer, and the nests remain on the bushes during winter. The caterpillar is blackish with white spots, and is nearly 1½ inches long when full grown. A spray or dust of pyrethrum will control these webworms if applied thoroughly when the caterpillars are small. Some of the nests can be handpicked and destroyed.

Other pests include the APHID, JAPANESE BEETLE, and SPIDER MITE.

## Diseases

**Rust** attacks European barberry and other varieties having sawtoothed leaves. (*Berberis koreana, B. mentorensis,* and most of the evergreen varieties are not susceptible even though they have sawtoothed leaves.) Japanese barberry is immune, but crosses of Japanese and European barberries may be susceptible. The fungus causes bright orange spots on leaves in spring from which come spores that infect wheat, oats, rye, barley, and some grasses. The disease rarely does much damage to the barberry shrub. In many states it is unlawful to grow susceptible varieties of barberry. If there is a susceptible variety of barberry growing on or near your property, destroy it.

**Wilt** is caused by a soil-inhabiting fungus that apparently enters the plant through roots. Leaves of affected plants turn yellow or red, followed by browning and premature defoliation. The sapwood of affected stems and roots is streaked with reddish brown lines. Older infected wood shows a brownish green discoloration.

Control by removing and destroying affected plants. Avoid replanting the location with barberry unless a volume of soil three feet square by one foot deep is replaced with new soil.

# BARRIERS

See BORDERS AND BARRIERS.

# BASSWOOD

See LINDEN.

# BEAN

## Insects

The tiny black bean **aphid** occurs on broad beans and, less frequently, on green and snap beans. They cluster on stems and under leaves, and cause leaves to curl and thicken. Plants become yellow and unthrifty. Aphids spread the virus of common bean mosaic (see below). Eggs are laid in autumn and, as typical with aphids, there are many generations each year. They overwinter on various shrubs. This species occurs throughout the United States, although infestations are local. See APHID and TIMED PLANTING.

The **bean leaf beetle** [color] is an occasional pest. Adults vary greatly in color and markings, but are typically reddish to buff in color, with a

Bean leaf beetle

black band around the outer wing margin and three or four black spots along the inner edge. They are ¼ inch long and overwinter in the adult stage. Adult beetles feed on the undersides of leaves and on stems of seedlings. The slender white larvae work belowground, attacking roots and stems. See MEXICAN BEAN BEETLE for controls.

Southern gardeners may run into the **bean leafroller,** an inch-long, chartreuse caterpillar that rolls leaves to feed within. Trichogramma wasps parasitize this pest, and are commercially available.

**Bean thrips** are barely visible, thin, dark insects with long, narrow, black-and-white wings that appear to be fringed. Groups of them feed to cause blanched, wilted leaves, spotted with dark specks of excrement. Control by ridding the garden area of weeds, especially prickly lettuce and sow thistle. Lacewings prey on thrips, and can be purchased for biological control. See THRIPS.

Thrips

The **bean weevil** [color] is ⅛ inch long and colored brown with thin lengthwise stripes on its mottled brown wing covers. The legs and antennae are reddish. The adult lays eggs in holes chewed along bean pod seams. The small, fat, white grubs enter and feed in young seed, and will emerge when beans are in storage. The exit holes are small and round. The beans can be cured by pulling up plants with pods on them at harvest time, and putting them on stakes to hold them off the ground. The curing process, which takes at least six weeks, involves certain biological changes, including fermentation and the production of heat. The cured beans may then be shelled and stored in a dry place without danger of weevil damage. Stored beans also may be protected by either heating dry beans at 135°F. for three to four hours, or by suspending seeds in a bag in water, heating to 140°F., and then drying rapidly.

Several species of leafhopper attack beans. The **beet leafhopper** [color] (commonly called whitefly in the West) is dangerous as a carrier of curly top, a virus disease. This pale greenish or yellowish pest is found west from Illinois and Missouri, excepting the fog belt along the California coast, which is relatively free from infestation. It is ⅛ inch long, wedge-shaped, and becomes darker toward winter. Spending winter as an adult in uncultivated areas, the leafhopper lays its eggs in March, and the first generation matures on these wild plants. The adults of this new generation swarm to other areas from early May to June, spreading curly top virus, of which they are the only known carrier. The symptoms of this disease are: warty pronounced leaf veins; kinked petioles; rolled leaves that sometimes take on a cupped or bell-like appearance, becoming brittle; and stunted growth, often resulting in death. The disease is not transmitted by seed.

The best control is the use of resistant varieties. Lacewing larvae feed on leafhopper nymphs. Rogue out afflicted plants as soon as the symptoms are spotted. Eliminate winter weed hosts from nearby. Since these insects do not breed in perennial grasses (they actually tend to avoid them), replace the neighboring weeds with grasses, such as *Eleusine indica* and *Leptochloa filiformis*. Once established, protect these grassy areas by watering or, if on a farm, by regulating grazing so that the areas are not destroyed.

Lima bean foliage is susceptible to the **cabbage looper,** a pale green measuring worm (inchworm) with light stripes on its back. See CABBAGE.

The **clover root curculio,** a small robust beetle with a short blunt snout, may visit your beans. They vary in color from light to dark brown with shadings of gray and brick red, and cut regular notches in the leaf margins. The larvae are tiny, gray, legless grubs with light brown heads; they attack roots. Large-scale problems may be solved by crop rotation.

All types of beans are host to the **corn earworm**[color]. The caterpillars grow to nearly two inches long and vary in color from yellow to green to brown, with lengthwise stripes and a yellow head. These pests help to keep their own numbers down—they're cannibalistic. Also in your favor are egg parasites, birds, moles, and cold, moist weather. Curiously, a border of marigolds or summer savory will increase earworm damage on snap beans. See TIMED PLANTING.

The yellow-and-green-banded **cucumber beetle** shows three bright green stripes running across the wing covers. They grow no larger than ¼ inch, and feed on bean leaves throughout the South from South Carolina to California. For control, see CUCUMBER.

The **cutworm** is familiar to almost every gardener. These are the smooth, fat, soft larvae of night-flying moths. They feed close to the ground on the succulent bean stems. See TOMATO for more information.

The larvae of the **European corn borer** occasionally bore into the stems and pods of beans. They are flesh-colored caterpillars, up to an inch long, with rows of small, dark brown spots. The female moth is yellow brown with wavy dark bands, and the male is slightly darker in color. These adults have a wing spread of one inch, and usually fly at night. See CORN.

The **fall armyworm** [color] is an occasional pest. The larvae vary greatly in color, from light tan to green to almost black, and have three yellowish white hairlines down the back from head to tail. On each side is a dark stripe paralleled below by a wavy yellow one, splotched with red. The head is marked with a prominent white V or Y. They eat leaves, stems, and buds of plants. See armyworms under CORN for controls.

Small bean seedlings may be attacked by the **garden springtail,** a tiny dark purple insect with yellow spots. They have no wings, but use tail-like appendages to hurtle themselves into the air. You will find springtails eating small holes in leaves near the surface of the soil. Clear the area of weeds, and spray soapy water or a mixture of garlic and water on leaves.

Garden springtail

The adult form of the **grape colaspis** (or clover rootworm) is a tiny, pale brown, oval beetle that is marked by rows of punctures. Telltale signs of this beetle are long curved or zigzag feeding marks on leaves. Plowing or spading in fall, rather than in spring, may give some control.

The **green cloverworm** is an occasional pest of beans from the eastern states to the Plains. The wriggling green caterpillars riddle bean leaves for about one month, but control is rarely necessary.

The **green stink bug** does indeed put off a foul odor, but is more easily identified by its appearance: large (⅝ inch long), oval, and bright green with a yellow or reddish margin. It has a special liking for beans, and sucks sap from shoots and leaves, causing pods to fall and distorting seeds. Control measures are rarely necessary for this pest. Dust with sabadilla in emergencies.

The **lima bean pod borer** can be a headache, particularly in the southern part of the United States. The borer is a white to greenish or reddish caterpillar, has a pale yellow head, and grows up to one inch long. It wriggles violently when disturbed. The borer pierces through lima bean pods and eats the seeds within, passing from one to another. While it usually devours many seeds from a small-seeded plant, it may obtain its entire nourishment from a single large seed. When the larva has finished feeding it goes to the ground and may crawl some distance to find a place to pupate. Its manner of crawling may attract your attention—larvae move rapidly and have been observed crawling up and down the walls of nearby houses. If the soil can be penetrated, the larvae attempt to spin their cocoons in the ground; otherwise, they enter leaves, trash, or other debris.

Eliminate these winter homes by cleaning the garden area thoroughly. Crop rotation is recommended for large-scale growers. Because the borers aren't difficult to spot, it should be easy to handpick them from leaves and pods. Damage is usually greater to beans planted late, so a crop planted early may escape a good deal of damage from the borer.

Irregular sections chewed from bean leaves may be the work of the **margined blister beetle;** it is black with a thin gray or yellow margin around the wing covers. Handpick with gloves. Protect plants with mosquito netting.

The voracious **Mexican bean beetle** [color] feeds on leaves, pods and stems the entire season as both a larva and adult and has a reputation as the number-one enemy of eastern vegetable gardeners. The latter form is ¼ to ⅓ inch long, broad-domed, pale yellow to coppery brown, and oval, with eight small black spots on each wing cover. It will attack any garden bean, but

Mexican bean beetle

has a particular fancy for the lima bean. The Mexican bean beetle spends the winter in rubbish or weeds, appearing late in March in the South and in June in the northern United States. Two weeks later, the adults deposit yellow clusters of eggs on the undersides of leaves. The larvae are ⅓ inch long, yellow, with six rows of long, branching, black-tipped spines covering their backs. After skeletonizing bean leaves, the larvae attach their hind ends to an uninjured leaf and the pupae emerge from their crushed larval skins. In the North, there is one generation and a partial second; near New York City there are two generations; and further south, three to four.

Control of this pest becomes easier as the season progresses and the beetle moves to the uppersides of leaves. See MEXICAN BEAN BEETLE. Keep an eye out for telltale skeletonized leaves, and destroy the yellow orange egg clusters found on the undersides of leaves. Sanitation is also a help, depriving the beetle of a winter home; clean up all plant debris after harvest. Time of planting can save your plants: early plantings have been found to be freer of the pest, so plant a heavy early crop for canning and freezing; near New York City, snap beans planted in early June will mature in July between generations of beetles. See TIMED PLANTING. Interplant rows of potatoes between every two rows of green beans; not only do the potato plants repel the Mexican bean beetle, but the green bean plant is noxious to the Colorado potato bug. The bug is not, however, put off by lima bean plants. Alternate rows of nasturtiums also are effective. A clove of garlic, planted in every hill of beans, will chase away this and several other bugs.

Try companion planting to keep the Mexican bean beetle away. Savory has a reputation as the "bean herb," and young seedlings spaced at intervals in the furrow in place of a bean at planting time will protect the bean patch. Build protection into the bean patch at planting time by placing young savory seedlings in place of beans at regular intervals. Because savory is fine-seeded, it is best to sow it inside and then transplant seedlings at the time beans are planted. Strong-smelling marigolds also do a good job of keeping away these beetles. Cedar-based sprays will repel them.

Bean varieties vary in their susceptibility to the Mexican bean beetle. See RESISTANT VARIETIES.

Root-knot is a condition caused by **nematodes**, whitish, translucent worms barely visible to the naked eye. See NEMATODE.

The **palestriped flea beetle** damages foliage by perforating the leaves. It is $1/16$ inch long and has a broad white stripe on the center of each pale to dark brown wing cover. See controls for flea beetle under POTATO.

The tiny ($1/16$-inch-long) black **potato flea beetle** is so named because it jumps like a flea when disturbed from its meal of bean leaves. See POTATO for control methods.

The **potato leafhopper,** considered the most serious pest to potatoes in the East, also visits bean patches, especially in the South, where it is called the bean jassid. It is small ($1/8$ inch long), wedge-shaped, and green with small white spots on its head and thorax. It causes bean leaves to whiten and may stunt, crinkle, and curl them as well. For control measures see POTATO.

The **seedcorn maggot** immigrated to New York from Europe more than a century ago, and is now found all over the country. The yellow white, $1/4$-inch-long maggot has a long, sharply pointed head, and bores into bean seeds in the ground so that they will not produce good plants. Damage is greater in cold, wet weather, and when seed is deeply planted. The adult form, a grayish brown, $1/3$-inch-long fly, emerges in July to lay its eggs in soil or on seeds and seedlings.

Seed should be planted shallow when the ground is warm enough to enable the seedlings

Seedcorn maggot and dark puparium

to get a good start. If maggot damage is heavy, the area should be replanted immediately.

Of the several species of **spider mite,** the twospotted spider mite is probably the most common. These pests are tiny, oval, yellow or greenish specks, and but $\frac{1}{16}$ inch long at

Twospotted spider mite

maturity. They live on the undersides of leaves and branches and suck plant sap. Leaves take on a sickly appearance, with yellow or reddish brown blotches. Centers of infection often arise near roads or other areas where plants may be covered with dust. The mite population increases rapidly in hot, dry weather. Control is often not necessary, but see SPIDER MITE if things get out of hand.

Bean growers east of the Rockies may be bothered by the **stalk borer.** When young, this caterpillar has a brown or purple band around its cream-colored body, with several stripes running lengthwise. The fully grown borer loses its stripes, and may take on a grayish or light purple color. For control, see ASTER.

The **tarnished plant bug** will likely retreat to the opposite side of its stem as you approach, but you may catch a glimpse of a flat, $\frac{1}{4}$-inch-long insect, colored green to brown with yellow, brown, and black markings, giving it a tarnished appearance. A very close look will reveal a yellow triangle, with one black tip, on each side of the insect. The tarnished plant bug is found all over the country, and is a pest of more than 50 crops. For habits and control, see ASTER.

Tarnished plant bug

The **whitefringed beetle** grub [color] lives in the soil and feeds on the roots of beans and a number of other vegetables and ornamentals. This larva is yellowish white, legless, curved, and grows to $\frac{1}{2}$ inch long. It eats tender outer root tissues and may sever roots, turning plants

Whitefringed beetle adult and larva

yellow and limp, and finally killing them. This wormlike pest invaded Florida in the 1930s and has spread through the South from Texas to Virginia. Along the way it acquired a taste for almost 400 plants. Less of a danger to the garden is the whitefringed adult, a $\frac{1}{2}$-inch-long beetle, dark brownish gray in color, with a broad, short snout. It is covered with short hairs and has a light band along its side. Because its protective set of wing covers are fused together, this beetle cannot fly.

Whitefringed beetles are particularly fond of legumes, and large-scale plant growers shouldn't plant more than a quarter of their crop land in annual legumes; neither should they plant the same land with these crops more than once in three or four years. On a small scale, the home gardener can do much to control this pest. Be certain not to plant its favorite foods, the legumes, together. Clean cultivation can go a long way toward the elimination of this beetle. Fertilize the soil heavily so that the plant may become strong enough to fight off the insects' attack; if possible, turn under a winter cover crop, and add the advantages of the crop's nutrients to the soil.

**Wireworms** may attack seeds and kill seedlings by eating underground stems of young plants. These worms are shiny, hard-jointed, smooth, yellow or brown, and grow to one inch long. For control methods, see WIREWORM.

## Diseases

**Anthracnose** is a fungal disease. The most obvious symptoms are dark brown sunken spots on the seed coat that may extend through to the cotyledons. The spots darken and take on reddish brown margins. Under humid conditions, the centers of the spots may become pinkish. Brown to black ovals develop on the stems of the seedlings and may girdle and kill them. Veins on the lower surfaces of the older leaves appear dark red or dark purple, and in severe cases angular dead spots appear on the upper surfaces of the leaves. Bean plants are most susceptible in cool, moist summers. The fungus is carried

Anthracnose

on infested seed and lives on the remains of diseased plant refuse in the soil; it occurs in the central northeastern, and southeastern states.

The use of disease-free, western-grown seed has nearly eliminated the threat of anthracnose in some areas. Do not plant seeds that are discolored or that come from spotted pods. Stay out of the garden when plants are wet, as the disease spreads more rapidly on wet foliage. A number of resistant varieties are available; check with your seed store or state experiment station for the variety best suited to your area. As a final precaution, rotate crops and plow down old bean tops as soon as harvested.

If **bacterial blight** strikes your bean patch, you should have little trouble recognizing the symptoms. Look for leaf spots with a water-soaked or greasy appearance. One species causes a yellow halo to form around the spots (halo blight). The affected areas later turn brown and die. Similar grease spots form on pods, but these marks may turn reddish brown in drying. If a plant is afflicted early in the season, it may be severely stunted or even killed.

Sprays will not help you to combat this blight, as the bacteria occur *under* the seed coat. The above methods of control for anthracnose apply to blight. Use of California-grown bean seed has eliminated this disease from much of the Deep South. Sidestep halo blight by delaying planting. See RESISTANT VARIETIES.

**Bacterial wilt** is seed-borne and overwinters on both the bean seeds and on diseased bean plant refuse in the soil. This disease usually attacks and kills seedlings; if a plant becomes infected when taller than three inches, it may continue to live for some time. Leaves become limp, wilt, and then die. The wilting is at its worst during the warm part of the day, and if the disease is not too far advanced, the wilted leaves may temporarily regain some of their vitality in the cool of the night or during humid cool spells. The causal bacteria invade the plants' vascular elements and plug or injure the vessels which carry water from the roots to the plant tops. Plants are often afflicted with both blight and wilt concurrently.

To avoid problems with bacterial wilt, plant only seed that is certified wilt-free. Control measures are the same as for bean blight, as described above.

High on the list of bothersome diseases that plague the bean grower is **bean rust.** Its symptoms are easily recognized: look for numerous small, orange to brown pustules, found most often on the undersurfaces of leaves, but also occurring on petioles, stem, and pods. Badly affected leaves turn yellowish, dry up, and drop to the ground.

Control of bean rust involves cutting down or discing vines as soon as the last picking is made; this serves to prevent formation of the overwintering stage. Don't leave overwintering spores on stakes, and be sure to use new stakes if the disease was particularly developed during the previous growing season. Use wire instead of stakes to support bean plants. Posts can be treated by spraying or dipping them with one part lime sulfur in ten parts water. Also, there are several rust-resistant varieties available, including White Kentucky Wonder 191, U.S. No. 3 Kentucky Wonder, and Dade. Avoid susceptible varieties, such as Blue Lake, McCaslan, and Kentucky Wonder. This disease is not seed-borne, but the wind can carry spores a great distance; therefore, a long crop rotation is advisable. When choosing a planting site, remember that prevailing winds may blow spores in from an infected field. The fungus overwinters in old stems. If you have had serious rust infections, dust plants with sulfur before you expect the first symptoms to appear.

**Common mosaic** is caused by a virus carried by aphids. Plants from seed that is infected with this virus and those that become infected in the seedling stage are severely stunted and set relatively few pods. The leaves turn a mottled light and dark green, become crinkled, and curl downward at the edges. Plants that become infected as they approach blooming time or later in the season may manifest mottling and a downward curling of the leaves only at the terminal growths, while in other cases the symptoms may not be at all apparent. Regardless of how affected the plants may seem, a few of the seeds in each pod may carry the virus inside.

Because mosaic viruses are carried in the sap of the plant, an afflicted plant cannot be saved. Prevent trouble by planting certified seed. See RESISTANT VARIETIES.

**Curly top** is a virus carried by the whitefly (or beet leafhopper) and is not seed-borne. The disease dwarfs young plants, and leaves pucker and curl downward. Leaves may be cupped, or take on the appearance of small green balls. Plants infected when very young will usually die, while mature plants do not show the typical symptoms and usually live.

The best method of control is the use of resistant varieties such as University of Idaho, Great Northern, and Red Mexican.

**Downy mildew** attacks lima beans, and does its worst in the Middle Atlantic and North Atlantic states. This fungal disease causes irregular, white, cottony patches on bean pods; some patches may be bordered with purple. Affected pods shrivel, wilt, and die. The disease occasionally attacks young leaves, shoots, and flower parts. Wet weather, with cool nights, heavy dews, and fairly warm days, causes downy mildew to spread rapidly. During the growing season, it may also be carried to healthy plants by insects.

Avoid seed from an infected crop; to be safe, use seed from the Far West. Large-scale growers should use a two- or three-year crop rotation. Thaxter is a resistant lima bean variety.

**Fusarium root rot,** a fungal disease, first appears as a slight reddish discoloration inside the taproot. Gradually the entire taproot turns a deep red, then brown, and finally decays. Under favorable conditions, developed lateral roots may keep the plants alive.

Control is difficult. In serious cases, cultivation should be shallow and the crop should be irrigated more frequently, if possible.

**Seed rot** and damping-off (seedling rot) are caused by fungi which are present in nearly all soils. Seed rot is a semidry rot that occurs at the time of germination; it is apt to be a problem in cool, moist weather. Damping-off is a watery, soft rot of the young seedlings at or below the ground line, resulting in wilting of the unfoliate leaves and death of the seedling.

Although plants can be affected under a variety of moisture and temperature conditions, root rotting is not likely to occur in cool, moist soils. Grow the beans in soils warm enough to encourage rapid germination.

**Southern blight** (or wilt) occurs in southern states and is considered a minor disease. However, it attracts attention because of its striking symptoms: yellowing, wilting, and shedding of the leaves, and sudden wilting and death of the vine. Control shouldn't be necessary.

It goes without saying that **weeds** make trouble in the bean patch. Care should be taken not to hoe too deeply, as bean roots lie close to the surface and deep or extensive cultivation may result in undesirable root pruning. A heavy mulch can do the job of suppressing weeds, and also serves to preserve moisture in times of drought.

The first signs of **white mold** are dark watery spots on pods, stems, and leaves. These spots enlarge rapidly under cool, moist conditions. The stems may be rotted through, causing wilting and death of part or all of the affected plant. Infected pods turn into a soft, watery mass, those that touch the ground being most susceptible. Very soon after the rot appears, you will notice a white, moldy fungal growth on the rotted area. These areas have the appearance of small patches of snow, unless darkened by soil. In a few days, the mold darkens and produces black kernels, called sclerotia, and may be seen in the drying fungal mass. The fungus carries over in the form of these sclerotia, which are resistant to drying and changes in temperature. Sclerotia in the soil may perpetuate the fungus for at least ten years.

Use crop rotation; do not follow lettuce with beans. Monitor humidity and soil moistness around the plants; a field known to have white mold fungus should not be irrigated more often than necessary, and rows should be spaced wider to allow air to circulate around the plants.

**Yellow mosaic** is carried by aphids to bean plants from infected gladiolas, crimson, and red clover, as well as from other bean plants. Leaves of affected plants show much contrast between the yellow and green areas. The plants become dwarfed and bunchy. Follow control measures listed above for common bean mosaic. In areas where yellow mosaic is widespread, avoid planting beans very close to fields of host plants.

# BEECH

## Insects

White **beech scale** is responsible for introducing a serious fungal disease to beech trees. As oil sprays may harm this tree, use a lime sulfur and water spray, mixed one part to ten.

## Diseases

**Leaf mottle,** or scorch, appears in spring on young unfolding beech leaves as small, semi-transparent spots surrounded by yellowish green to white areas. These spots enlarge, turn brown and dry, and are very prominent between the veins near the midrib and along the edge of the leaf. Later, entire leaves become scorched and premature leaf fall follows. In July, a normal set of leaves develops to replace fallen leaves. However, severe scalding of the bark on branches may occur as a result of premature defoliation exposing the branches to direct rays of the sun.

The causal agent of this disease is unknown. As a preventive measure, provide adequate fertilization. Where valuable trees have been defoliated, wrap exposed branches with burlap to protect them from the sun until the second set of leaves develops.

# BEET

## Insects

The **beet leafhopper** [color] (commonly called whitefly in the West) carries curly top, a viral disease, to table beets and a number of other vegetables and flowers. It is pale green or yellow, ⅛ inch long, wedge-shaped, and becomes darker toward winter. The symptoms of curly top are: warty, pronounced leaf veins; kinked petioles; rolled leaves that take on a cupped, or ball-like, appearance and become brittle; masses of hairlike rootlet growths on tap roots; stunted growth; and in many cases, death of the plant. See beet leafhopper under BEAN for control measures.

The **beet webworm** is found throughout the United States, but is especially harmful to beet crops in western states. It varies in color from yellow or green to nearly black, with a black stripe down the middle of the back and three black spots at the end of each segment. It feeds on leaves and constructs a protective hideout by rolling and folding a leaf and tying it together with webs. Sabadilla and pyrethrum are effective controls, but handpicking should usually suffice.

Another damaging insect that gives headaches to beet growers is the **black blister beetle,** also called old-fashioned potato bug, yankee bug, and just plain blister beetle. They are fairly long (½ to ¾ inch), slender, and have soft, flexible wing covers. The entire body is black or dark gray, and there may be thin white stripes

Blister beetle

on their wing covers. Blister beetles are usually found in swarms or colonies feeding on the blossoms and foliage of any of a number of garden and field crops—vegetables, vines, trees, and flowers. They are so named because a caustic fluid in their bodies may blister the skin if the beetles are crushed upon it.

To prevent damage, handpick the beetles as soon as they are discovered in the garden; because of the powerful, caustic fluid that they discharge, you should wear gloves. The blister beetle may be controlled with a safe insecticide made from sabadilla seeds. Sabadilla is available commercially from many seed and garden dealers. Perhaps because of the caustic liquid, hens will eat one blister beetle and no more.

Some gardeners literally chase large swarms away with branches and shouts.

The **carrot weevil** may injure the roots of beets. It is a coppery colored or dark brown beetle measuring ¼ inch long. The larvae are dirty-white, legless, curved grubs with brown heads. They grow up to ⅓ inch long, and tunnel into the roots of affected plants. For habits and control measures, see under CARROT.

Small seedlings may be attacked by tiny, dark purple, yellowspotted **garden springtails.**

Garden springtail

They have no wings, but use a tail-like appendage to hurtle themselves into the air. You will usually find springtails eating small holes in the leaves near the surface of the soil. Keep weeds down around the garden. A garlic spray on the vulnerable lower leaves may keep spring-tails away.

The tiny (¹⁄₁₆-inch-long) black **potato flea beetle** is so named because it jumps like a flea when disturbed. See under POTATO for control measures.

The **spinach flea beetle** emerges as an adult in early spring. This is a greenish black beetle with a yellow thorax and black head. It averages ⅕ inch long, and feeds on exposed beet leaves. The larval form is a dirty grayish purple, warty grub, ¼ inch long, that feeds on the undersides of leaves. The grubs drop to the ground when they sense danger. For control, see SPINACH.

The **spinach leafminer** is another spinach pest that attacks beets. The adult is a gray,

black-haired, two-winged fly about ¼ inch in length. It lays white, cylindrical eggs on the undersides of leaves in clusters of from two to five. In a few days these eggs hatch into ⅓-inch-long, pale green or whitish maggots that eat leaves. For control, see SPINACH.

## Diseases

A fungal disease that strikes gardens east of the Rocky Mountains is **leaf spot.** Small, circular, tan or brown spots with reddish purple borders appear scattered over leaves and stems. The spots later turn gray with brown borders. Heavy infection causes the leaves to become yellow and either scorch or drop off. Infection and dying of leaves and the production of new leaves result in a pyramiding effect of the crown. This fungus overwinters on seed balls and lives in the soil or on diseased remains of plants. Although not usually a serious threat to beets, it may be advantageous to ward it off by using a crop rotation of three years or more. Be certain either to cultivate deeply or to overturn beet refuse in the soil. This deep tillage will destroy the plant remains that usually harbor the fungus. The selection of resistant varieties depends upon your location; check with the county agent or extension service.

Beets will have a better chance of avoiding **yellows** virus if kept out of the wind—aphids are a prime vector and are light enough to be blown about in a breeze. Oil-emulsion sprays may keep aphids from transmitting this disease. Corn-oil emulsions stymie the aphid vector of beet mosaic virus.

# BEETLE

Here are descriptions of several beneficial beetles. For pests, see plant entries, and also JAPANESE BEETLE and MEXICAN BEAN BEETLE.

## BLISTER BEETLE.

While these slender, ½- to ¾-inch-long insects are most often thought of as pests, the larvae of some species travel through the soil to feed on the egg pods of grasshoppers.

## CHECKERED BEETLE.

Bark- and wood-boring beetles must contend with this brightly colored family of ⅛- to ½-inch-long beetles [color]. They are covered with a thick coat of short hairs, and bear clearly visible patterns on their rounded bodies. The larvae are hairy, yellow to red, and horned. Adult checkered beetles prey on adult wood borers, while the beetle larvae take care of pest larvae and eggs.

## FIREFLY.

Also called lightning bug, this insect is well known for its summer pyrotechics. The larval stage, called the lampyrid beetle [color] or glowworm, is short on looks but helps gardeners by feeding on slugs and snails. It is flat and has a toothed outline. Its jaws are suited for violent work: the larva ambushes snails by climbing over the top of the shell and digging in with its jaws when the mullusc sticks its head out. The snail is paralyzed so quickly that it isn't able to retract its head. A digestive enzyme is secreted to make prey easier to eat. The larva lives in the ground and hunts after dark. The adult is attracted to low vegetation.

## GROUND BEETLES.

Don't confuse these beneficial beetles with the big, bungling June beetles. The good species are less rounded. They are shiny, dark, and live in or near the soil by day. When disturbed in their daytime hiding places, these long-legged bugs run for cover; they don't often fly.

One ground beetle, *Calosoma scrutator*, is aptly named the fiery searcher for its caustic secretions (handle them with gloves) and voracious appetite for caterpillars. They are dark, with a greenish tinge to the thorax and a purplish tinge elsewhere. Another species, the European ground beetle *(C. sycophanta)*, was imported to North America and is now the most important predator of gypsy moth larvae, eating up to 50 of these pests to sustain its development. You'll have little trouble spotting this one — it is iridescent blue and green with a gold cast. Where it can't find gypsy moth larvae, it climbs trees to stalk cankerworms. The large (over an inch long) and attractive (black with a border of blue around the forewings and thorax) *Pasimachus depressus* attacks armyworms and cutworms east of the Rockies; by day it hides beneath rocks and logs.

## LADYBUG.

Likely the best known of beneficial insect predators, the ladybug (or ladybird or ladybeetle, if you will) [color] has become somewhat of a symbol for the entire process of biological control. The ladybug's appearance is paradoxical, in that its rotund, cheerfully colored adult stage looks anything but ferocious. On the other hand, the appearance of the spring larva matches its reputation for extended feats of avarice. This creature is flat-bodied, tapered to the tail when viewed from above, and shows blue or orange spots on a dark gray background. Even hungrier than the adult, the larva begins feeding the moment it leaves its orange egg, and can handle 40 aphids in an hour.

The species used most often to control aphids is the convergent ladybug *(Hippodamia convergens)* [color]. It is unfortunate that this beneficial beetle and the destructive Mexican bean beetle [color] should be near look-alikes. But the latter has 16 spots, and the ladybug usually has 12, sometimes fewer. Also, the ladybug has the converging marks responsible for its Latin name — they form a broken V on the dark thorax that points to the rear.

Some ladybug species are identified by the number of spots they carry. The 16-spotted Mexican bean beetle may look like a benign friend, but don't be fooled. It has no markings on its thorax, unlike the beneficial species, and it is larger as well.

For more information, see LADYBUG.

**ROVE BEETLES.**   With their long, dark, flat bodies, rove beetles look somewhat like earwigs. These predators are credited with eating mites, bark beetle larvae, and cabbage maggots. They lack the earwig's long, curved appendages at the posterior, and are equipped with stink glands that are capable of propelling a noxious fluid whichever way the tail is pointed. When alerted to danger, rove beetles typically lift their tails in readiness.

One member of the family is known as the red spider destroyer, a small, hairy, black beetle whose yellow larvae hatch from orange eggs to eat larvae, at a clip of 20 or so a day. Another small black beetle, *Aleochara bimaculata,* digs tunnels in the soil to deposit eggs. The larvae seek out and destroy cabbage maggots and larvae of the cabbage fly. The beetle's brown, horned larvae may eat up to 80 percent of the cabbage maggots in a field.

**SOLDIER BEETLES.**   Members of this family look much like fireflies, but lack the fire. The adults typically are found on flowers, while larvae are helpful predators. The downy leatherwing *(Podabrus tomentosus)* is found across North America; its black body bears short hairs, and thorax and head are pale. East of the Mississippi, gardeners are befriended by *Chauliognathus pennsylvanicus,* known as the Pennsylvania leatherwing; this beetle has yellowish wings, a large black spot on its thorax, and spots at the back of both wing covers.

**TIGER BEETLES.**   These ¾-inch beetles are distinguished by their flashy, iridescent hues of blue, green, bronze, and purple. The larvae spend their time at the bottom of tunnels, waiting for an insect to stumble in.

# BEGONIA

## Insects

The melon **aphid** (and likely a number of other species) occasionally infest the begonia; see APHID. Other pests include the cyclamen mite (AZALEA), flower thrips (PEONY), fuller rose beetle (ROSE), MEALYBUG, and SNAIL AND SLUG.

## Diseases

**Bacterial wilt** often strikes begonias. Small light spots appear on the undersides of leaves and gradually enlarge and merge into water-soaked areas. The leaves then turn brown and die. Also look for wilting and browning of stems and leaf areas. A puslike bacterial slime fills the stems and midribs of plants and leaves.

Plants severely affected by bacterial wilt must be pulled out and destroyed. Those that aren't so bad may survive if the infected areas are cut out with a clean knife. Since this disease is highly contagious, it can be spread by direct contact with tools, boxes, pots, and even sprinklers that have been used on infected plants. The disease thrives best in periods of high humidity, so if possible, keep the indoor area dry and well ventilated. Remove diseased leaves, stems, and plants from the growing area, as they are the springboards for the propagation of the disease. Set only healthy plants.

One trouble that may confront the begonia grower is **dropping of buds,** a malady with a couple of possible causes. Buds may drop if the begonias are exposed to high temperatures. Because tuberous-rooted begonias and gloxinias require temperatures between 65° and 75°F., they should be planted in a cool, shady spot of the garden. Indeed, begonias are a desirable ornamental for gardens that do not get a lot of sun.

Both begonias and gloxinias like well-drained soils, and if the center buds drop first, it is usually a sign that the soil has poor drainage. This does not mean that the soil should not be watered, for dried-out soil can also cause dropped buds, especially once the plants have formed buds. The best way to prevent such problems is to add plenty of well-composted organic material to the soil. This vital addition will allow the soil to soak up more water and nutrients, while enabling it to drain well without quickly drying out.

Begonias sometimes have too many leaves for the amount of flowers. **Excessive foliage** may result if too many shoots have been allowed to remain on the tuber. Leave just one shoot per tuber, and break off any others before they reach a height of two or three inches. Another possible cause of excessive foliage arises from soil that is too rich in nitrogen, often as the result of super-soluble chemical fertilizers. Too much nitrogen keeps plants in a vegetative state.

Gloxinias and tuberous-rooted begonias are both affected by **mildew.** The first evidence of this disease is a soft, grayish growth on leaves. Later stages of mildew are indicated by a soft, dark brown discoloration of the leaves. Mildew often occurs when plants get too much water. Water only as needed, provide good ventilation, and remove diseased plants.

**Rotting plants** are sometimes caused by contact with decayed matter that hasn't been well composted. Also, be certain to remove all flowers when they have finished blooming.

**Stem rot** results from overwatering and the use of unsterilized soil. You can escape trouble by clearing the area of old plant rubbish and by being careful not to draw soil or mulch up too close to stems. As soon as a rotted spot is noticed on a stem, cut the infected area out with a sharp knife or razor blade.

# BIOLOGICAL CONTROL

Biological control is defined as the restraining influence of parasites, predators, and pathogens on populations of insect pests and weeds.

This force goes on all around us, all the time, without our having to engineer it. But we do discourage biological control with monoculture (growing just one species of plant in a yard or garden or field). We also discourage it by spraying strong chemical insecticides and herbicides.

Humans can initiate and encourage *applied biological control,* which has been practiced in this country since the 1880s. The importation of natural enemies to control an exotic (imported) pest is known as *classical biological control.*

Biological control in the United States was given a boost by the spectacular success in controlling California citrus pests with natural enemies. Today that state's biological control program is the biggest in the United States. Another major catalyst was Everett Dietrich of the Rincon-Vitova insectary. He broadened the applications of biological control for the state, then split off to do his own work.

Backyard growers won't forage abroad for biological control agents, of course. You can encourage the beneficial organisms already in and around your plants, and attract others with plantings that serve as alternate food sources for predators and parasites. Beneficials may also be lured by spraying simple foods. Commercial nectar-and-pollen substitutes are available. These are mixed with water and sprayed.

Insectaries in the United States and Canada offer an increasing variety of natural helpers— insects, mites, and nematodes. Although these beneficials are marketed with the commercial grower in mind, small-scale growers find them of use in the garden, orchard, greenhouse, and even on windowsill plants.

Buying beneficial bugs or pathogens isn't necessarily a cheap way of managing pests on a small scale. But it is certainly safe and satisfying, even fascinating, to the grower whose interest in plants goes beyond what may be reaped from them.

See also MICROBIAL CONTROL, NEMATODE, PARASITIC INSECTS, PREDATORY INSECTS, and SPIDER MITE.

# BIRCH

Unless necessary for controlling a pest, prune birches as little as possible, and then only at a time when they are in full leaf, since these trees bleed profusely. The best time for doing the necessary pruning is late summer, for large wounds heal over rapidly in July and August when the tree is running less fluid in preparation for the coming winter season.

## Insects

Two species of **aphid** are common on birches, the yellowish birch aphid and the waxy birch leaf aphid. Migration of winged aphids in swarms occurs in severe infestations. Control these aphids with a forceful spray of water. Also see APHID.

The adult **birch leafminer** is a small black sawfly that measures $3/16$ inch long. It attacks most North American birches except the yellow birch, causing blotch mines and brown spots on the leaves, which come to appear blighted or scorched. Tender terminal leaves are most susceptible. The miner adult lays its eggs in newly growing leaves; it will not lay eggs in the older, hardened leaves. The larvae develop in mines in the leaves. At maturity they fall to the ground to pupate and overwinter as pupae in the soil. Adults emerge early in May (late May in Canada) and may continue to emerge for a two-week period. Depending on the length of the growing season, there may be three or four generations, the first two of which are the most injurious. The vitality of the tree is reduced because of loss of foliage, making it more susceptible to the bronze birch borer and other pests.

Sanitation is important in controlling the birch leafminer. Gather and burn or compost all leaves that fall to the ground prematurely. Because the leafminer takes a liking to new foliage, it is a good idea to prune back all unnecessary new growth. This is especially true of water sprouts, which are not necessary to the proper growth of the tree. Pruning is especially important when new broods of the insect are seen to emerge early in the growing season.

The larvae of the **birch sawfly** sometimes feed around the margins of birch leaves, but as they seldom do serious damage, control measures have not been necessary. The larvae of another sawfly, the birch leafminer, is much more important; see above.

The biggest enemy of birch is the **bronze birch borer,** a slender, bronze-colored, ½-inch-long beetle. The upper portion of the tree is often infested first, and shows spiral ridges on the bark of the branches. Although it consumes a bit of foliage, the adult causes little injury itself. After emerging in the latter part of May, the female lays her eggs upon or in the bark, and tunneling borers begin work at the top of the tree, giving it a stag-head appearance as the insects continue downward. Rusty red patches appear on the bark. When the tunnel network has expanded enough to cut off the flow of sap in the main trunk, death occurs.

Unfortunately, the birch borer often becomes entrenched before its presence is discovered. The small larvae hatch and tunnel through the bark to the sapwood, where the insect overwinters in a dormant state. The only method of safe control is to keep an eye out for infestation.

To control these insects by natural means, slice out and burn all infested parts and suspect debris before the adult emerges in May. This will prevent a profuse hatching of young beetles. Birches should be properly nourished and watered in dry spells.

The **casebearer** is a light yellowish green caterpillar, about $\frac{1}{5}$ inch long, that mines in the leaves and causes them to shrivel. Small cylindrical cases are formed on the undersides of the leaves. The adult moth is brown, and the insect overwinters in cases on the bark. Kill the overwintering larvae by spraying with lime sulfur and water, mixed one to eight, before growth commences in spring.

Caterpillars of the **gypsy moth** feed on birch. See TREES.

**Oystershell scale** is a common pest of young birch sprouts; see LILAC.

## Diseases

Paper and yellow birch are attacked by the **European canker fungus.** Cankers are formed on branches near forks, sometimes appearing as concentric rings of callus growth. The main trunk, if infected, may be flattened and bent.

Pruning and burning of affected wood will help your birch trees. Cut out small cankers, and sterilize the wounds and paint them.

**Rust** is seen as reddish yellow pustules on the undersides of infected leaves. It sometimes causes defoliation, but generally rust is not serious enough to warrant control measures.

**Willow crown gall** may strike birches. It is caused by a bacterium that invades the roots, trunk, or branches, stimulating cell growth so that tumorlike swellings with irregular rough surfaces are formed. The galls may be the size of a pea or larger, those on the trunk sometimes growing to one or two feet in diameter. Growth of the tree may be retarded, leaves often turn yellow, or branches and roots sometimes die.

To control crown gall, destroy affected trees and nursery stock. Do not replant the area with willow, poplar, chestnut, sycamore, maple, walnut, or fruit trees. Avoid wounding the stems of healthy trees, since infection occurs through wounds.

# BLACKBERRY

The best prevention against diseases and pests is a clean berry patch. Start with healthy certified stock from dependable nurseries. Keep the canes thinned out to let air and light circulate. Any leaves or fruit that drop out of season should be removed at once, since they may harbor disease or pests. All old canes should be removed as soon as they have finished bearing. When pruning, cut the canes all the way back to the soil level and destroy the prunings. Be sure to pick all the berries, even the few that turn small and dry toward the end of the season.

Since wild plants harbor many diseases and insects, it is best to eradicate those near cultivated plants. Rake and burn all nearby apple and cherry leaves. Keep a close watch on your plants for symptoms of serious troubles such as viral diseases or orange rust. At their first appearance, remove the entire affected plant, including all the roots, and burn it.

## Insects

The **blackberry psyllid** is a jumping insect that occasionally injures cultivated plants, and is usually to be found on wild blackberry. The adult is yellow brown, about ⅙ inch long, and is marked with three yellow brown bands on each wing. The adults live through the winter in protected places and appear on the plants soon after growth begins. They lay eggs in leaf stems and tender shoots. Both nymphs and adults puncture the stems and leaves, causing stunted or distorted growth (sometimes called galls).

As damage from psyllids is usually not very great, control measures are seldom needed. They collect and breed on lycium, so elimination of this plant will help curb the psyllid population. This pest has many natural enemies, including ladybugs and about fifty species of gall midges. If further control is required, spray a soap solution on the plants. Diatomaceous earth should give some control.

Occasionally, blackberry leaves are devoured by the larvae of the **blackberry sawfly.** Adults appear in late May, and the females lay white oval eggs, placed end to end and fastened to the larger veins on the undersides of leaves. The larvae roll the leaves, fasten them by a web, and feed within. In July they are fully grown, ¾ inch long, and bluish green in color.

When control is necessary, use pure pyrethrum or rotenone, applied early in summer to the undersides of leaves. If the eggs are removed and destroyed in spring when first laid, sawflies should not be a serious problem. Sticky traps will snare adult flies; see TRAPS.

The larvae of **leafminers** sometimes infest blackberry and dewberry leaves. There are two generations, the first in May and early June and the second in August. Eggs are laid in blisters on the undersides of leaves. The larvae mine the leaf edges, making blotched areas that give plants a scorched appearance. The adult is a blackish two-winged fly, about ⅙ inch long. So far, a satisfactory organic control has not been worked out for this insect. Infestations can be reduced somewhat by pinching feeding larvae.

The common brown **May beetles,** of which there are more than a hundred species, sometimes cause serious damage by eating blackberry leaves. In the larva or white grub stage they feed on the roots of bluegrass, timothy, corn, soybeans, and several other crops, as

May beetle and white grub

well as on decaying vegetation. Their pearly white eggs are deposited in spring, one to eight inches deep in the soil, and hatch three to four weeks later.

Populations of white grubs, and consequently of May beetles, can be reduced by rotating blackberries with deep-rooted legumes such as sweet clover, alfalfa, and other clovers that are unfavorable to them. Legumes are most effective if planted in the years of major beetle flights. The renovation of bluegrass pastures badly infested with white grubs may be necessary; the sod is thoroughly torn up in spring or late fall, treated with organic fertilizer, and sown in spring with a seed mixture consisting mainly of legumes. If you plant blackberries in fields that have been fallow for several years or on land that has been in pasture or grass sod, be especially watchful for white grubs. If they are in the soil, at least one intervening crop of legumes should be grown. Be sure to use clean cultural practices before the berries are planted. The winter before cane fruits are planted, disturb belowground stages by plowing deeply.

The **raspberry cane maggot** causes the tips of new shoots to wilt and take on a purplish discoloration. Galls may appear on the canes. For more information and control, see RASPBERRY.

The **rednecked cane borer** causes blackberry canes to develop enlarged swellings, ½ to 1½ inches in diameter and extending for several inches along the cane in a cigar shape. If these swollen places are cut open in winter, slender creamy white grubs will be found, usually in the pith. In May, bluish black beetles with coppery red thoraxes may be found on or in these swollen areas. Eggs are laid in June in the shoots, and the larvae hatching from them bore into the bark to cause the peculiar growth. These galls or swellings develop in late July or August and prevent the shoots from developing properly.

You can control the rednecked cane borer during fall or early winter simply by cutting out all canes having these abnormal swellings.

**Rose scale,** a common pest of both blackberries and raspberries, gives the canes a whitish, scaly, inflamed appearance. The scaly shell is secreted by a small reddish insect that covers itself with a waxy coating. Eggs are laid beneath the scale in fall, and hatch in spring to suck the sap from the canes. Rose scale is sometimes mistaken for anthracnose, but unlike anthracnose it appears only on the surface.

Control by spraying with a dormant oil. Remove and burn infested canes during the dormant season.

If the blackberry leaves are dull and pale, with yellow dots on the upper surfaces and webbing underneath, they may be infested by the **spider mite.** These pests are more prevalent in dry weather, but unless there is extreme drought they should not bother well-mulched berry bushes. Give your plants enough water, and dislodge the mites with water, as described in the RASPBERRY section. See also SPIDER MITE.

**Tree crickets,** especially the blackhorned tree cricket, may be a problem for blackberries. These greenish yellow insects, with feelers projecting from the front of the black head, will attack red and black raspberries, trailing brambles, wild shrubs, and fruit trees. Since they usually lay their eggs at the edge of a field, they do less damage to a square field than a long, narrow one. Canes injured by crickets show areas of bark split in an irregular line. Inside these splits are a series of small holes extending into the pith. These holes or punctures are usually made in rows of up to 75 or 100. Frequently the canes break off at the injury. The eggs remain in the canes through winter, hatching into pale green, slender young in spring. The nymphs grow rather slowly, reaching maturity in late summer, mating in fall, and laying their eggs in the maturing shoots as the weather turns cold.

If infested canes are not too numerous, they can be removed and burned to destroy the eggs. Where infestation is heavy, many of the eggs can be destroyed by cutting back the canes to half their length. This should save much of the crop. The prunings must be collected and burned before the eggs hatch in spring. It helps to keep down weed growth as well as any wild raspberry, blackberry, and dewberry plants growing nearby. If a further measure is needed, an application of rotenone should help.

## Diseases

**Anthracnose,** a fungal disease of a number of garden plants, is also a major enemy of blackberries. The most serious damage is to the stems of the fruit and to the berries, which wither early. Leaves may become infected and drop. The first symptoms appear in spring as small, purplish, slightly raised spots on the young plants. As the plants grow, the spots enlarge and their centers become gray and slightly sunken. Other spots appear on the leaves and stems. The lesions are often so close that irregular spots are formed over the cane surface. During the second year, when the fruit is to appear, the bark may crack, causing the plant to lose water and dry up.

Spores of this fungal disease are carried by rain, wind, and insects. As the buds open on the fruiting cane, new tissue becomes exposed and infected. The spores germinate and the fungus penetrates, often extending into the woody section of the plant.

Anthracnose is best controlled by careful sanitation practices. Remove old canes after fruiting and the stubs or handles of the trailing brambles after the plants are set. Cut canes low when planting them, and discard any showing severe, gray bark lesions. Although not entirely resistant, erect types such as Humble, Texas Wonder, Dallas, and Lawton are less susceptible. Since wild raspberries are usually heavily infected, it is best to eliminate them from the immediate vicinity. Avoid planting anthracnose-prone plants near your blackberries: this includes beans, cucumbers, grapes, sweet peas, and raspberries.

Moisture on the canes favors spore germination and infection, so encourage air circulation to speed drying. Eliminate weeds and thin out weak and spindly canes. If necessary, use a spray of lime sulfur, mixed $4/5$ pint per gallon or 10 gallons per 100 gallons of spray, in spring when $1/4$ inch of green leaf is exposed from the buds. Do not use this spray after the delayed dormant period, or severe foliage injury may result.

Trailing varieties of blackberries may be affected by **cane dieback,** which causes buds to fail to break early in the growing season. The plant may produce both normal buds and some that grow only a few inches before they wither and die. The leaves may be mottled and small, and scorched areas often appear on the leaf surface.

Maintain adequate soil moisture to a sufficient depth throughout the year. It is ideal to plant in sandy soil that runs eight feet deep and has a very high rate of water penetration.

Bramble fruits are vulnerable to **crown gall,** a bacterial disease that weakens the plant and ultimately may kill it. It is recognized by the presence of galls or knots on the roots and sometimes on the canes. The bacteria invade the plant tissue through open wounds and cause galls to form; even after the galls decay, the germs live for some time in rotted tissue in the soil, and may reinfect the plant. Plants with numerous galls may be stunted and produce dry, poorly developed berries. Latham raspberries are particularly subject to crown gall.

Destroy all plants that show symptoms of gall. Order plants from reliable nurseries and carefully inspect each plant before it goes into the soil. Once a planting is badly infected, the only recourse is to plow up the entire field and grow corn or grass for several years.

Blackberries stand the best chance where grain crops have been grown for several years. Stay away from land where brambles have been recently grown and land receiving drainage from old bramble fields.

A beneficial bacterial, Gall-trol, is marketed as a preventive root dip. See CROWN GALL.

A fungal disease that causes double blossoms from which no fruit sets is called, appropriately enough, **double blossom.** If you spot any blossoms with crinkled multiple petals among the healthy blossoms of your plants, inspect all buds before they open. Buds affected by double blossom have chubby contours and are fleshier and redder than normal buds. Pick off and destroy all such buds before they can open and release the spores of the fungus. After the harvest, cut all canes close to the ground and destroy them.

**Dwarf,** or rosette, is a viral disease causing marked reduction in growth and productiveness. Affected plants usually become unproductive within a year, with a downward cupping and reddening of the leaves and dwarfing of the cane growth.

Clear out any patches of wild blackberry in the vicinity, as they may contain the virus. Himalaya and Advance are very resistant to dwarf; Boysen and Young are relatively resistant.

One of the most serious diseases of commercial berry plantings is **orange rust.** Once infected with orange rust, a plant never recovers, as the rust reappears year after year. Usually the leaves are so heavily infected that the diseased plant will not bear a crop. Orange rust affects the leaves of the plant, causing yellow dots to appear on both sides early in spring. Two or three weeks later, light-colored areas develop on the undersurfaces of leaves, and in a few days the epidermis is ruptured, exposing large, bright orange red, powdery masses of spores. These spores blow about and infect other plants. Infected plants are stunted and worthless for fruit production. The orange rust fungus winters in the roots and stems of old plants. In spring it grows up in the pith of the developing canes and out into the leaves, where it produces the spore masses.

To combat orange rust, the infected plants must be dug out and destroyed, including all the roots and suckers. Since orange rust is a systemic disease that spreads through the plant, not even one cane should remain. If there are any wild berries nearby, be sure to inspect them too for disease—it is pointless to root out all of your own infected plants and ignore those just over the fence. Better still, eradicate wild berries near your plants. Avoid touching healthy plants with diseased materials, and do not cultivate diseased plants when the leaves are wet. Give the berries a heavy mulch of straw and leaf mold and, when the season is over, a heavy dressing of compost. Add some extra potash and phosphate to your compost, because fungal diseases are often encouraged by malnutrition. Try to plant resistant varieties, such as Eldorado, Boysen, Lawton, and Snyder.

**Septoria leaf spot** is a fungal disease that can cause serious defoliation. In early summer, small dots appear on the leaves of infected plants, purplish on the edge and tan or light brown in the center. When the spots are large enough, you may see several blackish dots in the center. These are the fruiting bodies, producing millions of spores that infect surrounding leaves and cause them to drop. In seasons of heavy rains, the leaves of an entire planting may be diseased.

Prevent weed growth and train the canes to give plants free air circulation. Remove old leaves and cultivate early. Remove and burn all diseased canes immediately after harvest.

Blackberries, like raspberries, are susceptible to **viruses.** There is no control as yet. Buy good stock from reliable nurseries. Some nurseries offer registered virus-free blackberries and raspberries.

For botrytis fruit rot, cane blight, and powdery mildew, see RASPBERRY.

# BLUEBERRY

Small plantings, especially new ones, are seldom troubled by the pests and diseases that hit large-scale plantings. Blueberry bushes are benefited by mulch because they are not deeply rooted and can be harmed by drought. Oak leaves, peat, sawdust, pine needles, or even spent hops from a brewery are all good mulches for these acid-loving plants. Careful cultivation will also help keep the bushes pest-free.

## Insects

The **blueberry budmite** is invisible to the naked eye. An infestation will distort flower buds or even prevent buds from developing. These mites are rarely a problem in the North, although they sometimes cause trouble in the South. Control with a dormant-oil spray in late September or early October.

The **blueberry budworm**, a type of cutworm that eats fruit buds, may develop in weeds under the bushes. Clean cultivation will control them.

The **blueberry fruit fly** is about the size of a housefly, with black bands on its wings. This fly lays its eggs in ripe or overripe berries. In just a few days, tiny colorless blueberry maggots burrow into the fruit and cause it to drop. Early varieties appear to be less susceptible to the blueberry maggot than late ones.

Pick up and dispose of all dropped fruit, because it may contain the eggs for another generation. Cultivation around the plants also helps, as larvae drop to the ground under the bush and plowing exposes them to predatory ants and birds. Some growers have controlled the blueberry fruit fly with a flat, paddle-shaped piece of board that is brushed with a commercial sticky compound (try Tanglefoot or Stikem) and hung on each bush. Clean up all old trimmings and pick ripe berries before the fruit fly can lay its eggs in them. Commercial growers have found a 2 percent rotenone dust to be an effective control.

The **blueberry leafminer** may eat a few leaves or roll up a few more, but it is not a serious pest and will usually not affect fruit yield. It has been effectively controlled by a parasite, *Apanteles ornigis.*

The **cherry fruitworm** hatches in late May in the North, and the pinkish larvae bore into the fruit and feed till mid-June. It is controlled by a parasitic fungus, *Beauveria bassiana,* which also keeps the cranberry fruitworm in line. A tiny wasp parasite, *Trichogramma minutum* Riley, is effective against both the cherry and cranberry species. The adult fruitworm is an ash gray moth, marked with black spots, that lays eggs on berries in mid-July. They hatch as green caterpillars that eat into berries near the stem, shriveling them. A web is woven around a cluster of berries.

Commercial growers are forced to flood the growing area after harvesting, but the cranberry fruitworm is usually controlled in small plantings by handpicking, keeping the bushes free of weeds and trash, and frequent cultivation of the soil between plants. Ryania and rotenone can be used against fruitworms.

The **plum curculio** [color] is a brown-snouted beetle, ¼ inch long, that damages many stone fruits. It overwinters in protected areas around the garden and in early spring lays its eggs in the young fruit, later feeding on the fruit. The mark of the plum curculio is a small, crescent-shaped depression in new fruit. The

Plum curculio

larvae enter the soil to pupate. Late-ripening berries are less frequently bothered by this pest. See TIMED PLANTING.

The best control is sanitation. Pick up and destroy any dropped fruit as soon as possible. Clean up rubbish heaps and other potential winter hiding places, such as old boards. Try to avoid planting near wooded areas, since the beetles are more numerous there than out in the open.

There are two varieties of **scale** which may become a problem for blueberries. Putnam scale is the more common. It often occurs on forest trees and will infest older blueberry bushes, clustering on the rough wood and feeding on the fruit. It is dark gray to black and circular, and produces a reddish secretion. Terrapin scale is a dark reddish brown insect with dark bands radiating from its center back to fluted edges. It secretes a honeydew on which a sooty fungus develops. When scales are numerous, there may be a putrid odor.

Prune to remove dead, excess, and infested wood. Dormant-oil sprays also keep down scale infestations.

The **stem borer,** or tip borer, is a long-horned, ½-inch-long, slender beetle that girdles the tips of blueberry and other acid-loving plants in spring. It lays eggs in the stems, and the grubs bore into them and down into the trunks of woody plants. The stem borer is controlled by cutting out and destroying all dead and wilting tips.

**Tent caterpillars** are more damaging to large commercial plantations of blueberries than to small plantings. Wild cherry trees are their favorite host, so be cautious if there are any in your vicinity. See TENT CATERPILLAR.

Other occasional pests of blueberries are the European corn borer (see CORN), JAPANESE BEETLE, and rose chafer (ROSE).

## Diseases

**Bacterial** or **stem canker** affects the canes of blueberry bushes, causing reddish brown or black cankers. All buds in the cankered areas are killed and sometimes the stem itself is girdled and dies. In other cases, the stem above the canker continues to grow, but it may be weak and unproductive. Eventually the plant will die.

Stem canker was formerly a serious problem in the South, but resistant varieties have been developed, such as Murphy, Wolcott, Angola, Croatan, and Morrow. The new rabbiteye varieties are also immune.

**Botrytis blight** affects growing tips, flowers, and newly formed berries. On canes, infected areas are brown or gray, extending down a few inches. The fungus, sometimes called gray mold blight, covers the flower with a dense, gray, powdery mass that quickly spreads to adjacent flowers and fruits. Young berries shrivel and turn purplish. Varieties that drop the corolla promptly are less susceptible.

Because succulent growth is most seriously affected, fertilizers heavy in nitrogen should be avoided. Prune out blighted twigs as soon as possible and burn them. Keep your pruning tools clean to avoid spreading the disease.

**Cane gall,** or stem gall, appears as rough, irregular, warty growths along the stems of plants. Older canes are more susceptible than young growth. Infected canes should be pruned out and destroyed; you're best off buying healthy nursery stock.

**Crown gall** is caused by a bacterium. See CROWN GALL.

The fungal disease **mummy berry** causes berries to shrivel and dry out on the bush just before they should ripen. Under the dry skin of the fruit is a thin black layer. Leaves may wilt and blacken. The mummy berry fungus overwinters on the ground beneath the bushes, so kill it with an early spring cultivation. Like many fungal diseases, it is nurtured by overfertilization of the plants. Mummy berry is seldom a problem with heavily mulched plants.

**Powdery mildew** is very common with blueberries. This whitish mold on the topsides of leaves appears after the harvest and so has no effect on fruit yield. Berkeley, Earliblue, and Ivanhoe are resistant varieties.

**Stunt** is a viral disease more common in the South than in the North. It is less troublesome than it used to be, although potentially present wherever there are wild blueberries. Stunt is spread by a leafhopper, and the reduced vigor of the plant results in abnormally small leaves that are curved upward and yellowish. The plant has a dwarfed appearance, with clusters of berries much smaller than normal. The plant may turn reddish in mid or late summer. Stunt-resistant varieties include Harding, Stanley, and Jersey.

## Wildlife

**Birds** may take most of your blueberry crop unless you find a way to protect your bushes by screening or enclosure. Posts eight feet tall can be set around your planting and through it in rows, with wire screening nailed around the sides and on top. Fish netting also works well and gives better ventilation than screening. The frame should hold the netting at least a foot away from the berry bushes.

# BORDEAUX MIXTURE

Organic gardeners sometimes use a bordeaux mixture as a last-ditch measure against potato blight. A mixture can be bought readymade or prepared at home by stirring eight ounces of copper sulfate into six gallons of water until the chemical is dissolved. So that the mixture will adhere to plants, stir five ounces lime into enough water to make a liquid with a creamy consistency. Add the quicklime solution to the copper sulfate solution. Prepare bordeaux just before spraying, and stir frequently as you spray.

# BORDERS AND BARRIERS

Instead of applying repellents and insecticides right on vulnerable crops, you can set up a miniature Maginot line around patches of certain plantings or even an entire garden. Winged insects may aviate right over such barriers, but those that walk to lunch can be stopped.

A perimeter of wood ashes scattered around the bases of plants will block all sorts of crawlers and walkers. You can also sprinkle lines of bone meal, lime, powdered charcoal, coal ashes, diatomaceous earth, or any of several vegetative dusts such as sabadilla. A ring of mothballs placed on the ground will repel both bugs and warm-blooded animals. Orchardists can keep deer from eating fruit by hanging small cloth bags of blood meal from branches. You don't need many of the bags—once the meal gets wet it releases an odor, and while the smell isn't noticeable to humans, to deer it signals trouble ahead. Hang mothballs in mesh bags from trees to ward away deer, squirrels, skunks, and other animal pests of the orchard.

The female cabbage moth doesn't care for the smell of tar, and you can capitalize on this dislike by cutting out square of disc-shaped collars of tar paper for the bases of young cabbage plants. Slit the paper so that it can be slipped around the plants, and the cabbage will be safe from the moth's eggs. To protect cucurbits from earthbound pests, try soaking newspapers with turpentine and laying them over the ground as the plants vine out.

Egg-laying, winged adults can be excluded from vulnerable garden plants by temporary shelters of clear plastic with cheesecloth ends to allow air to circulate.

If the birds have been beating you to the berries and fruit each year, consider investing in a roll of mesh. Nylon mesh is light in weight and impervious to mildew and rot. To protect thorny or low crops with delicate stalks, the netting can be suspended on stakes. Netting will keep fruit-robbing birds from trees.

Crawling insects, such as chinch bugs, armyworms, and wingless May beetles, are halted by plowing deep furrows between fields or around garden patches. The insects will fall into postholes dug at intervals along the bottom of the furrow, where they can be dispatched with a shot of kerosene, crushed with a stick, or buried.

Sticky bands can be applied to tree trunks and stems of large plants to halt the advance of a number of crawlers, including fall and spring cankerworms (you may know them as inch-worms) on fruit and shade trees. The adult female moths are without wings and have to crawl up trees to deposit their eggs. Put a strip of cotton batting two inches wide around the tree trunk (to protect the bark), wrap a strip of tar paper around this, and apply a sticky substance to the paper. Put bands on in the middle of October to thwart the fall female moths, and again in early spring to catch the spring moths. Chances are good that you'll trap a few tussock and gypsy moths too.

What sort of sticky stuff should you use? Commercial preparations, Tanglefoot and Stikem, are made for this purpose. Roofing tar is an old-time answer. You can even make your own goo by mixing pine tar and molasses, or resin and oil. Such sticky preparations may damage tree trunks, so make certain to protect bark with a layer of cotton batting or heavy paper before applying them.

Sticky bands are also used to block wingless aphids, rose beetles, and ants on roses; codling moth caterpillars on fruit trees; the whitefringed beetle on pecan and other susceptible trees; and any other walking insects that use stems and trunks as aerial highways.

Barriers don't have to be messy to work. Burlap bands thwart the gypsy moth, which rests or tries to pupate in the folds of cloth. Bands of corrugated cardboard or paper, wrapped in several thicknesses around a trunk, will attract hibernating codling moth larvae. This barrier should be put up in spring, and disposed of late in fall.

See TRAPS.

# BOXWOOD

## Insects

The **boxwood leafminer** is the most serious insect enemy of boxwood in the United States. This pest weakens the plant by removing tissue from the leaves, causing the foliage to become yellowish green and stunting new growth. Twigs may die, giving the plant a ragged appear-

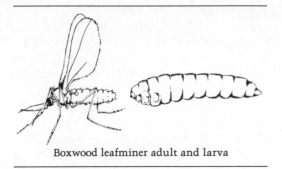

Boxwood leafminer adult and larva

ance. Most varieties of boxwood are susceptible to attack.

The larvae are yellow, wormlike insects, many of which may be present within a single mined leaf. Rake up and compost fallen leaves, and prune back succulent new growth, especially water sprouts, in early spring. The dwarf, or English, variety of box is immune to the leafminer.

Boxwood leafminer damage

The **boxwood mite** is a common pest in most boxwood plantings. The adults are $\frac{1}{64}$ inch long and yellow green to reddish brown in color. The overwintering yellow, rounded flattened eggs on the undersides of the leaves hatch shortly after the middle of April. The young nymphs feed on adjacent tissue. During the second instar, the mites feed on both surfaces of the leaves. On the upper surfaces, small patches are rasped from the outermost layer of cells, appearing as fine scratches. This gives the leaves a mottled appearance and serves as an early indication of the presence of mites. The third instar larvae go from leaf to leaf to feed. The adult mites feed mostly on the tender shoots and on the uppersurfaces of the leaves. As mites increase in number, the damage becomes more apparent. Leaves become bronzed, wither, and sometimes drop to the ground, leaving the plant looking scraggly.

Mites can be controlled by squirting the boxwood with water several times during spring and summer. Apply the water by hose late in the afternoon.

The **boxwood psyllid** causes a cupping of terminal leaves. It is a small, grayish green sucking insect, the nymphs of which are covered with a whitish waxy material. Plants are seldom seriously injured, although their appearance becomes unattractive. If infestations are severe, growth may be retarded. Control with a pyrethrum spray at the first evidence of leaf cupping.

## Diseases

**Decline,** or dieback, often follows winter injury or occurs on plants that are growing in poorly drained soil. Nematodes may also encourage this trouble. In spring the foliage of affected plants changes from the normal dark green color to gray green or bronze and finally to a straw color. Shoot growth is weak and old leaves drop prematurely. Entire branches die, especially in the middle or top of the plant. Sunken areas or cankers may appear in the bark of the trunk just above the ground line or on the branches in crotches where dead leaves and debris accumulate. The bark on the sunken areas may be brown or streaked with brown and often peels or splits away from the gray, discolored wood beneath.

Plant boxwood in well-drained soil and protect the plants from drying winter winds and severe low temperatures. Remove affected branches by cutting them back to healthy wood several inches below the cankers. On large branches, remove small cankers by cutting back to healthy wood on all sides of the canker and shape the wound to a point at either end. Paint all wounds with shellac and, when they dry, cover with a tree paint. Before growth starts in spring, remove leaves and debris lodged in the crotches. A strong stream of water will do an excellent job of removing this debris.

**Leaf spotting** on box is quite frequent and may be caused by any of several fungi. Infected leaves turn a strawy yellow color and their undersurfaces are thickly dotted with small black fruiting structures. Damage is limited to plants lacking in vigor, and it is important to enrich the soil and ensure good soil drainage.

Exposed boxwoods may suffer **winter injury** from severe cold and drying winds. Leaves may become grayish green or brown with dead twigs. Protect plants from exposure. Let the season's new growth harden off before the onset of winter, which means you should not overfertilize late in the season.

# BROCCOLI

## Insects

In general, the pests that injure cabbage also attack other members of the crucifer family, including broccoli, brussels sprouts, turnip, rutabaga, Chinese cabbage, mustard, and cauliflower. See CABBAGE for these pests of broccoli: cabbage aphid, cabbage looper, cabbage maggot, diamondback moth, flea beetle, and imported cabbageworm.

## Diseases

Broccoli is plagued by the same diseases that afflict other crucifers. Look under CABBAGE for these diseases: black leg, black rot, club root, downy mildew, rhizoctonia disease, seed rot (damping-off), watery soft rot, and yellows (or wilt).

# BRUSSELS SPROUTS

## Insects

Almost all the enemies of brussels sprouts are shared by broccoli, cabbage, and the other crucifers. Under CABBAGE, see cabbage looper, diamondback moth, flea beetle, and imported cabbageworm. Other possible pests are APHID and WHITEFLY.

## Diseases

Brussels sprouts share many diseases of the other crucifers, including black leg and club root; see CABBAGE.

# BUG

Here are descriptions of several beneficial bugs. For pests, see plant entries, and also MEALYBUG. Bugs, entomologically speaking, are members of just one order of the insect world.

AMBUSH BUGS.   The appearance of this small group of predators is variously described as "grotesque" and "odd-shaped," and is best learned from an illustration [color]. The front legs are suited for snatching up prey. You're most likely to find them hiding behind flowers or leaves, where they wait for passing insects. Although only ⅜ inch in length, they will attack large fliers, including some beneficial bees and wasps.

ASSASSIN BUGS.    Assassin bugs [color] are voracious predators in the garden, assaulting caterpillars, aphids, Mexican bean beetles, Colorado potato beetles, leafhoppers, Japanese beetles, an occasional honeybee, and even gardeners' fingers. The bite feels painful at first and then itchy. A very few people may be allergic to the substance these bugs inject through their curved beaks. After reading this, no doubt you'll want to know what these fierce predators look like. They are about ½ inch long, black or brown, have large eyes set in a narrow head and large front legs for grabbing insects or whatever.

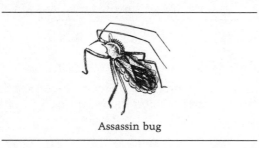

Assassin bug

BIGEYED BUGS.    These beneficial insects are supported by weedy areas. They eat mites, leafhoppers, aphids, Mexican bean beetles, and insect eggs. While bigeyed bugs feed on plant juices at times, they rely on meat to reproduce and grow. They grow to ¼ inch long, have large eyes, vary in color from gray to tan, and are oval in shape. See the photograph of the western bigeyed bug [color].

DAMSEL BUGS.    These small, dark, rounded bugs [color] are entirely predacious, and are responsible for the untimely deaths of many aphids, leafhoppers, treehoppers, psyllids, plant bugs, mites, and small caterpillars. They are about ¼ inch long, with the body narrowed toward the head; the large front legs are well suited for grasping small prey. Low-growing plants are damsel bugs' favored habitat.

# BUG JUICE

A garlic spray likely repels bugs for the reason it repels some humans. A pyrethrum spray works because it is toxic to plant pests. But bug juice—a slurry of mashed pests and water—is something of a mystery. Gardeners have used bug juice to protect corn and cabbage from aphids, squash and beans from pillbugs, and pepper plants from slugs. Other successes have been scored against Mexican bean beetles, armyworms, cutworms, cabbage loopers, velvet bean caterpillars, root maggots, grape skeletonizers, Colorado potato beetles, grasshoppers, ants, stink bugs, fungus gnats, and mountain-ash sawfly larvae.

For whatever reason, bug juice has worked for many gardeners since the USDA in California told people how to make it back in 1960 (12 years later, the Environmental Protection Agency put a stop to this recommendation on the ground that bug juice wasn't registered as a pesticide).

Mike Seip, a researcher in biological control and one-time head of a Florida pest management company, has used bug juice commercially. He first read of the method in a reader's letter to *Organic Gardening* magazine, tried it himself, and found it worked. He next tried it in farmers' fields, again with good results. At one point, he treated 500 acres with spray made from just three quarts of insects.

Seip notes that, curiously, the spray remains effective over a long period, up to two or three months. How can this season-long immunity be explained? He suspects the plant is changed by what amounts to an inoculative effect. Once the mechanism of immunity is found, companies might pick up the idea and go commercial with it through mass rearing of the bugs. Seip foresees the possibility of a range of inoculants for different plants. (At the University of Kentucky Joseph Kuć has successfully inoculated plants against diseases. See the discussion of beneficial diseases under MICROBIAL CONTROL.)

Other reasons have been suggested for the success of bug juice. Perhaps the ground-up bugs either repel the living or attract beneficial insects. Or the spray may broadcast a disease present in some of the ground-up bugs (*Bacillus thuringiensis* was first used in this manner). For this reason, Seip says it is better to use one-tenth of a solution containing 300 individuals than all of a batch of 30: your chances are better of including at least one sick insect, and the pathogen will be dispersed throughout the juice. Look for sluggish and dead pests as you collect bugs for your spray. After spraying, store infected insects for another season's sprays by freezing them. The responsible pathogens are likely host-specific, so you aren't apt to control caterpillars with a spray containing sick aphids.

Even if you are not at all squeamish, you shouldn't grind up bugs in a blender used for food preparation. If you can't spare a blender for bug juice and nothing else, try reducing the bug bodies with a mortar and pestle. Some people seem to be allergic to insects, and you would be prudent to wear a filter mask, long-sleeved shirt, long pants, and rubber gloves when spraying.

To ½ cup bugs, add 2 cups water and liquefy. Strain the juice through cheesecloth so that the sprayer won't become clogged. Before pouring the juice into the sprayer, cut it with 4 to 8 parts water. Leftovers can be frozen until the next growing season. Treat both sides of foliage, and spray again after a rain.

You also might experiment with a plant juice spray. Make a water spray of plant leaves that a particular pest doesn't seem to bother. Select smooth-leafed plants, as fuzzy ones may be unappetizing because of their surface rather than their taste or scent.

# BUTTERNUT

## Insects

For insects, see WALNUT.

## Diseases

For **anthracnose** of butternut, see TREES.

The fungus *Melanconis juglandis* infects small branches of butternut trees, producing a dark discoloration of bark and causing a **dieback** which progresses down the branches to the trunk. Severely affected butternut trees eventually die. Dead branches are covered with small, black, pimplelike fruiting bodies of the fungus. In wet weather, spores ooze out in a black, inky mass from the fruiting bodies and are splashed and blown about, infecting other branches.

Control by destroying severely affected trees. Where infection is light, remove diseased branches promptly, pruning back to sound wood.

Several of the wood-rotting fungi may attack butternut trees as wound parasites, causing **root rot,** or heart rot, which contributes to the decline and death of trees. The only control for these fungi is the proper care of tree wounds by cleaning, shaping, and painting them.

# CABBAGE

The several members of the cabbage, or crucifer, family have two things in common—insect enemies and diseases. Therefore, these plants will be treated as a group and included under the cabbage entry: cauliflower, broccoli, brussels sprouts, turnip, Chinese cabbage, rutabaga, and mustard. Where differences do occur, they are so indicated. Generally, mints and monarda (bee balm, bergamot, and Oswego tea) make excellent companions for cabbage plants. Not only do these perennials repel various pests, but they attract bees and hummingbirds to the garden as well. Growers have found that pests are less partial to red than green cabbage. The number of different cabbage enemies may be discouraging, but most of them are minor.

## Insects

Foremost among insects that attack cabbages is the cabbage **aphid** [color]. This tiny green to powdery blue insect has a soft body covered with a fine whitish wax. Aphids cluster on the undersides of leaves, causing them to cup and curl. They may also occur on flower heads. Broccoli is particularly vulnerable. Affected plants are stunted, and seedlings may die.

Aphid

When cabbage aphids appear in cool weather, they can usually be discouraged by a fine, forceful spray from the garden hose. Try pulling up old, woody mint plants in late summer and laying them as a mulch under broccoli and cabbage. Mint can also be planted among crucifers to repel the aphids. Fine-ground

limestone, sprinkled over cabbage and the heads and foliage of broccoli, can drive them away. While this powder makes aphids uncomfortable, sprinkling diatomaceous earth will actually kill them. If the damage has already been done, remove and destroy the plants as early in the season as possible, for they may be infected with mosaic, a viral disease that is often spread by aphids. You may be able to save later plants by removing only those leaves that have been affected. Because aphids winter in the North as small, dark eggs on the stems and old leaves of crucifers, a thorough garden cleanup at the end of the season could save you trouble next year.

All of the crucifers are open to damage from the **cabbage looper** [color], a large, pale green measuring worm with light stripes down its back. It grows to 1½ inches long, and it's so named because it doubles up or loops as it crawls. The round, greenish white eggs are laid singly, usually on the upper surfaces of leaves. The looper overwinters as a green or brown pupa, wrapped in a cocoon attached by one side to a plant leaf. See TIMED PLANTING for ideas on avoiding this pest.

Cabbage looper larva

Cabbage looper moth

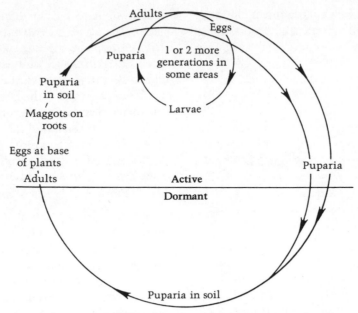

Life history of the cabbage maggot

Reprinted, by permission, from Pyenson, *Fundamentals of Entomology and Plant Pathology*
(fig. 3.25), AVI Publishing Co., PO Box 831, Westport, CT 06881.

Biological controls have been effective in subduing the cabbage looper. The tiny trichogramma wasp is a parasite of several caterpillars, and is commercially available. Gardeners can also use the bacterium *Bacillus thuringiensis* to kill and infect the looper and other caterpillar pests that include the saltmarsh caterpillar, imported cabbageworm, beet armyworm, and diamondback moth. See *BACILLUS THURINGIENSIS.* If you raise large amounts of cabbage, especially in the South, loopers may be controlled by spreading a homemade virus potion. A solution made from just ten loopers infected with nuclear polyhedrosis virus (NPV) will give an acre complete protection. The trick is finding a virus-infected looper in your cabbage patch. Look for loopers that have turned chalky white and appear half dead. They may climb to the top of the plant and lie on the leaves, or hang down from the undersides. These affected worms

soon turn black and liquefy. Capture ten of them and run them through a blender with some water. Then dilute to give the mixture a sprayable consistency. Once infected, loopers die within three or four days. In some cases, the application of NPV may control loopers for an entire growing season. See NUCLEAR POLY-HEDROSIS VIRUS.

The **cabbage maggot** [color] infests the stems of early-set cabbage and other crucifers, and late in spring radishes and turnips may also be damaged. The larvae are small (1/3 inch), white, legless maggots with blunt rear ends. They attack stems just below the surface of the soil, riddling them with brown tunnels. The seedlings then usually wilt and die; most infested cabbage and cauliflower seedlings never produce heads. The maggot earns more bad points by serving as a vector for bacterial soft rot and the fungus that causes black leg. Protect plants by dusting with diatomaceous earth or a mixture of lime and wood ashes.

Cabbage maggot adult and larva

Newly hatched larvae of the cross-striped **cabbageworm** are gray with small black tubercules and large round heads; mature larvae are green to bluish gray with at least three distinct black bands across each segment. Long dark hairs grow out of the black tubercules, and the slender worms grow to ⅔ inch long. The adult is a small pale yellow moth, with mottled brown patterns on the forewings and partially transparent hindwings. See imported cabbageworm, below.

The fat, smooth, pulpy **cutworm** [color] is as troublesome as it is ugly. Of the many species, you are most likely to run into the black, bronzed, and dingy cutworms. The black cutworm (or greasy cutworm) is gray to almost black and has a broken yellow line down the center of its back with a pale line to each side. It grows to 1½ inches or even longer and has a shiny, greasy appearance. As its name suggests, the bronzed cutworm is bronze in color and

Cutworm

displays five pale lines from head to tail. It is a pest in the northern states. A northern cousin is the granulate cutworm, which has a rough, granulate-appearing skin and a dusty brown color. It burrows very shallowly in the soil, often exposing its back. The dingy cutworm, a northern species, is also aptly named—it is a dull, dingy brown, and has a wide, pale stripe running the length of its back, flanked by a thin dark stripe on each side. It sometimes crawls up the stems of plants to feed. For control measures, see TOMATO.

Although the larva of the **diamondback moth** is a relatively minor pest, it at times can cause considerable damage to the cabbage family by eating small holes in the outer leaves. These caterpillars grow only to ⅓ inch long, are greenish yellow with black hairs, and when disturbed will wriggle and drop to the ground. The adult

Diamondback moth larva

is a gray to brown, small (¾-inch-across) moth, with fringed back wings; when the moth is at rest, look for the diamond on its folded wings. See *BACILLUS THURINGIENSIS* for an effective biological control. Southernwood is an herbal repellent.

The **fall armyworm** [color] is so called because it travels in veritable insect armies to consume everything in its path and because it does not appear until fall in the North. The larvae vary greatly in color, from light tan to green to almost black, and have three yellowish white hairlines down the back from head to tail. On each side is a dark stripe paralleled below by a wavy yellow one splotched with red.

The head is marked with a prominent white V or Y. They eat leaves, stems, and buds of plants. Naturally occurring enemies include parasitic ichneumon wasps and, of the braconid wasps, *Chelonus texanus, Meteorus laphygmae,* and species of Apanteles. Other beneficial insects are tachinid flies, ground beetles, and the commercially available nematode *Neoaplectana carpocapsae.* For other control measures, see armyworm under CORN.

The very visible **harlequin bug** [color] is black with brilliant orange to red markings. It is shield-shaped, flat, and ⅜ inch long. Although attractive to the eye, the bug has a disagreeable

Harlequin bug

odor at all stages. It is among the most important pests of the cabbage family in the United States, as local infestations often destroy whole crops, especially in the South. The bugs spend the winter around old plants and other trash that accumulates in the garden. In early spring they lay their eggs on the undersides of early blooming plants. The distinctive eggs are easy to spot —they look like small white pegs with black loops, standing on end, lined in two neat rows. Look for and destroy these eggs whenever possible; if allowed to hatch, the immature nymphs drain the juices out of the plants, causing them to wilt, turn brown, and die.

Plant a trap crop of turnips or mustard greens near or around the cabbage patch to lure away harlequin bugs. Patrol the trap area,

remove the bugs, and drop them into a jar of water topped with kerosene. Insecticidal soap sprays give good control. Sabadilla and pyrethrum, two plant-derived insecticides, are effective.

The bright green, velvety smooth **imported cabbageworm** [color; see Cabbageworm] is covered with close-set hairs and grows to one inch or longer. A yellow green stripe runs down its back. The worm is found in every state.

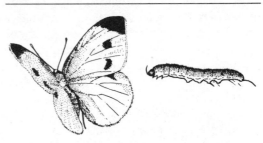

Imported cabbageworm adult and larva

No member of the cabbage family is safe from its appetite, although sea kale is rarely bothered. It eats huge, ragged holes in leaves and spots the feeding area with bits of excrement. It also bores into heads of cabbage. The adult is the well-known white cabbage butterfly, whose white to pale yellow wings have either three or four black spots; the tips of the front wings are grayish. In the early spring, these butterflies may be seen depositing their lemon yellow, bullet-shaped eggs at the base of leaves. The eggs usually take from four to eight days to hatch, depending on temperatures. They turn straw yellow just before hatching.

The cabbageworm may be attracted to your land by weeds of the cabbage family—wild radish, wild mustard, and wintercress. These wild plants should be cleared from the vicinity of the vegetable garden. Cole plants can be protected by a border of plants that are shunned by the cabbage butterfly; these include onion, garlic, tomato, sage, tansy, mint, southernwood, nasturtium, hemp, hyssop, and rosemary.

Insect predators and parasites claim a lot of these pests, but don't expect too much help from the birds around your garden; because of toxic body fluids, the caterpillars aren't favorite foods. Brownheaded cowbirds, song sparrows, and redwing blackbirds eat some cabbageworms. Yellow jackets have been found to thrive on these pests, and their presence should not be discouraged. The braconid wasps *Macrocentrus ancylivorus, Apanteles glomeratus,* and *Agathis diversa* are extremely helpful control agents. They are small insects with short abdomens, and rank as important parasites of aphids too. Numbers of dead aphids about the garden with a round hole in the back testify to the presence of these wasps. You can encourage *M. ancylivorus* by planting strawberries near the garden, as it also feeds on the strawberry leafroller, and by avoiding poison sprays. Another friend, the trichogramma wasp, is a commercially available parasite that attacks many worms of the Lepidoptera order, including the cabbageworm. This wasp lays its eggs inside the eggs of harmful insects by the use of a pointed egg-laying apparatus. When the trichogramma egg hatches, the young parasite proceeds to eat out the contents of the egg in which it lives, causing the egg to blacken and preventing one more insect from harming the garden. See TRICHO-GRAMMA. *Bacillus thuringiensis* was found to be far more effective than Sevin, a popular chemical insecticide. Nuclear polyhedrosis virus also works to control the larvae. The two pathogens can be sprayed at the same time for maximum effect; see individual entries for each.

Homemade remedies may help. Simply spoon a little sour milk into the center of each cabbage, say some gardeners, and the heads will be trouble-free. But a test at the University of Illinois showed that sour milk actually increases

damage. A repellent drench can be made by blending, in a mixer, spearmint, green onion tops, garlic, horseradish root and leaves, hot red peppers, peppercorns, and water. Add some pure soap, dilute, and pour on each plant. Or make a powder of ½ cup of salt and one cup of flour and shake it on the cabbages while dew is still on the leaves. Sprinkle up to an ounce on the worst heads. When the worms eat the mixture they bloat up and fall off dead. These two ingredients can also be mixed with water to produce a safe spray. Other gardeners get by with an occasional sprinkle from a saltshaker. This should be done right after rain or when there is dew on the plants. A sprinkling of straight rye flour (it's stickier than whole wheat) will gum up the worms if the plants are wet.

Protective canopies of lightweight polyethylene or nylon netting can keep egg-laying cabbageworm moths from tender young broccoli and cabbage plants. The netting is easy to stretch over rows and is available from most garden supply stores and mail-order houses. Simple frames of wood can be constructed to support the netting. Although the netting will keep butterflies from laying eggs on the plants, the pupae spend the winter underground and often surface around the plants to cause trouble. Till the soil several times in fall and early spring to expose these pupae to the air so that they will dry up and die. You should also relocate the patch to keep from perpetuating problems with these latent underground pests. Some gardeners trap cabbage loopers in sheets of colored plastic that are folded twice and placed on the ground every three days or so. The captives are then shaken into a bucket of water.

Not all of the many worm pests of the cabbage and its relatives can be discussed here, but a few of occasional or regional importance deserve mention. Control as for the imported cabbageworm, above. The yellow, purple-striped gulf white cabbageworm feeds on the leaves, stems, and the outer layers of leaves of cabbage heads; they also enjoy collards. In the Northeast and Utah, the larva of the purplebacked cabbage moth feeds from inside a silken web. The zebra caterpillar is a velvety black with two bright yellow stripes on each side and many thin, yellow transverse lines. The southern cabbageworm looks similar to the imported cabbageworm, but its appearance is distinguished by alternating longitudinal stripes of bright yellow and dark greenish purple, and it is scattered with black spots. The yellow woollybear caterpillar is bright yellow and very hairy.

You may find any of several species of the nocturnal **mole cricket** eating or flying about the garden at night or on very cloudy days. They may be drawn to lights near the garden. The northern and southern mole crickets are the species of most concern to gardeners. They are large (1¼ inches long) and have sturdy shovel-like forelegs that are adapted for digging into the soil. The northern species is brownish gray

Mole cricket

above and paler underneath, while the southern is a pinkish buff. Both species are a problem only in the South and are at their worst in warm, moist weather. Their tunnel cuts off the roots of seedlings, and some young plants may be totally uprooted. Moles may also chew off stems at the soil surface and pull the plants down into their tunnels. See GRASSHOPPER.

The **seedcorn maggot** also attacks young crucifers, especially cabbage and radishes. They are dirty-yellowish-white, ¼ inch long, and have pointed head ends. For control measures, see cabbage maggots, above.

The malodorous green **stink bug** [color] is smooth oval, bright green, and about ½ inch long. They are occasional pests of cabbage and mustard, piercing stems and leaves to suck sap and leaving tiny holes in the center of surround-

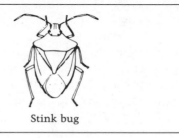

Stink bug

ing cloudy spots. Small plants are stunted and distorted. Control by keeping down weeds. Spray with soapy water. See SNAPDRAGON.

The **vegetable weevil** [color] is a dull buff color with a pale V on its wing covers. The head has a short, broad snout. They are ⅜ inch long, rarely fly, and attack cabbage, carrot, cauliflower, mustard, radish, and swiss chard. The adults devour foliage at night and hide close to the ground during the day. The green or cream-colored larvae are also destructive. They grow

Vegetable weevil

to ½ inch long, and their yellowish brown heads are patterned with brown dotted lines. Control by rotating crops, and destroy the underground pupae by cultivating; keep the garden clear of weeds and trash, as these provide homes for resting weevils, which become inactive as the summer progresses.

## Diseases

**Black leg** is a fungal dry rot that affects most crucifers. Like black rot, this disease is seed-borne and can live over in plant debris for one or two years. Spots appear on the leaves, starting as inconspicuous, indefinite areas and becoming well-defined dark patches in which many tiny black dots are scattered. Similar spots appear on cabbage stems, and the plant may wilt or even topple over as the head grows. Edges of the leaves of affected plants will wilt and turn bluish or red. Seed produced by such plants is affected. The disease spores are spread by the action of dew and rain; consequently, black leg damage is great only in humid, rainy weather. For control, see black rot, below.

Disease troubles are not unusual in the cabbage patch, and one of the more common is **black rot,** a bacterial affliction that appears on the leaves of any crucifer as yellow, wedge-shaped areas with darkened veins. The leaves then wilt, and severely affected young plants may not produce heads. Heads on more mature plants finally rot. There are some points of confusion brought about by the resemblance between black rot and yellows, described below. The chief points of distinction are: the veins and vascular bundles of leaves and stems in yellows tend to be brown while those in black rot are black and bad-smelling; and soft rot of cabbage heads commonly follows black rot but seldom yellows. Black rot is carried by insects, spattering rain, or surface water—it is primarily a wet-weather disease. The bacteria live over in the soil, in plant refuse, or in seed from diseased plants. After transplanting, infected plants may grow normally with little or no sign of disease for three to six weeks. Thereafter, in midseason, plantings of cauliflower or cabbage may suddenly show a rather large percentage of plants with black rot.

If black rot has been a serious problem, control by treating the seed with hot water before it is planted. Great care should be taken, however, for excessive heat is likely to somewhat reduce germination. Place the seed in a cheesecloth sack and immerse it in water held at a constant 122°F. for 15 to 20 minutes. Keep the water continuously agitated, and maintain the temperature by frequently adding hot water, being careful not to apply it directly to the seed. At the end of the treatment, immerse the seed in cold water, drain, and spread in a thin layer to dry. Seed that is produced in areas with little or no rainfall (such as California and some other Pacific Coast regions) is free of black rot infection, and need not be treated by the hot water method. You can also lessen chances of black rot infection by eliminating the conditions in which it can live over. Use clean seedbeds, removing plant refuse from the garden area. Because black rot can live in the soil for one or possibly two northern winters, grow noncruciferous crops for two seasons on a former cabbage patch, or sterilize the soil between cruciferous crops. If the seedlings are watered artificially, avoid sprinkling the foliage, since this helps to spread the black rot bacteria. Irrigate in the furrows between rows, if possible.

Plants affected with **club root** have yellowed leaves that wilt on hot days and may recover at night. Eventually, the plants may become stunted or die. The roots are enlarged and grotesquely misshapen with club-shaped swellings that later rot and become slimy. Such roots

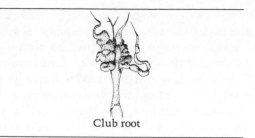

Club root

are unable to absorb nutrients and water, and this causes the above-soil symptoms. The most susceptible crucifers are cabbage, Chinese cabbage, brussels sprouts, some turnips, and mustard. Generally, most turnip and rutabaga varieties aren't troubled, but no member of the cabbage family is safe. The club root fungus is spread mainly on infected manure or plant refuse, and on soil clinging to shoes and tools. It's not seed-borne.

To keep club root fungus out of your garden, use only healthy seedlings for transplanting. Move the cabbage patch each year; the fungus persists in soil for at least seven years. Club root is most dangerous to plants growing in acid or neutral soil (pH of 5 to 7.2); add lime or other basic substances as needed. Weeds related to the cabbage family may carry the disease. Mix lime with wood ashes and sprinkle around plants. Badger Shipper, a globe-type cabbage, is resistant.

**Leaf spot** is a bacterial disease that causes small, gray to brown spots on cauliflower heads. These spots later develop into soft rot. Cabbage shows spots that first appear water-soaked and later grow in size and length, taking on a brownish or purplish gray color. Use seeds grown in western states where the disease does not occur. Seeds can be treated in hot (122°F.) water for 30 minutes, with fair to good results.

**Rhizoctonia** is a fungal soil disease that chiefly attacks young succulent tissue and dormant storage tissue. There are numerous strains, which vary in the hosts that they attack. Rhizoctonia has many effects on cruciferous crops, for it attacks them at different stages: damping-off and wire stem can attack young seedlings; bottom rot occurs in midseason; head rot develops between early head formation and maturity as a dark, firm decay at the base of the outer leaves and heads; root rot occurs before harvest and in storage. The important control measures are to sort out the plants that have wire stem and to avoid planting crucifers in short rotations where bottom, head, and root rot have been a problem.

Cabbage seedlings in seedbeds are also apt to be damaged by **seed rot** (or damping-off), caused by soil-inhabiting fungi. Stems of young seedlings will show a water-soaked appearance at the ground line, and young plants soon topple over and die. See DAMPING-OFF.

**Turnip anthracnose** is primarily a disease of turnip, although it affects other crucifers, including mustard, rutabaga, radish, and Chinese cabbage. Signs of this malady are many roughly circular spots on the leaves, first water-soaked and later drying up and becoming brown. The dead tissue in the center of the spots often falls out, giving a shothole effect to the leaves. The petiole may show elongated gray or brown spots. This fungus is seed-borne and starts from a few infected seeds. Plants set in cool weather are usually safe from turnip anthracnose. Southern Curled Giant is highly resistant.

**Viral infection** of cauliflower seedlings is minimized by growing a few narrow rows of barley as a barrier crop.

**Watery soft rot** is a fungal disease common to many garden crops. It causes cabbage and cauliflower heads to collapse with a very soft type of rot. The first symptom may be a white, cottony growth; this gives rise to black, hard bodies called sclerotia. These growths then fall to the soil, where they lie dormant until the next season. It is a cool weather affliction, and the spores are encouraged by very humid conditions.

While there is no truly effective means of control for watery soft rot, removal of affected plants before the sclerotia can fall should help. Rotation is of little avail, since this fungus attacks so many other vegetables. Destruction is rarely widespread, however, affecting individual plants and not whole fields.

**Yellows** disease (also known as fusarium yellows or wilt) is somewhat notorious and appears in many backyard gardens. Yellows is a warm-weather problem and is almost completely checked when the average soil temperature is below 60°F. As temperatures of soil and air rise, the disease becomes more virulent. However, when the average temperature reaches 90°F. or above, yellows development is retarded. It is caused by a soil-borne fungus, and affects cabbage plants at any age—seedlings wilt and older plants lose their vitality as leaves yellow. (Symptoms are similar for other cucurbits.) Lower leaves are the first to turn yellow, then brown, and finally drop, leaving the bare stem exposed. The heads may become stunted and bitter, but are usually not rotted. The plants become stunted, and the vascular bundles turn brown. The symptoms of yellows and the aforementioned black rot are very similar, and the two diseases may easily be confused: the most reliable distinguishing feature is that black rot turns leaf veins black instead of brown.

Yellows has been known to live in the soil for as long as 20 years. As a result, crop rotation is usually not a satisfactory control. The rapidity with which the disease spreads depends on the degree of susceptibility of the host plant and the favorableness of the environment. In the area where yellows is most severe (from Long Island to Colorado, including the southern parts of New York, Michigan, Wisconsin, and Minnesota and southward as far as cabbage is grown as a summer crop), the optimum conditions for the development of the disease usually prevail during the first three or four weeks after the transplanting of the main crop of cabbage. On resistant plants or susceptible plants in cool weather, the disease may only cause a leaf or two to drop, after which the plants may very well recover and produce normal heads. Early varieties, although very susceptible, often escape the disease. The right conditions for yellows may not appear until the early crop of cabbage is reaching maturity. Fortunately there are a number of resistant varieties, and yellows has ceased to be a problem in some areas. Use only resistant varieties that are appropriate for your region.

# CAMELLIA

Camellias are subject to many pests and diseases. If your camellias have had trouble in the past, look to the condition of the soil as both the source of and solution to your problems. These flowers thrive best in a well-drained soil to which manure or other humus material has been added. Mulching camellias encourages the vigorous growth that gives plants the strength to fight insects and diseases. It's a good idea to plant them in light shade, where there is little danger of damage from heat reflected off walls or sidewalks.

## Insects

Among the insects that attack camellias in the garden is the **fuller rose beetle** (or weevil), a grayish brown beetle with a cream-colored stripe on each side, growing to $1/3$ inch long. It feeds on the edges of leaves at night, and spends the day hidden in twigs, foliage, and flowers. Since this pest is wingless, sticky barriers around the base of the camellia trunk or branches may be enough to keep them from crawling to the leaves. Also see ROSE.

In flower gardens east of the Rockies, the small, shining bronze **rhabdopterous beetle** makes itself known as a nocturnal pest. Adults eat long, narrow holes in young foliage. If you tend to be nocturnal yourself, try handpicking with the aid of a flashlight. Otherwise, use a garlic-based botanical on camellia foliage.

The **root-knot nematode** is a microscopic, wormlike pest that is best combated by adding plenty of organic matter to the soil; some fungi that inhabit rich soil are lethal to this pest. Avoid growing susceptible plants near camellias. See NEMATODE.

Camellias suffer from a number of **scale insects,** including the camellia scale, parlatoria scale, and tea scale. When scales attack, foliage becomes weak and may drop from the plant. Some of the scale can be scraped off by hand, but the most effective control procedure is spraying with oil. Use any of the safe, refined oil emulsions at a strength of about 2 percent (2½ ounces to a gallon of water). It is best to spray in late winter and early spring. Be sure to cover all parts of the plant for thorough control. See SCALE.

Other pests include MEALYBUG and SPIDER MITE.

## Diseases

During winter, large numbers of buds may fall, with the edges of young petals turning brown and decaying. Some varieties suffer much more from **bud drop** than do other varieties. The cause has been attributed to severe frost in September and October, to severe freezing weather during the winter, and to an irregular water supply.

Make sure that the plants don't suffer from a lack of water. Cover them to give protection from sharp frosts and freezes. Avoid watering late in the season, so that little new (and vulnerable) growth will have to face the cold. Mulching will help to maintain uniform soil temperatures and a water supply through the winter.

**Camellia dieback** (canker) is a widespread disease that causes tips to die back and leaves to wilt and turn brown. Stems may turn brown and become girdled. Infection can occur on buds killed by freezing, so remove dead buds promptly. All dead twigs and stems with canker should be clipped off and burned.

**Flower blight** is potentially the most serious disease of camellias. The first symptoms are small brown spots that quickly enlarge to cover the entire petal. The veins are usually darker brown than the surrounding petal tissue [color]. Later, hard, dark brown or black bodies form in the bases of flowers or in individual petals. These sclerotia may resemble a petal or miniature flower in shape. They remain dormant during winter, and as the flowcring season approaches they produce saucerlike objects from which the spores are shot into the air. Sclerotia can lie dormant in the soil for at least three years before germinating.

This disease has sometimes been confused with frost injury, which browns only the outer or marginal portion of flowers or petals. In the South, areas known to harbor camellia flower blight are often quarantined. Northern growers should be on guard against purchasing diseased plants. It's best to buy barefooted plants from which all buds have been picked; on balled plants, discard the upper two or three inches of soil. Once this disease hits, remove all flowers, both from the bush and those on the ground, and burn them. If flowers have already decayed on the ground, remove and discard the upper three inches of soil or mulch.

**Leaf yellowing** may be either an environmental condition or a disease caused by a virus. In the former case, the variegations usually follow a uniform and rather typical pattern that is more or less similar on all leaves. In the latter, the yellow areas caused by virus are irregular and vary on different leaves, branches, and plants; affected plants grow rather normally, but the leaves are more susceptible to sunburn and frost injury than are normal leaves.

Rogue out any plant suspected of viral infection. Propagators should be extremely careful to use virus-free rootstock for grafting.

**Root rot** is a fungal disease that results in a gradual wilting, yellowing, and death of one or more branches. Affected leaves may fall prematurely. Small plants and cuttings succumb more rapidly than older plants, and varieties differ in their resistance. Diseased plants should be discarded and the soil prepared to provide good drainage.

Faded green or brown dead areas on leaves may be the result of **sunburn.** Symptoms are most common on the upper and exposed sides of bushes. If affected areas are subsequently attacked by fungi, they may change color and fall out.

Sunburn strikes when plants accustomed to partial shade are exposed to bright sunlight. This may come about by transplanting to a sunny spot, by turning bushes around so that the former north side is facing south, or by removing overhanging branches. In winter, bright days combined with cold soil may dry the leaves to produce similar symptoms; the problem here is that the leaves lose more water than is absorbed by the roots. You can help by providing partial shading in winter months.

# CAMPANULA

This flower is particularly susceptible to the foxglove **aphid,** a shining, light green insect with dark green patches around the base of its cornicles. Its feeding results in yellowed or blanched spots, leaf curling, and a distortion of leaves that resembles viral disease. See APHID.

Campanula may also be bothered by onion thrips (see ONION) and SNAIL AND SLUG.

# CANNA

## Insects

Two species of **canna leafroller** fold or roll leaves, tie them up with webs, and feed within them. Foliage often becomes ragged and brown, and leaves may die. One caterpillar is pale green, set off by a dark orange head, and reaches a length of 1¾ inches; the other is greenish white and measures 1 inch when fully grown. They are a problem in the southeastern states, south of Washington, D.C.

The caterpillars can be killed by pinching rolled leaves. Clean up and burn plant refuse in fall, and remove affected plant parts.

Other insects that may cause trouble are the corn earworm (see CHRYSANTHEMUM), and fuller rose beetle and goldsmith beetle (ROSE).

## Diseases

One of the most troublesome diseases affecting canna is **canna bud rot,** recognized by water-soaked leaves that darken and spread to form brown ragged areas. Flowers may be killed by the infection of young buds or the stem. Prevent this disease by buying healthy rootstalks, spacing plants, and watering carefully to keep new growth dry.

# CANTALOUPE

## Insects

Holes in fruit may be the work of the **pickleworm** [color]. This is a yellowish white to green, ¾-inch-long caterpillar, marked with a brown head. Look for sawdustlike frass at the holes.

Late crops are most susceptible to the pickleworm. A trap crop of squash will lure away these pests, but be sure the vines are disposed of before the worms leave the squash blossoms. Clear the melon patch of all plant debris immediately after harvest, and cultivate at this time to destroy pupae.

High on the list of pests that bother cantaloupes is the **striped cucumber beetle** [color]. In the adult stage it is yellow with three black stripes down the back, and grows to about ¼ inch long. The larva is white, slender, brownish at the ends, and about ⅓ inch long. Adults not only feed on leaves, stems, and fruit, but are responsible for spreading bacterial wilt as well. Larvae bore into the roots and also feed on the stems below the soil line; they usually attack young plants, causing them to wilt and sometimes die. The beetle is most common east of the Rocky Mountains.

Radishes interplanted among melons will discourage the beetle. See also CUCUMBER.

Other pests include those that infest SQUASH, and APHID, the grape colapsis (see BEET), potato flea beetle (POTATO), and SPIDER MITE.

## Diseases

**Curly top** of muskmelon is caused by the same virus responsible for curly top of beets. The young leaves of diseased plants are stunted and curled, but may be normal in color; the older leaves are often yellow. The stems are shorter than normal so that the plants are stunted and bushy. Infected plants usually fail to produce fruit, and under some conditions they may be killed. The virus is transmitted from beets to muskmelons by the beet leafhopper (see BEET). Where this virus is known to have occurred, keep down nearby weed hosts. Rogue out infected plants. Avoid planting muskmelons near sugar beets.

**Fusarium wilt** causes the vines to suddenly wilt away, sometimes almost overnight, and for no apparent reason. The disease is caused by a fungus that is spread by the yellow-and-black striped cucumber beetle; if the plants are protected from the beetles, they will be protected from wilt. Tender shoots can be protected by placing hot caps over the hills when the seedlings are young. Tents of cheesecloth or mosquito netting may be spread over the hills until the plants are large enough not to tempt the beetles. Resistant varieties include Golden Gopher and Iroquois.

**Leaf spot** is a fungus living on diseased remains of vines in the soil. It causes numerous brown spots on the leaves, many of which may be killed. There is no spotting of the fruit.

Be certain to destroy diseased vines so that the viral organism will not perpetuate itself. It is good practice to grow melons or cucumbers no more often than once in three years in the same soil.

**Mosaic** can be brought on by a host of viruses to cause chlorosis, stunting, and reduction of yield. The most commonly observed symptoms are mottled leaves which have dark green, light green, or yellow areas of irregular form. The leaves are often irregular in shape and portions of the veins may be killed. The vine tips are often stunted and yellowish. Under certain conditions, stem growth may continue while that of the leaves does not, producing a condition known as rat tail. Less readily observed is the abortion of flowers of all ages. Flowers that appear normal to casual observation often have aborted anthers and scant pollen. Fruit may appear normal, or may be mottled, small, or misshapen. Fruit not affected directly by the virus are often low in quality or sunburned, or they may have poorly formed net. These indirect effects result from lack of sufficient foliage to supply food and to cover the plants. Losses are slight if the plants are not infected until melons reach harvest maturity, but severe losses may result if infection occurs earlier.

Although many varieties of muskmelon and cantaloupe are susceptible, Honey Dew and Honey Ball are less affected than others. Control in the absence of other sources of infection can be maintained by using seed that is free from the virus or by carefully roguing at the thinning stage. The virus can be transferred from one plant to another by your hands. Keep down weeds in the vicinity.

Powdery mildew grows on the surfaces of the leaves and stems, sending minute feeding organs into the host tissues. This disease can be recognized by the powdery white growth on the stems and both surfaces of the leaves. Infected leaves and stems may remain alive for indefinite periods, or they may become brown and die. Severely attacked plants become weak and nonthrifty and produce few or no melons. The melons may be small or malformed or they may be normal in shape. Large melons often have poor net and poor flavor, and may be sunburned. Resistant varieties are available; consult your extension agent for an up-to-date list. Dusting with sulfur will take care of mildew, but only certain varieties can be dusted without an adverse reaction. Cantaloupe varieties tolerant to sulfur have been developed.

# CAPE-JASMINE

Southern growers of this plant are often bothered by the **Florida wax scale,** distinguished by a thick, waxy covering on a reddish brown body; the covering is sometimes tinted pink. This scale lives on stems, and causes plants to become stunted. A dormant-oil spray, applied before growth starts in spring, should take care of this pest.

Other pests include the fuller rose beetle (see ROSE) and WHITEFLY.

# CARNATION

## Insects

A number of **aphids** infest young leaves and buds. They do the most damage in greenhouses. See APHID.

The **variegated cutworm** is a common pest of greenhouse carnations in fall, when they climb stems and eat holes in the buds. The adult is a moth with a wingspan of up to two inches, and displays brownish gray forewings. The caterpillar is light brown, mottled with darker brown. Normally there are two annual generations outdoors, but there may be more in the greenhouse environment. For control measures, see CUTWORM.

Other insect pests include the cabbage looper (see CABBAGE), celery leaftier (CELERY), flower thrips (PEONY), fuller rose beetle (ROSE), MEALYBUG, SNAIL AND SLUG, and SPIDER MITE.

## Diseases

Whitish spots on leaves may mean a fungal infection, one that is capable of invading the branch and causing death of the plant above the infected part. Known as **alternaria leaf spot** and branch rot, this disease is controlled by watering the soil, not the plant, to keep foliage dry. Soil that is rich in organic matter will mete out water as it is needed.

**Bacterial wilt** causes wilting of plants and shoots, accompanied by splitting of the stems at the joints and by yellow streaks in the vascular system. Tissue just under the epidermis is sticky, and cutting across the stem shows discoloration and a bacterial ooze. The disease appears to spread rapidly at high temperatures. Although the parasite works internally, it is spread by splashing. Use clean cuttings, rogue affected plants, and water carefully.

**Botrytis flower rot** causes a brown spotting or rot of flowers. The gray mold you may see growing over these brown spots is the fruiting stage of the fungus. Since the fungus fruits on dead plant material at the base of the plants, it is nearly impossible to eliminate the sources of infection. Keep moisture from condensing on flowers by turning on the heat one-half hour before sundown, especially during late winter.

The fungal disease **bud rot** is carried by small mites. It causes buds to shrivel, and the resultant blooms are unhealthy and malformed. As this disease has been found to flourish under cool, wet conditions, it's advisable to hold off the planting of susceptible varieties until the warm weather has somewhat alleviated the danger of cold nights. Avoid overwatering. Remove and destroy any diseased buds, so that the disease does not have an opportunity to propagate itself. Buds should also be removed from the ground, because they carry spores that are responsible for the spread of disease.

The first symptom of **fusarium wilt** is a dull green color to plants, which then wilt and eventually turn the color of straw. Cutting across the main stem reveals brown discoloration, sometimes accompanied by a dry, punky rot. A pink mass of spores may be seen on the roots. This wilt is caused by a soil-borne fungus which enters through root wounds and then plugs the water-conducting system to kill the plant. Plants that have been outside are very apt to become infected. The disease may be carried in infected cuttings. It can show up at any time, but is particularly noticeable after rooted cuttings have been planted in flats or benches and in later winter when flower production draws on the strength of the plant.

Soil sterilization controls the disease. Use only clean cuttings and sterilized sand in the cutting bench.

**Leaf rot** is one of the most common diseases of carnations, causing discolorations on the stems and both sides of the leaves. These discolorations appear most often on the tips of leaves. When most of the plant is afflicted, death usually follows.

Any plant that appears to be suffering from leaf rot should be removed immediately to avoid infecting healthy plants in the vicinity. Use cuttings from healthy plants only, and give the plants plenty of room to allow air to circulate. When watering, avoid wetting the foliage, as wet leaves render a plant vulnerable to the disease.

Pale tan **leaf spots** with purple borders are symptoms of fungal disease. Look for small black fruiting bodies on the surface of the spots to make sure of the cause. The disease is spread by splashing water; if water is to be syringed on plants, do so early in the day, and use a mulch to discourage splashing. Stick to healthy plants for cuttings.

**Mosaic, streak,** and **yellows** are viral diseases that may show as light mottling in the leaves (mosaic), red or yellow spots or streaks (streak), or yellows (a combination of both diseases). Streaks in the flowers are either due to mosaic or yellows infection. Use only healthy stock. Take care in disbudding, as these viruses are transmitted by mechanical contact. Keeping down the aphid population will discourage the spread of the viruses; see APHID for means of control.

Carnation **rust** is the agent behind numerous brown pustules that appear over various parts of the plant. The pustules usually start as water blisters on the lower leaves, and burst to discharge a brownish powder that spreads the disease to other parts of the plant as well as to healthy plants.

Rust-resistant varieties are available. But only those plants you know to be healthy, as one sickling can quickly spread its misfortune to neighbors. See to it that plants don't have excess moisture on their leaves—it's best to water the ground, not the plants, and to do so early in the day. A soil rich in organic matter will retain moisture around the roots without making puddles. Generally, carnations and dianthuses should be planted in a sunny part of the garden, far enough apart so that air can circulate between them.

# CARPENTERWORM

The carpenterworm grows to 2½ inches long and is pinkish with a brown head. It makes galleries in trunks and larger branches. Remove and destroy affected branches, and dress wounds. Light traps have been used to attract the gray moths; these adults have a wingspread of about three inches, and the wings show brown or black markings. A recent control employs nematodes, which transmit a lethal bacterium to the carpenterworms. The nematode *Neoaplectana carpocapsae* is available commercially; see NEMATODE.

# CARROT
## Insects

Carrots are prone to injury from several **aphids**: among these are the bean, green peach, and corn root aphids. Control is not usually necessary; if you do need help, see APHID.

The **carrot rust fly** [color] is particularly destructive in the northeastern states and in coastal Washington and Oregon, and also is found in parts of Idaho, Utah, Wyoming, and Colorado. The larvae are yellowish white, legless, and grow up to ⅓ inch long. These maggots hatch from eggs around the carrot crowns, and burrow down into the roots. Their tunnels are rusty in color from the maggots' excrement, which explains the name. Plants may be stunted, and the infestation leads the way for soft-rot bacteria (see below). If this damage is slight, affected parts can be cut out and the carrots used for cooking. Prevent maggot troubles by putting used tea leaves in with the carrot seed. If the infestation becomes serious and cannot be controlled by rotating the crop, skip the early planting and do not plant until early summer, after the maggots have died of starvation. This procedure requires that celery plantings also be postponed, as the maggots will otherwise live happily on the celery roots while they wait for the carrots. In the New York City area and similar climates, main crops planted after June 1 will escape the larvae if harvested in early September, before the second brood has hatched. To repel the slender, shiny green carrot fly, try sprinkling pulverized wormwood around plants. The flies will not lay their eggs at the crowns of plants protected with this homemade pesticide. They also are repelled by adjacent plantings of leek, onion, pennyroyal, rosemary, sage, black salsify (or oyster plant), and coriander.

From Colorado east to Georgia and New England, the **carrot weevil** [color] produces pale, legless, brown-headed grubs. The larval worms grow up to ⅓ inch long, and tunnel into carrot roots. Early carrots are usually the most severely injured. The weevil spends the winter in the adult stage in weedy or grassy areas immediately adjoining old carrot fields. The following May, these adults travel to the nearest host plants.

The adult usually reaches its goal by walking, and thus travels only a short distance; this suggests rotation as a means of control. Avoid replacing carrots with other crops that are attractive to the weevil, such as parsley, celery, and parsnips. If debris, high grass, and weedy, fallow fields are cleaned out, the pest will have a hard time finding a snug winter home. In spring, destroy any grubs in the soil by cultivating deeply around the area where the carrots are to be planted.

The **carrotworm** (also called parsleyworm and celeryworm [color]) is the larval form of the black swallowtail butterfly. The larval form is as bizarre in appearance as the winged adult is beautiful, and both are easy to spot. The two-inch caterpillar is green with a yellow-dotted black band across each segment; when disturbed, it projects two orange horns and emits a strong, cloyingly sweet odor. The well-known black swallowtail has large, black forewings with three rows of yellow markings parallel to the wing edges, the rear wings also having a blue row, an orange spot each, and the projecting "swallowtail" lobe.

Carrotworm

Because the carrotworm population is rarely very great, early morning handpicking should work well enough. *Bacillus thuringiensis* gives biological control. Rotenone dust is sometimes used in serious cases.

A serious disease, aster yellows, is spread by the **sixspotted leafhopper,** also known as the aster leafhopper. The first symptom you are likely to spot is a yellowing of young leaves as they emerge from the crown. Side shoots also take on a jaundiced appearance, and older outside leaves may become bronzed or reddened and twisted. Finally, near the end of the season, the crown becomes blackened and dead. The responsible leafhopper is greenish yellow with six black spots, and grows to ⅛ inch long.

Because this leafhopper seems to prefer open areas, the planting of carrots near houses or in protected areas may help to prevent infestation. If this is not possible or practical, try enclosing your plants in cheesecloth, muslin, or plastic netting, supported on wooden frames. Pyrethrum can be dusted onto the carrot tops. Avoid planting carrots near lettuce or asters.

If southern gardeners find holes appearing overnight in carrot foliage, a prime suspect is the **vegetable weevil** [color]. This pest is a dull buff color and has a pale V on its wing covers. The head has a sharp, broad snout. It is ⅜ inch in length, rarely flies, and does its dirty work at

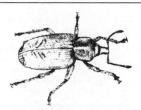
Vegetable weevil

night. The green or cream-colored larva is also destructive. It grows to ½ inch long and its yellowish brown head is patterned with brown dotted lines. Control by rotating crops, and destroy the underground pupae by cultivating. Keep the garden clear of weeds and trash, which provide homes for nesting weevils as they become inactive as the summer progresses.

Many species of **wireworm** dine on carrots, puncturing and tunneling roots. They are so called because they resemble jointed wires. Color varies from white to yellow to brown—they get darker as they grow longer. For control of this ubiquitous pest, see WIREWORM.

Carrots also fall prey to a number of relatively minor pests: MILLIPEDE, onion bulb fly, THRIPS, webworms (see parsnip webworm under PARSNIP), and yellow woollybear caterpillar (imported cabbageworm under CABBAGE).

# Diseases

In the South, carrots are grown during the cool months of the year when most of their parasites are not very active, and southern gardeners should therefore not have any serious problems with diseases of this crop.

Peculiar shapes can be the result of poor growing conditions. Forked carrots are often found in compacted soil. Soil rich in clay may benefit from the addition of sand. Too much fertilizer and other agricultural chemicals may be responsible for hairy carrots. Stunted carrots are found below soil that has crusted after sowing. Try interplanting radishes; they crumble the soil as they emerge, and can be pulled to give room to the carrots.

Carrot seeds are vulnerable to **damping-off**, which causes the watery soft rot of stem tissue at the ground line. Roots may become infected, turn black, and rot, and the seedlings themselves later become yellowed and die. See DAMPING-OFF.

**Fusarium root rot** primarily affects cucurbits, but may also cause a spongy rot of carrots that are stored at temperatures above 45°F. The rot fungi enter through wounds and bruises in the root, causing shallow, scabby spots on the taproot. Try crop rotations omitting root crops; avoid injury to the roots at harvest; make sure storage rooms or areas are clean; and if possible, store carrots at approximately 32°F.

**Leaf blight** on carrot foliage first appears as yellow to white spots, irregular in shape, on the margins of seedling leaves. These spots quickly become brown, and take on a water-soaked appearance. Spots may also girdle slender taproots, resulting in the death of the root below the infected spot. Affected roots may appear water-soaked, and sometimes have lesions, dark spots, craters, or pustules. The flower cluster of seed plants may have both blasted seeds and seeds capable of germination that will transmit the infection. Infected parts of the flower cluster become discolored and may die.

To avoid blight, plant only seeds that have been treated by the hot water method, at 126°F. for ten minutes. There is no dust or disinfectant that will kill the bacterium of the blight inside the seed without injuring the embryo. A four-year schedule of crop rotation, good garden cultivation, and sufficient spacing between plants will all aid control. Also, good healthy soil encourages the root system to develop and feed nutrients that are used for fighting off the disease symptoms.

Tiny nematodes in the soil cause **root knot.** The small knots or galls form on the lateral roots and rootlets, and small pimplelike swellings in the taproot enlarge and roughen the surface. Plants are often stunted and chlorotic (yellowed or blanched in appearance), but the knots are the only dependable symptom.

Rotation of carrots with nematode-resistant crops will sufficiently reduce the nematode population so that at least one crop of carrots may be successfully grown. Such resistant crops include small grains, some cowpeas, and velvet beans. Sometimes they must be grown for three or four years before a satisfactory decline in nematodes is noticed. See also NEMATODE.

**Vegetable soft rot** is an infamous bacterial disease that affects many vegetables, causing heavy damage. The disease is characterized by a soft, slimy rot of the taproot that is distinguished from other rots by a foul, sulfurous odor. Infected tissues are discolored gray or brown, and either the tender lower end or the crown may be affected. Leaves often wilt or turn yellow in the late stages of the disease. Soft rot is favored by hot humid weather, and is caused by bacteria in the soil.

Rotation of the rows often prevents vegetable soft rot in the garden, but the disease can also occur in storage. Discard at harvest those roots which are bruised or otherwise damaged, and store the crop in a cool, dry, clean place. Carrots should only be packed in containers that allow some ventilation.

# CASTOR BEAN

The castor bean plant is valued by gardeners for its lushly tropical appearance, as a means of discouraging moles from tunneling in the garden, and as a companion plant that keeps insects from vine crops. Because of its size (8 to 12 feet high), the castor bean is better suited for field plantings of vine crops than small gardens. The highly poisonous seeds should not be planted if accessible to children. Castor oil con-

centrates the noxious properties of the plant. To make a repellent solution that can be sprinkled from a watering can, mix a tablespoon of the oil with two tablespoons of liquid detergent in a blender until the product has the appearance of shaving cream, and add to warm water in a sprinkling can. Sprinkle the liquid wherever moles are causing trouble. It penetrates the soil best after a rain shower.

Although the castor bean is used as a repellent, it has a modest list of its own pests. The hairy **garden webworm** spins light webbing over plants and feeds on the foliage underneath. It is a light green, inch-long worm with small, dark green spots, and hides within a silken tube on the ground. Growers in the Midwest and Southwest are particularly bothered by the garden webworm. Cut off and destroy webbed branches; handpick feeding webworms, and step on those hiding in their silken tubes.

The black-and-yellow southern armyworm may bring castor beans to grief; see CELERY. Other possible troublemakers are the corn earworm (see CHRYSANTHEMUM), potato leafhopper (POTATO), serpentine leafminer (PEPPER), and SPIDER MITE.

# CATALPA

## Insects

Trees may be defoliated by the **catalpa sphinx**, a brown or green caterpillar about 2¼ inches long. There are two broods of the sphinx, one in June and another in August or September. If you find any caterpillars crawling on the limbs of trees, handpick and crush them. Generally, this is all that is needed, for there are many parasites of the sphinx that keep the insect under control naturally.

Presence of the **comstock mealybug** is often made known by patches of white in cracks in the bark. See MEALYBUG.

**San Jose scale** occasionally hits catalpa. See LILAC.

## Diseases

Certain fungi may cause brown circular **leaf spots** on tree foliage. Infected tissue within the spots often drops, leaving holes in the leaves and causing premature leaf fall to follow. Good garden housekeeping is needed to aid materially in control. Collect, remove, and burn or compost all dead leaves that may harbor disease-producing fungi. If leaves are composted instead of burned, they should be piled in alternate layers with soil. Be sure that any leaves left exposed are covered completely with soil.

**Verticillium wilt** may cause one or several branches to wilt. Purple to bluish brown streaks in sapwood of wilted branches indicate the presence of this disease. For control, see TREES.

# CAULIFLOWER

In general, the insects and diseases that harm cauliflower are also injurious to cabbage; see CABBAGE. Insect pests seem to prefer green to purple cauliflower. Although the purple variety doesn't sell as well at the roadside stand because of its odd appearance, many have found its taste is superior to that of the green.

# CEDAR

When a white cedar suffers **foliage browning**, all the leaves on the inside of the tree turn reddish brown and fall to the ground either in spring or fall. This natural shedding often occurs after a dry season, or it may happen every second or third year, varying with the individual tree and conditions. The discoloration is distinguished from fertilizer burn in the way it hits the inside of the entire tree, whereas fertilizer burn or sunburn due to root injury start at the tip of the branch and work back, generally progressing up the tree from the lowest branches. Red spider mite injury differs too, causing a sudden bronzing of all the leaves on a branch during summer.

Foliage browning or winter bronzing is apparently due to physiological changes in the tree, and is tied in with the tree's nutrition in a way not yet understood. It disappears with the advent of warm weather.

Two types of **gall rust** may infect white cedars. One causes a good-size spindle-shaped gall; the branch beyond the gall usually dies. The life cycle of this fungus is completed on the leaves of the serviceberry plant, from which come the spores that infect the white cedar. A second type of gall rust, *Gymnosporangium hyalinum*, causes a smaller spindle-shaped gall on white cedar twigs and branches. The infective spores come from the leaves of pears and hawthorn. Cedar-apple rust gall [color] may also damage apple trees in the area.

Pruning and burning of affected parts help control these fungi. If possible, eradicate the alternate host plants—serviceberry, pear, and hawthorn—within a half-mile radius.

**Nursery blight** is a condition causing tip dieback. It occurs occasionally on white cedar but is much more common and serious on red cedar. A small canker girdles the twig, which eventually dies. Control by cutting and burning affected twigs.

A fungus may cause **witches' brooms** to appear at the ends of the branches of white cedar. On young trees the entire top may be broomed. Branches beyond the infection point usually die as soon as the fungus girdles the branch. Cross sections of infected wood show wedge-shaped brown spots in the wood. In spring, brown horns grow out of the infected area and produce spores that infect bayberry and sweet fern, which in turn produce spores to reinfect the white cedar. The only control known is the eradication of the bayberry and sweet fern.

# CELERY

## Insects

Of several **aphids** known to feed on celery, the green peach aphid is possibly the most common. It is yellow green and bears three dark lines on its back; see APHID.

The larva of the **carrot rust fly** is an occasional celery pest, feeding on the roots of early plants. These ⅓-inch-long worms are yellowish white, legless, and leave rust-colored tunnels as they eat their way through the plant. The adult is a slender, shiny green fly. For habits and controls, see CARROT.

The major pest of celery is the **celery leaftier** (or greenhouse leaftier), ravaging a variety of vegetables and ornamentals throughout North America. These caterpillars are initially pale green in color, with a white stripe running the length of their backs and a pale stripe centered within. As they grow to their full length of ¾ inch, their color changes to yellow. The worms eat holes in leaves and stalks, and have the peculiar habit of folding leaves and tying them together with webs. The adult is a small, brown, nocturnal moth with wings crossed by dark wavy lines. The eggs are laid on the undersides of leaves and look like fish scales.

Since the leaftier does not usually appear in gardens in great numbers, gardeners have relied on handpicking to prevent trouble. It's also a good idea to handpick and destroy any leaves that appear damaged or rolled, as a leaftier may well be sequestered within. *Bacillus thuringiensis* is a commercially available pathogen of this pest. In serious cases, pyrethrum can be applied in its pure state; this botanical poison makes the larvae sick and they then leave their webs. A second dusting should kill them.

The **celeryworm** [color] is hardly a picky eater, which explains why it is also known as the carrotworm and parsleyworm. The adult is

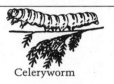

Celeryworm

the well-known black swallowtail butterfly. Its black wings are marked with three rows of yellow markings parallel to the wing edges, the rear wings having a blue row, orange spot, and characteristic "swallowtail" lobe. The larval worm is as bizarre in appearance as the winged adult is beautiful; reaching a length of two inches, it is green with a yellow-dotted black band across each segment. When disturbed, the worm projects two orange horns from its head and emits a sweetish odor.

Because the celeryworm population is rarely very great, most gardeners find that early morning handpicking works well enough. *Bacillus thuringiensis* gives biological control. If your celery gets into bad trouble, consider a dusting of rotenone.

If your celery plants are stunted and yellowed and you can find no obvious cause, the plants may be victims of **nematodes** (eelworms).

These tiny pests, some of which are too small to be seen with the naked eye, live in the soil and suck juices from celery roots. Roots develop root knots, or galls, and affected rootlets turn brown. Finally the plants become chlorotic and may die.

Seedbeds should never be located on soil that is known to be infested with nematodes. If it is impossible for you to obtain clean soil, infested soil can sometimes be rendered safe by starving out the nematodes: keep the land fallow for several months, and if practicable, flood the fallow land for two or three months. Nematodes that hatch in such plots will starve to death in a few weeks. An alternative method involves planting the land to resistant cover crops, but this requires that the land be kept free of weeds, as a few weeds can sustain a considerable number of the tiny pests. For more information, see NEMATODE.

The **parsnip webworm** is a small black-spotted caterpillar that webs together and feeds upon unfolding blossom heads of celery. It grows to about ½ inch in length, and varies in color from greenish yellow to gray. When mature, the webworm leaves its web and burrows inside a flower stem to pupate. The gray moths appear late in summer. Damage is usually not serious; control by cutting off infested flower heads.

Gardeners and farmers from South Carolina to Florida may run into the **southern armyworm** at work on their celery. These caterpillars are black or yellow with black markings; luckily, they are susceptible to many natural predators and parasites, including birds, skunks, and toads. The armyworm is so named because in some seasons they attack plants in huge numbers, destroying everything in their path.

Encircle the garden with a ditch, plowed so that the side nearest to the crops is perpendicular; the worms will be trapped in the entrenchment and may then be crushed, buried, or cremated.

Tiny, oval yellow or green specks swarming over celery leaves are likely twospotted **spider mites.** They live on the undersides of leaves and suck plant sap; their feeding gives a sickly, blotched appearance. Control is often not necessary, but if you feel the plants need help, see SPIDER MITE.

The **tarnished plant bug** is a nuisance to more than 50 plants, including celery. It is shy and will likely hide as you approach, but you may catch a glimpse of a flat, ¼-inch-long insect, colored green to brown, with touches of yellow,

Tarnished plant bug

brown, and black that give the wings a tarnished appearance. A very close look will reveal a yellow triangle, with one black tip, on each side of the bug. The nymph is very small, greenish yellow, and marked with black dots on the thorax and abdomen. As it feeds, the adult liberates a plant toxin that may result in black celery joints. Also, look for brown sunken areas on leaves and shoots near the top of the plant. For control methods, see ASTER.

Celery growers should also keep an eye out for the cabbage maggot (see CABBAGE), the reddish brown carrot beetle, the white, brown-headed carrot weevil, which burrows into celery hearts (CARROT), various cutworms (see cabbage cutworm under CABBAGE), the flesh-colored European corn borer (CORN), the potato flea beetle (also jumpy; see POTATO), jumpy, aphidlike sixspotted and beet leafhoppers

(ASTER, BEAN), THRIPS, and the yellow woollybear caterpillar (see imported cabbageworm under CABBAGE for controls).

## Diseases

Perhaps the most serious diseases of celery are the early and late **blights.** Early blight first appears as small, pale green or yellow spots on seedlings in the plant bed or on transplants in the field. These spots enlarge, turn brown or gray, and gray or pale lavender spores are produced on the upper leaf surfaces. In severe cases, the spots grow together and kill the leaves. Sunken, tan, elongated spots may mar stalks just before harvest, and growth is stunted.

The fungus is sometimes seed-borne; it doesn't live in the seed for more than two years, and newer seed should be disinfected by dipping for 30 minutes in water at 118°F. It has been found that the longer plants are left in crowded seedbeds, the worse blight is apt to be in the field.

Late blight begins as small water-soaked spots on the leaves of seedlings. The spots later darken and become dotted with tiny black dots which are the fruiting bodies of the fungus. Thousands of gelatinous threads of spores may be exuded from a single dot in wet weather. Infected seed can be treated in hot water at 118°F. for 30 minutes; while this treatment could be used on all seed sown late in fall for a spring crop, it is not so necessary for early plantings, since temperatures in early fall are probably too high to permit infection. If the disease cannot be controlled in the seedbed, it is better to destroy the seedlings than to plant them in the field where control would be very difficult and other plantings would be exposed to the disease. Select a blight-resistant variety.

**Damping-off** is the principal cause of losses occurring in seedbeds. It's particularly serious in beds sown in summer, and at times may destroy as many as 95 percent of the plants. Usually the first sign of damping-off is wilting or toppling over, caused by a watery soft rot that attacks the plants at the ground line. If the affected plants are scattered about the seedbed, damage may go unnoticed, but the disease may also attack groups of plants. During periods of high humidity and high temperatures, the threadlike mycelium of the causal fungus may lie on the soil like a spider web. It is during these periods that damage is greatest. Losses are highest in August and September, and are reduced sharply as weather becomes cooler and drier in fall.

Crowding plants aggravates damping-off, as the dense mat of vegetation keeps the humidity high. Use double shades over the beds during the first month or until the first leaves unfold. One shade can then be removed, and the plants are further strengthened by lifting the other shade a few minutes longer each day. Soil that has been used previously for seedbeds is apt to harbor the fungus.

Stunted, one-sided growth and yellowing are signs of **fusarium wilt,** a fungal disease. Vascular strands become reddish brown from the roots to the leaves. Avoid this problem by planting seed in clean, healthy soil and by discarding diseased plants. Green celeries are generally resistant.

When **pink rot** occurs, the stems become water-soaked in spots, and a telltale white to cottony pink growth appears at the base of the stalks. These spots of infection usually start in growth cracks or in wounds left by blight or insects. The stalks rot and take on a bitter taste. This fungus lives in the soil for many years, and may cause damping-off in the seedbed.

To control pink rot, rotate the growing area whenever possible, avoiding successive plantings of celery, lettuce, or cabbage in the same soil. Remove and destroy sick plants so that healthy ones will not become afflicted.

# CHALCID WASP

Mealybugs, aphids, scale, and larvae of beetles, moths, and butterflies are parasitized by these small ($\frac{1}{32}$-inch-long) wasps [color]. Some chalcids are a metallic black, while others are golden. They are rated as even more important than ichneumonid and braconid wasps as biological control agents, and attack all life stages. The egg parasites of the trichogramma family are the best known to growers, and are commercially available (see TRICHO-GRAMMA). A more recent wasp to appear on the market is the golden chalcid, *Aphytis melinus.* It is used to control California red citrus scale, San Jose scale, and ivy scale. The parasites are shipped as adult wasps. Upon release, they spend the 10 to 15 days remaining to them laying eggs under the scales' protective covers. Adults may sustain themselves by feeding on the scales. The wasp larvae feed inside the scales and destroy them.

In warm climates such as that in southern California, *A. melinus* may overwinter. In cold climates, new releases are needed to reestablish the wasp in spring. These beneficials are highly vulnerable to most chemical insecticides. Less damaging are light, supreme petroleum oils, and insecticidal soaps. The wasps are further discouraged by ants, which feed on scale honeydew and attempt to protect their food source. Ant baits, traps, and barriers are recommended.

# CHARD

## Insects

Because chard shares most of its insect enemies with the beet and other vegetables, none is dealt with fully here. See BEET.

Tiny black dots may be bean **aphids.** See APHID.

The long, slender **black blister beetle** should be handled with care—that is, with gloves. If a bug is crushed during handpicking, a toxic substance may blister the skin. See BEET.

An occasional pest of chard is the **European corn borer,** a flesh-colored caterpillar with dark brown spots. See CORN for means of control.

The **garden springtail** is a tiny, dark purple insect with yellow spots; it attacks chard seedlings. The springtail has no wings, but uses a

Garden springtail

tail-like appendage to hurtle itself into the air. It is usually found eating small holes in leaves near the ground line. Clear weeds from the area; a garlic-and-water spray can be applied to the lower leaves.

The tiny ($\frac{1}{16}$-inch-long), black **potato flea beetle** is so named because it jumps like a flea when disturbed. See POTATO for control measures.

The **spinach leafminer** occasionally makes tunnels in chard leaves. The adult is a gray, black-haired, two-winged fly about $\frac{1}{4}$ inch in length. It lays white, cylindrical, reticulated eggs on the undersides of leaves in clusters of from two to five. In a few days, these eggs hatch into the $\frac{1}{3}$-inch-long, pale green or whitish maggots that tunnel the leaves. For control see SPINACH.

As the **tarnished plant bug** feeds, toxins in the plant are liberated and chard leaves may be deformed. The adult is small, mottled with

Tarnished plant bug

Aphid

white and yellow, and has a tarnished appearance. Look for a yellow triangle on each wing. See ASTER for control measures.

The nocturnal **vegetable weevil** does its dirty work at night. It grows to ⅜ inch long, is a dull buff color, and has a pale V on its wing covers. The head has a short, broad snout. See CABBAGE for controls.

## Diseases

Chard may be spoiled by **chard blue mold,** which appears as large yellowed spots on top and bluish purple mold underneath. This disease also occurs as spinach downy mildew. A week of humid, cool weather is needed to introduce blue mold. Plant in fertile, well-drained soil, avoid crowding, and use a two- to three-year rotation with other crops.

# CHERRY

For an effective spray program, see FRUIT TREES.

## Insects

**Aphids** [color], particularly the black cherry aphid, may be seen massed two deep, standing on each other's backs as they try to stick their beaks into the leaves. They are seldom a serious problem except on young trees, and favor sweet cherries.

Nature provides natural enemies that handle aphids, particularly in the case of large infestations, but in some orchards it may be necessary to introduce ladybugs or lacewings to supplement low native populations. Strong sprays of soapy water have been used to kill aphids, and heavy local infestations can be controlled by dipping whole branches into soapy solutions. Dormant oils will kill many overwintering eggs. If you can tolerate some wrinkled leaves for a month or two, aphids will migrate in midsummer to plants of the mustard family. See APHID.

The **buffalo treehopper** [color] and other related treehoppers may be serious pests of cherry trees. These light greenish insects have a distinctive triangular shape. Two horns protrude at the shoulders. The damage, much like that of the periodical cicada, is caused by laying eggs in the bark of the tree through a curved slit. Infested trees often appear rough and cracked, and the slits made by the treehopper's ovipositor can give entry to a variety of fungi and diseases.

Do not grow alfalfa, sweet clover, or other legumes as cover crops, since treehoppers favor these plants and can increase to damaging numbers. A thorough spray with a 4 percent dormant oil will kill many of the eggs overwintering in (not on) the twigs.

The **cherry fruit fly** [color] is usually to blame for wormy cherries, whether sweet or sour. It looks something like a small housefly, but is distinguished by bold bars on the wings.

The maggots develop from eggs laid in the cherries and eat until fully grown, when they drop to the ground to pupate two or three inches belowground.

European cherry fruit fly and larva

Once eggs are laid, the damage has been done, so the adults are the object of control measures. You should spray trees with rotenone as soon as the flies appear, and repeat the application seven to ten days later if trap counts remain high. They can best be monitored with a commercial or homemade trap. A simple trap can be made by painting a ten-by-six-inch plywood board a bright yellow and hanging a screen-covered jar below, to which has been added a mixture of half household ammonia and half water. The board is then covered with a commercial sticky material, and the contraption is hung up at a height of six to eight feet, preferably on the tree's south side. If attached to young limbs, the traps can be pulled down for easy inspection. A minimum of four traps should be used in small orchards and about one per acre in larger orchards. Traps seem to work best when surrounded by foliage, in that more places are provided for the flies to alight as they approach the traps. Renew the ammonia-and-water bait each week, and check the board for stickiness from time to time. Also effective is a red, apple-size ball coated with sticky material.

Since they overwinter as pupae under the trees, the pests will be destroyed by cultivating down to two inches. Similar species, the black and western cherry fruit flies, cause similar damage. Monitor and control as for the cherry fruit fly (see wing marking illustration).

The **cherry fruit sawfly** occurs only on the Pacific coast. It occasionally attacks plum, peach, and apricot. The overwintering pupa may be caught in the ground and destroyed by the shallow (two-inch) cultivation used against fruit fly.

The **cherry fruitworm** is reported in New Jersey, Iowa, and Wisconsin, but rates as a serious pest primarily from Colorado northward and west to Washington and Oregon. The larvae overwinter in bark crevices. They may be distinguished from fruit fly larvae by their black head and caterpillarlike (rather than maggotlike) appearance. A thorough spray of *Bacillus thuringiensis,* applied the first week in June, will stop many of the worms before they enter fruit.

The **cherry** (or pear) **slug** often defoliates young cherry trees, and resembles a shiny tadpole. See pear slug under PEAR.

The **tent caterpillar** develops webs in spring and will eat every leaf in reach before dropping to the ground—unless you take action. Wild cherries are particularly vulnerable; in fact, cherry growers are often better off if nearby wild cherry trees are chopped down. While it seems a shame to destroy such beautiful trees, they may

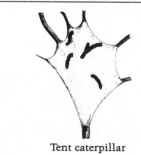

Tent caterpillar

host infestations that could ruin a harvest. Once caterpillars have already attacked, a strong soap spray may help, and a spray of *Bacillus thuringiensis* will stop a developing colony if applied before the caterpillars are ¾ inch long. See TENT CATERPILLAR.

Other insect pests of cherry include leafrollers and scale (see FRUIT TREES), peachtree borer (PEACH), and plum curculio (PLUM).

## Diseases

**Alternaria fruit rot** is seldom a problem if the cherries are harvested on time and have not been cracked due to rain or hail. The lesions begin as circular to oblong sunken areas, and later become hard, darkened, and wrinkled. They may cover one-third to one-half of the fruit and in a few days a fluffy, grayish growth develops as the fungus produces spores. The sulfur applied for brown rot will reduce this disease, but the first line of control is to pick the fruit on time and avoid breaks in the skin.

**Bacterial canker** is primarily a disease of sweet cherry, although it affects sour cherry, peach, and plum. Blighted shoots and limbs exhibit a heavy oozing of gum in spring and fall. Leaves may wilt and die in summer and, if the disease is severe, the entire limb may die. Fruit infections are evidenced by a darkened depression on the surface; this in turn may increase the severity of brown rot. Prolonged periods of cool, windy, moist weather favor spread of the disease. Bacteria overwinter in buds, cankers, and the tree's sapstream. Promptly prune out any blighted shoots as soon as they are noticed. After each cut, disinfect pruning shears in a solution of one part household bleach in nine parts water. There are no known cures for canker. The variety Hardy Giant seems to be the most susceptible, with Schmidt and Windsor slightly less so.

**Cherry leaf spot** is sometimes known as yellow leaf, because bright yellow foliage develops soon after the purple spots appear. It has also been referred to as shothole disease because the dead areas of the spots often drop out. No matter what the name, this is the most common and serious of cherry leaf diseases. Early defoliation caused by this disease results in dwarfed, unevenly ripened fruit. The most serious cases occur in the season following extensive early defoliation. The results can be: cold injury (death of limbs or trees); small, weak fruit buds; death of fruit spurs; reduction of fruit set and size; and reduced shoot growth. The chief symptom of cherry leaf spot is found on the leaves, but symptoms can also be found on petioles (leaf stems), fruit, and fruit pedicels (stems). Small purple spots appear on the upper surfaces of leaves about 10 to 14 days after infection. They may be the size of pinpoints, or up to ¼ inch in diameter. On the undersides of the infected leaves during wet periods, whitish pink masses appear under the purple spots. These masses contain spores and a gelatinous substance. The spots usually dry, and a chlorotic (yellow) ring forms around the edge. The center of the lesion may shrink and fall out, causing a shothole effect; sour cherries are more prone to this damage than sweet. Older leaves turn yellow and usually drop prematurely. The fungus develops most rapidly under wet conditions and warm temperatures of around 65°F. Infection occurs at lower temperatures, but more slowly.

There are two stages in the life cycle of this fungus, each with different control measures.

In the *primary cycle*, the fungus overwinters in dead leaves on the ground. Fruiting structures called apothecia develop on these leaves in late April, and spores mature at various times, depending on temperature and moisture. Generally, most of the spores are discharged from bloom to four to six weeks after petal fall, during wet periods. The infection early in the primary cycle is limited, because the susceptible developing leaves are small and most of the air pores (stomata) are too immature to be infected. Other factors are low temperatures and drying out of spores. Prevention of this early infection is crucial to eliminating early spore production on the leaves.

Upon germinating, the fungus enters through the stomata, and the mycelium branches out inside the leaf to later form a fruiting structure (acervulus) on the underside of the leaf surface. Each acervulus contains thousands of spores (conidia). A spray of 95 percent wettable sulfur, applied for brown rot control at 6 pounds per 100 gallons, will greatly reduce this potentially serious disease.

In the *secondary cycle,* the acervulus appears as a pink to whitish pink mass on the undersides of the leaves during humid conditions. In dry periods this mass is brown in color. The conidia are spread from leaf to leaf by water. The rapid spread of leaf spot fungus in summer and fall is usually due to rapid increase and spread of the fungus during wet periods by means of repeated generations of conidia.

Try rotary mowing the orchard after leaf drop in fall. A fall application of a nitrogenous fertilizer will hasten the decay of the leaves and thereby reduce the number of those on which the fungus can overwinter. A slight discing around the trees just before bloom also serves to decrease the spore discharge of the cherry leaf fungus, as well as the brown rot fungus.

Even as sweet cherries near harvest after a picture-perfect season, all can be lost in one day if heavy rains occur when the fruit is in its final swell and it cracks (sour cherries seldom crack). The tender flesh is opened to fungal attack at the worst or simply dries out, in either case leaving the cherry useless. An old remedy that will provide good control is to spray the trees with a dilute solution of nondetergent dishwashing soap *before* the rains begin, and allow the soap to dry. Use two teaspoons per gallon of water. Crack tolerant varieties include May Duke, Hardy Giant, Rainier, Jubilee, Sam, Victor, Corum, Venus, Lamida, Emperor Francis, Ulster, Gold, and Hudson. Varieties susceptible to crack are Black Tartarian, Yellow Spanish, Napoleon, Schmidt, Bing, Chinook, and Lambert.

**Foliage browning** may be traced to several causes. A natural browning and shedding often occurs after a dry season. In this instance, all the leaves of the inside of the tree turn reddish brown and drop either in spring or fall. The discoloration is distinguished from fertilizer burn or sunburn due to root injury because the discoloration takes place on the inside of the tree; fertilizer burn and sunburn start at the tip of the branch and progress up the tree from the lowest branches. Red spider mites bronze leaves, but they cause most of the leaves on a branch or tree to suddenly discolor in summer. Winter bronzing is due to some physiological changes brought about by the freezing of the tree's circulatory system and is tied in with the nutrition of the plant.

A deficiency of zinc in a cherry tree's diet brings about a condition known as **little leaf.** Leaves are crinkled, mottled, and may be bunched together at the tips. Fruit is sometimes misshapen. On a small scale, growers can get around this problem simply by driving galvanized nails or metal strips into the trunk.

**Verticillium wilt** is a soil-borne fungal disease affecting cherry, peach, nectarine, apricot, and plum trees. The disease first appears as a wilting or drooping of the leaves during midsummer or later. Leaves later may yellow, curl upward along the midrib, and finally drop. These symptoms develop on the older part of the shoot first, and progress toward the youngest growth. A few normal green leaves sometimes remain at the tip of the shoot. In the following year, twig dieback may result, and shoot growth on the surviving twigs is shortened. It is somewhat difficult to distinguish verticillium wilt from winter injury. In the case of winter injury, the wood just under the bark of the larger branches is discolored a dark brown, with less streaking of color than found on trees affected by wilt. Also, winter injury is responsible for killing all the tissues of the affected areas, including the inner bark.

Stone fruit should not follow susceptible crops—tomatoes, potatoes, eggplant, peppers, strawberries, and raspberries—or such weeds as pigweed, lamb's-quarters, horse nettle (nightshade), and ground cherry. Once trees are planted, these susceptible weeds must be kept down. Soils high in organic matter tend to reduce the levels of the fungus, and well-nourished trees have a good chance of recovering from infections. It is useless to replace a diseased tree with another susceptible tree, as the wilt fungi inhabit the soil.

When **virus yellows** or sour cherry yellows comes to the orchard, leaves show green and yellow mottling with waves of defoliation starting three to four weeks after petal fall. Trees infected for several years will develop willowy growth, abnormally large leaves and few spurs, and bear small crops of large fruit. Susceptibility to winter bud damage increases, and bloom may be prolonged, leading to uneven ripening; the tree gradually deteriorates. Yellows is transmitted by budding and through seed. Specify virus-free trees from the nursery. Sweet cherries can be infected but show no symptoms. Avoid replanting young cherries in a block of older, likely infected trees, since the virus will quickly infect the replants.

Trees infected with **X-disease** on Mazzard rootstock make poor growth. Bloom is delayed, leaves often become duller than normal, and some of the fruit never matures. This immature fruit is light-colored and insipid in flavor. It tends to be triangular in cross section, and has less flesh in proportion to the pit than does normal fruit.

On infected red-fruited trees such as Montmorency, the fruit is pale pink or dull white. On a black sweet cherry such as Windsor, the fruit is usually red in color, and develops other symptoms similar to those on Montmorency. A sweet cherry such as Napoleon with light-colored fruit usually has dull white fruit when infected. On all varieties, the fruit is worthless.

Late in the season, bronze discoloration occasionally develops along the midveins of a few leaves. After several years of poor growth, dieback may develop on infected trees. The disease progresses more rapidly in peach and sour cherry trees than in sweet cherry trees; it may be five or more years after infection before large sweet cherry trees become unprofitable.

It is often difficult to identify X-disease on cherry trees propagated on Mahaleb rootstock because fruit symptoms rarely develop. In June or later, infected trees often suddenly wilt and nearly die by the end of the season. Other trees do not wilt but may decline slowly without characteristic symptoms. If no cause of decline can be found, such as girdling by mice, verticillium wilt disease, or dead roots from wet soil, then check the immediate neighborhood for diseased chokecherry. If chokecherries are found nearby, it can be assumed that the wilted trees are infected with X-disease. Trees on Mahaleb rootstocks that produce roots above the bud union may develop X-disease symptoms characteristic of trees on Mazzard rootstock.

See PEACH for control suggestions.

The **yellows** and ring spot virus complex causes the most serious reduction of sour cherry yield in some regions. The disease may be known as tatter leaf (from a characteristic symptom) or necrotic ring spot. Fruit set is reduced as soon as infection occurs; growth is reduced because leaf efficiency is impaired; and, as the yellows disease develops, buds that normally produce lateral shoots and fruit spurs produce flowers instead. Once these buds have blossomed, no additional growth occurs at the nodes where they are borne, and long, bare twigs and branches eventually develop. The only fruit on such branches develops on the terminal shoots. The final result is a small number of blossoms and an equally small number of leaves, many of which drop in June because of the disease. In addition, the weakened buds are more susceptible than normal to winter injury, and the twigs finally become so weakened that dieback occurs.

Ring spot symptoms are most prominent during the two-week period following petal fall. Typically, they are produced by a given tree for only one or two years, when the tree is in the initial "shock" stage. Etch-type symptoms, characterized by darkened, depressed fine lines and rings, develop on the upper surfaces of the leaves. Dead tissue may develop on leaves on this stage of symptom expression. Sometimes the lines are broad, chlorotic, and not depressed. Yellows symptoms develop later in the year, commonly beginning three to four weeks after petal fall. Leaves turn yellow in different patterns, and drop along with other leaves that do not turn yellow.

Sweet cherries also are infected by cherry yellows viral disease, although leaves on affected trees do not turn yellow. Most varieties of sweet cherries appear to be less susceptible than sour cherries, but they are damaged to a significant degree. Follow the precautions given for purchasing and planting of sour cherries. It may not be wise to rogue out affected sweet cherry trees, however, because of the lack of diagnostic symptoms.

The practice of replanting trees in old orchards exposes new trees to large sources of nearby inoculum. The interplants become diseased quickly and, even before they show symptoms or suffer serious damage, provide inoculum for infection of subsequent replants.

Spread of the viruses can be reduced by making new plantings in solid blocks isolated from other cherry orchards. Many old orchards that were planted in isolation are producing much better crops than those in which constant replanting has been practiced, or in which a high percentage of trees were diseased at planting. Do not replant in old orchards. Remove all trees from old orchards at the same time. The block should not be replanted until all large roots have been completely removed and the soil has good tilth, proper nutrient level, adequate organic matter content, and suitable pH; adequate soil preparation requires one year or more. Any factor unfavorable for growth will add to viral damage when the trees eventually become infected. The trees should be as virus-free as possible; this is the only means of excluding serious forms of the disease.

There appears to be a delay of a few years before the viruses begin to spread from trees that were diseased at planting to neighboring healthy trees, since infection will occur via infected pollen and the tree must bloom to be susceptible. Therefore, any tree showing yellows symptoms during the first years after planting should be removed at once. Also remove those that show such severe symptoms of ring spot that twigs are killed, that develop unusually poor fruit spur systems, and that grow very poorly without explanation. Once this is done, you can plant groups of trees in which a low percentage are infected by the less severe viral strains. Such trees are those most likely to be available in nurseries at the present time; because of the widespread use of budwood from isolated virus-free foundation plantings, trees have been greatly improved during the last ten years.

Use the following steps to reduce spread of the virus.

For orchards *less than ten years of age,* first remove all trees that grow poorly, fail to set up a good spur system, or show yellow leaf symptoms characteristic of the viral disease. This practice will remove a potential source of infection to other young trees and will eliminate border trees. Replace trees only if the original planting is less than five years old. Replant with the best trees available, preferably with those certified to be free of viruses, available from all reputable nurseries.

For orchards *more than ten years of age,* start by destroying all trees that show many yellow leaves in June, produce a poor spur system, or have many dead twigs. Such trees do not pay for the care they receive and they endanger other trees in the orchard. Do not replace old trees, as new ones will soon be infected by the remaining diseased trees. Bulldoze complete blocks if the trees are so diseased that a normal crop has not been harvested for a few years. Such orchards cannot be returned to profitable production.

In *planting for the future,* isolate new plantings, setting them as far as possible from existing cherry blocks and other stone fruits. At least 100 feet is necessary, and distances of 500 feet or more are preferable. Open fields between plantings of different ages will provide suitable isolation, but may not be economical in areas of intensive fruit growing. An alternative is to plant apples or pears between cherry blocks or between cherry and other stone fruit blocks. Occasionally it may be necessary to reduce the size of individual plantings and to have more blocks of each kind of fruit. Smaller plantings will increase the cost of spraying and other orchard care operations, but increased cherry production will more than offset the added costs. Make solid plantings, and avoid interplanting among older trees. When an orchard is prepared

for replanting, all existing trees in the old planting must be removed to eliminate this certain source of yellows virus. Purchase the best trees available. These should carry a certificate stating they were produced with virus-free buds. Successful cherry production depends heavily on careful selection and preparation of planting sites. Soil condition and cultural care also are important phases of a prevention program.

Because one aspect of the damage caused by cherry yellows viral disease is to increase susceptibility to winter injury, other factors conducive to winter injury must be taken into account. While one growth factor alone might not result in trouble, it may cause damage when compounded with viral yellows. In new plantings, follow the roguing programs outlined above.

Other diseases that may attack cherry include brown rot (susceptible varieties are Seneca, Black Tartarian, Bing, and Schmidt, and controls are given under PEACH), cytospora canker, rhizopus rot, and X-disease (PEACH), and black knot (PLUM). Crown gall is caused by bacteria that enter the plant through wounds. See CROWN GALL.

# CHESTNUT

## Insects

Several types of **aphid** may attack your chestnut trees; see APHID.

The most serious insect pest is the **chestnut weevil,** of which there are two species, both attacking the nuts of Asian chestnut trees. In years of heavy infestations the feeding of the larvae inside the nuts may destroy practically the entire crop.

The larvae in the nuts can be killed by immersing them in water kept at 120°F. for 30 to 45 minutes, depending on the size of the nuts. If the nuts are gathered from the ground daily, few worms will enter the soil. It helps to let chickens run beneath infested trees.

The white flat-headed grubs of the **twolined chestnut borer** make tortuous and interlacing galleries under the bark of chestnut and oak. The grub is about ½ inch in length, and the parent beetle is ⅜ inch, black with two narrow converging longitudinal pale stripes on the wing covers. The beetles appear in May and June and deposit eggs. The grubs work in the inner bark and outer sapwood, pupating in cells in the wood. There is one annual generation each season. Mutilated, weakened, or dying trees are especially vulnerable to attack. Grubs interfere with the flow of water and nutrients through the tree, and leaves wilt in that portion of the tree beyond the injury. Leaves turn completely brown and remain on the tree; oaks may also be affected by oak wilt, but this disease causes leaves to pale at the edges and curl, then fall from the tree. If you lift the bark from a tree infested with borers, you may see their galleries and slender, inch-long white larvae. Look for the dark projections at their tail end.

Stressed trees are most vulnerable. Help them with fertilizer and watering in spring. Remove infested parts of trees or the beetles will spread throughout the tree, killing it, and then travel to neighboring oaks and chestnuts. Beetles may leave piled firewood to infest nearby trees (the adults can fly well). The woodpile can be shrouded snugly with a tarp to keep beetles from escaping and laying eggs on trees.

Cankerworms, gypsy moths, lecanium scales, mites, and twig girdlers may occasionally be pests of chestnut trees. See TREES.

# Diseases

By far the most destructive disease afflicting chestnut trees is **blight.** Within 15 years after it was first noticed in the United States, blight had spread across the country. Few native chestnuts remain today. Even sprouts that arise from old stumps are hit, as the disease is still with us. On young wood, cankers appear as swollen, yellowish brown, oval or irregular areas. On older wood they are brownish, circular or irregular areas with slightly raised or depressed edges. Partial or complete girdling of stems by cankers causes leaves to yellow and brown. Dead leaves and burrs cling to diseased branches long after normal leaf fall. The surface of older cankers is covered with minute pinpoint fruiting bodies of the blight fungus. During wet weather, yellowish spore masses ooze from fruiting bodies. These spores are carried by rain, birds, and insects. New infection develops following the entry of spores through wounds.

As soon as the disease is definitely identified, cut down the tree and burn the branches and trunk. Debark the stump to prevent sprout growth, and replace the tree with resistant hybrid chestnuts. Chinese varieties, such as Crane and Orrin, and some Japanese varieties are resistant.

Though Chinese and Japanese chestnuts have some resistance to blight, they are susceptible to the fungus that causes **twig canker,** also called twig blight and dieback. This fungus enters the tree through a dead twig. Once in the wood, it progresses rapidly, growing lengthwise in the wood and eventually girdling the branch. The trouble may be traced to winter injury in varieties not hardy enough for the area, to a lack of compatibility between stock and scion in grafted trees, or to some other adverse environmental factor.

Control this disease by planting varieties well adapted to the climate. Provide good air circulation. Trees should be protected from injury as far as possible, and should be periodically inspected. If any branches appear to have cankers, prune them one to two feet below the site of injury. Or, you can cover cankers with mud, and wrap the site with a plastic sheet or bag, taped securely so that the mud won't dry out. This won't cure the tree, but merely arrest the disease. Keep an eye out for new cankers.

# CHINESE CABBAGE

## Insects

Chinese cabbage is preyed upon by a number of insects discussed under CABBAGE. The potato **aphid** (see APHID) and the yellow-striped **purplebacked cabbageworm** (see imported cabbageworm under CABBAGE) have a particular craving for Chinese cabbage.

## Diseases

Two crucifer diseases likely to strike this crop are **black leg** and **black rot;** see CABBAGE.

# CHRYSANTHEMUM

Despite the awesome list of bugs and diseases that follows, chrysanthemums are relatively trouble-free. Remember that mums should not be planted in wet, shady places, and that it is best to water early in the day so that leaves have a chance to dry before nightfall, as fungal diseases thrive on wet leaves. Plants should be spaced well and staked to keep branches well off the ground.

## Insects

The shiny, dark chrysanthemum **aphid** is a tiny speck that is often seen on stems, leaves, and flowers of this plant. They congregate in clusters on tender shoots of growing plants, stunting growth and causing a slight leaf curl. Plants sometimes die in serious cases. When cut mums are brought into the house, the aphids abandon them and crawl about. When crushed they leave dark stains. See APHID for control measures.

The **chrysanthemum gall midge** is a tiny, slender, orange, two-winged fly which lays eggs in the leaves and tender shoots of greenhouse plants. Each larval maggot forms a cone-shaped gall. A number of galls together often form knots, twisted stems or distorted buds. This tiny insect is so prolific in reproduction that it is difficult to tell just how many generations there are in a year. The midge is usually more abundant in spring and fall than during summer.

Damage can be reduced by bringing only clean cuttings or plants into the greenhouse. Because greenhouse plants are most bothered, think about moving your plants outdoors if you have trouble with this pest. Pick up foliage that appears galled or knotted and burn it, as the maggots are probably inside the leaf tissue.

The **chrysanthemum leafminer** is a yellow maggot that feeds in petioles and between leaf surfaces, causing large, blisterlike mines. Leaves often dry up and drop off. Look for telltale specks of excrement in the tunnels. Control this pest by discarding affected leaves and stems.

At night, the mottled-brown **climbing cutworm** leaves the soil to climb chrysanthemums and eat tip leaves, flower buds, and blossoms. If only a few plants are involved, just scratch the soil surface and destroy the exposed worms; if you're in trouble with a lot of mums, see CUTWORM.

The **corn earworm** [color] (also known as the tomato fruitworm and cotton bollworm) is better known for its ravages in the vegetable than in the flower garden. Nevertheless, it is a pest of annual and perennial flowers and attacks abutilon, ageratum, amaranth, canna, carnation, chrysanthemum, dahlia, geranium, gladiolus, hibiscus, mignonette, morning-glory, nasturtium, phlox, poppy, rose, sunflower, and sweet pea. The caterpillars show a marked preference for the opening buds and flowers of chrysanthemum, calendula, dahlia, gladiolus, and rose, although they also feed on leaves and may tunnel in stems of certain plants. Their injury to the buds is not unlike that caused by climbing cutworms, with which they are often confused. The earworms gouge out the inner part so deeply that the flowers are a complete loss. When fully grown they are about two inches long, and their color may change as they mature from reddish brown to green, with stripes. The parent is a fawn-colored moth with dark spots on the forewings.

If the flower buds have been bored into, there is little that can be done except to remove and destroy those that have been infested. Screen choice and valuable plants with cheesecloth to prevent them from becoming infested. See CORN.

The translucent, microscopic **cyclamen mite** infests new leaf and blossom buds, causing them to become swollen and distorted. It is a pest of both garden and greenhouse. Plants may be stunted. See CYCLAMEN for the mite's life history and control measures.

The **foliar nematode** is a serious pest both in the greenhouse and garden. The first symptom you'll notice is darkened spots on the undersides of leaves. After a few days, veins on the top surfaces of the leaves are discolored, and wedge-shaped spots appear between the veins until entire leaves become blackened and withered.

Once nematodes get inside leaves, there's no way to get them out, and control is aimed at preventing their trek from soil to foliage. They are so tiny that they can swim vertically in the water film on mum stems, so see to it that stems stay dry. Don't water from overhead, and use a mulch to prevent splashing on the plants. Remove and burn affected leaves, and discard badly infected branches. Avoid crown divisions, and take tip cuttings instead. While most varieties are susceptible, Koreans are reported to be especially so. Be careful in selecting plants from nurseries, checking for damaged foliage.

Round, depressed tan spots on new leaves may be the work of the **fourlined plant bug** [color], so named because of the black lines

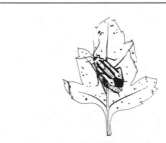

Fourlined plant bug

running down its greenish yellow back. The nymphs are red in color. Handpicking should take care of most problems; rotenone is sometimes used as a last resort.

Lacy wings are the key characteristic of the chrysanthemum **lace bug** [color]; the nymphs are similar to the azalea lace bug. These pests bleach foliage and stems of mums, and leave dark spots of excrement on the undersides of leaves. They breed on weeds, especially golden-rod, and are common wherever mums are grown. Soap sprays have been used to good effect against this pest. If not many bugs are about, it may suffice to crush them by drawing leaves between the thumb and forefinger.

Several species of **spittlebug** [color] suck plant juices from chrysanthemums, causing stunting and distortion. The adults look some-thing like leafhoppers, are straw-colored or brown, and measure ¼ inch long. Pinkish to yellow green nymphs hatch from eggs in spring, and are usually found under the masses of white froth or spittle that they produce. It is this stage that causes damage; adults apparently cause no visible trouble. Spittlebugs are found through-out the northern states, and are particularly serious in regions of high humidity.

As for controls, you're on your own with this one. Almost every insect pest reference recommends a shot of chemical, with no sug-gestions for prevention or cultural control. Rotenone is effective against this pest, but you might concoct a milder botanical before turning to this powerful agent.

Several species of **thrips** feed on mums, especially the flower thrips. They feed in the developing flowers, causing deformity and mot-tling. The adult of the flower thrips is brownish yellow and displays feathery wings; the young are lemon yellow. See ROSE.

The larva of the **whitemarked tussock moth** [color; see Western tussock moth] is a strange-looking, hairy caterpillar that grows to 1½ inches long. It has a red head from which sprout two hornlike tufts of long black hair. Another tuft projects from the tail. A black stripe is centered down the back, flanked by a wide yellow line. Usually a number of caterpillars feed together, skeletonizing leaves of many plants including geranium, German ivy, rose, and several fruit and deciduous shade trees.

Pick off and destroy infested leaves and groups of larvae. Scrape off the egg masses or paint them with creosote; these are very visible, lathery, and about an inch long. Fortunately, birds aren't put off by the worm's looks, and devour great numbers of them. This pest is preyed upon by many parasites, but nature has seen to it that the parasites have to contend with their own hyperparasites.

Another conspicuous pest is the **yellow woollybear caterpillar,** a hairy, yellow, two-inch-long food tube that attacks a great number of plants. Look for the black lines running down its back. Handpicking will take care of modest infestations; see imported cabbageworm under CABBAGE for other controls.

## Diseases

**Aster yellows** is caused by a virus that is carried to chrysanthemums by aphids and leafhoppers. In general, the infected plants are yellow or mottled, and stunted. They tend to develop distorted flowers and many small, weak shoots. Flowers are deformed and often are colored green rather than their normal hue.

Control of aster yellows must involve the control of the aphids and leafhoppers that transmit the disease. See APHID and ASTER. Yellows can also be carried on garden imple-ments or your hands. Take out any suspicious-looking plants that may serve as a source of virus; no matter how valued a plant may be, it should be destroyed as soon as it is infected.

**Dodder** is a parasitic vine that is distributed by birds or in the seed. It appears as many strands of orange, stringlike growth, without apparent roots, that wind around mums. The little vines are not readily detached. Small clusters of whitish flowers produce abundant seeds for another year's crop. When seeds germinate, the dodder climbs up the nearest plant and sends little penetrating knobs into the host, through which it draws its nutrients. The dodder roots are dropped as soon as this attachment is made, and the plant becomes truly parasitic.

If the dodder is noticed early, the portion of the flower that has been attacked can be cut out and destroyed with the dodder. However, once this parasite has had an opportunity to become firmly established, it's necessary to remove and burn the infested plant.

**Foot rot** and **wilt** cause lower leaves to turn yellow and die, along with a general wilting of the plant. Near the ground the stem is greatly discolored and may show masses of pink spores; when cut across, it usually shows discolored vascular bundles.

Prevent foot rot by using cuttings from healthy plants. Any plants that appear to be affected should be destroyed. In the garden, it may be a good idea to grow mums in a new location.

**Gray mold** appears as flower spots that enlarge and rot the entire flower, particularly those of light-colored varieties. The fruiting of the fungus appears as a gray mold. This disease is particularly prevalent during wet or cloudy weather, and in greenhouses during late winter.

Avoid humid conditions by improving ventilation. In a greenhouse, turn on the heat an hour before sundown to prevent a sharp drop in temperature, which might condense moisture on the flowers and allow the fungus to grow.

**Petal blight** of chrysanthemums is a fungal disease. It may be expected during cool weather when excessive moisture remains on flower heads for several hours.

Ventilation is the key to preventing this disease, as it ensures that moisture can evaporate from the flowers even in cool, wet weather. Plants under cloth will be less likely to suffer from blight if intermittent openings are left in the roof to enable air to circulate.

**Powdery mildew** appears as a whitish powder on the leaves, and spreads rapidly during hot dry weather.

**Rhizoctonia stem and root rot** hits the base of stems and roots, and causes plants to wilt slowly. The disease is encouraged by high soil moisture, and in warmer areas it may live on in the soil from one year to the next.

Plant mums in soil that has been free of disease, if possible. Where rotation is impossible, some growers have replaced soil. Composted soil discourages the rot fungus.

**Stem rot** (cottony stem blight) is a wind-borne disease that is hard to control in most gardens. In the soil-borne phase, stems near the ground level are attacked, showing a white surface mold that causes the plants to wilt and suddenly die. In the wind-borne phase, the leaves, blossoms and stems are attacked by spores blown either from surrounding infected plants of many kinds, or from small mushrooms formed on the garden soil by the disease.

Be certain to plant in disease-free soil, even to the point of replacing infected soil with good soil. Soil around plants should not be too wet. Wide spacing between plants allows air to circulate, reducing the moisture on plant sufaces. Diseased clumps should be removed and burned to prevent spread of the disease to healthy plants.

**Stunt** appears to affect part or all of mum plants, in both the greenhouse and garden. Infection is characterized by stunting of the entire plant or one or more branches, and early production of off-color flowers. Leaves on infected branches may be smaller than normal, flowers are small, and pink and bronze varieties are bleached. Greenhouse stunt appears to be mechanically transmissible and is often spread by handling, so break off buds instead of pinching or cutting them with fingernails or a knife. Take care in selecting healthy propagating stock, and rogue infected plants as soon as possible. Keep plants free from leafhoppers, as these insects also serve as vectors. For the first year, segregate new stock if at all possible.

The several species of **verticillium wilt** may bring about a number of symptoms. Leaves wilt at the margin, turn brown, and die progressively from the base of the plant upward. A cross section of the stem near the ground may show a discoloration of the woody tissue, and many plants grow lopsided. Once plants are visibly stunted, they are usually beyond hope. Occasionally, wilted plants can be saved by providing very little soil moisture, but this is an emergency measure that is recommended only for valuable plants. Temperatures of 60° to 75°F. are most favorable for this wilt, but once started, the fungus can persist outside this temperature range.

Because of the nature of verticillium wilt, protection measures are concerned more with prevention than control. Avoid planting chrysanthemums in the same spot continuously, as this increases the soil's susceptibility to the disease, especially if the soil is kept fairly moist and at a temperature of 60° to 75°F. For garden and field planting, do not use soil where susceptible crops (such as cotton, tomatoes, and strawberries) have grown, and avoid soil that has previously hosted the disease. If space is limited and the same soil must be used yearly, select resistant varieties.

# CINERARIA

## Insects

Both the adult and nymph leaf-curling plum **aphid** are flat-backed, a shiny pale green in color, and are found in the crevices of plant tips. They cause severe distortion and stunting of growth. Infested leaves are covered with white, crystalline droplets of honeydew. The winter host is typically plum; summer hosts include aster, mum, heliotrope, dahlia, erigeron, eupatorium, gerbera, cynoglossun, lobelia, mertensia, marguerite, and sunflower plants. See APHID.

**Foliar nematodes** cause water-soaked spots that appear first on the undersides of lowest leaves and then work through to the uppersides, where brown blotches become evident. Avoid splashing water from leaf to leaf or from soil to leaf; see CHRYSANTHEMUM.

The **greenhouse leaftier** is a translucent, greenish white caterpillar that grows to about ¾ inch long; it is marked with a dark green stripe. When disturbed, they wiggle violently and often drop to the ground. They usually feed on the undersides of leaves, but may web several leaves together. The adult is an inconspicuous, rusty brown moth with a wingspread of just under an inch. See celery leaftier under CELERY.

The adult **greenhouse whitefly** feeds in both the adult and immature stages by sucking plant juices. Heavy feeding gives leaves a mottled appearance, and may cause them to yellow and die. The adult is a tiny mothlike insect with a mealy appearance that's due to wax secretions. It also secretes sticky honeydew that glazes the lower leaves and encourages black, sooty mold. The nymph is oval, flattened, light green, and about the size of a pinhead. It looks somewhat like a small, soft scale insect, and spends its time attached to leaves until maturation. Don't bring infested cuttings or plants into the greenhouse. For controls, see WHITEFLY.

Other likely pests include the cabbage looper (see CABBAGE), CUTWORM (TOMATO), MEALYBUG, sixspotted leafhopper (ASTER), SNAIL AND SLUG, and SPIDER MITE.

## Diseases

**Powdery mildew** gives cineraria leaves a whitish, powdered appearance.

# CITRUS FRUIT

## Insects

Brown, $1/12$- to $1/8$-inch-long Argentine **ants** are responsible for carrying around aphids, mealybugs and scale insects. The ants give off a peculiar, musty odor when stepped on.

Black citrus **aphids** are usually kept down by a good variety of natural enemies and diseases. Growers who spray trees with nothing more severe than dormant-oil sprays will receive maximum benefits from this free help. These aphids are carried to leaves by Argentine ants (see above); see also APHID.

The **citrus mealybug** feeds on the juices of the plant, causing eventual death of the affected parts after they wilt and lose their color. The mealybugs coat the foliage with a sticky honeydew on which an unsightly black mold grows.

Honeydew is also the natural food of certain ants that tend the mealybugs and carry them from plant to plant.

Mealybugs are usually found in clusters along the veins on the undersides of leaves and crevices near the base of leaf stems. Since they multiply rapidly, all stages may be present at the same time. They have amber-colored bodies with short, waxy filaments. The eggs are laid in a protective cottony mass or sack that resembles a small puff of cotton. Each mass may contain 300 eggs or more that hatch in 10 to 20 days.

The first step in the control of mealybugs is to eliminate ants in and about the planting area. This can best be done by scattering bone meal on the ground. On plants that will not be damaged by frequent watering, syringe frequently with a forceful stream of water. You can buy a mealybug predator, *Cryptolaemus montrouzieri*. Rotenone and pyrethrum are effective botanicals.

The **citrus red mite** (known also as purple mite and red spider) is about $1/50$ inch long, and varies in color from deep purple to rose. This species infests leaves, fruit, and new growth. Injury appears as a silvery scratching or etching of the upper leaf surfaces. Leaves may turn brown and drop, and fruit becomes yellowed. Use a magnifying glass to check for the tiny mites and their red, stalked eggs on upper leaf surfaces, especially along the midrib, and in

Red citrus mite

angular crevices on leaf stems and young, tender twigs. Although citrus red mites are most numerous from November to April, they can be seen at any time of the year. Regular dormant-oil sprays should control populations of this species, but natural enemies may well make spraying unnecessary. You can purchase a predatory mite, *Amblyseius californicus.* The drop in the mite population observed yearly around April is due in part to viral diseases.

**Citrus rust mites** are present most of the year, and their injury, while not materially affecting fruit quality, often results in rusty, rough patches on the skin. These pests are about $\frac{1}{200}$ inch long, and cannot be seen by the

Citrus rust mite

unassisted eye; in fact, they're even hard to observe with a 10× magnifying glass. They appear on green fruit and both sides of leaves. The mites can be biologically controlled by a commercially available fungal miticide, *Hirsutella thompsonii.* A lime sulfur preparation, diluted 1 to 50 or 1 to 100 with water, will knock out this pest in serious cases. Sulfur dusts also work well. As with the citrus red mite, this mite is very vulnerable to naturally occurring disease.

The **navel orangeworm** is an important pest of almond, walnut, fig, and citrus, especially in California. The worms are colored yellow to dark gray with dark heads. The moth is distinguished by a line of crescent-shaped markings along the outer margins of its light gray wings. Plant and orchard sanitation, along with an early harvest, will prevent most damage. Two navel orangeworm parasites, *Goniozus legneri*

and *Pentalitomastix plethoricus,* are on the market. They affect orangeworm pupae, and releases should be made all at once after harvest or in early spring.

**Nematodes** are responsible for two serious diseases of citrus, spreading decline and slow decline (see below). Soil that is rich in organic matter has been found to encourage fungi that prey on nematodes.

Californians have found the **orange tortrix caterpillar** to be of much trouble throughout the growing season. A host of other types of plants may also be subjected to damage from this brown-headed, brownish white caterpillar. The larva rolls leaves into a web and feeds within. If oranges drop before their time, suspect it of boring into the rinds of the orange or scarring them near the bottom.

If gray moths with dark markings are seen in the vicinity of your citrus plants, chances are that the tortrix is laying its eggs on leaf surfaces. Look for these eggs (they are cream-colored discs, found on both leaf tops and bottoms) and destroy them. *Bacillus thuringiensis* kills the larvae. Pyrethrum has been found effective against this pest when either sprayed or dusted. Should you notice a caterpillar with white eggs on its back, do not pick it, as these may be the eggs of a parasite that will hatch to produce beneficial wasps. At least a dozen parasites of this caterpillar have been identified. Light traps will attract the moths.

**Scale** insects can be extremely bothersome to the citrus grower. There are several varieties of this small pest to contend with.

*Black scale* is most easily recognized as a female which is dark brown or black, hemispherical, and about ⅛ to ³⁄₁₆ inch long. Most specimens have ridges on the back that form the letter H. This pest may be found on leaves, twigs, and sometimes fruit. It produces honeydew that serves as a medium for the growth of a sooty mold; this mold reduces the effectiveness of the leaves and is difficult and expensive to remove from the fruit. A parasite, *Metaphycus helvolus,* is now commercially available.

*Citricola scale* produces honeydew, which causes sooty mold to grow on the fruit. The female is gray, oblong, and about 3/16 inch long. As is true of other scale species, this pest is the victim of a variety of parasites.

*Purple scale,* found on leaves, twigs, and fruit, produces a toxic substance that kills the more heavily infested parts of the tree. Its presence can often be detected at a distance by these dead areas. The scale can be recognized by the narrow, oystershell-shaped female, which is about 1/8 inch long and purplish in color.

*Red scale* is found on leaves, twigs, and fruit and produces a toxic substance that causes a more-or-less general killing of leaves and twigs. The scale is reddish brown, almost round, and a little over 1/12 inch in diameter. On green fruit it causes a yellow spot somewhat larger than the scale; on leaves the spot extends through to the other side. Four chalcid wasp species give good control.

Although lemons may be sprayed with petroleum oils in April and May, the preferred timing of oil sprays on most other varieties of citrus is late summer or fall. The timing on navel oranges is critical, as this variety is most susceptible to injury. Oil treatments to maturing fruit are to be avoided—coloring of the rind and internal quality may be affected and fruit drop can occur. Timing of oil treatments varies for citrus varieties and areas. For specific information on timing, consult your local extension agent. High or subfreezing temperatures, low soil moisture, and low humidity can increase the susceptibility of trees and fruit to injury by oil sprays.

In California, discontinue spraying when it is evident that temperatures will reach 90°F.

in the coastal areas, 85°F. in the intermediate areas, and 95°F. in the interior areas. Oil-sprayed trees may be injured if spraying is followed by hot, dry winds. Spraying should be discontinued if the relative humidity is 35 percent or lower in coastal areas, 30 percent or lower in intermediate areas, or 20 percent or lower in interior areas. You should spray as soon as possible after an irrigation so that the foliage will be as turgid as possible.

Red scale has been controlled biologically in a substantial number of orchards, but certain conditions are necessary: favorable weather, control of ants, a minimal amount of dust or dirt on the trees, and choice of control measures that are least harmful to the red scale parasites. In California, weather conditions are most favorable for parasites in the coastal and intermediate climatic areas, and the possibility of biological control of red scale is good there if the other conditions are provided. The most important enemies of red scale are the parasites *Aphytis lingnanensis* and *Prospaltella perniciosi* in coastal areas, and *A. melinus* and *Comperiella bifasciata* in interior areas. The advice of a qualified entomologist should be sought in order to evaluate the possibilities for biological control in a given orchard. *A. melinus,* the golden chalcid, can be supplemented with two to six releases. *C. bifasciata* is also marketed, with two to four releases called for.

*Yellow scale* is often found mixed with red scale. It lives on leaves, where it produces a marked yellow spotting, and on twigs. It may cause heavy leaf drop, but not the serious diebacks that result from red scale. A chalcid wasp imported from the Far East handles yellow scale populations, as long as ants aren't bringing in great numbers.

Control scale insects with an oil-emulsion spray. This may be applied from petal fall in spring through September, but the preferred time is June 15 to July 15 in Florida, and June through August in California. Do not spray trees that show signs of wilt or apply dosages less than six weeks apart. Although this spray is primarily for scale control, it will also remove sooty mold from the foliage and provide protection from other pests that attack citrus during midsummer. Important natural enemies include the Australian ladybug, twice-stabbed ladybug, and several predacious mites and parasitic wasps.

**Sixspotted mites** are about the same size as purple mites but are yellow white to sulfur yellow in color. Adults usually have six dark spots that are barely visible on the back or adbomen with a 10× magnifying glass. These mites live in colonies on the undersurfaces of the leaves only, especially along the veins and midribs. Injury appears as yellow spots, often cupped toward the top of the leaf. Sixspotted mites prefer grapefruit, but are found on other types of citrus. They usually disappear with rainy weather. Oil sprays are effective. See SPIDER MITE.

The **whiteflies** [color] that infest citrus are not considered serious pests in commercial groves but are disliked by growers of ornamental citrus plants. The nymph, which is seldom recognized by the grower, infests the undersides of the leaves and sucks sap from them, resulting in injury to some trees. Ornamental growers object to the dark sooty mold fungus that grows on honeydew given off by the immature stages of the whitefly. If this pest is causing intolerable damage, consider an oil spray, as recommended for scale insects (see above). It may help to remove chinaberry trees from the area. Given favorable weather conditions, red-and-yellow fungi grow on and destroy many of these pests. These colorful beneficial organisms are seen as brightly colored pustules on the undersides of leaves. *Encarsia formosa* is a commercially available parasite of the greenhouse whitefly.

## Diseases

**Anthracnose** is a catchall term for three citrus maladies—wither tip, anthracnose stain, and anthracnose spot. While it was originally thought that these conditions were caused by a fungus, the fungus is actually either a secondary invader of weakened tissue or a latent infection that only becomes noticeable after the fungus takes hold in dead tissue. Control of wither tip and anthracnose spot depends on maintaining tree vigor, and not on eliminating a fungus.

*Wither tip* describes the gradual yellowing, parching, and dropping of leaves, followed by dieback of twigs and branches. In some cases, the wilted leaves stay on the tree, giving portions of the canopy the appearance of having been scorched by fire. If you look closely at affected twigs, you may see a light gumming and a distinct line of separation between healthy and dead tissue. Tiny black fruiting bodies of the secondary fungus appear as speckles on dead terminals.

Now that it is known that the fungus acts as a secondary problem and is not alone responsible for seriously damaging trees, wither tip is no longer held to be a serious problem. Dieback usually results when injured roots can no longer support the leaves and twigs, which then wilt and die back until there is a balance between roots and tops. Root troubles are usually the result of deep plowing, fluctuating water tables,

Citrus whitefly

drought, fertilizer burn, soil deficiencies, oil spillage, and insects and diseases. Conditions up top may also lead to dieback—hurricanes, long spells of hot and dry wind, salty sea spray, and winged pests.

*Anthracnose stain*, a russeting and tear-staining of rinds, was also attributed to anthracnose fungus. Researchers have found that these symptoms can be traced to the citrus rust mite and to melanose.

*Anthracnose spot* is the third disease to be blamed on fungus. Lesions on rinds are round, brown or black, sunken, hard, and measure ⅛ to ¾ inch in diameter, with tiny black fruiting bodies sometimes visible. These spots later serve as entrances for rot organisms, but pathogens are a secondary cause of damage and do not initiate spotting. Anthracnose spots usually develop at sites of mechanical injury or in fruit that is overripe. Grapefruit is especially vulnerable.

In light of what is now known about so-called anthracnose diseases, control is a matter of maintaining tree vigor, as well as preventing mechanical injury to fruit.

If you have inherited an orchard in which chemical controls were practiced, trees may show signs of **arsenic toxicity.** Arsenic is sometimes sprayed on grapefruit trees to lower the acidity of fruit that is harvested early, and an overdose of the chemical shows up as chlorosis of the leaves. The effects are cumulative, and several years of spraying cause the most serious damage.

In light cases, chlorotic spots appear in a mottled pattern, extending to the leaf margins and often crossing veins. This symptom could be confused with manganese deficiency except that arsenic injury is first seen on mature leaves in summer and fall, usually on the south and southwest sides of trees. In serious cases, leaves become yellow throughout and drop, and fruit is misshapen, dwarfed, hard, and spotted within.

Such symptoms on fruit are similar to those caused by boron deficiency, and it is likely that overdoses of arsenic are linked with boron deficiency. You can help trees out by supplying boron in the form of ground glass, commercially available as FTE (fritted trace elements). This supplies the soil with the raw stuff necessary for natural chelation.

**Blight** is an umbrella term for several diseases, all having in common blightlike wilt and dieback in the tops. These diseases are orange blight, limb blight, go-back, wilt, dry wilt, leaf curl, roadside decline, and Plant City disease. Just as confusing as this maze of names is the disease itself—no one seems to know its cause. As most people think of it, blight of citrus crops involves: foliage wilting despite sufficient soil moisture; dieback that is apparently independent of trunk or root damage; and the appearance of water sprouts after the onset of decline.

This mysterious disease is terminal—once affected, trees never recover. Infection in the orchard does not spread from tree to tree, but jumps around without apparent reason. You'll first notice an unusual dullness of the foliage that precedes wilting. Leaves may revive somewhat at night, but wilting becomes progressively worse, and rains and irrigation are of no help. Leaves then roll, droop, and drop, and dieback soon follows. An unusual number of water sprouts appear with the rainy season, and the new foliage is soon killed by the disease. Trees appear to have normal root systems. This characteristic distinguishes blight from disorders having both similar aboveground symptoms and roots obviously damaged by water, excess fertilizer, nematodes, gophers, fuel oil spillage, or mushroom root rot. Also, blight seldom hits trees of less than 12 years.

While researchers debate the causes of this disease, your only recourse is to take out affected trees and stick in new ones. Chemicals, transplanting, and pruning are to no avail.

**Brown rot gummosis,** also called foot rot, is caused by a fungus that attacks and kills the bark at or near the ground level. Large quantities of gum are usually produced when the infection is aboveground. Under some conditions, the fungus may attack the upper part of the trunk or even limbs and twigs at the top of the tree. Lime and lemon varieties are most susceptible to this fungus; orange, grapefruit, and mandarin varieties are somewhat resistant. Of the common stocks, smooth lemon, sweet orange, rough lemon, Cleopatra mandarin, Carrizo and Troyer citrange, and trifoliate orange are most resistant. Highly susceptible rootstock varieties include smooth lemon, Page, sweet orange, and grapefruit. Infection is favored when wet soil remains in contact with the bark.

To avoid gummosis, plant trees high enough so that, after settling, the point at which the first lateral roots branch out is at ground level. On heavy soil it is especially important to use stocks that are resistant to the disease. With older trees, all soil should be removed from around the trunk down to, or even below, the first lateral roots. Never apply water around the base of the tree, for this area should be as dry as possible. Nursery growers can treat seeds for ten minutes in 125°F. water.

Treatment consists of exposing the crown roots and cutting away the dead bark and about ½ inch of the live bark. No wood should be removed. Cover the wound with a tree paint. If the disease has killed the bark more than halfway around the tree, remove the tree and replant.

Lemon trees grown on sour orange stock are particularly susceptible to **citrus scab,** the principal lemon disease in Florida. The scab is caused by a fungus that attacks the new growth and young fruits of citrus trees. It seems to attack the growing parts of the treetop only while the shoots are very young and tender. Once the tissues have become developed and hardened, the disease cannot affect them; leaves more than ½ inch wide and fruits more than ¾ inch in diameter are immune. The scab develops its worst attacks in cool weather, when the foliage and fruit are growing slowly and it has an opportunity to gain a foothold. Dampness also favors the disease.

Scab is not a problem in dry, arid regions of the country. California plantings never suffer from it, though the disease has doubtless had many chances to reach California orchards in stock from Florida. For that reason, lemons have been grown with greater success in California than in Florida.

Scab fungus is carried over from year to year in infected twigs and leaves. It does not live over from season to season in ripe fruit. The fungus finds the most congenial conditions in sour orange sprouts that happen to arise from the rootstock and in water sprouts from rough lemon stock. These sprouts should be destroyed as soon as they appear in any citrus tree. Also, where the scab has appeared, prune all infected growth before new growth starts in spring. Remove and burn fallen twigs and leaves. A hybrid lemon, the Perrine, is very resistant to scab.

**Green and blue mold** first occurs as watersoaked spots of up to ½ inch across that rapidly enlarge to cover most of the fruit. Harvested fruit is most often affected, but may be occasionally struck while on the tree. Blue or green spores soon appear on the water-soaked area. There's no way to avoid the two species of causal spores, as they are always present in the air. Fruit should be handled carefully to prevent the injuries to the rind through which the spores enter. The blue species can spread from infected to healthy fruit by physical contact, and this means even greater care during harvest.

The false spider mite is responsible for spreading **leprosis,** a canker affecting fruit, leaves, twigs, and branches of citrus trees. Damage is worst on early and midseason varieties of sweet oranges, and sour oranges are also troubled. The disease first appears as chlorotic spots on green fruit. The spots eventually enlarge and turn brown, earning the alternate name of nailhead rust, and become cracked in the center. Spots on the leaves are somewhat long and ragged, not round as on fruit, and most are found near the margins. On the bark, thick reddish brown scales develop from tiny yellow spots.

Control of leprosis is dependent on control of the casual mites. A spray of wettable sulfur, mixed 10 pounds to 100 gallons of water, will do the trick.

**Lightning** is a spectacular and frequent phenomenon in citrus groves of Florida. Lightning damage usually covers an area in the shape of an irregular circle or oval. The effects of lightning on citrus are much less dramatic than on other trees; there are no fires, no branches wrenched from the tree, and no bark torn loose. Usually the first symptoms are a permanent wilting of foliage, yellowing of the midveins and leaf veins, and defoliation. Wilting may start within four to six hours of the strike. The most characteristic symptom is a green bud surrounded by dead bark. Lightning typically hits one or two trees hard; those suffering a direct hit show narrow strips of injured bark running down the trunk. Moisture-deficient trees suffer more than those well supplied with water. Deep-rooted trees are damaged more than shallow-rooted ones, and trees on clayey soils are more resistant than those on sandy soils. But there seems to be no way to avoid damage entirely, unless each tree is equipped with lightning rods.

You can best discourage **mushroom root rot** as land is being cleared for the orchard by removing stumps and roots of the previous inhabitants.

**Psorosis** is the common name for a number of viral diseases that share the symptom of vein-banding chlorosis in immature leaves. The chlorosis may either appear as thin, pale bands bordering veins and veinlets, or as a leaf-shaped chlorotic area centered on the affected leaf. Other causes may generate these symptoms, but psorosis leads to patterns that are bilaterally symmetrical. In lemons the virus may cause uneven growth in the leaves and fruit; this results in rough fruit and crinkly leaves.

As sweet orange, grapefruit, and mandarin trees mature, the virus causes local areas of outer bark to break away in small, irregularly shaped scales. Only the surface of the bark is killed; new bark forms under the old as it continues to scale off. Frequently, pockets of gum are formed under the bark and beads of gum appear in the affected areas. As the areas of infected bark enlarge, the wood becomes stained, the top begins to die back, and the trees gradually cease to produce. The virus may also cause yellowish, circular areas to develop on the fruit and mature leaves, occasionally resembling symptoms of ringworm.

The heartbreak of psorosis is transmitted mainly by means of buds taken from infected trees, and occasionally by natural grafts that take place between roots of adjacent trees. You can avoid the disease by propagating nursery trees with buds from virus-free stock. Remove the bark on the outer, affected layer. This may be done by carefully scraping the bark with a garden tool. However, trees that develop bark symptoms of the disease within five to seven years of planting should be removed.

Nematode-free nursery stock should guard you from the effects of **slow decline,** caused by injury of the citrus root nematode to the root system. Tree growth subsides, foliage turns dull green or bronze, and leaves become cupped. Later, leaves can be seen to be smaller and tops thinner than those of healthy trees, and fruit yield is affected adversely in both size and number. To make sure that nematodes are responsible, a visual inspection is necessary. Severely attacked feeder roots are typically more stubby, misshapen, and brittle, and they lack the usual yellow or white color of normal roots. Nematode nests and eggs are visible under 10× magnification. See spreading decline, below.

Nematode problems do not come about suddenly, but build up over a period of years. Even after affected trees are removed, the soil may harbor nematodes for several years, so replanting healthy stock in infested areas is useless. Be sure to plant nematode-free trees in soil known to be safe.

Use resistant rootstocks—Trifoliate orange and Troyer citrange in some locations—but be aware that citrus nematode races exist, and it is important to determine if Troyer citrange roots are susceptible to the race of nematodes present in the soil.

**Sooty mold** is caused by aphids, mealybugs, certain soft scales, and particularly, immature whiteflies. All secrete a sweet, syrupy material known as honeydew. This falls on leaves and fruit and supports sooty mold fungus. Sooty mold can blacken an entire tree, including the fruit. Control sooty mold by controlling these insects. Oil sprays will usually cause the mold to flake off, making the leaves bright and shiny.

Burrowing nematodes are responsible for **spreading decline.** The limits of infested soil are clearly demarcated in the case of this disease, whereas the borders of slow decline (see above) are less evident. Another difference is that the progress of spreading decline is clearly visible, taking in about two trees in a row each year. The roots of affected trees show half the normal number of feeder roots, and inspection with

the aid of a lens will show the nematodes in the roots. Avoid tangling with this nematode by using only nematode-free stock in uninfested soil.

**Water spot** occurs during long wet spells, late in the season when water is absorbed by the white portion of the rind through openings caused by injuries. The resultant swellings cause minute splits in the rind, permitting more water to be absorbed. This condition, together with the oil liberated from the injured rind, causes the cells to break down. If dry weather soon follows, depressed brown scars result. Affected fruit is usually subject to decay and is not suitable for shipment. If moist weather continues in the orchard, various fungi attack the fruit and cause rapid decay.

Susceptibility to water spot increases as fruit matures. It is greatly increased by mechanical injuries and by ice that forms on the fruit in cold weather. Early picking will help.

# CLEMATIS

The **clematis borer** is a dull-white worm with a brown head, growing up to ⅔ inch long. It hollows out stems and tunnels the crown and roots, causing vines to become stunted and branches to die. Larvae spend the winter in the roots. To control, cut out and burn infested stems, and dig out larvae in crowns.

Other pests with a penchant for clematis include blister beetles (see ASTER), NEMATODE, and SCALE.

# COLD WEATHER

See FROST INJURY.

# COLLARD

Collard is afflicted by many pests and diseases of the members of the cabbage family; see CABBAGE.

# COLUMBINE

## Insects

The columbine **aphid** is one of several species seen on this plant. They cause stunting of foliage in late summer. See APHID.

The **columbine borer** feeds in the petioles and stems of both wild and cultivated plants in April and May; it later bores into crowns and roots. Infested plants wilt and die. Look for sawdustlike castings near entrance holes. The larva is a caterpillar, salmon colored with a pale stripe down its back, and grows to 1½ inches long. To keep down the population of these borers, cut off infested stems or destroy infested plants. Clean up beds in spring, and scrape away surface soil or mulch to remove any overwintering eggs.

The **columbine skipper** is a butterfly with purplish wings. While pretty to look at, its larval form is a ¾-inch caterpillar that chews holes in leaves and rolls them up when not feeding for protection. This worm is green and has a black head. Control by handpicking feeding worms or crushing them when at rest.

Two species of **leafminer** attack columbine. One makes serpentine mines and the other makes blotch mines. The pale, yellow green maggots eat their fill and then pupate on the undersides of leaves. Tiny brown, two-winged flies emerge in about two weeks. In the fall, fully grown miners drop to the ground and crawl into the soil to spend the winter. When

infestation is light, pick off and destroy leaves; hoe the ground around plants in fall to expose the miners to the elements and birds.

MEALYBUG, stink bugs (see SNAP-DRAGON), and the WHITEFLY may feed on columbine.

## Diseases

**Wilt** disease causes leaves and stems to wilt and dry up, resulting in death of the entire plant. Inside the stem you may be able to see sclerotia (resting bodies) that look like mustard seeds. Cut out and dispose of affected plant parts, being careful not to let infected material come in contact with the soil.

# COMPANION PLANTING

Plants seem to do better in the neighborhood of certain other plants. The practice of making use of these happy combinations is known as companion planting.

While the mechanics of repellent planting and mixed cropping aren't all that mysterious, the various benefits claimed by practitioners of companion planting aren't all that easy to explain. Seemingly random combinations of plantings produce crops that not only grow better, but are relatively free from insects and diseases as well. Many gardeners who follow this method just accept that it does work, without trying to come up with the reasons why. Some see companion planting as the product of witchcraft and wishful thinking.

But it's likely that companion planting is based on something more mundane than alchemy. The protection may be due to the increased health of vegetables growing near their friends; insects typically pass by healthy plants, preferring to pick on weak or sick plants. Why are companionate plantings healthier? Perhaps because the plants make complementary demands on the environment. For instance, the compatibility of celery and leek can be traced to the upright-growing leek enjoying the room and light near the bushy celery plant. Or the roots of two friendly plants may occupy different strata of the soil, as do those of swiss chard and beans. Lettuce and kohlrabi get along well because the lettuce is harvested about the time the kohlrabi comes to need all the space in the row.

Unbeknown to the casual observer, the root exudates of a plant can affect the well-being of its neighbors. Marigolds put off a substance that keeps nematodes at bay, and interplanting the flower with susceptible crops will show positive results over a year or two. Mustard oil released from the roots of mustard family members will sweeten an acid soil, helping out adjacent plants that suffer when the pH is too low. The secretions also inhibit the hatching of potato nematode cysts. Root exudates of some asparagus varieties resist nematodes. The roots of both oats and flax excrete compounds that inhibit the growth of some harmful soil fungi. Many deep-rooted plants, including certain weeds, have the ability to bring toward the surface many essential minerals, making them available to their neighbors. This is important to plant protection because plants with deficiencies are especially prone to disease and insect feeding. Associations like these can strengthen the vegetables, giving you better taste, better nutrition, and less insect damage.

When you plant a companion garden, it's good to get the companions right in there together. One way is to plant zigzag rows, with the zigs and zags of the beets and onions tucked into one another. Another method is to use the technique of intercropping, and plant several companions in the same row. You'll find that your companionate garden breaks down into loosely defined sections. The corn, squash, cucumbers, pumpkins, etc., might be in one section. The strawberries, spinach, beans, etc., might be in another. It's best to place paths between these sections rather than between companions. Plant herbs in borders and scattered throughout the garden. See MIXED CROPPING and REPELLENT PLANTING.

# COMPOST

Leaves, grass, vegetable debris, wastepaper, table scraps, manure, and a good number of other possible ingredients all combine to make this an excellent fertilizer for flower beds, orchard, garden, trees, and ornamentals. The ingredients will decompose naturally, so there's no need to add lye or other chemicals. To speed up the process, you can add materials high in nitrogen and protein, such as dried blood, bone meal, and manure. Avoid putting fat or meat in the compost pile, as they don't break down very well; soapy water is of no benefit either. If young children play anywhere near the areas to be composted, don't empty the cat box on the compost pile—a dangerously toxic by-product may be formed.

Aside from these cautions and limitations, just about any organic odds and ends will do. It's a good idea to shred materials before adding them to the pile, using either a rotary mower or a compost shredder. Generally, compost should be applied about a month before planting. If the mixture has evolved only to a half-finished state that is still quite fibrous, you can still use it right away or apply it in fall so that it will decompose sufficiently by planting time. If the compost is ready in fall and won't be used until spring, cover it and store in a protected place. If the finished material is to be kept for long periods during summer, water it from time to time.

In the garden, compost should be applied freely, from one to three inches thick a year. Flower beds need just a scattering. In the orchard, cultivate between the trunk and the drip line, and work in a bushel of compost per tree; trees that bear heavily may be able to use more than this allotment. Shrubs and ornamentals are usually composted according to their size.

The physical condition of the soil has an important bearing on the prevention and control of diseases, and compost is the most important contribution you can make to your soil: a healthy dose of compost will keep tomatoes relatively free from blight and rots; fruit trees seem to become more resistant to disease when well fertilized with compost; and sowing seeds in a mixture of half compost and half sand will discourage damping-off of young seedlings. Root rot fungus has a relatively short existence in soil with a high organic matter content because of organisms which produce antibiotic substances.

The heat of a compost pile can be sufficient to kill weed seeds, grubs, and diseases on discarded vegetable matter, but the pile must be turned several times so that plants near the surface get a chance on the inside where peak heat is generated. It is all right to compost most diseased vegetable matter if the pile reaches temperatures high enough to inactivate the organisms in its core.

Wireworms don't like composted soil, as organisms in the compost give the pests a hard time.

Because compost contains so many elements in proper amount, it gives growers insurance against diseases prompted by any of several deficiencies. For example, dieback has been attributed to copper and potassium deficiencies, too much fertilizer, acidity, and so on. When properly made compost is used regularly, these problems are all alleviated.

Compost also tends to balance soil that is too acid or alkaline, since it itself is generally neutral. This is important because, if bound up in too much acid, nutrients cannot reach a plant. The optimum pH level for purposes of nutrition is between 6.0 and 8.0. Counter an overly acid soil by adding ground or crushed dolomite limestone (not burnt lime), wood ashes, or ground oyster shells to the compost pile. Alkaline soil will be improved with compost buffered by acid material, such as leaf mold, pine needles, or acid peat. (It is notable here that well-fed plants can tolerate a wider range of pH.) If you can't arrange to have a compost bin or heap, these additives can be mixed right in with the soil around the plants in need.

There is evidence that some plants fertilized with their own composted residues may eventually become strong enough to resist disease. Some gardeners have found this to be true of black rot on grapes. They compost infected clusters of mummified fruit and then apply the decomposed material around the base of vines, digging some into the soil. Though not universally applicable, the principle behind this practice is valid: grape tissue is known to be able to build up resistance to disease in much the same way the human body does.

# CORN

## Insects

Just as corn is a favorite crop for many gardeners and a major crop for many farmers, it is an important food for an impressive, or depressing, number of pests.

Of the several **aphids** [color] known to damage corn, two are of particular concern to gardeners and farmers. The small, green blue corn leaf aphid infests the tassels, upper leaves, and upper part of the stalk. It is most prevalent in the South, and recently its numbers have increased in the American Corn Belt and Ontario. As the aphid feeds, it secretes a sticky substance called honeydew, which serves as a medium for a black fungal growth. This secretion also attracts corn earworm moths. Severe aphid infestations may prevent pollination and as a result the ears will have no kernels.

Growers have several natural predators on their side, including ladybugs and lacewings. Plant corn as early as possible. Some hybrid varieties are less susceptible to aphid damage. See APHID.

Corn root aphids are similar in appearance to the corn leaf species. They feed on corn roots,

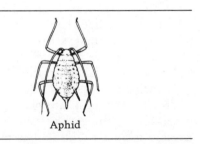

Aphid

but often not of their own volition—they are carried to their meals by ants. Winged aphids that try to fly away are snapped up by the ants and taken belowground. Because these aphids would be of little harm without other insects to lug them to your corn plants, efforts should be directed at controlling the ants.

In some years, the **armyworm** is a fairly serious pest of corn. The larva is a green or brown caterpillar with several longitudinal dark stripes and one broken light stripe. It grows to 1½ inches long and feeds chiefly at night and during cloudy days.

Predators and parasites usually keep the armyworm in control—tachinid flies, egg parasites, birds, skunks, and toads. In most years, the outside rows along hay, pasture, or grain fields will absorb most of the damage. If vast numbers of worms make you feel helpless, trap them in a ditch around the corn patch, with the side next to the corn plowed so that it is perpendicular. You can reduce armyworm populations with the old-time practice of planting alternate rows of sunflowers. You can purchase a beneficial nematode parasite, *Neoaplectana carpocapsae.*

Several species of **billbug** attack corn, occasionally causing serious damage east of the Great Plains in the United States and Canada. They are shaped like slender footballs and range in color from reddish brown to black. Billbugs typically play dead when disturbed. The larvae are humpbacked white grubs with a yellow or brown head. The adults eat small holes in the stems of very young plants and perforate unfolding leaves; the grubs attack roots and may tunnel up into stems.

To avoid problems with billbugs, don't plant corn on land that hasn't been cultivated for several years; these pests favor old pastures and reclaimed swamplands. Fall plowing, good drainage, crop rotation, and clean cultivation all help minimize danger.

The **chinch bug** is tiny, black, and has white wings and red legs. Nymphs are reddish in color. One positive way of identifying the chinch bug is by stepping on a few—they will put off a bad odor. They do their worst in the Mississippi, Ohio, and Missouri river valleys, in Texas and Oklahoma, and in extreme southwestern Ontario. Chinch bugs and the nymphs climb the corn plants and suck juices from the stem.

Chinch bug

Corn earworm adult and larva

Keep the area around the garden free of weeds. Well-fertilized soil and early planting also help. Remember that small grains also appeal to this pest; do not plant wheat next to corn. Interplant soybeans to protect corn; the broad soybean leaves shade the base of the corn stalk and make the stalks less appealing to the chinch bug. There are several resistant varieties available.

Chinch bugs come on strong in hot and dry stretches. Warm and humid weather favors the naturally occurring green muscardine fungus, *Beauveria bassiana*. The disease may become commercially available, but it is effective only when temperature and humidity are optimal. Some entomologists believe artificial dissemination is unnecessary because the spores are common where chinch bugs feed.

If you grow sweet corn, you probably are familiar with the **corn earworm** [color]. It is found throughout the United States and eastern Canada. The adult is a strong-flying moth that feeds on flower nectar. Early in the season, the eggs—dirty-white, ribbed domes—may be seen glistening singly on host plants. These hatch to produce one of the most destructive of insects, a striped caterpillar that varies in color from light yellow or green to brown. They grow to almost two inches in length. (Because of this pest's omnivorous habits, it is also known as the tomato fruitworm and the cotton bollworm.) Early in the season, the earworm attacks buds

and eats ragged patches out of tender, unfolding leaves. The most serious damage occurs later, when the larvae attack the ears. At first they feed on the fresh silky tassels, and may sever them. As the silks dry, the earworms shift to the kernels, starting from the tip. They frequently penetrate to the middle of the ear, leaving behind moist castings. Overall, trouble to the crop occurs in several ways: chewed silk prevents pollination; damaged kernels are prone to several diseases; and marred ears must be trimmed before they can be marketed, lowering value while increasing labor cost during production.

Timed plantings may sidestep this pest. In the South, early and late plantings are most vulnerable; in the North, the borer hits early corn and the earworm hits late plantings. See TIMED PLANTING.

Their population increases and decreases in cycles, because mating increases with the new moon. The worms are most numerous two weeks or so after the full moon. Ideally, corn should silk during the full moon.

If you find earworms in the first corn that you harvest, prevent an outbreak by applying mineral oil to the silk just inside the tip of each ear—half a medicine dropper's worth for a small ear, three-quarters for a larger ear. Large-scale growers may find that an oilcan is easier to use. Do not apply oil until the silk has wilted and begun to turn brown at the tip, as earlier treatment will interfere with pollination and result in poorly filled ears. Mineral oil acts by suffocating the worms; it is tasteless and will not affect the flavor of the kernels. However, oil may spoil the ears in hot, dry weather. Another method of dealing with earworms already at work is to pull back the sheath a bit and gouge the pests out. This should be done after the silk begins to turn brown and when you are sure that pollination is complete. A third method of control is easier but not quite so effective. Simply walk through the corn patch every four days and cut the silk off close to the ear. Some growers plant marigolds around and among the corn.

Any corn variety with long, tight husks is physically safer from the earworm; see RESIST-ANT VARIETIES. You can clip a clothespin on the silk channel at the early silk stage to increase the effect of husk tightness and keep the worms from damaging the ears. Luckily, egg parasites, birds, moles, and cold, damp spells all are in your favor; even earworms, if hungry enough, will eat other earworms. The adult, a non-descript grayish brown moth, is attracted to light and its numbers may be reduced with blacklight traps.

Soldier beetle larvae enter the pest's tunnel and eat the worm. Small black flower bugs eat both eggs and larvae; the beneficial's red nymphal form is also very helpful. Other insects that prey on earworms are the green lacewing and damsel bug [color]. Blackbirds, grackles, English sparrows, and woodpeckers eat larvae. Birds may damage ears in the process. Moles eat pupae in the soil.

Trichogramma wasps parasitize earworm eggs. You can increase their effectiveness by spraying your corn plants with a watery emul-sion of amaranth (or pigweed) leaves. See TRICHOGRAMMA. Releases of the wasp can be complemented by dusting with the bacterial pathogen *Bacillus thuringiensis*. It is more lethal to the younger larvae that work close to the surface than to older larvae that are on their way into the ears. Pollination is not affected by this pathogen. See *BACILLUS THURINGIENSIS*. Another disease, one you can collect from nature, is NUCLEAR POLYHEDROSIS VIRUS. The earworm is also controlled by Elcar, a viral insecticide.

**Corn rootworm** problems respond to plant-ing soybeans, alfalfa, or small grains after a season of corn. See TIMED PLANTING for ideas on avoiding this pest's annual appearance.

Several species of the fat, smooth, pulpy **cutworm** [color] attack corn, usually at night. You can often catch them in the ray of a flash-light. When disturbed, cutworms curl into a ring, and they spend their days resting in this position just below the soil around affected plants. The black cutworm (also known as the greasy cutworm) is black, or nearly black, and has a greasy appearance. Look for a broken

Cutworm

yellow line on the back with a pale line to each side. They grow to 1½ inches or longer, and move restlessly about from plant to plant, cutting them off at ground level. The dingy cutworm is a dull brown and is marked with a pale, triangular pattern on its back. They some-times will climb to eat leaves, and become fairly inactive as summer progresses. Other species that feed on corn plants include: the variegated and spotted cutworms, which climb to the leaves at night; the glassy cutworm, a greenish white, translucent worm that feeds underground; and the claybacked cutworm, which is distinguished by a broad, pale stripe running down its back. For control measures, see TOMATO.

Another very important insect invader of corn is the **European corn borer** [color], a flesh-colored caterpillar that grows to one inch long and has brown spots and a dark brown head. You will likely find this borer on other crops as well—it is known to attack more than 200 plants. It has yet to spread to the Far West or Florida, but farmers and gardeners elsewhere in the United States are vulnerable. The borer is a serious pest of market sweet corn and canning corn from Quebec eastward; it is no longer a major pest of grain corn in southwestern Ontario, thanks in part to recent resistant varieties. The larva passes winter in corn stubble and pupates in spring. The nocturnal yellow

European corn borer adult and larva

brown moth appears in May, or early June in Canada. Populations are reduced by unusually cold winters and dry summers. The male is much darker than the female, and dark wavy bands cross his wings. The eggs are laid in masses of 15 to 20 on the undersides of leaves, overlapping each other like fish scales. These hatch early in June, and the borers immediately crawl to protected places on the corn plant, eating their way through the stalks and into the bases of ears. Larvae may be found inside corn kernels. Broken tassels and stalks, and sawdust castings outside of small holes are signs that borers are feasting on your corn.

Shred and plow under cornstalks in or near fields where borers overwinter. This should be done in fall or early spring, before the adults emerge. Generally, plant as late as possible, but be certain to stay within the normal growing period of your locality. Another caution is that late plantings are especially vulnerable to the corn earworm, and if this pest is an important one in your area, corn should be planted to come to silk when other plant hosts are up. In New York through southern New England, corn set in the latter half of May will usually mature between the two generations of borers. In Ontario, very early and very late plantings are particularly susceptible—moths of the overwintering generation are attracted to early-planted corn, and midsummer moths lay their eggs on late-planted corn. In Maryland and Delaware, corn growers avoid borers by planting when forsythia and star magnolia bushes are in full bloom.

Although there are no immune strains currently available, there are plant hybrids that are resistant or tolerant. Consult your county agent or state experiment station for help in selecting the best hybrid for your locality.

Handpicking is the oldest and simplest of remedies, and it can be very effective in controlling the borer. On plants showing external signs of borer activity, split the stalk a little below the entrance hole with a fingernail and pluck out the worm. For growers north of the Mason-Dixon line, shredding corn stalks from the ground up by April 1 and covering them with soil will go far toward discouraging a borer infestation. Clean up grassy or weedy areas adjoining the corn field.

Swallows can be attracted to corn and corn borers with bird houses and ponds. If cornstalks are left in the field after harvest, woodpeckers and flickers may hunt among them for these pests.

Parasites have been used against corn borers with considerable success. Since the corn borer lives from 10 to 24 days of its life as a nocturnal moth, it can be attracted to light traps. A daylight-blue lamp will draw hundreds of the moths to their death. The USDA has imported a parasitic tachinid fly, *Lydella stabulans grisescens*, and this ally is now well established in southern New England and the Middle Atlantic states, and in Illinois, Indiana, Iowa, Kentucky, and southwestern Ohio. The adult fly deposits larvae, not eggs, at the entrance of borer tunnels.

Other parasites of the borer include the gardener's good friend, the ladybug, which can consume an average of almost 60 borer eggs a day, and the trichogramma wasp, a commercially available egg parasite. *Macrocentrus grandi*, a braconid wasp, parasitizes as many as 35 percent of the corn borer larvae in the Corn Belt and 20 percent in Connecticut. Borers are also susceptible to the protozoan diseases *Perezia pyraustae* and *Nosema pyraustae;* these may be the most important naturally occurring control in some areas. You can buy the bacterium *Bacillus thuringiensis,* which is effective in controlling the borer and other lepidopteran pests (see *BACILLUS THURINGIENSIS*). The fungal disease *Beauveria bassiana,* which turns larvae a pink shade and eventually white with fungal growth, may soon be marketed.

The small, dark **flea beetle** is so named because it jumps quickly when disturbed. Although it eats small holes in leaves in early summer, it is more important as a vector of

Flea beetle

bacterial wilt of corn, known also as Stewart's disease. This malady stunts growth of young corn plants and can kill them. It is most likely to occur after a mild winter and a cool, wet spring.

Plant varieties that offer resistance to wilt; in Canada, North Star and Northern Belle are particularly vulnerable varieties of sweet corn. For other ways of combating this pest, see POTATO.

Corn on the margins of fields is vulnerable to damage from **grasshoppers.** To help keep leaves, tassels, and stalks from being devoured, mow fields early in the season and keep them cut short. For other control methods, see GRASSHOPPER.

Chainlike markings on the surfaces of leaves are probably the work of the **grass thrips.** Lowermost foliage may be killed, and larger leaves may take on a silvery sheen. Developing kernels are sometimes destroyed. Control by cleaning up plant litter where the pests hibernate.

The **lesser cornstalk borer,** a blue green worm with brown stripes, bores into lower parts of corn stalks. Although this insect is generally distributed, control is only necessary in the South. The borers make dirt-covered, silken tubes that lead away from stem tunnels in the soil. Stalks of young plants become distorted

Lesser cornstalk borer adult and larva in stem

and may not produce ears. Because the larvae spin their cocoons in plant debris, fall and winter cleanup of the garden area should reduce their numbers. Early planting and rotating corn with a resistant crop also are effective measures.

Several species of **sap beetle** attack corn. The adult is small (³⁄₁₆ inch long), usually black, and active, and feeds on the ears of sweet corn, frequently making small round holes through the husk to enter the ear. The maggotlike larva is also harmful, eating into kernels of corn that is already damaged. It is white or cream-colored, grows up to ¼ inch long, and scatters over the ear when exposed to light.

There will be fewer overwintering beetles if rotting plant material is kept from the area. The mechanisms that prevent damage to ears by corn earworms appear to operate somewhat against sap beetles, although corn earworms are rarely found in ears damaged by sap beetles. A number of resistant varieties are available.

The **seedcorn maggot** burrows its way into planted seeds, injuring them so that they produce sickly plants; also, soil organisms may thus gain entry to the seed and cause it to rot. The maggots are yellowish white, ¼ inch long,

Seedcorn maggot and dark puparium

and have pointed head ends; the adult is a small, grayish brown fly. Damage is greatest in cold, wet weather and with shallowly planted seed. See BEAN for control measures.

The **stalk borer** is distributed generally east of the Rockies. It is a very active, slender caterpillar that grows up to 1½ inches long. The young insect is creamy white with a dark purple band around the body and several brown or purple stripes running down the back. The mature caterpillar loses the lengthwise stripes and takes on a grayish or light purple hue. It bores into the stalks of corn and works unseen.

Plants around the periphery are particularly apt to be damaged.

Because the borer works out of sight, by the time you spot damage it may be too late to come to the plant's rescue. You can try slitting the stalk with a fingernail, plucking out the borer, and binding the stem. Keep weeds and grasses down in the garden area, since these serve as winter nurseries for eggs and newly hatched stalk borers.

Several similar species of **white grubs,** which are the immature forms of the common, large brown May or June beetles, may harm corn by feeding on the roots. They are white with brown heads and are found in the soil in a

May beetle and white grub

characteristic curved position. Because eggs are usually laid in grasslands, damage to crops is greatest when they are planted on newly broken sod land, especially if the soil is sandy or sandy loam. These beetles have a three-year life cycle, and outbreaks of white grubs occur every third year in the eastern United States and most of Ontario (in 1975, 1978, and so on); north of Montreal Island and around St. Jean, damage will be heaviest one year later in the same cycle, while south of Montreal Island and in the Eastern Townships, white grubs will be troublesome one year earlier.

Most problems can be avoided by not planting corn on new sod land. Plowing in summer or fall kills or exposes the grubs to their predators.

Any one of several species of **wireworm** may cause poor development of plants. They are hard, wirelike worms, varying in color from white to yellow to brown and darkening as they

Wireworm

grow longer. They may bore into corn seed and eat the germ, or enter the underground stem to cause young plants to wilt, discolor, or die. Damage is worst when corn is planted on soil that is the first or second year out of sod. See WIREWORM.

## Diseases

Missing kernels may be a sign of poor pollination, which in turn results from too few plants sown too far apart. Try to plant at least four rows, no more than 2½ feet apart.

Wilted leaves on dry days may be a sign of **bacterial wilt,** or Stewart's wilt, a common disease of corn in the central, southern, and eastern states. Sweet corn is the most susceptible, especially the early, yellow, sweeter varieties. The wilting is caused by blockage of vascular bundles by the developing bacteria, and increased watering is of little use in saving plants. Leaves take on long pale streaks; when a severely infected stalk is cut after a rain or heavy dew, a sticky yellow ooze of bacteria exudes from the severed bundle.

The flea beetle (see above) is an important vector of this disease, especially after mild winters. Most white late-maturing varieties are resistant to wilt. Destroy infected refuse and deeply plow stubble in fall.

**Ear rot,** or diplodia ear rot, usually first occurs in the shank and then spreads into the ear. Rotting may also begin at the tip of the ear. Husks tend to become dry and bleached, and are often stuck together. You will see minuscule, black fruiting bodies embedded in the moldy kernels and husks; these create the spores that transmit the disease.

Use diplodia-resistant hybrid corn seed and don't follow corn with corn. Be sure to plow cornstalks under deeply so that the disease is not spread.

**Helminthosporium leaf spot** is a fungal disease that appears late in summer after wet weather as a sudden leaf scorch. Affected leaves look as though damaged by an early frost, and the fungus makes black threadlike growths on the injured tissue. To reduce the chances of infestation, avoid late planting and fertilize the plants properly.

**Root rot** is characterized by plants that are easily blown over and stalks that show a reddish tint when split open. The only known control measures are crop rotation and use of soil that is properly drained.

Corn **smut** [color] is a common fungal disease that appears as large, irregularly shaped blisters or galls on all parts of the plant. At first the galls are grayish white, darkening as the season progresses until they burst to release masses of black fungal spores. Ripened spore masses have an oily appearance. The disease is favored by a dry spring followed by a wet spell.

Institute a three-year crop rotation; use disease-resistant seed (see RESISTANT VARIETIES); turn under garden trash or manure, and remove old stalks in fall. After the fact, the only course left is to remove and destroy the galls. If plants become too badly affected, they should be removed and burned. Don't till the plants into the soil. Do not use diseased plants in making compost, since the fungus may affect the entire compost heap.

Another disease caused by diplodia fungus is **stalk rot.** After pollination, look for reddish purple or dark brown blotches on leaf sheaths and stalks. Corn ears are rotted or chaffy, and many stalks will break after maturity.

Resistant varieties are available. A four-year (or longer) crop rotation will help you avoid this disease. Because the causal organism overwinters on injected stalks, crop residue should be plowed well under.

## Wildlife

**Birds,** especially crows, are fond of pulling up small corn plants in search of the tender kernels below. You can plant untreated seeds through a permanent mulch to give plants a chance to get off to a good start before the crows can spot them. By that time, any plant pulled with yield disappointing results to the average crow, and the flock will move on to another stand. Crows can be further discouraged by planting corn a bit deeper than recommended on the packet.

After plants are well established, pull the mulch back. Crows may not bother germinating corn when three small twigs are stuck in the ground so that they lean together at the center of the hill (but note that these birds do a good job of cleaning out cutworms early in the season). A straight row of newly planted seeds can be protected by placing a stake at either end of the row and running a dark sewing thread between

them, about two inches above the soil. Some growers have diverted the crow's attention from the corn patch by setting out other food. You might try scattering feed corn about the plants. This practice could have the negative effect of encouraging more hungry crows to visit your area, and even to overwinter and greet you the next growing season. Netting around the corn plants gives protection at two periods in the season: until seedlings are well on their way, and for the two weeks preceding harvest.

Blackbirds raid corn plantings at dawn and dusk, stripping the ears. These birds do their worst when corn is in the silk stage.

**Raccoons** are a serious pest for some corn growers. They may be fenced out of the garden area with chicken wire stapled to posts so that the wire projects a foot or so above the posts—raccoons climb fences rather than dig under them, and they will find themselves caught as their weight brings the slack top over onto them. Any that manage to get beyond that point may be kept from doing serious damage by tying a paper or plastic bag over each ear you want to ensure, covered in turn with a piece of wire screening. The wire prevents the coon from tearing the bag, while the bag prevents it from reaching the corn through the wire. Although this scheme is virtually coon-proof, fencing is a lot easier.

You might set up an electric fence. String 20-gauge wire at 8 and 16 inches above the ground, and use either a 6- or 12-volt battery or knocked-down house current for juice. Turn the fence on at night, since raccoons sleep during the day.

A final suggestion for outsmarting the coons: try planting pumpkin seeds with your corn, setting the seeds about four feet apart. The idea is that coons won't strip the ears because they like to be able to stand up and look around while they eat, and the big pumpkin leaves make that difficult. The animals will also avoid a corn patch bordered with squash, because they don't like the prickly leaves of the squash vines.

If **squirrels** have been stealing corn seeds as fast as you can plant them, mix a teaspoonful of pepper to a pound of seed.

# COSMOS

## Insects

Among the insect pests of cosmos are the European corn borer (see CORN), root APHID, sixspotted leafhopper (ASTER), and SPIDER MITE.

## Diseases

Sudden wilting of plants may be an effect of **bacterial wilt,** and is usually accompanied by a soft rot of the stem at the ground level. When stems are cut, a yellowish ooze appears. Infected plants should be destroyed and cosmos planted elsewhere, since the bacteria are soil-borne.

**Stem canker** causes dark brown spots to appear on stems. The spots later turn gray, rapidly enlarge, and girdle the stem to kill the plant. Burn old stems at the end of the season to reduce chances of the fungus overwintering.

# CRYPTOLAEMUS MONTROUZIERI

Known as the mealybug destroyer, this predator is commercially available. See MEALYBUG.

# CUCUMBER

## Insects

Melon **aphids** [color] like nothing better than a meal of sap from the leaves of a cucumber. They are usually green but their color varies from yellow to black. Likely before you spot

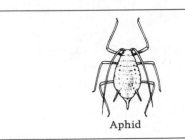

Aphid

them feeding on the undersides of cucumber leaves, you will notice the wilting and curling of the leaves. You may also see an unusual number of ants, bees, and flies swarming around, as they are attracted by the aphids' honeydew. If the aphids are numerous enough, the vine may become too stunted to produce a crop or even die. The melon aphid is also dangerous as a vector of bacterial wilt.

Control is often not necessary, as parasites and predators do a good job of keeping populations down. These allies may have more than they can handle, however, if a cool, wet spring is followed by a hot, dry summer; for control measures, see APHID.

A particularly harmful pest of cucurbits in the Gulf States is the **melonworm,** a slender, active, greenish caterpillar that is identified by two narrow white stripes running the full length of its back. It feeds mostly on foliage, but may burrow into fruit. Control by handpicking.

Cucumbers in much of the country may be damaged by the **pickleworm** [color], which at first is a pale yellow with black spots that

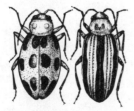

Pickleworm

disappear as it matures. The full-grown worm is ¾ inch long and colored green or copperish. The pickleworm is especially serious in the Gulf States, but late in the season it is found as far north as New York and Michigan. They cause damage by burrowing into the buds, blossoms, vines, and fruits. Although very early spring plantings are seldom damaged, late crops may be ruined.

Any hibernating pupae may be killed by plowing early in fall. At the end of the season, clean up all plant refuse from the garden. Pickleworms are attracted to squash blossoms, and squash may be planted as a trap crop. See RESISTANT VARIETIES.

The striped cucumber beetle's partner in destruction is the **spotted cucumber beetle** [color] (also known as the southern corn rootworm). Like the striped species, this pest is rather

Spotted cucumber beetle (left) and striped cucumber beetle

slender, measures about ¼ inch long, and has a black head. The wing covers are yellow green, and 11 or 12 black spots are scattered on its back. This beetle not only has a big appetite, but carries the bacteria that cause cucumber wilt as well.

Control measures are generally as listed below for the striped cucumber beetle. The larvae, known as rootworms, are easily controlled by rotating crops and prove troublesome only to growers who insist on growing continuous corn. In California, the western 12-spotted species has been kept in line with a trap crop of alfalfa. See RESISTANT VARIETIES.

The yellow, black-spotted **squash beetle** looks something like a ladybug, and in fact is a member of the same family. It is a minor pest, and you shouldn't find control measures necessary.

Although the **squash bug** prefers the taste of other cucurbits, it is an occasional pest of cucumber. The adults are dingy brownish black insects, are about ⅝ inch in length, and give off a very disagreeable odor when crushed. Immature bugs first have reddish heads and legs and green bodies, and later, black heads and legs and gray bodies. The southern squash bug, or horned squash bug, is a similar pest, and attacks cucumber north to Delaware. For control, see SQUASH.

East of the Rocky Mountains, the **squash vine borer** may tunnel in stems of cucumber and other cucurbits; you may see bits of yellow sawdustlike excrement around the holes or about the base of the plant. For control, see SQUASH.

East of the Rockies, the star insect enemy of the cucumber is the **striped cucumber beetle** [color]. The adults are distinguished by yellow wing covers with three black longitudinal stripes, have a black head, and grow up to ¼ inch long. They enter the garden as soon as cucumber seed germinates, and you can usually find them on the undersides of leaves. Other vine plants are also attacked; winter squash is particularly susceptible to damage throughout the growing season. The slender white larvae are up to ⅓ inch long, have brown ends, and damage plants by feeding on roots and underground parts of stems, sometimes destroying the entire root system. These wormlike larvae may also inoculate plants with bacterial wilt. The first and greatest damage results from the adults feeding on stems and leaves when the plants are just pushing through the ground and before they have developed true leaves. The beetle also is a vector for bacterial wilt and cucumber mosaic.

Perhaps the best control is the time-tested procedure of heavy mulching. The beetles also may leave your crops alone if goldenrod is growing nearby. Cucumber will fare better if interplanted with catnip or tansy. Rotenone or neem extract can be used in serious cases. A spray made from geranium leaves and stalks is less powerful.

Late planting is one of the easiest ways to avoid the worst beetle damage, although it is by no means foolproof. If practicable, plant seed right after the first horde of beetles lays its eggs. See RESISTANT VARIETIES and TIMED PLANTING.

Nature provides a good measure of free control. The most effective enemies are soldier beetle, tachinid fly, braconid wasp, and nematode.

## Diseases

**Anthracnose** is a seed-borne disease caused by a fungus that lives in the soil. Leaves develop small, dark spots that may combine and destroy entire leaves. Fruit on young plants may turn black and drop off, and older fruit may be covered with irregularly shaped, sunken pits filled with salmon-colored growth. The disease is favored by warm, moist conditions.

A three- or four-year crop rotation is the best means of preventing anthracnose. None of the host plants—including watermelons and muskmelons—should follow each other.

**Bacterial wilt** is sometimes a problem to cucumber growers, particularly those living in north central and northeastern states. Plants with this disease wilt so quickly that leaves may be both dried up and still green. The bacteria are spread by spotted and striped cucumber beetles, and grow to plug the water vessels of stems and leaves; if you cut across a stem you may notice a sticky, white ooze.

Affected plants found early in the season should be promptly removed and destroyed. You might plant extra seed to ensure that there will still be enough plants left for a good crop. Follow the recommendations for the control of the spotted cucumber beetle discussed above. Cover other plants with cheesecloth. There are several varieties with some resistance to scab.

Hot, dry spells and disease-ravaged foliage may lead to cucumbers spoiled by **bitterness.** Older varieties tend to be more susceptible. The bitter flavor is worst at the stem end and at or near the surface. Cut away these portions before eating, and prevent further trouble by watering plants thoroughly in dry weather. The bitter flavor can be eliminated by heating cukes for one minute in boiling water.

**Downy mildew** is a fungus common in the Atlantic and Gulf states. The first symptom you will see is spots on leaves, yellow on top and downy purple underneath. The spots spread, killing the foliage and sometimes whole plants. Mildew is encouraged by cool, wet nights, and warm, humid days; if such conditions prevail in your area, use a variety resistant to mildew; see RESISTANT VARIETIES.

**Mosaic virus** is characterized by mottled or curled leaves that turn yellow and die, and fruit (often called white pickles) is warty and light colored. It overwinters in perennial weeds that grow near the garden, such as milkweed, ground cherry, catnip, ragweed, pigweed, burdock, various mints, and horse nettle. The virus cannot overwinter in garden trash or annual hosts. It is usually spread by aphids, and occasionally by cucumber beetles or on implements.

The area surrounding the garden should be kept as free as possible from perennial weeds. This no man's land for weeds should be 100 yards wide in most seasons, although in cool, wet summers a 10-yard zone could be sufficient.

Because the disease is spread by aphids, a reflective mulch that repels the bugs will help guard the plants from mosaic. In an experiment at the University of California, Riverside, a mulch of 30-inch-wide strips of aluminum gave an 85 percent reduction in incidence of virus over a growing season. The plants grew through holes in the strips. A 63 percent reduction was brought about with white polyethylene strips of the same size. In the same test, an insecticide failed to give any protection, because the aphids migrated in from elsewhere. There are several mosaic-resistant varieties available, including Spartan Dawn, SMR, and Salty; check with your county agent or experiment station to learn which varieties will do best in your area.

**Scab** is a fungus found chiefly in the northeastern and north central states. Symptoms include dark, oozing spots on the fruit that look like insect feeding punctures. The spots later dry up, leaving small pits covered by an olive gray fungus growth. Small brown spots may appear on leaves and stems. This fungus lives in the soil and does its worst damage in cool, moist weather.

To avoid scab, do not grow cucumbers or squash in the same soil more often than once in three years. See RESISTANT VARIETIES.

# CURRANT

## Insects

The currant **aphid** [color] is the most troublesome pest of currants, stunting the whole plant and diminishing fruit production. Look for distorted and yellowish or pale foliage that is rolled up and wrinkled by the yellow green pests feeding on the undersides of leaves. Later in the season the injured leaves may drop, and this hinders fruit development. The eggs of the currant aphid are shiny black, cucumber-

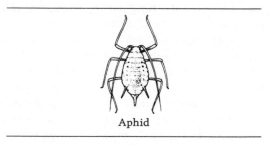

Aphid

shaped, and attached in October to the bark of new growth near the nodes. Soon after the leaves open in spring, the eggs hatch and the aphids crawl to the leaves, feeding on them by sucking the sap.

The currant aphid is not easy to control. Eggs are often transported on certified nursery stock, and the sprocketlike cavities of curled leaves protect the aphids and hinder control. It is important to pluck off affected foliage if serious curling is evident. The shiny eggs can be rubbed off canes and destroyed in spring, or washed off with water when they are laid in October. See APHID.

**Currant borers** can often be found and destroyed before they become a serious problem. In larval form the borer is a yellowish caterpillar measuring about ½ inch long. Adults are moths with yellow-and-black markings on their wings. Whenever you see a cane with wilting leaves at its tip, suspect the borer. Usually the damage occurs in spring.

Cut damaged canes back to ground level and quickly burn them before the moths can emerge. Light traps may be effective, and should be used before the moths lay their eggs in the canes, in June or July.

The **currant fruit fly,** or gooseberry maggot, is about the size of a housefly, with a yellow body and dark-striped wings. It emerges from the soil in April or May and lays eggs in the fruit. When the fruit drops, the maggots enter the soil. Early varieties are less susceptible. Control by dusting with rotenone as soon as the blossoms have wilted, in two or three applications at weekly intervals.

In spring, the **currant sawfly** (or currant stem girdler) lays an egg in new shoots which are then girdled or cut partly off above the egg. After hatching, larvae burrow several inches down into the cane, where they will stay until the following year. The sawfly can be controlled by handpicking on small plantings. Prune and destroy all affected canes.

The **fourlined plant bug** [color] is a pest of many plants, including currant and gooseberry. Like many berry pests, it winters in the canes, lays eggs there in spring, and hatches in mid-May or early summer to injure plants by sucking on the young leaves. Affected foliage shows spotted sunken areas and, later, holes. The nymph is bright red with black dots on the thorax; the adult is greenish yellow with four wide, dark stripes down the wings. If necessary,

control this pest by dusting with rotenone at weekly intervals. A soap spray may also be effective.

Fourlined plant bug

The **gooseberry fruitworm** is a pest of currants and gooseberries that winters in the ground under dead leaves and trash. The following spring, the adult lays its eggs in the flowers and the larvae hollow out the fruit. The moth is ashy colored with dark markings; the larva is ¾ inch long and yellow green with a pinkish cast. Control by spraying with rotenone at seven- to ten-day intervals. Destroy all the infested berries.

Currants are sometimes troubled by the **green currant worm** (or imported currant worm), which is black with yellow markings and devours foliage. Fortunately, they are rare.

Early in spring, check your plants for the white eggs on the undersides of leaves. Unless you pick off and destroy them before the larvae hatch in late May, you may have to resort to a summer treatment of rotenone or pyrethrum.

**Spider mites** are an occasional pest of currants. For control, see SPIDER MITE.

Crinkly dwarfed leaves may be the work of **thrips.** See THRIPS.

## Diseases

In the home garden, currants are often grown in the partial shade of trees, buildings, or fences. This practice makes them more vulnerable to disease than if grown in full sun with good ventilation.

Currants are susceptible to **anthracnose,** a disease of leaves and sometimes fruit that may occur if you work on the bushes while they are wet. You will see small reddish brown or purplish spots with masses of pink spores on the undersides of leaves. There is not yet a cure for anthracnose of currants, but some varieties are resistant to it.

Both currants and gooseberries are host plants to the **blister rust fungus,** which is not so much a danger to them as it is to white pine trees. Blister rust is so destructive to white pines that there are federal and state regulations about planting gooseberries and currants in the vicinity of white pine nurseries or forests. You may need a permit to grow them. Currant and gooseberry bushes should be planted at least 300 yards away from white pine trees. For more information, check with your local state agriculture extension service, the nearest USDA office, or the nursery from which you buy your plant stock. The rust disease shows as bright orange spots on currant leaves in midsummer.

Sometimes a fungal invasion will cause **cane blight:** the canes suddenly wilt and die while loaded with fruit. Small black cushions on the canes will identify the disease. The best remedy is to remove all affected stalks and burn them.

**Crown gall** may be caused by ordinary soil bacteria. It is not always serious, and often the lesions or galls are not too severe and the bush continues to produce.

**Mildew** is a common malady of both currants and gooseberries. You will notice white cobweblike growths on the top surfaces of leaves. Mildew is most prevalent in wet seasons or in moist areas. Let your plants have good exposure to the sun and plenty of air circulation to help keep them dry. Promptly rogue out mildewed sections to prevent its spread.

# CUTWORM

There is little about cutworms [color] to commend them. They are ugly to the eye and unpleasant to the touch. Worse yet, they attack just about every plant in the garden. Cutworms can be classified into four groups, according to the damage they do. Fruits, buds, and leaves are vulnerable to climbing cutworms. Tunnel-making cutworms do their work near the soil line, and may cut off and topple younger plants. All parts of the plant are vulnerable to armyworms. Subterranean cutworms eat out of sight.

Cutworms are plump, soft-bodied larvae, usually dull colored and scattered with coarse

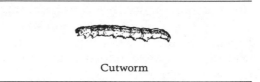

Cutworm

bristles or hairs. When disturbed, the worms commonly coil their bodies. They burrow down several inches into the soil to pupate, and reappear as night-flying moths.

Many animals find cutworms a good source of food. Natural enemies include firefly larvae, meadowlarks, blackbirds, toads, moles, and shrews. They also fall prey to the parasitic braconid wasp, tachinid flies, and nematodes. The black cutworm larva alone is kept in check by 15 species of bacteria, viruses, fungi, and protozoans.

The garden should be kept clear of weeds and grass through the fall months, in order to discourage egg-laying cutworm moths. Crops planted on sod land are especially apt to be infested, unless the land is plowed during late summer or early fall and then kept free of weeds and grass thereafter. Sunflowers can serve as a trap crop if grown around the border of the garden. Tansy repels cutworms.

If you have the time, a labor-intensive control is to place leaves of wild onion or garlic in planting holes, and then to tie the leaves around the plants. To kill resting worms, cultivate lightly or probe by hand around the base of the plant. Place a stiff three-inch-high collar of cardboard around stems of young plants, pushing it down one inch into the soil (toilet paper tubes will work, and they eventually disintegrate).

Cutworms can be kept from vulnerable plants with barriers of crushed eggshells, damp wood ashes, chicken manure, or a mulch of oak leaves. They can be done in by their appetite for cornmeal, a food they cannot digest. Scatter cornmeal where the worms have been a problem, and they may die from indigestion. A commercial product combines cornmeal with a bacterial pathogen. Cutworms also like molasses, and it can be mixed into a lethal bait. Combine equal parts of hardwood sawdust (pine sawdust repels cutworms) and wheat bran, and add enough molasses and water to make a sticky substance that can be scattered around each plant at dusk. As the cutworms crawl around in the sweet bait, it clings to their bodies and soon hardens, rendering them helpless.

Cutworms are controlled by *Bacillus thuringiensis.* Adult moths will fly into electronic light traps. If you aren't squeamish and have a flashlight, you can handpick the worms after dark.

On a large scale, all fields and gardens should be plowed or disced as soon as crops are harvested. This will leave no grass for the moths to lay their eggs upon. Large fields that had been in crops or grass the previous September should be plowed again in spring. A flock of chickens will follow in the furrow and snap up any worms that are brought to the surface. Hogs are capable of rooting up and eating large quantities of cutworms, grubs, and other destructive insects.

# CYCLAMEN

The translucent, microscopic **cyclamen mite** often infests new leaf and blossom buds, causing them to become swollen and distorted. Infested plants give no satisfactory blossoms. If the mites hit early, no blossoms will form at all; later infestation leads to streaked or blotched blooms that fall before their time.

The mite is difficult to control when cyclamen is badly infested. Provide ample space between plants, and don't touch healthy plants if you've just been around ones in trouble: you won't notice the tiny pests on your hands. Rogue out plants that are in a bad way. Plants can be ridded of the mites by immersing them in a hot water bath—110°F. for 15 minutes (or 100°F. for 20 minutes for strawberries).

**Foliar nematodes** cause leaves and stems to collapse, and the plant may not produce buds properly. Nematodes are sometimes carried in the dormant buds of the corms (underground stem bases) or in the soil; use clean corms and clean potting soil.

# DAFFODIL

**Blast** is fairly widespread and is characterized by flower buds which turn brown and dry up before opening. The cause is unknown, and bulbs behaving in this way year after year can be dug up and discarded.

**Ring disease** can cause plants to fail to grow in spring. Leaves may be abnormal and twisted with small yellowish lumps called spikkels, which contain many nematodes. Flowers are not produced. This disease is so called because the bulbs show discolored rings when cut across. Yellowish pockets in scales contain numerous nematodes. Hot-water treatment of bulbs gives reasonably good control (110° to 115°F. for three hours for small bulbs and four hours for large ones). Plant in clean or sterilized soil and remove infected plants and bulbs, along with some of the soil around them.

Because of bulb and **root rots,** plants develop poorly or don't even grow at all. An examination of bulbs shows rotted areas. Roots are brown and rotted, and you may find white or pinkish mycelium or spore masses. Rogue infected bulbs and remove soil from the area where the bulbs were planted. Narcissus should not be planted in an infected location for two or three years.

# DAHLIA

## Insects

The **cocklebur billbug** is a weevil with a long, curved snout. It is ⅓ inch long and has 13 black spots on its reddish back. The larvae are legless, curved, and white, with brown heads. They bore into the pith and hollow out stems for a foot or more near the base of the plant. Serious infestations are restricted primarily to the southeastern United States.

Stake dahlia plants to prevent breaking of stems. Postpone planting until June in order to escape the adult billbugs which emerge in May. In the garden area, destroy such wild hosts as ragweed, thistle, cocklebur, and joe pye weed.

The second generation of the **European corn borer** lays its eggs on dahlias and sometimes injures them seriously. See CORN.

When the **potato leafhopper** sucks plant juices, foliage is blanched. See POTATO.

The slender **stalk borer's** presence is usually first made known when its feeding causes dahlias to wilt, but by then it's often too late for help. The young borer is creamy white with a dark purple band around its body, and several brown or purple stripes running the length of its body. The full-grown larva is creamy white to light purple, and has no stripes. When young borers hatch early in the spring from eggs laid on grass and weeds, they first feed upon the leaves of the nearest plant and then gradually work their way to larger, stemmed plants.

Remove and destroy nearby weeds. Infested plants can sometimes be saved by slitting stems, destroying the borer, and then binding the stems and keeping plants watered.

The **striped cucumber beetle** is attracted to the pollen and petals of dahlia. These beetles are elongate, yellow green, and marked with three black stripes. See CUCUMBER.

The **tarnished plant bug** is small, flattened, and brownish, with a somewhat brassy appearance to its wing covers. It causes deformity by puncturing buds and new shoots. See ASTER.

## Diseases

**Bacterial wilt** is characterized by sudden wilting and death of plants. Cut stems will reveal a yellowish oozing from vascular bundles. Because bacteria overwinter in the soil, gardeners have found it important to sterilize soil and remove infected plant parts.

Wilting of a single branch or entire plant, followed by recovery at night, is symptomatic of **fusarium wilt.** The plant may remain alive for a short time, but it eventually succumbs. The fungus lives in the soil and in the tubers, so that once this wilt has struck, you must get new, healthy tubers and plant them in disease-free soil.

**Gray mold** causes flowers and buds to rot, especially in prolonged periods of cool, wet weather. There is no known control for this disease.

**Mosaic** (stunt) is a viral disease that appears as a mottling in leaf tissues, or as pale green bands along midribs and larger veins. Leaves are generally dwarfed or distorted, and infected plants are stunted, small, and bushy. Tubers are shorter and thicker than normal and show reddish brown necrotic spots inside.

Promptly destroy infected plants to avoid spreading mosaic to other plants. Dwarfing may also result from the feeding of many leafhoppers, aphids, European corn borers, or thrips. In this case, plants should recover after the insects have been controlled.

Conspicuous light green or yellow concentric rings on leaves are a sign of **spotted wilt;** plants are not dwarfed. Rogue out affected plants, and control thrips, aphids, and leafhoppers, which serve as vectors of the disease.

# DAMPING-OFF

This disease of seedlings is caused by fungi that live in the soil. It may strike either seedlings in plant beds or those seeded directly in the garden. The tiny plants are attacked by a watery

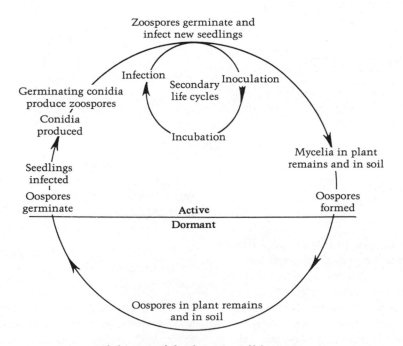

Life history of the damping-off fungus

Reprinted, by permission, from Pyenson, *Fundamentals of Entomology and Plant Pathology* (fig. 9.10), AVI Publishing Co., PO Box 831, Westport, CT 06881.

rot at the soil line or just below. Tissues shrink rapidly, and the seedlings topple over.

The disease is encouraged by humidity so you shouldn't sow seed too close or too deep in the soil. You might add sand to seedbed soil so that it will dry quicker after watering. Do not water more than is necessary, and expose plants to sunlight. Cultivate soil if it is compacted.

Soil can be pasteurized in the oven. Place it in metal or glass pans no more than four inches deep, and cover with aluminum foil. Stick a meat thermometer into the center of the pan, through the foil, and bake for 30 minutes at 180° to 200°F. You can disinfect tools and pots by submerging them for 30 minutes in water heated to at least 160°F. A half-hour bath in diluted household bleach (one part bleach to nine parts water) will have the same effect.

Traditionally, emerging plants have been protected by steeping seed for several hours in brine or lime and wood ash solutions.

# DECOLLATE SNAIL

A predatory snail, commercially available for biological control of the brown garden snail. See SNAIL AND SLUG.

# DELPHINIUM

## Insects

The shiny orange delphinium **aphid** lives on the undersides of leaves and between buds in flower spikes. Heavy infestations cause curling and cupping of leaves and dwarfing of shoots. Flower spikes become stunted, and florets fail to open. Damage is similar to that caused by the cyclamen mite, but is less severe and lacks the blackening symptoms of mite injury. See APHID.

The most important pest of delphinium is the **cyclamen mite,** a microscopic insect that causes distorted, brittle, thickened leaves and black, deformed buds that fail to open. This condition has become known as blacks, but the cause for the malady is not a disease pathogen but an insect.

Once established in a planting, mites are difficult to control. Destroy heavily infested plants. Mites on valuable plants can be destroyed by digging plants and immersing them in water at 110°F. for 15 minutes. Cut out badly infested shoots. Tools and hands are often the means of transportation for the insect—it rarely goes very far under its own power. Keep the garden free from weeds, and remove old plant parts after the season. Burn them on the spot, because digging and pulling may spread the mites. When

Cyclamen mite

planting delphiniums, give them enough room so that they will not be in danger of touching each other. Should mite damage become serious, repeated dusting with rotenone will give effective relief.

Delphiniums are vulnerable to the **red spider mite.** Flower growers have used a garlic spray with good results. See SPIDER MITE.

Roots are often damaged by **slugs** and **snails,** and in some parts of the country plants are susceptible year-round. In Delaware, for example, January sees the worst damage to delphinium crowns. During late fall, before the first frost stiffens the ground, remove all soil above the plant crowns. Then cover the entire plant and fill the excavation up to ground level with coarse river sand. The sand is sharp and painful to the tender bodies of the molluscs, making an effective barrier. When compost and pulverized rock fertilizers are added to the soil in spring, the sand will merge with the soil and improve drainage. Seedlings can be protected by placing the flats on inverted flower pots that are surrounded with ashes or sand. See SNAIL AND SLUG.

Other common pests are black blister beetle (see ASTER), cutworm (TOMATO), fourlined plant bug (CHRYSANTHEMUM), flower thrips (PEONY, ROSE), MILLIPEDE, and potato leaf-hopper (POTATO).

## Diseases

**Bacterial blight** causes an irregular black discoloration and softening of leaves and stem which may extend nearly to the ground. When the stem is split, masses of bacteria ooze out and black internal streaking is visible. The bacteria appear to dissolve the cell membranes and the affected tissues, and give off a foul odor. The plants may be stunted or killed back severely. The causal bacteria appear to overwinter in the soil on dead leaves and in the crowns of infected plants. In spring, the bacteria are splashed by rain to new leaves, where they produce the characteristic black spots. Starting on lower leaves, the disease will quickly spread upward during periods of wet weather until it finally affects the topmost leaves and the lower spike.

To reduce chances of trouble, clear away and burn all plant refuse in spring before new shoots emerge. In fall, cut off and burn dead plants in order to eliminate them as a source of infection for the next year's garden. If the disease

is a real problem, obtain plants that are free from leaf spot and plant them in soil which has not grown delphiniums for many years.

**Bacterial leaf spot** is a common disease which causes the appearance of irregular, tarry black spots on leaves and sometimes on stems

Bacterial leaf spot

and buds. When infection hits young leaves, they frequently become distorted. This disease is carried both in the seed and in the soil, but is usually lethal only under very moist conditions. Excessive watering or irrigation intensifies the disease. Cracks in the stem at ground level also provide a means of entry for the bacteria.

If leaves and flowers are badly deformed or if leaves are brittle and thickened with black or brown spots or streaks, suspect a condition known as **blacks,** brought on by cyclamen mites (see above).

A number of different fungi can bring about **crown rot** (or root rot) of delphiniums. The first sign of the disease is a discoloration of the lower leaves, which is quickly followed by a wilting of the young shoots. Within a few days the entire plant is dead. Roots of infected plants are black or dark brown, rotten and covered with white threads of the causal fungus. In wet weather, small cream-colored, egg-shaped bodies known as sclerotia are found on the crown and roots of infected plants and in the adjacent soil. These sclerotia are the durable seedlike bodies of the crown rot organism that carry the fungus through winter. It's not necessary to figure out which fungus is responsible, as these diseases occur infrequently and control measures are the same: remove and destroy diseased plant parts.

When **mildew** strikes, plants become small and deformed, and stems and leaves are covered with a mealy, white coating. Generally, the flower yield is low, buds are stunted, and the plant wilts. A plant that has been attacked by mildew should be cut down to ground level to prevent the disease from becoming established. Resistant varieties are available.

**Stunt** (or greens) is caused by the aster yellows virus. Plants may be dwarfed, with small leaves showing mottling or chlorosis. Flowers are characteristically green, and a bunchy flowerhead is typical. This virus is spread by leafhoppers and has many plant hosts, including aster, ragweed, and chrysanthemum. Rogue out infected plants; controlling the sixspotted leafhopper will contain infections (see ASTER).

# DERRIS

See ROTENONE.

# DIANTHUS

See CARNATION.

# DIATOMACEOUS EARTH

About 30 million years ago, countless one-celled plants called diatoms lived in the sea and built little protective shells out of the silica they extracted from the water. When a diatom died, its tiny shell drifted to the sea floor; over the course of centuries, these shells built up into deposits, sometimes thousands of feet thick. When the ocean waters receded to leave the continents, the shells in these deposits were fossilized and compressed into a soft, chalky rock that came to be called diatomaceous earth. Today, the stuff is prepared for commercial use by quarrying, milling, grinding, screening, and centrifuging. The result is a fine, talclike powder that can be handled safely without gloves and can even be eaten; it is used to filter wine and beer. However, it may cause silicosis if inhaled. Wear a cotton face mask.

The seemingly innocuous powder also will kill insects on contact. Diatomaceous dust is not a poison, but works mechanically. Proper milling cracks apart the diatom skeletons to expose microscopic needles of silica that are razor-sharp daggers at the insect's scale of life. If these daggers pierce an insect's exoskeleton, the unfortunate creature dehydrates and dies. If the dust is eaten by an insect, the daggers interfere with breathing, digestion, and reproduction.

Fortunately, the tough internal and external tissues of warm-blooded animals make them relatively invulnerable to the microscopic edges. Still, you should wear a filter mask while applying diatomaceous earth.

The use of protective dusts is not original with man. Birds and mammals have been taking dust baths, with the same effect, for millions of years. The Chinese picked up the idea over 4000 years ago, but the method was then apparently lost, to be rediscovered in this country in the late 1950s.

Diatomaceous dust also works inside animals, controlling worms and other internal parasites. Dairymen, cattlemen, horse owners, dog owners, and poultry men find that this method is gentle and natural. One interesting benefit of feeding animals the dust is that their manure seems to have a degree of built-in fly control.

Some bugs are more vulnerable to diatomaceous earth than others. It usually means death to the gypsy moth, codling moth, pink boll weevil, lygus bug, twig borer, thrips, mite, earwig, cockroach, slug, adult mosquito, snail, nematode, fly, cornworm, tomato hornworm, and even mildew. Also, on fruit trees the dust can handle aphid, spider mite, oriental fruit moth, codling moth, and twig borer; brown rot can be controlled by dusting a day or two before harvest.

If diatomaceous earth sounds like an easy answer to your problems, it's not. If you don't know the life cycle of the insects on your trees, you might blitz the beneficials that are all-important to getting unpoisoned fruit through the season. Ladybugs, praying mantises, honeybees, and the larvae of parasatic wasps are especially vulnerable. Do not apply dust to blossoms. This insecticide works about as well as the farmer or gardener using it, and can't be regarded as a panacea. One point against the dust is that it performs less effectively in hot, humid weather because the outside moisture balances an insect's internal moisture—and the insect doesn't dry up. Another problem is that a healthy and thorough application is needed to effect good control; unless an insect comes in contact with the tiny particles, it will go on its merry way. On top of this, death comes slowly, and pests may remain active for 12 hours. The dust will adhere best if applied after a light rain or after plants have been misted. Treat the undersides of leaves as well as the tops. It can be applied wet. Mix 1 teaspoon flax soap (available from paint supply stores) to ¼ pound dust and 1 quart warm water. Pour in a sprayer and top with water. Keep the mixture agitated. A viscous preparation of diatomaceous earth and flax soap can be painted on tree trunks to kill larval pests.

You might also sprinkle the dust on the ground surrounding the trunks. To kill grubs, cutworms, and other larval lawn pests, dust four times a year, using 25 pounds per 1500 square feet. (Diatomaceous earth is a mineral, and offers the grass some nutrients.) Seeds can be dusted before planting.

# DISEASES

Plant diseases are hard to diagnose. A particular symptom may be triggered by any number of problems—fungal, bacterial, or viral organisms, weather, pollution, bugs feeding out of sight, bugs so small that they are all but invisible, and deficiencies or overdoses of nutrients. Whereas insect pests can be identified (given patience and a good field guide), diseases elude naming even by experts. On the positive side, pests are vulnerable to countless diseases. Recent research has shown that plants can be inoculated against diseases by making them ever-so-slightly sick, intentionally. See BACTERIA, FUNGI, MICROBIAL CONTROL, and VIRUSES.

The following checklist of leaf symptoms may help you narrow the lists of suspects, if not to pinpoint the problem. Also consult the color photographs in this book.

## LEAF SYMPTOMS

> CHLOROSIS (YELLOWING) OF LEAVES
> *all leaves:* lack of nutrients, extremely bright light, high temperature.
> *youngest leaves:* lack of iron or manganese, not enough light.
> *older leaves:* lack of nitrogen or potassium, soil needs aeration.

*(continued on next page)*

CHLOROSIS (YELLOWING)
OF LEAVES — *Continued*

*leaf edges:* lack of magnesium and potassium, spider mites.

*between leaf veins:* lack of iron or manganese, spider mites, sulfur dioxide in the air.

*round spots:* bacterial or fungal disease.

*irregular spots:* cold water, bacterial, fungal, or viral disease.

*small dots:* spider mites, planthoppers, leafhoppers, thrips.

*mosaic pattern:* cold water, viral disease.

### DEAD AREAS ON LEAVES

*edges or tips:* potassium deficiency, boron excess, fluoride, heat or cold, insufficient water, spider mites, bacterial disease.

*spots or sections:* cold water, leafminer larvae, bacterial or fungal disease, foliar nematodes.

*edge and inner sections:* too much light, cool temperatures, cold water, foliar nematodes, insect damage, air pollution.

### OTHER

*water-soaked* or *greasy appearance:* heat or cold, cold water on foliage, bacterial or fungal disease, foliar nematodes.

# Bacterial diseases

Bacteria are plants, invisibly small ones, that cause trouble when they live in animals and other plants. They lack chlorophyll and do not benefit from photosynthesis. Some bacteria live only on dead animal or plant remains, and are termed saprophytes. Those that cause disease in living organisms are parasites or pathogens. Many kinds of bacteria can live in either role, and thus plant pathogens are often able to overwinter or maintain themselves between successive crops by living saprophytically on plant refuse.

Each bacterial cell is an independent plant. Some even wiggle little hairs, or flagella, to move about. Although bacteria can't propel themselves very far, they are carried to plants in flowing or splashing water or in transported soil. They can enter a plant through wounds or squeeze through the tiny natural openings in the skin, or cuticle. Once inside, bacteria flow with or swim short distances in the tides of sap.

Fortunately for the backyard diagnostician, bacterial diseases are often easily identified by symptoms. *Rots* attack leaves, stems, branches, and tubers; because of a bacterial enzyme that dissolves cell walls, the affected tissue becomes soft, slimy, and malodorous. Soft rots often follow other, less serious, diseases. For example, black rot of cabbage and late blight of potato would be far less serious if it weren't for the subsequent soft rot. Also grouped with rots are necrotic blights and leaf spots, which are areas of dead tissue that sometimes spread to kill leaves.

The second group of bacterial diseases is characterized by *wilting,* caused when pathogens block a plant's vascular system. The blockage may lead to such diseases as black rot of cabbage, ring rot of potatoes, and tomato canker.

The third group, *galls,* results from an overgrowth of the affected plant's cells. Problems begin when the galls interfere with the flow of food and water through the vascular system.

Bacterial diseases are generally favored by wet soil, high humidity, and high temperatures. A plant's nutrition may make it susceptible to disease—for example, too much nitrogen in the soil encourages bacterial wilt pathogens—so be sure your plants have the advantage of gradually available nutrients in the form of compost and mulch. Problems can also be avoided by using disease-free seed and resistant varieties, and by crop rotation. If you notice a few wilted plants in the garden or field, remove and destroy them immediately, as bacteria multiply by cell division and their numbers increase very quickly. There's no saving a plant once it is infected, and susceptible crops should not be returned to the immediate area of infection, as bacteria can lie in the soil for years.

Why hasn't someone come up with a safe, effective spray or dust to fight plant bacteria? Consider the unique power of the enemy: the continual natural selection of millions of little disease organisms makes pathogenicity—the ability to cause disease—possibly the most stable characteristic in nature. In other words, avoid the bacteria rather than cast about for a cure.

## Fungal diseases

Like bacteria, fungi are plants that take their energy entirely from organic matter. Neither has the ability to turn the sun's energy into food, as both are without chlorophyll. If a fungus feeds on live matter, it is classified as parasitic; if the matter is not alive, the fungus is called saprophytic. It is the former that causes you trouble in the garden and orchard; the latter is beneficial, helping to break down wastes in the compost pile.

Fungi that affect plants are known and named for their appearance. *Downy mildew*, also known as false mildew, grows from within a plant and sends branches known as sporangiophores out through the plant's stomata to create pale patches on leaves. True or *powdery mildews* live on the leaf surface. They send hollow tubes into the plant to suck out nutrients. *Rust* fungi are named for the color their pustules impart to leaves. But a rusty area is not necessarily the work of a fungus; weather and spray injury are likely to be the cause. *Leaf spot* fungi cause round, yellow to yellow green spots that darken with time. If you've seen the grayish, downy patches that form on berries that don't

get eaten soon enough, you know what *gray mold* looks like. This fungal disease affects many fruits, flowers, and vegetables. *Soil-inhabiting fungi* cause damping-off, the dread disease of seedlings, and various root rots.

Conventional chemical sprays and dusts are widely used to fight fungal diseases, but the organic grower can safely rely on sanitation, eradication of diseased plants, and the use of resistant varieties.

## Viral diseases

Viruses are neither organisms nor inanimate molecules, and it's not even certain whether they are living or nonliving. In some ways viruses behave like living organisms, and in other ways like nonliving chemicals. They can increase their numbers, but only when in the living cells of the host plant or animal.

Viruses are spread from plant to plant in a number of ways. Bugs often act as vectors, especially aphids, leafhoppers, mealybugs, and whiteflies. Aphids are the worst offenders—the green peach aphid can carry more than 50 different plant viruses. Parasitic plants known as dotters pass on viruses through the graft unions they make with their hosts. You, too, can transmit some viruses just by handling plants or brushing them with infected tools. Occasionally, infested soil may be responsible. And because viruses can invade all parts of a plant, a new plant propagated from infected tubers, bulbs, seeds, or scions will usually be infected, too.

Although virus particles can only be seen with the aid of an electron microscope, there are several characteristic symptoms of viral infection to look out for. Typically, yields are small and of poor quality, and some strains can bring a quick death; spotted wilt and curly top on tomato are examples.

The largest group of viruses includes those that cause a yellow and green mottling of leaves, called *mosaic,* and spotting of leaves. Mottling results from chlorosis (the death of chlorophyll in cells) and it may be displayed on stems and blossoms, as well as leaves. Chlorophyll manufactures food for the plant through photosynthesis, and chlorosis therefore leads to stunting and poor yields.

The second group includes those diseases that cause yellowing (called *yellows*), leaf curling (or *leaf curl*), dwarfing, or excessive branching, but little or no mottling or spotting. Yellows and leaf curl block up a plant's vascular system (analogous to our circulatory system), thereby restricting the flow of water and nutrients. The task of identifying a virus is complicated when plants are infected with two or more viruses concurrently; the symptoms of one may be added to those of the other, or a new symptom may evolve.

Methods of control are directed at eliminating one or more of the conditions that enable viruses to spread. Once a plant has been infected, there is little you can do to restore it to health.

## Nematode diseases

Nematodes are classified as pathogens because, like a disease, these little worms cause a continuous irritation rather than a transitory injury.

Nematodes are tiny parasitic worms that either stick their heads in a plant to suck the sap or actually spend their lives inside the plant. See NEMATODE.

## Environmental diseases

When gardeners behold a sick-looking plant, they tend to jump to the conclusion that a parasitic disease is the agent. But a more probable cause of trouble is a physiological disorder, one that results from a less-than-ideal environment for the plant. Of course, weather causes its share of problems: the vagaries of wind, rain, sunlight, and temperature combine to create such conditions as dieback, blasting, leaf scorch, hollow heart, and sunscald.

All too often, well-intentioned gardeners are their own worst enemies. This is especially true of those who use chemical insecticides, herbicides, and fertilizers, as these extremely concentrated substances are easy to misuse. Organic gardeners have a built-in advantage: unless chemicals have blown or washed in from elsewhere, there is no need for them to worry about such things as bordeaux injury and copper injury.

Growers are increasingly aware of the effects of polluted air and water on their crops. Ozone, soot, sulfur dioxide, chlorine, and escaping piped gas are as detrimental to plant health as they are to ours, and the symptoms are shown graphically on the greenery of our gardens, lawns, and trees. The cures for environmental diseases are often out of the hands of the individual, but you'll find some suggestions for controlling them under AIR POLLUTION and SOIL POLLUTION.

## Resistance to diseases

Plants discourage disease organisms in a number of ways that aren't fully understood. For one, leaves may carry an electrical charge. Wax layers are thought to repel water-borne organisms. Beneficial bacteria and fungi may take up residence on plant surfaces, and fight diseases as they come along with antibiotics. Other factors include thickness and hairiness of skin, and toxic substances within the plant.

Cells of attacked tissue may either become thicker or die promptly to prevent spread of the disease. And then there are the plants that can just sit there and take it. This is known as tolerance. A tolerant plant supports the enemy without serious damage.

A list of many resistant varieties is given under RESISTANT VARIETIES.

# DOGWOOD

## Insects

The most serious pest is the **dogwood borer.** Very young trees are frequently killed, and older ones are left reduced in vitality and with dead and dying branches. Borer-infested trees begin to show swollen, knotty, callused or gall-like areas on the trunk, just at or immediately below the surface of the ground, or between the level of the soil and the branches above. Injury may also occur at the union of the trunk and principal branches. Young dogwood trees are attacked mostly at the crown.

The larva is white with a pale brown head. The adult dogwood borer is a clear-winged moth with a one-inch wingspan, colored blue black with some yellow markings. There is only one

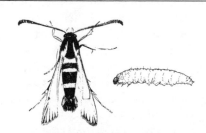

Dogwood borer adult and larva

generation a year. The adults begin to emerge late in May and continue to do so throughout the remainder of the spring and summer months. They may appear as late as the end of September. The eggs are deposited on both smooth and rough bark, frequently near an injury. The larvae wander around aimlessly until an opening in the bark is reached, and then tunnel in. An infested tree four inches in diameter may be killed by a single borer in the course of a season, although under most conditions several borers would be required to kill a dogwood.

Once these borers are in trees, they are difficult to control, and prevention is a far more successful approach. Wrap the tree, especially if a new transplant, in kraft paper, and keep it wrapped for two years or until well established. Any entrances or wounds on the tree should be coated with tree paint. Maintain the vigor of the tree by fertilizing, watering, mulching, and pruning out deadwood. Trees planted more than 300 yards away from an established infestation should not become seriously affected.

**Dogwood club gall** is a grayish tubular swelling that contains a larva. The growths occur on the small twigs, often killing them back for several inches. The orange larva overwinters on the soil under the trees, protected by the sod and decayed grass and leaves. The adult is a small fly, or club gall midge, that emerges in May or June to lay eggs at the base of leaves.

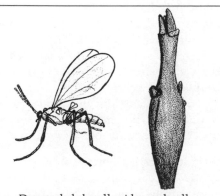

Dogwood club gall midge and gall

Upon hatching, the maggots work their way into the twigs and cause galls to develop. Usually the fastest growing and most vigorous twigs are heavily infested, as are water sprouts or sucker growth nearest the ground. Most of the flower and leaf buds that develop beyond the apex of the galls will die. While serious infestation of dogwood club gall stunts the tree, a light infestation hardly affects development.

Cut off and destroy the galls soon after they have formed, before the maggots leave them in late summer.

Trunks and limbs heavily infested with the **dogwood scale** have a whitish, scaly appearance. The female scales are grayish, pear-shaped, and about $1/10$ inch long; the males are narrow and pure white. Eggs overwinter under the female's protective shell. Infestations weaken trees and may kill encrusted branches.

To control, apply a dormant-oil spray in spring before new growth starts. When the trees begin to leaf out, the young scales hatch and can be killed by brushing them with turpentine or, in serious cases, applying two or more doses of a nicotine sulfate and soap solution, spaced ten days to two weeks apart.

The larvae of the **dogwood twig borer** infest the terminal twigs of dogwood. Eggs are laid in June and July, hatch in about ten days, and the larvae tunnel downward in the twig, making several openings through the bark. The twig is girdled above and below the place where the eggs are deposited. There is apparently one generation each year.

Cutting and burning the infested twigs will help control this insect. Clip off the tips several inches below the girdle soon after wilting occurs.

The **flatheaded borer** deposits its eggs on the bark of dogwoods, and the larvae make irregular tunnels under the bark. When feeding is completed they bore into the wood to pupate.

Larvae are white, have a flat head, and measure about an inch long when fully grown. The $1/2$-inch-long adult is a wedge-shaped, metallic-sheened beetle. Control as for the dogwood borer. Since these beetles attack seriously weakened trees, the best preventive is to keep the trees growing vigorously.

Other insect pests of dogwood include cottony maple scale (see MAPLE) and the oyster-shell and San Jose scales (LILAC).

# Diseases

A generally unhealthy appearance of the tree may be the first sign of infection by **bleeding canker** fungus. Look for sunken areas in the bark, from which may ooze a reddish brown fluid. The bark eventually sloughs off, revealing wood with blue black or reddish brown streaks. Twigs and branches on the side of the tree above the canker die first, but the entire tree will die if the canker has girdled the trunk. Cut out infected wood back to live, normally colored sapwood, and seal the wound with tree paint.

**Crown** or **trunk canker** leads to trees with an unhealthy appearance. The leaves are smaller and lighter green than normal and turn red prematurely in late summer. Later, twigs and even large branches die. Cankers are found on the lower trunk near soil level at spots where the fungus has invaded bark, cambium, and outer sapwood to cause a discoloration of infected tissues. The cankers enlarge slowly for several years until they extend completely around the base of the trunk or root collar, killing the tree. For a few years before their death, infected trees often bear large flower and fruit crops.

Since the causal fungus appears to enter only through injured tissue, avoid wounding the trunk during transplanting and when mowing. In case of an injury, trim the bark around the wound, shaping it to a point at the top and the bottom. Paint the edge of the wound at once with orange shellac and then cover the injured area with a wound dressing. If a small canker is found, cut out the infected area and all diseased bark. The edges should be painted with orange shellac and then covered with a wound dressing. If the canker is very large, destroy the tree and do not replant any dogwoods in that spot for several years.

**Twig blight** may be confused with the more extensive dieback of shrubs affected by crown canker. Control is usually not too difficult. Prune out the dead twigs until you reach sound wood. Improve the vigor of the tree by fertilizing and watering more heavily than usual.

Other diseases of dogwood include leaf spot and spot anthracnose; see TREES.

# DOUGLAS FIR

**Needle cast** is a fungal disease that infects developing needles. Mottled yellow spots appear, but may go unnoticed. In spring, the spots change to reddish brown, and elongated brownish orange fruiting bodies of the fungus develop in May or June. By July, spores from the fruiting bodies infect new needles and old needles fall. Trees that have been infected over a period of years have a thin appearance with usually only one year's needles at the tips of branches. It's best to locate Douglas fir seedbeds away from older fir trees. Remove fallen needles.

For other diseases, see EVERGREEN and FIR.

# DUSTS

See SPRAYS AND DUSTS.

# EARWIG

These brown, beetlelike insects are distinguished by a pair of sharp pincers at the tail end [color]. Although the appendages are fierce-

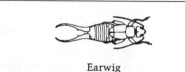

Earwig

looking, they do no harm to humans. Earwigs are usually beneficial in habit, acting as scavengers on decaying matter and predators of insect larvae, snails, and other slow-moving bugs. They are nocturnal feeders, spending the day under bark, stones, and garden trash. At times they may enter homes and feed on stored food, and they occasionally feed on foliage, flowers, and other parts of many plants, including dahlias, zinnias, butterfly bush, hollyhock, lettuce, strawberry, celery, potatoes, and seedling beans and beets. They damage sweet corn by feeding on the stalks.

The earwig is primarily spread by man, in bundles of plants and shrubbery, in cut flowers, and in florists' equipment. Left to its own resources, this insect cannot travel very far or fast. To fly at all it must take off from a high place, as its wings are not strong enough to lift it from the ground.

Earwigs are victimized in some areas by a parasitic tachinid fly brought to this country from Europe; *Bigonicheta spinipennis* is established on the eastern seaboard and in Idaho, Utah, and Washington. Groundhogs and chickens are earwig enemies.

Earwigs fall for traps. Sections of bamboo can be set horizontally through the garden or flower bed; check the traps early each morning and dump them into a bucket of water topped with kerosene. Or try crumbling up paper into an inverted flowerpot that's set up on a stake.

# EGGPLANT

## Insects

Among the insects most troublesome to the eggplant is the **Colorado potato beetle.** Both adults and larvae defoliate plants and are particularly destructive for the home gardener with small plantings. The adult is a yellow beetle with a broad convex back and is about ⅜ inch long. Running the length of both wing covers are five black stripes, and the thorax has black spots. The larval form is a humpbacked, red grub. For control measures, see POTATO.

Several **flea beetles** feed on the leaves of the eggplant. They are small, brown or black and jump like a flea when disturbed. For control, see potato flea beetle under POTATO.

The eggplant **lace bug** [color] is grayish to light brown, flat, and has transparent, lacelike wings. They are a pest to crops in the South as far west as New Mexico and Arizona, feeding in groups on the undersides of leaves, which they cover with brown spots of excrement. The leaves take on pale, mottled patches and plants often die.

If you observe either symptoms or the lace bugs themselves, draw affected leaves between the thumb and forefinger to kill any bugs present. This should be done as early in the season as possible, before a destructive infestation can build up.

Eggplant fruits may be marred by the feeding of **pepper maggots** [color], which at first are translucent white and turn yellow as they feed. The adult is a yellow fly with three brown bands across both wings; it lays eggs in fruit.

Keep the flies away with dusts of talc, diatomaceous earth, or rock phosphate during the egg-laying period in July and August. If an infestation has already occurred, remove and destroy affected fruit.

The **tomato hornworm** and tobacco hornworm eat eggplant leaves and occasionally fruit. They are light or dark green with white stripes, and can grow to four inches long. The former has a black horn at the back end, and the latter's horn is red. Both species are often found in the same garden. See TOMATO.

Other possible culprits include APHID, black blister beetle (BEET), cutworm, (TOMATO), harlequin bug (CABBAGE), SPIDER MITE, stink bug (SNAPDRAGON), WHITEFLY, and yellow woollybear caterpillar (imported cabbageworm under CABBAGE).

## Diseases

**Fruit rot,** also known as phomopsis blight, may cause brown and gray spots in leaves, and large, ringed, tan or brown spots covered with small pustules on fruit. Damage is worst in warm, wet weather. Select rot-resistant varieties. A four-year rotation of crops is helpful.

**Verticillium wilt** may cause leaves to wilt during the heat of day and then recover toward night. These leaves eventually dry up and fall off. The inside of the stem is discolored.

Avoid growing eggplant in soil that has recently grown tomatoes or potatoes, especially if the soil has previously harbored the disease, and rotate crops so that fungi cannot become established in the soil. Be sure to use only clean seed. Recent research with inoculants has shown that wilt can be combated with a soil-borne fungus. A beneficial microorganism has yet to be marketed, however.

# ELM

## Insects

The worst pests of elms are the **bark beetles** that transmit Dutch elm disease (see below).

The **elm borer** and the flatheaded apple-tree borer lay their eggs in bark wounds on elm trunks. The larvae then burrow down to the cambium layer and may cause girdling of the tree. Borers are likely to attack trees weakened by drought, insects, and disease, and a good maintenance program will greatly reduce chances of trouble. Keep elms in a healthy, vigorous growing condition. Trees should be watered and organically fertilized. Repair bark wounds promptly, and remove dead, dying, or injured branches before April 1. Borers can sometimes be dug out of the bark by means of a wire.

The **elm leaf beetle** [color] is a yellowish to olive green insect, about ¼ inch long, with a black stripe along each wing cover. It devours

Elm leaf beetle larva

elm leaves, especially those of Chinese or Siberian elms, reducing the tree's vigor. These beetles may become a household nuisance by migrating indoors in fall; normally, they overwinter in tree bark and outbuildings. In spring they fly to the elm trees and begin feeding on the new leaves, laying their eggs on the undersides. When the ½-inch-long, yellow-and-black larvae

hatch, they feed for several weeks and then drop or fall to the soil at the base of the tree where they pupate. Successive generations continue throughout summer and into fall.

To prevent their entry into the house, caulk cracks and other openings where they might enter. *Beauveria bassiana,* an important naturally occurring fungal pathogen, takes care of many of the beetles. Elms in California have been protected by releases of a tachinid fly and a chalcid wasp, both of which have become established.

**Elm scurfy scale** is found most often on the American elm, although it often kills other types of trees. The white scale is very small and may be found on both the leaves and bark. Scales spend the winter on the tree, and shelter purple eggs.

A dormant-oil solution gives the best control results, as it does with most other scale species that spend the winter on the tree. Apply a 3 percent spray late in winter before the growth of the tree has started.

The **European elm scale** is easily seen on the bark of elm trees during early summer when the scales reach maturity. The scales are oval, not over $\frac{1}{16}$ inch long, reddish purple, and surrounded by a fringe of white, waxy secretion. They are commonly found on the undersides of limbs and branches. They are soft, and leave a reddish stain when crushed with the finger. Infested elm trees display a yellowing of the leaves on lower branches during July. Later in the season, seriously infested branches may become yellowish brown. When scales are extremely abundant, the foliage turns gray green and wilts.

Often it is possible to control the scale by merely washing the trees off with a forceful stream of water from the garden hose. Should a stronger control be needed, apply a mild dormant-oil spray early in spring. See SCALE.

# Diseases

A fungus causes **dothiorella wilt,** or dieback, by infecting trees through wounds to cause a wilting and yellowing. When cankers form, they cause a dieback of infected branches. Diseased bark becomes shrunken and reddish brown with black, raised pustules appearing in the dead areas. A brownish discoloration of the sapwood may be confused with that caused by Dutch elm disease or verticillium wilt.

To control dieback, prune out and burn infected branches, making the cuts at least a foot below any evidence of brown streaking in the wood. Fertilize trees in low vigor so that they will have the strength to ward off the disease.

The greatest enemy of elm trees is **Dutch elm disease,** so named because it was first described by plant pathologists in the Netherlands. It first came to the eastern United States from Europe about 1930, in elm logs intended for veneer, and has since spread rapidly to most parts of this country and Canada. The West Coast was hit in the 1970s. Many millions of dollars have been spent on control measures and on removal of diseased trees, not to mention aesthetic and other losses impossible to measure in terms of dollars.

Diseases are one form of nature's vast variety of checks and balances. The diversity of nature tends toward stability, whereas man is inclined to modify nature and reduce this diversity. The loss of street after street of elm trees is a result of this lack of diversity. Planting and replacement programs should consider interplanting of different species in order to avoid similar tree losses in the future.

Dutch elm disease control is a community problem. The wider the area over which controls are carried out, the better will be the results. Once an elm tree is infected with the disease, it never recovers. Therefore, as soon as infection is discovered, control measures must be promptly applied to keep the disease from spreading in a community. It is imperative to completely destroy all debris from diseased elms. A town in Massachusetts dumped its diseased elms in a river only to have the beetles survive the dunking and infect a community downstream.

The fungus that causes Dutch elm disease is *Ceratocystis ulmi.* Spores of this fungus germinate in the water-conducting tissues, causing the tree to wilt and die. Sometimes the tree is killed within a few weeks, and only a very few live longer than the second or third season following infection. All elms are susceptible, the American elm most of all. Chinese and Siberian elms are highly resistant to Dutch elm disease, as are some hybrid elms advertised as immune or resistant. However, some Asiatic species and hybrids are either not winter-hardy in northern climates or are very susceptible to diseases other than Dutch elm disease.

The danger of infestation is less common for young vigorous trees than for older specimens. Dutch elm disease grows rapidly in the direction of the length of wood fibers, and moves very poorly across the grain of the wood as it constantly encounters thick growth rings. In a young tree where a thick growth ring is formed every year, the fungus is walled off by a layer of healthy wood. But the fungus can easily keep pace with the slow growth rate of larger trees and readily infects them.

The disease may be introduced through the holes made by the native and the European bark beetles as they feed on twigs and branches. The smaller European beetle is about ⅛ to ¼ inch long; it has reddish brown wing covers and a darker thorax. The female lays her eggs under the bark of dead or recently cut elm wood, in galleries that run lengthwise with the grain of the wood. The hatching larvae bore small tunnels around the trunk or branch and away from the centrally located egg gallery. Adult beetles emerge in May or early June, fly to healthy trees, and feed in the crotches of the twigs. It is during this time—if they have emerged from the bark of diseased trees or wood—that they may introduce the fungal spores from their bodies into healthy trees. The fungus fruits abundantly on diseased wood and the sticky spores frequently adhere to the beetles. The habit of feeding on healthy trees after emerging from the bark of diseased elm trees makes this insect a major carrier of Dutch elm disease. It overwinters as a grub in the bark of unhealthy elm trees or of recently cut logs or firewood.

The native elm bark beetle is smaller, brownish, with a moderately stout body and wing covers coarsely punctured with small depressions. The female lays her eggs in galleries that run across the grain of the wood (or around the circumference of the tree).

The most noticeable symptom of Dutch elm disease is the wilting and yellowing of one or more branches of infected trees. There are several wilt diseases that attack elms, all of which produce symptoms that look very much alike, but if you see oval-shaped, depressed feeding holes at the crotches of one- and two-year-old twigs, you can be fairly sure of Dutch elm disease. Internally, a brownish staining appears in the annual rings, showing as discontinuous streaks when the bark is peeled away from the wood, or as small, shiny black or brown dots (or a partial to complete ring) when the branch is viewed in cross section.

The only way to be sure your trees have Dutch elm disease is to take twig samples for laboratory examination by plant pathologists. First, cut six twigs or small stems about 7 inches long and ½ to 1 inch in diameter from the diseased branches of each tree. Carefully mark the twigs or stems from each tree, and wrap securely in a cardboard box for mailing. Do not send material that has been dead for some time or that doesn't show the discolored ring under the bark. Send the samples to the laboratory of your state's agricultural extension service. It is very important that you mail the samples immediately after collecting, as the longer you delay in getting the twigs to the laboratory, the harder it is to identify the disease. Remember that a letter giving tree location, city, county, and date of collection should accompany each sample.

Controls are of four kinds: sanitation, prevention of root grafts, use of insecticides to kill the beetles, and injections of fungicides. Sanitation involves winter pruning of old and dying branches. Promptly remove diseased or dying elms, regardless of the cause of their problem. Infested trees should be removed before May.

Root grafts are likely to interconnect large elms within 30 to 50 feet of each other. Elms infected in this way tend to show first symptoms in their lower branch sprouts. Once a tree is ailing, it is likely too late to save its neighbors. Trees can be isolated by trenching to sever roots. The other alternative involves a chemical soil fumigant, poured into holes dug in a line between healthy and diseased elms.

Insecticides will not render elms safe from Dutch elm disease but can only slow its progress. Trees can be injected with a systemic fungicide. The chemical is injected into small holes drilled in the trunk, and is carried upward to the crown by the flow of sap. If infection enters through root grafts, the treatment is not likely to be effective; also, trees with more than 5 percent of their crown infected are probably beyond help. This fungicide is a chemical treatment, and will be ruled out by some people for this reason. Other drawbacks are its price, the skill required to inject the trees, and the fact that some strains of fungus are resistant. Contact your state extension. Recently, an antagonistic bacterium, *Pseudomonas syringae,* has been used to combat the fungus. Research is under way to find a resistant elm that will do well in North America.

Elms, like all living things, should be kept in vigorous growing condition. Keep them fertilized and watered. Trees have heavy nitrogen needs, so use a high-nitrogen fertilizer such as blood meal or cottonseed meal. Repair bark wounds promptly; remove dead, dying, or injured branches before they attract beetles.

**Phloem necrosis** is a viral disease transmitted by leafhoppers that affects a considerable part of the root system to eventually kill the tree. It is common in the midwestern and south-central states, and seems to affect primarily the American and winged elm. After the tree is infected, it takes a year for symptoms to appear. In midsummer, leaves of infected trees roll and turn yellow, then wither and fall. In later stages of the disease, the thin layer of inner bark that is in contact with the sapwood becomes butterscotch in color and may be flecked with brown or black. A faint wintergreen odor can be detected when the bark is placed in a closed container in a warm place for a few hours. Once the external symptoms appear, the diseased tree will die within a period of from two weeks to two months.

Since diseased trees cannot be cured, protect healthy trees by following rigid sanitation measures. As soon as infection is noticed, diseased trees should be destroyed.

**Slime flux,** or wetwood, afflicts the wood of elms and other hardwoods such as maple, birch, oak, poplar, sycamore, and willow, causing an increase in internal sap pressure. The sap seeps out of infected areas through cracks, wounds, and pruning cuts and flows down the trunk, soaking large areas of bark. The seepage often becomes contaminated with bacteria and yeasts, resulting in a foul-smelling substance called slime flux. The slime flux may prevent healing of the wounds and kill bark and wood over some time. Toxic sap from infected wood is sometimes carried to branches to cause wilting and defoliation, and the tissues between veins and along edges of leaves are browned. Wood under the bark of wilted branches often shows grayish brown streaks similar to those caused by Dutch elm disease.

At the present there is no cure or preventive for wetwood. Drill holes into infected wood to relieve sap pressure and correct the fluxing condition and prevent the spread of toxic sap into the branches. The holes should be drilled 6 to 14 inches below the fluxing area and slanted upward into the heartwood to within a few inches of the other side of the tree. It may be necessary to drill several holes at various locations on the trunk or branches before the infected wood area is located. A short piece of pipe screwed into the hole will carry the dripping sap clear of the tree.

When tapping and drainage of infected wood areas has reduced sap seepage, healing of wounds may be hastened by cutting the bark around the wound to a football-shaped form, pointed at the top and bottom. Shellac exposed bark around the wound, and paint the whole area with a tree dressing.

Other diseases of elm trees include black leaf spot, various twig cankers, verticillium wilt, and wood decay. See TREES.

# ENDIVE

Endive is especially vulnerable to aphid damage; several species are responsible. For control methods, see APHID. Other possible pests are dealt with under LETTUCE.

# ENGLISH IVY

See IVY.

# EUONYMUS

## Insects

Euonymus is chiefly bothered by **scale** insects, including greedy, San Jose, and euonymus species. The euonymus scale is perhaps the most serious pest of euonymus, often killing entire branches. It also affects bittersweet and pachysandra. The female shells are gray and pear-shaped, and the male shells are smaller, narrower, and whiter. There are two broods each season, and the winter is passed in a nearly mature condition. Eggs are formed during May and hatch later in the month.

To control scale, cut and burn all infested and injured branches, and apply a dormant-oil spray in late winter or early spring, before growth starts. See SCALE.

Euonymus is also attacked by the bean APHID and cottony maple scale (see MAPLE).

## Diseases

A bacterium causes rounded galls with an irregular rough surface on the stems or roots. This condition is known as **crown gall.** The bacteria enter through wounds made during cultivation. *Euonymus fortunei radicans, E. f. vegetus,* and *E. patens* are susceptible to this disease.

To avoid crown gall, do not plant euonymus having galls on the roots or stems. Avoid wounding stems or roots in cultivating. If the disease appears, prune and destroy affected parts. Badly affected plants should be removed and destroyed.

A powdery appearance of the leaves means that one of the **powdery mildew** fungi is present. Blast plants with a forceful water spray.

# EVERGREEN

Many evergreens are troubled by the same pests. **Aphids** are among the most universal. See APHID.

**Bagworms** appear frequently on arborvitae and junipers. These insects pass the winter in the egg stage within baglike cocoons on trees. A fully developed bag is about two inches long.

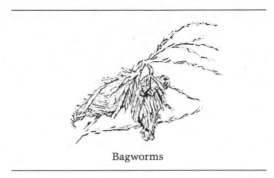

Bagworms

To prevent bagworms from chewing up the evergreens next year, you should pick the bags during winter and burn them. Burpee sells a pheromone trap designed specifically to control this pest.

If evergreens appear to be dying, and fine wood dust or shredded wood particles are forced from holes in the bark, **bark beetles** or wood borers have probably struck. Since the insects usually attack dead or dying trees, you should try to keep plants in a healthy, thrifty condition and prevent mechanical injury. If the tree is dead already, cut and burn it so that it does not provide a home for other insects.

**Spider mites** infest evergreen and cause severe damage. The mites are small creatures, scarcely visible, that suck sap from the needles, weaken the tree, and cause needles to turn yellow and die. Older and lower branches are usually attacked first, but eventually the entire tree

Spider mite

and narrower. A white-oil spray will suffocate this pest.

The **Florida fern caterpillar** is 1½ inches long and starts life colored a pale green that

Florida fern caterpillar

may become infested. The mites' webbing appears around the base of the needles, and the branches take on a dusty or dingy look.

A good way to check for spider mites is to hold a sheet of white paper under a branch and rap the branch sharply. If mites are present, they will fall onto the paper where they are seen as tiny reddish specks moving about. Wash the tree or shrub with a forceful spray from the garden hose. Squirt the water up through the center of the tree. Inspect frequently through the hot months for an infestation. See SPIDER MITE.

# FALSE HELLEBORE

A botanical insecticide. See SPRAYS AND DUSTS.

# FERN

The fern **aphid** [color] is a tiny black speck that often infests the undersides of Boston fern fronds, but it seldom causes much harm. Spray with soapy water; if the plants are in small pots, just dip them in the mixture.

The armored **fern scale** infests Boston fern in greenhouses and dwellings. The female scale is about 1/12 inch in length, pear-shaped, and light brown in color. The male is smaller, white,

turns to velvety black as it matures. It hides by day along the midrib or in the fern crown, and feeds at night. The chief hosts are nephrolepis and adiantum ferns. The caterpillar eats young leaves and strips leaflets from old growth. Although a tropical species, widely distributed in Florida, it has been introduced into northern greenhouses on infested plants. These caterpillars are quite visible, making handpicking a practical means of control.

Other pests include flower thrips (see PEONY, ROSE), MEALYBUG, and WHITEFLY.

# FERTILIZERS AND MULCHES

Here are descriptions of some organic soil-builders—what they are, where you can find them, and how to use them. Unless otherwise noted, these fertilizers can be worked into the soil in spring or fall, top-dressed around growing plants, added to the compost heap, or used as mulch.

How can you tell what your soil needs without relying on dying plants to signal the problem? A soil test will tell you just what nutrients you have and have not. You can take some assurance in knowing that nearly all the nitrogen and sulfur, and more than one-third of the phosphorus available for plant use, is supplied by organic matter. Many other nutrients are present in organic matter, too.

## ALUMINUM FOIL.

This unlikely mulching material serves three purposes: it retains moisture in the ground, reflects an extra dose of light onto plants, and repels aphids. When aphids take off to fly, they head directly toward the sky, and when they decide to land again they reverse their direction *away* from the sky. It is believed that the reflection of the sky on aluminum foil confuses the insects into constantly reversing their direction so that they never land on plants. Aluminum is most effective for the first foot or so of space above the ground. Controlling pests on tall roses may require structures to hold the foil well above the ground.

All types of aluminum foil can be used for insect control, but a special paper-backed kind, sold in many lumber yards as insulation material, will do the best job of mulching directly on the ground.

## BLOOD MEAL AND DRIED BLOOD.

Although the plants in your garden aren't likely carnivores, they will benefit from sparing applications of these slaughterhouse by-products. Blood meal and dried blood contain 15 and 12 percent nitrogen, respectively, and 1.3 and 3 percent phosphorus. Blood meal also has 0.7 percent potash. These materials can be used either directly in the ground or added to the compost pile. They should be used sparingly because of high nitrogen content—a sprinkling is enough to stimulate bacterial growth. Both are excellent for breaking down green fibrous matter in the compost pile. An added benefit is that deer will stay clear of gardens and orchards fertilized with blood meal or dried blood.

## BONE MEAL.

Years ago, mountains of buffalo bones were collected from the Great Plains for use as fertilizer; nowadays the main source is the slaughterhouse. Bones aren't all alike, and bone meal differs too—phosphorus and nitrogen content depends mostly on the kind and age of the bone. Raw bone meal has between 2 and 4 percent nitrogen, and 22 to 25 percent phosphoric acid. The fatty materials in raw bone meal somewhat delay its breakdown in the soil, and steamed bone meal (the most commonly used form) is made from green bones that have been boiled or steamed at high pressure to remove the fat. The steamed bones are easily ground into meal and are usually in better condition for mixing with the soil. Steamed meal contains between 1 and 2 percent nitrogen and up to 30 percent phosphorus. It is available in hardware stores, local nurseries, and wherever garden products are sold. You can also buy bone black, or charred bone, which has a nutrient content similar to that of steamed bone meal.

Bone meal is most effective when used with other organic materials. It generally acts more quickly when applied to well-aerated soils, and works to counteract soil acidity. Gardeners report that freshly cut seed potatoes will have little trouble with potato bugs when rolled around in bone meal so that a little adheres to the damp cuts. Others have found that bone meal can serve as an insect repellent; sprinkle handfuls of it in borders and between plants to chase aphid-carrying ants from the lawn, flower beds, and garden.

## COMPOST.

Probably the best organic fertilizer, compost can include almost any of the other materials listed here. See COMPOST.

COTTONSEED MEAL.    This meal is made from cottonseed that has been freed from lints and hulls and then deprived of oils. Its low pH makes it especially valuable for acid-loving crops. Cottonseed meal contains 7 percent nitrogen, and 2 or 3 percent potash. It is available commercially.

GRANITE DUST.    This is another excellent source of potash, containing an average of 3 to 5 percent (up to 11 percent from one Massachusetts quarry). Granite dust is also rich in potassium. Commercially processed, "fortified" granite dusts are now widely sold as complete natural soil conditioners. The added ingredients may include chicken droppings, fish meal, castor pomace, blood meal, greensand, or any of several other fertilizers.

GRASS CLIPPINGS.    Because they're fairly rich in nitrogen, grass clippings are useful as a green manure to be worked into the soil, for adding to compost heaps, or for mulching. Most lawns produce over 1 pound of nitrogen and 2 pounds of potash for every 100 pounds of clippings in the dry state.

GREENSAND.    This interesting mineral fertilizer comes from deposits laid down in what was once the ocean; at one time in short supply, it is now mined in New York, New Jersey, Maryland, and Ohio. Greensand contains more potash than does granite dust—6 or 7 percent, as compared with from 3 to 5 percent. Because it's an undersea deposit, it contains most if not all of the elements occurring in the ocean; some of the major constituents are silica, iron oxide, magnesia, lime, and phosphoric acid.

Greensand does a good job of absorbing and holding water, won't burn plants, and is especially valuable in conditioning sandy and hard soils. It stays put better than rock powders, and you can spread some around plants as a sort of combined mulch and compost.

HERBS.    Any plant with repellent qualities, including herbs, peppers, and members of the allium family, can lend its repellency to a mulch.

LEAF MOLD.    To make this nitrogen-rich (up to 5 percent) fertilizer, shred the leaves if possible and compost them in a container made from wood, stone, or snow fencing. Keep the leaves damp and apply limestone to offset acidity, unless you plan to use the leaf mold around acid-tolerant plants only. Leaf mold from deciduous trees has been found to be somewhat richer in potash and phosphorus than that made from conifers.

LEAVES.    Just plain old leaves can work as a fertilizer; they are an excellent source of humus and mineral material, including calcium, magnesium, nitrogen, phosphorus, and potassium. Leaves, especially acid maple leaves and pine needles, are best used around acid-loving plants such as azaleas, rhododendrons, and hollies. (The leaves of a few trees, the sugar maple for one, are alkaline.) Leaves may be applied directly to the soil, decomposed in a leaf mold container or compost pile, or laid down as mulch. Some small-landowners find that a shredder is a very wise investment, while others make do with a power lawn mower. A mulch of oak leaves serves to control cabbage maggots on cabbage, radishes, and turnips.

LIMESTONE.    Several common plant diseases —including potato scab, beet rot, and tobacco, tomato, and cotton wilts—thrive in acid soil and are consequently reduced by the alkalinizing action of limestone. The microorganisms that produce penicillin, streptomycin, and other soil antibiotics, which in turn kill or make harmless the microbes causing these diseases, must have calcium and magnesium to do their work. Liming of acid soils supplies these nutrients, and therefore increases production of the antibiotic-producing microbes. Besides working to increase the fertility of the soil, crushed limestone can be built up around tree trunks to act as a shield against the tiny, sharp teeth of pine and meadow mice. Snails and slugs won't cross a line of lime.

**MANURE, DRIED.**   Manure is widely available in this form. Pulverized manure of sheep, goats, and cattle has about 1 to 2 percent nitrogen (N), 1 to 2 percent phosphorus (P), and 2 to 3 percent potassium (K), while poultry manure contains 5 N, 2 to 3 P, and 1 to 2 K.

**MANURE, FRESH.**   This has been a basic fertilizer for many centuries. Horse, hen, sheep, and rabbit manures are considered hot manures because of their relatively high nitrogen content; rabbit manure, for example, contains 2.4 N, 1.4 P, and 0.6 K. Because of all that nitrogen, it's best to allow these manures to decompose into compost before applying them directly to plants. Cow and hog manures are relatively low in nitrogen, and are called cold manures.

Animal manures make things grow in the garden because they stimulate and release a lot of energy in the soil. They have been called the keystone of the compost program. If you don't have livestock on your place, you may be able to get manure from a local dairy farm, egg farm, or riding stable.

Manure is just a bit too rich for some applications. For example, fire blight, the dread disease of pears, is encouraged by manure mulch or liquid fertilizers that would cause rapid or late growth.

**MULCH.**   A layer of mulch can moderate soil temperatures in all seasons, conserve moisture, provide a physical barrier to walking and crawling pests, add nutrients to the soil, produce antibiotic agents, and keep down weeds. See MULCH.

**NEWSPAPER AND MAGAZINE PAPER.** Newspaper serves as both a soil-builder and an animal repellent. It is mostly composed of wood cellulose, a natural product of the forests. Repeated use on the garden will tend to make the soil acid, and you should keep a record of soil acidity and apply lime at suitable intervals. Added nitrogen in the form of blood meal assists decomposition of newsprint, and blood meal serves as an animal repellent in its own right.

**PEAT MOSS.**   Though it doesn't contain any nutrients, peat moss serves to aerate the soil and improve drainage, and ultimately helps plants absorb nutrients from other materials. It is made up of the partially decomposed remains of plants that have accumulated over the centuries in relatively airless conditions. Established lawns can be top dressed with a ½-inch layer of peat moss twice a year, and an inch or more can be spread and worked into vegetable gardens and flower beds.

**PHOSPHATE ROCK.**   This is the commercial term for a rock containing one or more phosphate minerals of sufficient grade and suitable composition for use as fertilizer. It is often recommended as an antidote to onion rot and other underground diseases. Calcium phosphate is the most common of these minerals; the term also includes phosphatized limestones, sandstones, shales, and igneous rocks. Phosphate rock ranges from 28 to 39 percent phosphoric acid. Colloidal phosphate is a finely divided type of rock phosphate with an application rate about 50 percent higher than that of rock phosphate.

**SAWDUST.** This mulch material is often applied in a one-inch layer around plants that are about two inches high. Prior to spreading the sawdust, many gardeners side dress with a nitrogen fertilizer such as cottonseed meal, bone meal, or tankage. You should use well-rotted sawdust only, as raw, pale-colored sawdust will mat and cake, thereby preventing proper penetration of rainwater.

A sawdust mulch can keep down the weeds in your vegetable garden, too. In most instances, a one-inch layer applied when vegetable plants are three to four inches tall will be enough to eliminate most of the need for cultivation, besides conserving soil moisture. Sawdust also can be used to a five- to six-inch band over rows after seeding to prevent soil crusting, but cultivation will be needed to control weeds between rows. A good supply of organic matter in your soil, or some nitrogen-rich materials like cottonseed or soybean meal, dried blood, or bone meal, will prevent the nitrogen shortage which sometimes develops while the sawdust mulch decomposes.

Sawdust is especially useful as a mulch for strawberries because of its acidic nature. By holding each plant's leaves and berry-producing stems in one hand, you can work the mulch in and around the plant with the other. This layer also serves to hold the growing berries off the ground, and you'll find that the harvest is cleaner and damaged little by slugs.

**SEAWEED AND KELP.** Many seaweed users apply it fresh from the sea, while others prefer to wash the salt off first. In either case, seaweed is high in potash and trace elements. It can be used as a mulch, worked directly into the soil, or placed in the compost heap. Dehydrated seaweed and liquid preparations are available commercially. Some orchardists swear by seaweed sprays for pest control.

**SLUDGE.** Sludge is processed in two ways. Activated sludge is produced by agitating sewage with air bubbles. Bacterial action coagulates the organic matter, which settles out to leave a clear liquid. The nitrogen content of the product is between 5 and 6 percent, with from 3 to 6 percent phosphorus. Its value as plant food is similar to that of cottonseed meal, a highly recommended fertilizer.

Digested sludge settles out of sewage that is not agitated and has about the same fertilizer value as barnyard manure—2 percent of both nitrogen and phosphorus. It often has an offensive odor that persists for some time if it is put on the ground in cool weather. This odor may be avoided by storing sludge in a heap during warm weather.

Sludge brings excellent results when applied to lawns, trees, ornamentals, and flower beds. It's a first-rate compost ingredient, as well. (Note, however, that sludge may contain heavy metals—poisons that can be taken up by vegetables grown with it.) To get some for use on your soil, call up your city hall or local sewage treatment plant. Many plants will deliver truckloads at a nominal charge, and you might even consider renting a truck and doing your own hauling.

**TANKAGE.** Meat tankage consists of waste processed into a meal which contains 10 percent nitrogen and up to 3 percent phosphorus. Bone tankage has the same ratings.

**WOOD CHIPS.** In addition to doing a fine job of aerating soil and increasing its moisture-holding ability, wood chips have a higher nutrient content than sawdust.

A number of other soil-builders deserve your consideration: hoof and horn meal, dry fish scraps, feathers, brewery and cannery wastes, castor pomace, hay and straw, tobacco

FIG **151**

stems, lobster shells, molasses residue, and last but certainly not least, garbage. Mixing mature dried crop residues with nematode-infested soil will reduce plant damage. In one test certain crop residues, such as oat straw, clover, tomato stalks, and lespedeza hay, gave better control than sawdust and peat moss. A 95 percent reduction in number of galls per plant was obtained when oat straw was added at a rate of ten tons per acre.

# FIG

Figs enjoy a greater range than many people would expect, growing north into New Jersey, but in colder climates they must be planted either in sheltered places or on southern slopes with good air drainage.

## Insects

One rather prevalent pest is the **longtailed mealybug** [color]. This insect infests a good number of plants, including avocados. It has two long strands that extend from the rear, accounting for its name. These pests are usually brought into the garden by ants, and the control of ants will probably eliminate mealybug trouble.

Ants can be kept off fig trees by banding the trunk with cotton and roofing paper covered with Tanglefoot or Stikem. You can also control them by sprinkling bone meal on lawns, flower beds, and in the garden. Two chalcid wasps, one from Australia and the other from Brazil, were successfully imported in the 1930s. Ladybugs are also effective. For further controls, see MEALYBUG and PINEAPPLE.

Another serious pest is the **root-knot** (or garden) **nematode.** This minute worm penetrates the fibrous roots and causes small knots or swellings to develop. Root growth and functioning is impaired. Severely affected roots usually die, and infested trees are usually stunted or weakened. See NEMATODE.

Also look for black scale, citrus mealybug, and navel orangeworm under CITRUS FRUIT.

## Diseases

**Fig mosaic** shows as chlorotic mottling and severe leaf distortion. Light or rusty spots occur on fruit, which is sometimes misformed and drops from the tree. The fig mite is a vector of this virus. The Mission variety is especially susceptible.

**Fig rust** is a fungal disease that attacks young fig trees without injuring mature leaf tissue. Leaves affected by fig rust fall prematurely, and affected trees are more susceptible to cold injury than are healthy trees.

You can recognize fig rust by the small yellowish green spots that appear on leaves. These spots enlarge and turn yellowish brown and the leaves often become distorted. Rake up and burn all the old leaves.

**Smut** of figs is not a true smut but a mold. It is caused by a strain of the common black mold that grows on many kinds of spoiled fruits. The fungus produces spores in a black smutlike mass. Some of the spores are introduced into the healthy fig fruits through the eyes by the same insects that transport the yeast that causes souring (see below). The control measures for smut, like those for souring, are based on sanitation and control of the insects by destroying the waste fruit in which they breed.

**Souring** is used to designate spoilage of fresh figs that is accompanied by a sour odor or taste or by a dripping from the eye. Souring is caused by the action of yeast and bacteria that are carried into the fig by insects such as the fig wasp, thrips, mite, and driedfruit beetle. The condition is less of a problem on scattered trees than in fig orchards located in the vicinity of melon patches, tomato fields, mulberry, or other fruit trees in which the driedfruit beetle finds abundant breeding material. Mission, Celeste, and Kadota figs are somewhat resistant to this form of spoilage, as the eye is small and not easily penetrated by beetles. Destroy decaying fruit and vegetables near the fig trees.

# FILBERT

Immature filbert **aphids** are of a light green color, while mature or winged forms are darker. Heavy infestations may damage the nut crop. The harm caused by aphids is cumulative, and the benefits of control measures may therefore only become evident after two or more years of aphid control. Aphids overwinter as small, dark-colored, oval eggs attached to twigs and small branches of filbert trees. The eggs hatch in March, and the aphids move to the unfolding leaves to feed. For control, see APHID.

The **filbert bud mite** is microscopic in size but can be easily recognized by the damage it causes. The buds swell into an oversized, puffed shape, and deformed red flowers fail to produce fruit. See SPIDER MITE.

The larvae of the **filbert leafroller** feed on the leaves and roll them for protection. The most serious damage occurs to young fruit buds, which are often severely damaged or cut off entirely. Heavy infestations can cause a serious reduction in nut production.

Adult filbert leafroller moths have a wingspread of about ¾ inch and are usually buff colored with darker, irregular markings on the wings. The mature larvae are approximately ¾ inch long, light green to dark green, and have a dark head. This insect overwinters in the egg stage. Eggs are laid in silvery masses looking somewhat like overlapping fish scales. Each egg mass contains about 50 eggs, and they are found on limbs and trunks of filbert trees. The eggs hatch in spring, usually in early April. Mature larvae pupate within rolled leaves and emerge as moths during the latter part of June or in July. Control as for the obliquebanded leafroller, discussed under ASH.

The adult **filbertworm** moth has a wingspread of about ½ inch and is marked by two golden bands across each forewing. The larvae feed within the nuts. Filbertworms overwinter in silken cocoons found in the soil, leaves, and debris on the ground. Adult moths usually begin to emerge in early July, and may continue to appear through late August and early September. Soon after emergence, eggs are laid singly on leaves, usually on the upper surfaces, and hatch in about eight or nine days. The newly hatched larva moves about until it finds a nut to enter and devour. To avoid tangling with this pest, harvest nuts as early as possible and immediately dry them.

The larval **omnivorous leaftier** occasionally causes damage to young filberts by feeding on buds and leaves. When fully grown, larvae are about ½ inch in length and grayish yellow, with two light stripes and a darker central stripe on the back. Damage occurs during late April and May. For control, see the obliquebanded leafroller under ASH.

# FIR

## Insects

Firs are susceptible to various **aphids**, including the balsam twig aphid, which feeds on shoots of fir and causes the needles to bend upward. For control, see APHID.

Vigorous trees may be attacked by the **balsam bark beetle,** which causes a bleeding of the sap from the trunk and an eventual reddening of the needles and dieback of the upper parts of the tree. To control, prune and burn the affected parts.

Fir trees are preferred over spruce by the **spruce budworm** [color]. The young caterpillars feed upon needles of the new growth on terminal shoots and those of the preceding season, usually webbing the needles together and eating them off at the base. The webs holding the severed leaves and bud scales give the trees a sickly appearance, and in fact weaken the trees to such an extent that they become the prey of bark beetles and other secondary destroyers. At maturity the caterpillars are about ¾ inch in length, dark brown, and bear cream-colored tubercles. The adult moths have a wingspread of about ¾ inch and are brown, marked with gray spots. They are most abundant in June and July and the females lay clusters of pale green, flat eggs upon the needles. These eggs hatch in ten days and the caterpillars hibernate in a partially grown condition. There is one generation a year.

If you are planting a great many seedlings, plan on stands that are mixed with a reduced percentage of balsam fir to cut chances of infestation. Should trouble occur, cut off and destroy the infested tips. If the spruce budworm has been a problem, encourage birds by providing bird houses and alternate sources of food. Other natural enemies of the budworm have proven their value: wasp predators, red squirrels, spiders and spider mites (both very susceptible to pesticides and slow on the rebound), and the parasitic trichogramma wasp. *Bacillus thuringiensis,* when suspended in water and sprayed, effected an 80 percent mortality of budworm larvae in tests in eastern Canadian forests. Eight grams of spore dust were used per gallon, and full effects were reached in eight days. Because the preparation washes off trees with rain, it should be renewed from time to time. It is less virulent in cool weather, and sunlight reduces its effectiveness.

The best method of harvesting is to cut the oldest stands first, as these are most susceptible.

For control of the dark green or black, pink-legged **spruce spider mite,** see SPIDER MITE.

## Diseases

Among the diseases that afflict firs is **gray mold blight.** In cool, wet seasons it causes the new growth to curl, wither, and die. The grayish mycelium and fruiting bodies appear if the cool, wet weather continues.

It is best to improve ventilation to keep down the dampness that encourages this disease. Avoid crowded planting sites.

There are several **needle cast** diseases of fir trees. Needles affected by these fungi turn brown the spring following infection, and may persist on the tree. The infective spores are usually discharged about July. Prune and burn badly infected branches.

The orange yellow or white stages of **needle rust** appear on the upper or lower surfaces of the needles. The alternate hosts include blueberries, willows, and several ferns. The usual way to control needle rust is to remove alternate host plants in the area.

**Twig blight** causes needles of new growth to yellow and twigs to die. Prune and burn affected parts.

**Yellow witches' broom** is a stem rust infection that causes dwarfing and yellowing of needles beyond the point of attack, and dwarfing and browning of twigs. The alternate hosts are chickweeds. This fungus may kill saplings if the trunk is affected. To control the disease, prune and burn all infected parts, and eliminate all chickweed from the immediate area.

# FIRETHORN

## Insects

The **webworm**, prevalent in the southwestern United States, is a major pest of the firethorn. The full-grown webworm is a yellow green caterpillar, about one inch long, that is covered with numerous black spots along its back and sides. A chewing insect, the webworm eats holes in the leaves and stems of plants. It is usually protected by a thin silken web that it spins about the food plant, webbing together both leaves and twigs.

Keep down weeds and other wild plants along the margins of the garden to prevent injury. Should damage from the webworm become too extensive, consider investing in trichogramma eggs. An effective spray can be made from fresh cow manure and clay, diluted in water sufficiently to pass through a spray nozzle. If cows don't live in your area, dried cow manure is available from most supply outlets. Webworms are vulnerable to *Bacillus thuringiensis*.

## Diseases

**Fire blight** bacteria cause a sudden wilting, dying, and browning or blackening of new shoots. Leaves on these shoots die, hang downward, and cling to the blighted twigs. The variety *Pyracantha koidsumii* is susceptible and *P. crenulata* is moderately susceptible. *P. coccinea lalandii* is relatively resistant.

Destroy any nearby diseased and neglected pear, quince, and apple trees, since they may harbor the fire blight organism. Between November and March, cut off affected branches at least three inches below the affected area. Before each cut, disinfect the pruning saw or shears with alcohol.

# FLOWERING CRAB

See HAWTHORN.

# FLOWERS

Of the several types of plants dealt with in this book, flowering ornamentals are least likely to cause you trouble. Perhaps this is because bugs and diseases, like humans, find flowers less appetizing than vegetables, fruits, and berries. Or perhaps flowers as a group have come up with ways of dealing with their problems that other domesticated plants lack. But the best explanation probably lies in the way we use flowers. We look at them, but rarely eat them, and so we are less sensitive to their flaws. There is still beauty in a bug-tattered rose, but a worm-eaten tomato does little for the appetite. In other words, anyone but a florist can tolerate more damage on flowers than on plants raised for food, and this makes organic flower protection a relatively easy matter.

GARDEN SANITATION. Preventive maintenance goes a long way in the flower garden. Keep the immediate area free of plant debris—in most cases, you can put the stuff to work in the compost pile. Many growers clear the garden of dead stalks at the season's end. This makes good sense for several reasons: the material is more useful if composted, rather than left to sit as a mulch; there will be fewer places for pests to overwinter; and many hollow plant stalks serve as subterranean turnpikes that give bugs and disease pathogens easy access to the rootstocks and bulbs below. If yanking a stalk leaves a hole in the soil, just sprinkle a little dirt to seal it up.

It's best to rogue out diseased or infested plants as soon as they're spotted. While this may leave an unsightly gap in a chorus line of flowers, experience shows that sicklings can quickly pass on their ills to neighbors. However, experience has yet to bring growers to a consensus on what to do with the rejected plants. Should they be burned, or submerged in the hottest spot of the compost pile? As you'll find in the flower entries, the answer depends on the virulence and contagiousness of the particular disease.

### PROPER NUTRITION.

Perennials that show up year after year may suffer from the eroding effect of heavy rains. Keep an eye out for plants on high ground that might be in trouble and, if necessary, come to the rescue with an occasional top dressing of garden soil mixed with a bit of compost.

As is true of other types of plants, healthy flowers resist damage. Be sensitive to the particular needs of each flower variety you grow—an overdose of a nutrient can be as harmful as too little.

# FLY

Although gardeners probably would vote flies least likely to help out in the garden, their numbers include many valuable predator and parasite species. The predators outnumber the parasites, but the latter group has proven a more valuable biological control. For fly pests, see plant entries and also WHITEFLY.

### FLOWER AND ROBBER FLIES.

Of the predacious flies, robber and syrphid (or flower) flies are best known. The flower fly hovers, beating its wings so fast that they can hardly be seen and staying in one spot much like a hummingbird. These flies are often brightly colored, and some have beelike bands of yellow and black. Typically they are a bit larger than houseflies and more slender. Adult flies live on pollen and nectar (they serve importantly as pollinators), but the larvae's diet includes aphids, leafhoppers, mealybugs, and scale. Individuals have been clocked at one aphid per minute over extended periods. They hold the pests aloft while draining them. If you find these legless, tan or greenish, sluglike maggots on flowers, hold off with sprays and dusts.

The robber fly is beneficial to growers as both adult and larva. The big, powerful flies attack other winged insects (butterflies, moths, beetles, wasps, bees, and grasshoppers) on the wing; the subterranean larvae eat grubs and grasshopper eggs that they happen to bump into. It is in this immature stage that robber flies are most useful.

You can attract the adults to your garden by providing flowers that bloom throughout the growing season.

### TACHINIDS.

The most important parasite group, the tachinids, have little about their appearance to suggest they aren't houseflies: both groups of flies are drab in color and bristly. But you're apt to see tachinids around foliage and flowers, where they feed on nectar and insect honeydew. Some tachinids have the ability to inject the caterpillar hosts with active maggots. Others place eggs or maggots on leaves.

One particularly important fly, *Lydella stabulans grisescens*, parasitizes from 16 to 75 percent of European corn borers in the Middle Atlantic states, southern New England, and from Ohio through Iowa. Like other tachinids, the female *Lydella* has the ability to hatch eggs within her own body, but this trick can turn against the mother. If she can't find a place to lay her young, they will try to find their own way out and may exit through her eyes.

Other tachinids destroy cutworms, armyworms, the larvae of browntail and gypsy moths, Japanese beetles, Mexican bean beetles, and sawflies.

# FORSYTHIA

Though forsythia is seldom infected by diseases, **dieback** hits flowers and then twigs. Black sclerotia (fungal resting bodies) may develop on the surfaces or inside infected twigs. Control by pruning and burning infected branches and by improving aeration around the bush.

# FROST INJURY

In spring, a late frost may yellow and crinkle leaves. New blooms, shoots, and leaves are killed. The crop may be lost. Tree fruit may show russet frost bands. Look for dead bark on the southwest side of tree trunks. Cool weather may interfere with pollination. Soil in raised beds is warmer than typical garden soil. Fruit trees will be safer if their trunks are painted white on the south side.

# FRUIT TREES

Backyard fruit growers, like commercial orchardists, should first determine what is tolerable damage on fruit and then fashion a suitable pest control program. Learn to distinguish the three types of damage: indirect, on leaves, stems, roots, and trunk, such as that caused by borers, aphids, leafminers, and leafhoppers; cosmetic, on fruit that is damaged but not enough to affect it seriously (plant bugs, curculios in apple, scale, some diseases); and severe, meaning eating quality is reduced or lost (codling moth, birds, oriental fruit moth, fire blight). If you can accept fruit that is slightly blemished but otherwise sound, you may be able to cut down on spraying and other time-consuming controls.

Keep control in mind when you decide what variety to plant on which rootstock and where the tree will spend its fruitful years. Choose a deep, well-drained soil where water doesn't remain for more than a day after rainy periods. Trees planted on a hillside will be free from frost pockets, and benefit from improved air movement. A southeast exposure is best for all fruits (except apricots, which should face north to delay their early-blooming tendency); good sites will reduce risk of winter injury, meaning fewer troubles with peach canker, black rot in apple limbs, and the lesser peachtree borer. In choosing varieties and rootstocks with pest control in mind, consult nursery catalogs, your extension office, nursery workers, and orchardists. The connection between rootstock selection and a crop can be critical. An example is the Siberian C root for peach trees. It is very winter-hardy, as the name implies, but will break dormancy with a week of warm weather. If you live in the far North where warm spells are rare, Siberian C will serve you well; but if your area experiences January thaws, this root will break dormancy as though spring had arrived, with disastrous results when winter resumes the following week.

When planting trees, keep in mind the old adage, "Better a 50-cent tree in a 5-dollar hole than a 5-dollar tree in a 50-cent hole." Run a soil test before planting, as it's easier to correct a deficiency before the tree is growing. Dig the hole large enough to accommodate all the roots without cramping. Don't skimp when mixing peat moss, compost, and well-rotted manure with the native soil in the planting hole; chemical fertilizer may burn the roots. Set the tree slightly deeper than it grew in the nursery, pack the soil well to eliminate air pockets, and water thoroughly. If you leave a slight depression around the trunk for the first few months, the tree may catch some rainwater. To discourage mice and help anchor the young tree, spread gravel extending out 10 to 12 inches, and to a depth of several inches (plan on using a half-bushel per tree). Attach a hardware cloth mouse guard (18 inches high), trim back all sound shoots, and remove any dead or broken shoots. Early spring planting is preferred over fall, unless you live far enough south to avoid the severe freezes which can kill a seedling that has not had long enough to develop a root system.

An overfertilized fruit tree will produce bushels of leaves and dull-flavored fruit that ripens poorly; the tree will also be susceptible to aphids, mites, fire blight (in pears), and winter injury. Let your tree tell you if it's receiving enough by checking the length of the current year's shoot growth (12 to 15 inches for apple; 15 to 18 inches for peach, plum, nectarine, apricot; and 8 to 12 inches for cherry). Soil and leaf analyses will describe a tree's nutrient needs.

Wood ashes are an excellent potash source. A good mulch is old or spoiled hay, laid at least six inches thick in a three-foot band under the drip line. Allow the soil to warm in the spring before mulching. Keep mulch away from the trunk, as such a covering will enable mice and rabbits to sneak up to the trunk and nibble. If mice have been a serious problem in the past, try pulling the mulch out past the drip line in October.

Annual pruning is essential for healthy trees and good-quality fruit. Open spaces in the foliage will allow sunlight and sprays to penetrate and air to circulate. Remove dead or diseased limbs; limbs rubbing against another; upright sucker growth; and root suckers. To avoid winter damage, wait until just before spring bud swell to prune pears and apples. Stone fruits should be pruned after growth begins, to allow the pruning cuts to heal quickly and naturally resist cytospora canker infection; they can be pruned up to several weeks after bloom. Don't leave "shirt hooks"—stubs of branches that die back to the trunk and may serve as an avenue for wood-rotting fungi. Prune close without cutting into the larger limb. Burn all prunings.

So that you won't harvest baskets of egg-size apricots, nectarines, or peaches, you should thin the crop eight weeks after bloom, leaving one fruit every eight inches along a shoot. Though the results are not as dramatic, thinning also works on apples, pears, and plums. Cherries are little affected.

SPRAYING.    Organic growers use a variety of sprays and dusts: plain water, soap, oil, rotenone, ryania, diatomaceous earth, *Bacillus thuringiensis* (BT), liquid seaweed, and sulfur. If microfine, wettable sulfur is not available, you can substitute lime sulfur. Neither form is compatible with oil sprays. Diatomaceous earth will stick better if you add a pound or so of white pastry flour to 100 gallons of mixture. Here is a suggested regimen for fruit trees.

1.  At the delayed dormant stage (when the buds are beginning to swell), apply a superior dormant-oil spray at 2 gallons oil per 100 gallons water. Wet the tree well, and you should take care of most scale, mites, and aphids.

2.  Approximately one week later, apply microfine wettable sulfur, mixed 5 pounds per 100 gallons water; also apply diatomaceous earth, mixed 10 pounds per 100 gallons water.

3. If the weather is rainy and warm, repeat sulfur applications at weekly intervals.

4. When pink shows on the flower buds, apply the sulfur as before. Use liquid seaweed to help control mites and aphids, to feed trees through the foliage, and also to offer some frost protection. If you spot fruitworms, cankerworms, tent caterpillars, or other pest larvae, spray rotenone (3 pounds per 100 gallons water), BT powder (1 pound per 100 gallons water), diatomaceous earth (10 pounds to 100 gallons water), or 6 pounds 50 percent ryania (sold as R-50) per 100 gallons water.

5. At the pink stage (as the blossoms are just beginning to open), no insecticide should be needed.

This is just as well, as bees are starting to pollinate at this time. If you do have pests at the pink stage, use either BT or diatomaceous earth, which are harmless to bees. If the blossom period is a long one, it's best not to let the trees go for more than seven days without a sulfur spraying.

6. At the time of petal fall (when most of the petals are off), apply sulfur for scab, and diatomaceous earth or BT mixed with liquid seaweed (1 gallon to 100 gallons water).

7. Repeat step 6 after 12 days, plus 6 pounds of 50 percent ryania per 100 gallons.

8. After another 12 days, repeat step 7; if scab is not a problem, cut the sulfur rate in half. Omit ryania if monitoring traps detect no plant bugs or curculios.

9. Repeat step 7 if scab or insects are noticed.

10. Use commercial pheromone traps to monitor leafroller activity and apply BT if the weekly trap catch increases dramatically. Monitor codling moth flights with pheromones or molasses bait (see the apple maggot under APPLE) and apply ryania if traps catch more than an average of eight per week.

You can encourage populations of native beneficial insects by planting food sources for the adults, many of which feed exclusively on nectar and pollen. These beneficials may congregate in your orchard if dill, mustard, or buckwheat grow nearby. Meadows of native wildflowers also provide a good supply of nectar and pollen, so hesitate before mowing blooming meadows near your fruit trees. In one study, an unsprayed and uncultivated orchard had 4½ times as much larval parasitism as the average heavily sprayed orchard. One victim of clean cultivation is the braconid wasp, a parasite of the fruit moth; the braconid overwinters in other hosts that include the strawberry leafroller and the ragweed borer, and if you are to benefit from its natural control, you should neither plow under second-year strawberry beds after each harvest nor destroy nearby ragweed, smartweed, lamb's-quarters, or goldenrod. In general, by letting the weeds around fruit trees go, you keep the population of beneficials high.

You can make a substitute food source—a spray of brewer's yeast, sugar, and honey, mixed two to four to one. Add water, and spray foliage. Begin in early June and repeat only once two weeks later. This seemingly harmless concoction has been shown to damage grape leaves if used throughout the growing season. Commercial pollen-and-nectar substitutes are available.

A parasite of the white apple leafhopper overwinters as a pupa near the soil surface and is killed by discing in fall. Apple growers should instead cultivate in June or leave plots of uncultivated ground beneath the trees.

**OTHER REMEDIES.** If you've done everything right, and your trees bloom profusely but won't make any fruit, there are several possible cures.

Check to see that you have another nearby specimen of the problem fruit to provide pollenization. Consult nursery catalogs to make sure that you aren't trying to pollinate one fruit with another that is incompatible (for example, Bartlett pear cannot be pollinated by Seckel pear, Lambert cherry will not pollinate Bing cherry, and neither Stayman nor Rhode Island Greening apples can pollinate anything because their pollen is not viable).

If a tree makes excessive shoot growth each year but bears no fruit, it may be receiving too much nitrogen. You can encourage fruiting simply by cutting back on fertilizer. This is often a problem with trees planted in an old manure-rich livestock area.

## Insects

Several **borers** concentrate on young or recently transplanted fruit trees. The Pacific flatheaded borer is present only in western states; the flatheaded appletree borer (see below) occurs throughout North America but is most common in eastern and central states. Larvae of these beetles bore under the bark surface. A single larva can kill a tree by completely girdling it, but usually several larvae are required to kill a tree. Larvae grow up to ¾ inch, and are grublike with flat heads. Their tunnels are filled with frass, a dry powdery substance consisting of larval droppings and coarse sawdust. Adults are dark bronze with a metallic sheen. They emerge from oval holes (approximately ³⁄₁₆ inch wide) in bark in early summer. Eggs are laid in bark crevices.

Protect young or recently transplanted trees from winter sunburn, as this renders them more susceptible to attack. Wrap burlap or cardboard from the soil up to the lower branches, or apply a coat of diluted white latex paint. Alternatively, a board approximately six inches wide can be placed against the south side of the tree trunk to shade it. Use a sharp knife or moderately stiff wire to dig larvae out if only a few trees are attacked (but be careful not to damage the tree further in the process).

The shothole borer is a little reddish brown to black beetle that attacks fruit trees of any age and many broadleaved woody ornamental plants. It makes many tiny holes ($1/16$ inch in diameter) in bark; they resemble buckshot holes. Clear sap may ooze from the exit holes. The adult beetle is about $1/10$ inch long and first appears in early spring. There are two or three generations per year.

This pest seeks out injured and weakened trees. Prevent sunburn of young trees as for flatheaded borers (above). Keep trees healthy and vigorous. Maintain adequate soil moisture; extra irrigations may be needed in dry years. In winter, prune and burn any limbs that have shothole symptoms.

Problems with the **European red mite** (and other orchard spider mites) have been aggravated by the use of high-powered insecticides. This has become a serious pest of all tree fruits. The mature mites are very small and can best be observed with a magnifying glass. The adult female is dark red with white spots; other stages may vary from red to green in color. Eggs laid in summer are brown and those that pass the winter on bark are red. They are slightly flattened topped with a delicate central hair. There are several generations to contend with in summer, the peak of infestation usually occurring in the hot days of July or early August. Feeding on the chlorophyll causes bronzed foilage that may drop prematurely; fruit is likely to be small and low in sugar; and the bud set for the following year may be reduced if the infestation is early.

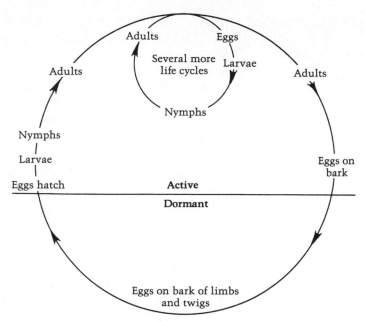

Life history of the European red mite

Reprinted, by permission, from Pyenson, *Fundamentals of Entomology and Plant Pathology*
(fig. 5.5), AVI Publishing Co., PO Box 831, Westport, CT 06881.

Superior dormant-oil sprays can be applied in early spring to kill the overwintering eggs. Many predators aid control in unsprayed trees, including ladybugs, green and brown lacewings, black hunter thrips, pirate bugs, and predatory mites. Mites occupy the unenviable position at the bottom of the food chain and will seldom cause serious damage if you leave natural controls intact. A thorough application of 2 or 3 percent dormant oil, just as the buds begin to swell, will prevent mite explosions for the season. Forcefully apply a soapy water spray during the growing season.

Spider mites can be kept in check more easily if trees are kept watered as needed. Overhead irrigation helps by washing dust off leaves, which in turn discourages mites.

The **fall webworm** [color] forms large silken tents or nests on the ends of branches of several hundred different plants. Eggs are laid in white clusters of 200 to 500 on the undersides of leaves. These hatch in ten days and the larvae from each egg mass leave together as a colony. Unlike tent caterpillars, which occur in the spring and feed outside their tent, the fall webworm is seen in summer and spring and feeds entirely from within the tent. This tent is extended to include more foliage as needed. When nearly full grown, the worms leave the tent and crawl about in search of a place to make their gray cocoons. They select rubbish, fences, crevices in bark, and other secluded sites. One or two generations occur each year. This caterpillar is often bothersome in cycles; several years of large populations will be followed by several years of relative scarcity.

The fully grown caterpillar is about 1¼ inches in length and is marked with a broad, dark stripe along the back and yellowish sides thickly peppered with small blackish dots. Each segment is crossed by a row of tubercules bearing rather long, light brown or whitish hairs. Keep down infestations by cutting off and burning the nests when they first appear, or spray with *Bacillus thuringiensis.*

The **flatheaded appletree borer** [color] is just that, but its appetite is much more catholic than the name indicates: this pest also mines the bark and wood of apricot, ash, mountain ash, beech, boxelder, cherry, chestnut, cotton-oak, currant, dogwood, elm, hickory, horse chestnut, linden, maple, oak, peach, pear, pecan, plum, poplar, prune, sycamore, and willow. Feeding tunnels show through the bark as sunken areas, and it is in these tunnels that the larva passes the winter. Once the borer reaches full size, it forms a cocoon. Tunnels are filled with a dry powdery substance known as frass, made of droppings and sawdust. The adults are dark bronzy beetles with a metallic sheen. They emerge in May and June and rest on the sunny sides of trees, where they lay their eggs in bark crevices. They will cause minor foliage damage with their feeding.

Because the beetles are drawn to sun-warmed trunks, shade the trunks of young trees with any sort of shield. Protect trees that have just been transplanted by wrapping with burlap or cardboard from the soil to the lower branches, or apply a good coat of white exterior latex paint. Keep young trees pruned so that they have a low profile. If you have but a few young trees, you can control borers that have already got a start by digging them out with a sharp knife or a moderately stiff wire. Since the borer most often attacks trees in poor condition, seal up wounds with tree paint and make sure that the orchard has an adequate supply of organic fertilizer.

**Fruit tree leafrollers** [color] are small green caterpillars, usually dark-headed, that damage fruit clusters, leaves, and buds. They roll up leaves with webbing and feed within. Pheromone traps can help you decide if controls are necessary. *Bacillus thuringiensis* has proved to be effective.

Two- or three-inch-long **gypsy moth** larvae ravage the foliage of apple trees. As is true with many other fruit tree pests, the gypsy moths may start life as overwintering eggs on bark, though many blow in from surrounding forests. The gray or brown hairy larvae feed nocturnally through June and July. For control methods, see GYPSY MOTH.

An infrequent but potentially serious pest of tree fruits, the **Japanese beetle** [color] may attack ripening peaches and devour whole fruit in an afternoon. If they're a problem in your

Japanese beetle

area, keep an eye on peaches and nectarines as they near harvest. Handpick the beetles if possible, or apply rotenone.

Milky spore disease will control beetle grubs in the area, but in years of high populations the beetles may fly in and cause damage. Almost as serious as actual feeding damage is the brown rot fungus that grows on damaged fruit. See JAPANESE BEETLE.

**Oystershell scale** [color] is elongate and enlarged at one end. It resembles a miniature oyster shell. Overwintering in the egg stage, the pest hatches in May and crawls about for a time before settling down on the bark. Eggs are laid under the scale, and a thorough scraping of the bark should reduce the population of this pest to an acceptable level. Dormant-oil sprays work to suffocate it.

The **periodical cicada** (mistakenly called a locust) can devastate new plantings with its egg-laying punctures. The "stitching" on small shoots or trunks of young trees will so weaken the shoots that they may break off under the slightest pressure or remain attached for a few years until a fruit load proves too much for the damaged wood. If possible, avoid planting in years when a large brood of cicadas is expected in your area; check with your county extension office. Fortunately, these cicadas appear just once every 17 years.

Various species of **plant bug** are destructive just before, during, and several weeks after bloom. They pierce fruit with their hypodermic mouthparts, and inject a toxic saliva that causes the fruit to distort around the injury as it grows. On apples, damage may be limited to a very acceptable dimple. Pears develop a small eruption where feeding is light and sunken areas where many plant bugs have had a snack. Peaches and nectarines react with severe catfacing damage—large knotty, sunken areas where the flesh is inedible. Monitor for plant bug activity with white sticky cards, or use a large sheet spread under the limbs to catch bugs as you knock the branches. Legumes and plants with large seed heads attract the pests; remove them from the orchard if plant bugs have been a problem.

**San Jose scale** sucks the sap from the wood, leaves, and fruit of apple, pear, peach, and other fruit trees. Serious infestations can lower the vitality of a tree, eventually killing branches

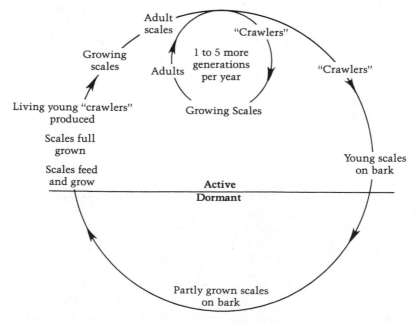

Life history of the San Jose scale

Reprinted, by permission, from Pyenson, *Fundamentals of Entomology and Plant Pathology* (fig. 3.12), AVI Publishing Co., PO Box 831, Westport, CT 06881.

and even the tree itself. The mature female insect is yellow and about the size of a pinhead. It lives under a protective covering that forms over it as it grows. Small, reddish discolorations often may be seen at the point of feeding, particularly on new, tender wood and fruit. Except for the first few hours of its life, and the short time that adult males are active, the scale remains in one place. The tiny, newly emerged young, called crawlers, move around on the tree and are sometimes blown or carried to trees some distance away.

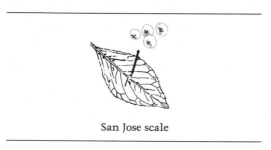

San Jose scale

To control, spray trees early in spring, just as buds begin to swell, with an oil emulsion; repeat one week later for better results. The commercially available parasite *Aphytis melinus* attacks this scale. Osage orange trees support this pest.

The larvae of the **whitemarked tussock moth** [color; see Western tussock moth] are very noticeable pests of fruits and foliage. They reach a length of 1½ inches, are striped lengthwise with brown and yellow, and are hairy, bearing four upright white tufts on the front half, two long sprouts of black hairs near the head, and a similar sprout on the tail. These hairs are irritating to some people. Look for a bright red spot just behind the head. The female is ash gray and has no wings. The male moth has prominent, feathered antennae and gray wings marked by patterns of darker gray.

Fortunately, birds and parasites help out the organic orchardist with this one. You can do your part by scraping away the egg masses that appear in fall. They may be found on bark, empty cocoons, or leaves.

## Diseases

A great many plants host **armillaria root rot.** The states west of the Rockies are most troubled, California in particular. Infected trees show a progressive yellowing and wilting of the foliage and stunted growth. Apricots are among the species that may live but one or two years after planting. Infected bark at the roots and lower trunk becomes moist and spongy. Characteristic whitish fan-shaped growths and dark stringlike fungi are formed in the bark and between the bark and the wood. After fall and winter rains, the mushroom stage may be seen around the bases of rotted trees. These honey-colored mushrooms appear for only a few weeks of the year. They grow to six inches or so in height, and the cap measures from two to four inches across. Avoid planting in an area known to be infected or in recently cleared oak forests.

The fungus *Phytophthora* causes **crown gall** in apple and pear. Most common in poorly drained soils, the disease first appears as an inconspicuous sunken area on the trunk at the ground line. It may expand and eventually encircle the trunk, killing the tree. Select resistant rootstocks. Avoid those areas where water stands for more than a day after a hard rain; the soil is poorly drained and favors the spread of crown gall spores.

## Wildlife

In many areas, the most serious four-legged pest of fruit trees are **deer,** who will spend the winter nipping off buds when other food is scarce, and may feed on tender growing tips in summer.

**Mice** can cause considerable injury to apple trees. Meadow mice feed on the trunks of trees at the ground level and may girdle trees. Pine mice work underground gnawing at roots, and damage is not so easy to spot.

Wire guards can partially protect young trees. Place a cylinder of hardware cloth (three to four wires to the inch), measuring 6 inches in diameter and 18 inches high, around each young tree soon after planting. The cylinder should be set at least 2 inches into the soil, preferably in coarse gravel. Mice are loath to leave protective cover, and it therefore helps to remove all vegetation within a three-foot radius of the trunk. Protective cover can also be reduced by mowing the grass and discing or chopping the sod.

**Rabbits** also like the bark of the young apple trees. Although they will occasionally try plum, peach, or cherry, their favorite is apple. It is important to protect the trunks of young trees until they can bear fruit. Although rabbits rarely feed on tree bark during summer, they can be quite destructive from November through winter, especially after a heavy snow. Mouse guards also foil rabbits.

# FUNGI

Naturally occurring fungi are rated by entomologists as being very important in keeping insect populations in hand, serving as common causes of population crashes during outbreaks of many insect orders. Only modest progress has been made in using fungi to control pests. One fungus is registered for use in the United States: *Hirsutella thompsonii* controls citrus rust mites, and is commercially available.

Fungal controls have been hard to develop because few species can be grown outside of the host insect. A couple of species have been in common use abroad. *Beauveria bassiana* is used against the Colorado potato beetle in the Soviet Union and against the European corn borer in China. In Britain the fungus *Cephalasporium lecanii* is commercially available as a biological control of aphids and the greenhouse whitefly. It soon may be registered for use in the United States.

Chances are good that a fungus is working for you right now, if you are troubled by aphids and if conditions are warm and humid. *Entomorphthora aphidis* makes aphids appear darkened or fuzzy. To spread this disease, you can collect infected aphids and distribute them throughout infestations of healthy pests.

Florida citrus growers enlist the help of a fungus to disable the citrus whitefly. The pathogen, known as red aschersonia, is obtained from infected whitefly nymphs. These sicklings are swollen, secrete more than the normal amount of honeydew, and after death show a fringe of fungal growth that bears red, spore-bearing pustules. The pathogen is prepared by inoculating sterilized slices of sweet potato. The slices are kept in bottles, and in from 30 to 40 days the cultures are ready and the bottles are sent out to growers, or kept in storage if not immediately needed. Growers simply add water, shake the container, and filter its contents through cheesecloth to leave a suspension for spraying. One pint proved sufficient for an acre of orchard in the moist period of June and July.

A fungus called white muscardine is considered likely to be the most important natural enemy of the chinch bug, and at one time the disease was propagated for distribution to large areas of the Great Plains. However, the disease is present naturally wherever there are chinch bugs, its effects waxing and waning with the weather, so there is little need for humans to supplement this pathogen. Keep a look out for muscardine—affected bugs are covered with a white, cottony growth. Another naturally occurring fungus that might catch your eye in the garden is *Spicaria rileyi*, a disease that holds down populations of the corn earworm (known also as the tomato fruitworm and cotton bollworm).

Fungi even have been used to combat other fungi. One-quarter of the chestnut trees in France have been guarded against the fungal chestnut blight with a beneficial strain. Some chestnut trees in Appalachia seem to be protected naturally by a combatant strain of fungus. Joseph Kuć of the University of Kentucky has vaccinated plants against fungal diseases by spraying on them a weak solution of fungi. The plants then produce antibodies, much as do inoculated humans.

# GARDENIA

## Insects

The **root-knot nematode** causes swellings or galls on the roots which prevent uptake of water and nutrients. Plants become weak and stunted, and are often yellowed. Young plants usually do not show the effect of nematode infestation, but older plants show marked symptoms.

Prevention is the best cure for root-knot. Burn infested plants or remove them to the compost heap. Revitalize the soil with a good helping of organic matter, as microorganisms in rich soil make life unpleasant for these pests. See NEMATODE for more information.

Other insect pests of gardenia are flower thrips (see PEONY, ROSE), fuller rose beetle (ROSE), MEALYBUG, SCALE, and SPIDER MITE.

## Diseases

The cause of **bud drop** is physiological, connected to a change in a plant's environment. Buds discolor and drop before opening. Troubles arise from moving plants, too much heat or dry air, or waterlogging or dryness at the roots. No cure is known, and until one is discovered, surround gardenias with constant high humidity and a temperature close to 70°F.

**Stem cankers** start as sunken brown areas on the stem. The bark then splits, exposing the wood. The leaves of infected plants turn yellowish green, then yellow, and finally drop. Bud drop is increased, and plants generally appear sickly and stunted, dying when the canker girdles the stem.

The fungus gains entry through wounds made by leaf pruning. Avoid wounds during handling, and always use a sterilized rooting medium. Badly infected plants should be rogued. If the cankers are on smaller stems, it may be possible to cut off the stems below the canker. Bathe the cut with alcohol immediately after cutting. Sometimes the plant will gain new roots and progress nicely if soil is heaped over the canker so that new roots can develop above it.

# GARDEN SYMPHILID

See SYMPHYLAN.

# GERANIUM

## Insects

A number of **aphids,** including the green peach and foxglove species, feast on plant juices. See APHID.

At night, **slugs** leave their hiding places under rubbish to eat notches in the tender leaves of greenhouse geraniums. They leave a telltale slimy trail. See SNAIL AND SLUG for control measures.

**Termites** have a widespread reputation as household pests, and their appetite for living plants has earned them a bad name in the garden, too. Worker termites that damage plants are sometimes known as white ants. They are wingless, about $\frac{1}{16}$ inch long, and are found in injured plants and in pieces of wood beneath the soil surface. Termite colonies can be found in old stumps, fence posts, or the structural timbers of buildings. The pests tunnel out the stems of geraniums both indoors and outdoors.

Clean up stumps, roots, and other woody material in contact with the soil. Use redwood or other termite-resistant wood for stakes and posts. Kerosene can be poured on colonies.

Several species of **whitefly** cause leaves to become pale, mottled, or stippled. Plants eventually loose vigor, turn yellow, and die. Leaves often are sticky with honeydew and may be coated with black, sooty mold. The adults are white, wedge-shaped, and fly about like snow-

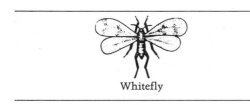

Whitefly

flakes when disturbed. The young are rounded, flat, and look much like scales. They have waxy threads on their spines, and stay motionless on the undersides of leaves. Whiteflies live outdoors in warmer parts of the country, and in greenhouses in the North. For control, see WHITEFLY.

Other pests to keep an eye out for include the celery leaftier (see CELERY) and MEALYBUG (both greenhouse pests), corn earworm (injures geraniums in fall in the North; CHRYSANTHEMUM), and cyclamen mite (occasionally curls leaves; AZALEA, DELPHINIUM).

# Diseases

**Bacterial leaf spot** and stem rot may seriously affect both greenhouse and garden geraniums. Bacteria invade the vascular tissue of cuttings or plants to cause a variety of symptoms. Leaf spots vary from scattered, minute, translucent dots to brown, papery, wedge-shaped areas. Leaves may wilt and drop, and branches turn black, collapse, and rot. The bacteria are very infectious, and plants can pick up infection from splashing water, contaminated soil, or handling.

Use only thoroughly cleaned and disinfected cutting tools and pots. Propagate only those plants known to be free of leaf spot. Lightly infected plants should be kept well spaced; with careful watering of the soil so that leaves and stems don't get wet, the plants may survive. However, heavily infected plants must be rogued.

**Botrytis blossom blight** and leaf spot describe a fungal disease that causes petals of entire florets to turn dark and wilt prematurely. If the fungus persists, a gray mold appears on infected tissues and falls on healthy leaves to spread the infection. Leaf spots are brown and irregular, becoming dry, wrinkled, and surrounded with yellow tissue.

In the greenhouse this disease is easily controlled by lowering the humidity and raising the temperature. Trouble usually occurs in late winter or early spring when a sharp drop in temperature at sundown causes moisture to condense on the plants. Turn on the heat a half-hour before sundown and open vents a bit to prevent this temperature drop and subsequent infection. Avoid wounding, crowding, or mishandling the plant. If the plants are planted outdoors, give them plenty of room. Rogue out infected plants.

The viral infection **crinkle** shows as irregular or circular spots on crinkled or dwarfed younger leaves. Symptoms show only during the spring months and disappear during summer. Even though plants have apparently recovered, cuttings taken from them will show viral symptoms the following spring. Rogue infected plants.

**Edema,** or dropsy, is a physiological problem. Small, watery swellings appear on the leaves, and corky ridges develop on the petioles, stems, and petals. This trouble occurs when overwatering or high humidity accompanies cloudy weather, often in late winter. Plants are improved by withholding water and providing good ventilation and light.

**Mosaic** is a virus that causes a light and dark green mottling of foliage in late winter or early spring. Leaves and plants are usually small. Symptoms are masked during summer, but cuttings taken from these plants will show the disease when temperatures are optimum. Control by roguing and insect control.

# GLADIOLUS

## Insects

The **bulb mite** attacks bulbs of gladiolus and other bulb flowers. It is very small ($\frac{1}{50}$ to $\frac{1}{25}$ inch long), whitish, and is often marked with two brown spots. The mites are reared on onions and potatoes, and later congregate beneath the scales of the gladiolus. Luckily, their feeding doesn't usually sap a plant's strength, but mites may enable decay organisms to get a start. The usual treatment for the bulb mite is to dip the bulbs in hot water for about 20 minutes. Clean up rotting bulbs and decaying plant material. It's interesting, if unessential, to note that under unfavorable conditions the mite goes into a stage in which it develops a hard shell, ceases to eat, and searches about for a free ride to greener pastures. This ride may be aboard a mouse or fly.

**Gladiolus thrips** can cause injury to gladiolus. The flower buds may be injured so that no good blossoms result. The minute insect hides in the sheaths of the flower stem, where it is difficult to reach with a spray. Adult thrips are milky white to dark brown; for a short time, the

Gladiolus thrips

younger thrips are light orange. The insects are very slender, with their lacy wings folded back over the abdomen. They are large enough to be seen crawling about on the palm of the hand.

Since thrips like to hide in a narrow space, you can massage the flower spikes to kill any thrips found beneath the leaves or bracts. However, such a method is of practical value only on a limited scale in a small garden. If home gardeners leave the bulbs in the ground from year to year, they may well be host to a population of gladiolus thrips. No corms should be left in the ground from one year to the next, particularly in an area where neighbors grow gladiolus.

Gladiolus thrips also feed on iris, calendula, lily, dianthus, sweet pea, potato, cucumber, dahlia, aster, delphinium, hollyhock, and white clover; the presence of these host plants nearby complicates control. Generally, trouble with gladiolus thrips is much less serious when bulbs are planted early than late. It helps to dig early in the fall, before the corms are quite mature, and to cut off the tops. The pests will be carried away in the cut portion. As a final measure, burn the decapitated tops. See also THRIPS.

Other prominent insect pests include APHID (hibernates on the stored corms and damages new growth), cutworm (see TOMATO), tarnished plant bug (ASTER), and zebra caterpillar (see imported cabbageworm under CABBAGE).

## Diseases

If you grow gladiolus for their beauty and not for a living, you shouldn't be troubled much by diseases. Select varieties resistant to fusarium yellows. **Dry rot**, recognized by small circles of decay on the lower section of the corm, can be

Dry rot

prevented by using soil with adequate drainage and removing husks before planting. Aphids are vectors of several diseases; see APHID. Infected corms should be burned and not composted.

**Fusarium yellows** is a fungal disease that causes a dark brown, dry rot of the corm (usually near the base) and yellow foliage. Plants may either remain yellow or darken to brown and eventually die. To confirm that yellowing is caused by the fusarium infection, uncover the corm and look for the characteristic dark spots; the corm may appear normal on the outside,

and the only sign of the disease will be a dark core, revealed by cutting the corm in half.

A generous application of nitrogen will aggravate fusarium yellows. Do not use infected corms, and consider varieties that are described as less susceptible. Remove infected plants, along with some of the surrounding soil.

A variety of **viruses** may infect gladiolus, with symptoms depending on the strain of disease and the variety of flower. The damage is not life-threatening to the plant, but affects appearance—even enhancing it in the case of attractive "flower breaks," or patterns of color. Leaves show mottling or streaking. Flowers may not open fully. Plants may be stunted. Red-flowering gladiolus is more susceptible than other colors; so are plants from late-planted corms and older corms.

Specialty growers may be able to provide you with disease-free corms. Remove plants that are seriously affected, and take care that you don't spread the virus to healthy plants by handling them soon thereafter. Aphids spread the virus, and this suggests alternating rows of gladiolus with other plants. See APHID and THRIPS for means of controlling these vectors.

# GLOXINIA

See BEGONIA.

# GOOSEBERRY

Gooseberries are subject to most of the diseases and insects of currants. See CURRANT for full information.

Gooseberry witchbroom **aphids** are yellow green insects that cause the leaves to curl tightly. Control with a spray of soapy water, or rotenone or pyrethrum if infestations are severe.

# GOPHER

See GROUNDHOG.

# GOURDS

Gourds are attacked by the same insects that plague CUCUMBER.

# GRANULOSIS VIRUS

See VIRUSES.

# GRAPE

You won't find every pest of grapes listed below—far from it. Most of the insects that trouble grapes are general feeders, with few classified as specific to grapes. Small plantings and backyard arbors are likely to host a diverse mix of insects, although recent grape plantings may go for several years without insect troubles if vineyards and wild vines aren't nearby.

Good sanitation goes a long way toward preventing disease. Inspect the vines regularly and often, and remove any grapes showing dark spots (indicative of fungus) or red or purple spots (entry holes of insects). Discolored or mined leaves should also be removed. After the fruit has set, prune back excess growth ruthlessly in order to facilitate air circulation and thereby discourage fungus. This pruning prevents a second fruiting, which will not ripen before frost and may keep the first bunches of grapes from reaching their full potential.

While hybrid varieties win favor because they yield the highest-quality fruit, most are less winter hardy than the native American species. Plant on a slope having a southeastern, southwestern, or southern exposure. Such a slope creates excellent air drainage in spring (thereby decreasing the risk of late frosts), tempers the cold winter winds out of the northwest, and provides more early- and late-season sunlight to speed the growth and maturity of the crop. The air circulation on such sites discourages fungal diseases such as downy and powdery mildew. Avoid flat or valley bottom sites.

In the first year, clean cultivation involves keeping the middles well pulverized and free from grass and weeds. You can use a cultivator, disc harrow, or Rototiller, and then follow up several times during the growing season by hoeing around and between the plants.

For a spray program, see FRUIT TREES.

## Insects

Several species of climbing **cutworm**[color] may damage grapes. These pests hide on the ground under weeds and trash by day, and climb the grape trunk on warm nights to eat the primary grape buds. The damage usually takes place in only a short period each year, as the vulnerable buds grow quickly. Serious damage is usually limited to areas of the vineyard in which there are weeds or grasses under the trellis. Destroying weed hosts at this time leaves only grape buds for the cutworms to feed on; a thorough reduction of the weed cover a month before bud swell will greatly reduce cutworm population. See TOMATO for other control measures.

The **grape berry moth** [color] is likely the primary pest of grapes. It occurs in almost all grape-growing areas east of the Rockies, including Ontario's Niagara Peninsula. Growers generally have to deal with two generations a year. Young larvae are white to cream in color at first, and become green and then purple if they feed on deep blue berries. They reach a length of ⅜ inch when fully grown, at which time those of the first generation form a cocoon by folding over a section of leaf and securing it with silk. Second generation larvae typically drop to the ground and form pupae in dead leaves. The adult moth is distinguished by tan forewings with sharply distinguished areas of dark brown.

As the larvae feed, they tie the berries and flowers together with a silken webbing. Red spots show where larvae have entered a berry;

Grape berry worm

later, the spots often turn purple and berries may split as they grow. About the time the berries touch in their clusters, the second brood of larvae is due. Affected green berries of blue varieties will sometimes turn red prematurely. Eggs for the second generation are laid from late July through August.

Because the grape berry moths pupate and spend the winter on the ground, an early cultivation (before mid-May in Pennsylvania) will bury them and halt the first generation of larvae. Pupae fall prey to several species of ant and ground beetle, and the eggs are parasitized by a tiny wasp.

Early in June, a row of tiny holes may be seen around grape shoots. These are chewed by the **grape cane girdler** with the object of placing a single egg in each hole. If the eggs are allowed

to hatch, the larvae feed within the cane and the shoot breaks and falls over, leaving the infested section still on the cane. The adult is a small black snout beetle that overwinters in trash in or around the vineyard.

Damage is not usually serious, as the stems are girdled beyond the grape clusters and there is seldom any loss of fruit. If infested canes are to be pruned, do so before the beetles emerge, in late July or August. The cuts should be slightly below the girdle. Damage is usually confined to vines closest to border or trashy areas. Clean the vineyard of long grass, rocks, and corrugated paper.

**Grape galls** are caused by several species of small adult flies or midges that lay their eggs either on or in leaves, stems, tendrils, and canes. After feeding within the galls, the full-grown larvae emerge and drop to the ground to pupate and emerge as small flies. While galls look like trouble, they rarely cause economic damage. If the larvae are still within (a small hole indicates that they are not), it may be to your advantage to cut the galls from the vine and destroy them.

In the home vineyard, you should be on guard against the **grape leafhopper** [color]. The adults of this insect pest are about ⅛ inch long and lay their eggs on the leaves. When the eggs hatch into nymphs, both adults and nymphs may be found feeding on the undersides of leaves, causing whitish spots that later turn brown. If this injury is extensive, fruit may be seriously affected. There is a tiny microscopic wasp parasite, *Anagrus epos*, which becomes established early in the growing season and can produce three generations in a summer. By the time that the last generation has developed, the leafhopper is all but eliminated from the vineyard. During the early part of the year, the parasites breed on a noneconomic leafhopper that feeds on blackberries. As the family grows, the first brood of leafhoppers beckons the parasites to the grape vines, where *Anagrus* feeds and continues to produce offspring. With the advent of winter, the leafhoppers find a home on the vines or in nearby weeds but the beneficial parasites must overwinter in blackberry vines, which are often a great distance from the vineyards. By the time summer arrives and the parasites find their way back to the vineyards, the leafhoppers are already established on the grapes and have begun to feed. It's good practice to grow blackberries near the grapevines so that the parasites are able to move quickly from their winter home to the vines.

Pyrethrum is effective against the grape leafhopper. Less disruptive to grapevine ecology is Safer Agro-Chem's insecticidal soap spray. You might try a control method used by grape growers of the last century. The vineyardist would saturate a sheet with tar or kerosene and carry it, stretched on a frame, along a row while a partner walked down the row and disturbed the grapevines so that the insects flew into the homemade trap.

Insect-electrocuting light traps, using a clear blue light, will draw significant numbers of grape leafhoppers to their deaths. Keep the area cleaned and raked, especially in autumn, so that the insects will be exposed to the weather. Avoid planting thin-leaved varieties such as Clinton and Delaware, as they are more susceptible to the leafhopper than are varieties whose leaves are more leathery.

The **grape phylloxera** causes leaf galls on the undersides of leaves. These growths are wartlike and about the size of a pea. Heavily infested leaves become distorted and may eventually die. The responsible phylloxera stage is a plump, orange yellow, wingless insect. Five hundred yellowish eggs are laid in the galls. Nymphs leave the galls to feed on new leaves at the shoot tip, and galls result.

While nearly all grape varieties are susceptible to this pest, some varieties fare better than others. Consult your local extension agent.

The adult **grape sawfly** is a small, black, wasplike insect that lays its eggs in clusters on the undersides of terminal grape leaves. The larvae line up side to side and strip leaves from the undersides, leaving only the heaviest of leaf veins. Full-grown larvae drop to the ground to form cocoons and pupate. Infested leaves should be removed and burned.

The iridescent adult **Japanese beetle** [color] feeds on the foliage and sometimes fruit of both wild and cultivated vines. The larvae (grubs) occupy themselves primarily with grass roots. The beetles seem to prefer vines with thin, smooth leaves to those with thick, pubescent leaves.

Since Japanese beetles are very mobile, absolute control is difficult. Your best defense lies in inoculating nearby grassy areas with milky spore disease. See JAPANESE BEETLE. On cool evenings, the beetles turn sluggish and can be shaken into coffee cans containing kerosene and water.

**Nematode** damage is not always easy to spot. Nematodes are slender, wormlike creatures that are often too small to be seen with the naked eye. A single gram of root may host from 2000 to 10,000 pests. They suck plant juices through a sharp mouthpart called a spear or stylet. Some nematode species that prey on grape work from within the roots and others feed externally. While their feeding naturally saps a plant's energy, the feeding wounds can aggravate problems by admitting pathogens. In addition, so-called root-knot nematodes produce swellings or galls on roots, which work to block the flow of water and food to the plant. These nematodes are especially severe in light, coarsely textured soils.

If nematode injury is so subtle, how can you tell what's going on underground? Root galls are concrete evidence that root-knot nematodes are at work, but other species are not as easy to detect. Laboratory examination of both roots and soil is usually necessary to establish what species of nematode, if any, are present. The species must be determined because a number of harmless or beneficial nematodes also reside in the soil. Samples should be collected from several spots within the vineyard, and roots and soil be taken from a depth of at least 8 to 12 inches below the surface. So that the nematodes do not die before reaching the microscope platform, place the soil samples in plastic bags and avoid setting them in the sun. Some states will examine the samples for you; in others you should take the samples to a commercial laboratory.

To prevent nematode problems, use resistant rootstocks and set them in planting material that is free of virus and nematodes. Crop rotation and fallowing of the land will help. Rootings can be purged of the pests by submerging them in water heated to 125°F. for exactly five minutes. Cool the rootings immediately after the treatment and keep them from drying out.

Nematode-resistant rootstocks may be an answer. See NEMATODE.

The ungainly, light brown **rose chafer beetle** measures about ⅜ inch long and looks something like a dull Japanese beetle. It flies in from surrounding grassy areas about the time of grape bloom, especially in vineyards located near areas of sandy soil. Chafers reduce the crop by feeding on grape blossoms or small grape berries, during an active span of only five or ten days out of the season. The larvae are grubs that feed on grass roots, and apparently do not damage the roots of grapevines.

Adult chafers rarely cause significant damage. Milky spore disease will kill off the grubs (see JAPANESE BEETLE). Heavy infestations can be removed from vines with a good shake and collected in a sheet spread below. The chafer is very partial to the Clinton variety, which may be planted as a trap crop to concentrate the beetles for easy disposal.

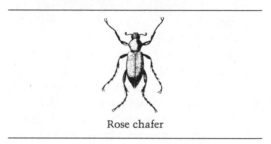

Rose chafer

The **snowy tree cricket** lays its eggs in a series of tiny holes that run the length of a cane for a short distance. The eggs are laid in these holes in early autumn and hatch into nymphs in spring. This stage feeds on other plants in the vineyard, and the only damage caused is by diseases that occasionally enter the site of injury. The adult is much like a cricket in appearance but is more slender, has very long antennae, and is a very pale green in color. Vinelands having overgrown natural ground cover or cover crops are the most susceptible to infestations, as they afford cover as well as food for the feeding nymphs. Control measures are not required if the ground cover is reduced or eliminated.

**Wasps** occasionally beat growers to the grapes. To foil these pests, tie or staple small paper or plastic bags around each cluster. Clear plastic bags aren't necessary, as the sugar stored in grapes is made in the leaves.

A naturally occurring predatory mite, *Amblyseius californicus*, preys on the **Willamette mite.** You may get better control if you allow Johnson grass to grow about the vineyard because this weed supports alternate prey for the beneficial mite. *A. californicus* is available commercially.

# Diseases

**Black rot** [color] is a fungal disease that attacks leaf, blossom, and fruit. About the time the berries are half-grown, light spots may appear on them. These spots enlarge rapidly until the whole berry is affected. The berry becomes a shriveled, hard black mummy bearing small black pimples that contain the disease spores. A few or all of the berries in the bunch may be affected. Grapes thrive best where there is good soil drainage and good ventilation.

Pick off the dried fruit in which the fungus overwinters and remove it to the compost heap. Compost made from diseased fruit can give vines the ability to resist the black rot fungus, as the grape tissue will build up resistance to disease much as the human body does. Varieties vary considerably in their susceptibility to black rot. Catawba, Concord, Niagara, and Duchess are among those most apt to be troubled. See RESISTANT VARIETIES.

Rounded, reddish brown galls on roots are symptoms of **crown gall.** Galls occasionally appear on aboveground stems as hard brown or black growths. When the galls are numerous, growth of the vine is adversely affected.

There is no known cure for this disease, and care should be taken not to injure roots or to plant grapes in soil known to be infected with crown gall. Do not plant infected stock in a vineyard. A friendly strain of bacterium, marketed in the United States as Gall-trol and elsewhere as Agrocin 84, is used to discourage the crown gall bacteria as a preplanting root dip for almond, apple, apricot, blueberry, bushberry, cherry, kiwi nectarine, peach, pear, pecan, plum, prune, walnut, and various ornamentals as well as grapes.

**Dead-arm** is a fungal disease that kills the shoots and branches (or arms) of grape plants. Control depends upon your ability to recognize symptoms so that affected parts can be removed.

Most new shoot infections occur before the shoots are 12 inches long. All parts of the shoots are subject to infection. New lesions are dark. The lesions eventually become elongated, and split to expose the inner fibers of stem tissue. Dark specks of the spore-forming pycnidia can be seen on the infected stems. The rotted tissue is not as dark as black rot and the pycnidia are larger but less numerous. Spores are dispersed early in the season and infections are usually limited to the first six or seven nodes. Symptoms of trunk infection are most easily detected in June. The shoots from infected arms and trunks that are not killed are very much stunted. The leaves are small, crinkled, and yellow—especially at the margins. They cup upward or downward and the margins are ragged and irregular. The fungus enters the trunk and supporting arms through pruning stubs or breaks caused by winter injury, and invades all portions of the wood. A cross section of the trunk reveals a V-shaped section of darkened wood that extends toward the ground line. Affected shoots occur on either or both sides of the vine, depending on the location of the infected trunk tissue. Affected canes may die at any time during the year, but most die during the winter months. If the disease girdles the trunk, all above portions die.

Dead-arm is most effectively controlled by removing infected trunk and arm cankers the first year that symptoms appear. This should be done before June by cutting off all affected parts and removing and destroying them to reduce danger of spread of the disease. When in doubt, cut off the trunks at the ground line.

Although **frost injury** does not come within the scope of this book, it is a danger worth mentioning—many growers find that spring frosts are as harmful as mildew and birds. In advance of heavy frost warnings, hang four or so thicknesses of newspaper over tender shoots or blossoms. The paper can be held in place by stapling. When the danger is past, the paper is easily removed in the morning when still damp. Avoid leaving it on more than a few days, as the foliage will become so tender that strong sun may burn it.

**Mildews** are a big headache for many grape growers. Downy mildew is particularly harmful to European varieties. Powdery mildew seriously affects the berries of French hybrids Rosette and Aurore in the East, and affects the foliage of most other French hybrids. Native American varieties are not very seriously damaged in the East, but the disease is considered a major problem in California. White patches are visible on leaves, fruit, and canes and turn brown or black toward the end of the season.

Commercially, mildew is blitzed with a variety of sprays, ranging from sulfur on powdery mildew to insidious chemicals on the downy species. But beyond being a disease in its own right, mildew serves as a sign that other problems are in need of the grape grower's attention. It has been suggested that the widespread use of sulfur aggravates mildew. Some growers get by with frequent wash-downs of water. Others are careful to avoid overfertilization, on the theory that it grows a vine that is especially susceptible.

Good air circulation will discourage mildew, and pruning is important here.

## Wildlife

It's no news to grape growers that **birds** have a penchant for fruit. Some growers use paper bags, slipped over maturing fruit bunches. The bottom corners are cut off to provide ventilation and let any accumulated moisture drain out. Keep a small pair of pruning shears handy to clip off any tendrils that get in the way. Staple the top of the bags, leaving just enough of an opening to allow ventilation without giving access to birds. As the season advances you should see to it that the bagged bunches are shaded by leafy vines during the hottest part of the afternoon, or else fruit on the sunny side may be cooked.

# GRAPEFRUIT

See CITRUS FRUIT.

# GRASSHOPPER

Grasshoppers are a concern to growers over most of the continent, but are most serious in areas having an annual rainfall of from 10 to 30 inches—particularly the midwestern United States and Canada, including the High Plains and Mountain States. Most troubled are rangeland, wheat, corn, barley, oats, rye, and alfalfa, but there are plenty of grasshoppers left over to attack vegetable crops.

The eggs are vulnerable to cultivation, and the nymphs, which start feeding within a day of hatching, are susceptible to weather, disease, predators, and parasites. Most species deposit their eggs in the ground late in summer and in fall. The common pest species spends six to eight months of the year in the egg stage in the top three or so inches of soil. Cultivation compacts the surface layer, and discourages the newly hatched hoppers from emerging into the world. Working the soil also exposes the egg pods to parasites, predators, weather, and dessication. Fall tillage, right after harvest, is preferable, as it makes the soil unattractive to egg-laying females, as well as stirring up any eggs already deposited. Large-scale growers should either plant such resistant crops as sorghum or

Grasshopper

border susceptible crops with a resistant planting. After reaching a height of eight to ten inches, such varieties as sorgo and kafir are practically immune to grasshopper attack. They can be planted rather late in the season to provide valuable feed for livestock.

Weedy margins, such as roadsides and fence rows, offer ideal places for overwintering grasshoppers. Replace them with perennial grasses, such as crested wheatgrass.

NATURAL ENEMIES.   A good layer of compost or mulch will keep many grasshoppers from surfacing in spring. Those that do make it are game for crows, sparrows, hawks, killdeer, flycatchers, bluebirds, mockingbirds, catbirds, brown thrashers, and meadowlarks—to name a few of the birds—and a wide variety of insect, mite, and nematode enemies. An impressive number of animals, including coyotes, skunks, cats (both domesticated and wild), snakes, toads, and spiders, also contribute their appetites. Guinea hens do a fine job of patrolling for hoppers.

Some enemies of grasshoppers attack the immature and adult stages, while others prey on the eggs. Some attack at one season, some at another. In combination they carry on a destructive action that never ceases. Flesh flies deposit active maggots on grasshoppers even while the latter are in flight. The tiny maggots work their way into the body and feed on its contents, leaving the more vital organs for last. When flesh flies are abundant, they frequently kill large numbers of grasshoppers by midsummer, and infest so many of those remaining that egg laying is greatly reduced.

A protozoan parasite, *Nosema locustae*, is now on the market. It has been used to control grasshoppers on rangeland throughout Montana and Idaho.

Tangle-veined flies are similar in color to honeybees but are slightly smaller. Unlike flesh flies, the adults pay no attention to grasshoppers. Upon emerging from the ground, the egg-laden females fly to the nearest wooden fence post or other upright object and start laying eggs in cracks and holes in the wood. Eggs are laid very rapidly and in great numbers: one female was observed to lay 1000 eggs on a single post in 15 minutes; another female, confined in a pillbox, laid 4700 eggs in seven hours. Eggs hatch in about ten days, and the tiny maggots are blown away by the wind. Just how they find grasshoppers is a mystery, but the large numbers of grasshoppers in which they are found proves that many are successful. When a maggot gets on a grasshopper, it bores into its abdomen, lives on the contents, and thus eventually kills it. Once fully grown, it forces its way out of the grasshopper and goes into the ground, where it changes to a pupa the next spring. The adult fly emerges several weeks later. There is only one generation a year.

Bee flies, blister beetles, and ground beetles lay their eggs in the soil close to grasshopper egg pods, or even in them. Larvae hatching from the eggs work their way into the egg pods and usually consume all the eggs of the pods they enter. These predators have been known to destroy 40 to 60 percent of grasshopper eggs over considerable acreages.

Red mites feed on grasshopper eggs early in spring. They burrow into egg pods and suck the eggs dry. Later in the season larval and adult mites attach themselves to immature and mature grasshoppers of almost any species. Infested grasshoppers may live normal life spans but their egg production is often greatly reduced.

In some states, nematodes are common parasites of grasshoppers. They are long, whitish, extremely slender, and frequently are found coiled within the body cavity of living grasshoppers. Grasshoppers thus infested may live one to three months but are retarded in their development, and the females are rendered sterile. When the worms complete their growth in a grasshopper, they kill it by forcing their way through the body wall. They then enter the ground.

Surprisingly large numbers of immature and adult grasshoppers are trapped by spider webs. Even the largest grasshopper is securely bound with silken strands within a few seconds after it becomes entangled in a web.

Ground squirrels, field mice, and other rodents eat grasshoppers and dig in the ground for their eggs. No figures are available on the percentage of eggs destroyed by rodents, but it must be high, since evidence of their digging can be seen wherever grasshopper eggs are abundant. All birds, except the strictly vegetarian doves and pigeons, feed on these insects. Some eat the eggs after scratching them from the ground. Birds are of value in holding grasshoppers in check when the pests occur in moderate numbers, but they cannot prevent outbreaks.

**CONTROLS.**    Seedlings can be protected by laying cheesecloth over the rows. If your crops are in trouble right now, a hot-pepper spray might come to the rescue. The ingredients are hot-pepper pods, pure soap, and water; the proportions are up to you. It's likely that there are other ingredients with repellent powers, but grasshoppers are not so easily turned away as other insects.

A few well-placed traps will do much to rid the garden of grasshoppers. Half-fill two-quart mason jars with one part molasses in ten parts water, and put several where the grasshoppers are at their worst.

Some traps are designed to catch the insects that hop or jump when disturbed. One trap, known as the hopperdozer, is merely a long, narrow, shallow trough of boards or metal mounted on runners that can be drawn across a field to catch grasshoppers. At the back of the trough a vertical shield, about three feet high, is partly filled with water. Sometimes enough kerosene is added to cover the water with a thin film. The grasshoppers fly up to avoid the hopperdozer, strike the vertical shield, and fall into the kerosene-coated water.

# GROUNDHOG

Growers at times wield guns as a last resort against groundhogs (woodchucks) and gophers. A .22-caliber rifle is a more humane means of dispatching pests than slow-acting poisons. Flushing their burrows with water may make them move out, but they'll likely return.

Gophers will shy away from plantings of scilla bulbs. Scillas (sometimes called squills) are flowering bulb-type ornamentals with grassy leaves and clusters of flowers at the tops of long stems. They grow readily with minimum care, adapt well to borders and rock gardens, and bloom in early spring. It's difficult to protect *all* vulnerable plants with scillas, as gophers nibble everything from vegetables to trees. Traps can be placed in main runways.

A fence will exclude gophers. Use ¼-inch hardware cloth, extending a foot aboveground and two feet below. The rodents can tunnel more easily in loose soils, and this condition may call for a deeper barrier.

Groundhogs are a persistent problem for many gardeners. Anyone who has tried to keep these pests away with a fence soon discovers that the groundhog climbs as well as it burrows. Again, the only sure way to rid your garden patch of this animal is by shooting or trapping.

If you feel compassion for the animals that would share your crops, traps are available for gophers, groundhogs, rabbits, squirrels, raccoons, mice and others. Wire traps come in many sizes, and neither the animal nor curious children can get hurt. Groundhogs are attracted to a bait of apples.

# GUAVA

## Insects

The **broad mite** is a pale pest of foliage, usually to be found on the undersides of leaves. Water sprays should dislodge most of the mites. They are vulnerable to sulfur dust.

The **guava whitefly,** a minute sucking insect that looks like a tiny white moth, attaches its eggs to the undersides of guava leaves. The eggs, ¹⁄₁₀₀ inch long, hatch in a short time and the young suck the sap from leaves. Use oil-

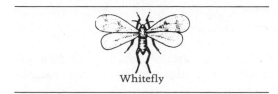

Whitefly

emulsion sprays in fall when foliage begins to harden, or after fruit set in spring. The spray should contain 1¼ percent oil. This spray will also control the numerous species of scale that attack guavas.

**Nematodes** sometimes attack and kill wild plantings of guava. They should not menace any planting fed and mulched with organic matter, however, for a fungus that kills nematodes is present in decomposing organic fertilizer. See NEMATODE.

## Diseases

**Crown** or **root rot** will sometimes destroy a planting that is made on soil containing tree stumps and old roots, particularly if the trees were oak. It is best to plant guavas where the land has been cleared long enough for all of the old roots to have decomposed.

# GYPSY MOTH

In a bad year, voracious gypsy moth larvae [color] can strip entire forests of their foliage. Sanding crews once were called out in Connecticut when roads covered with crushed larvae caused seven accidents in two hours. Deciduous trees can make a comeback, after the caterpillars turn into adults, by putting out another set of leaves; but the tree is weakened, seriously so after consecutive seasons of defoliation. Evergreens cannot grow a second set of needles, and fare worse. They are troubled by older larvae.

The northeastern United States is hardest hit. Susceptible trees include oak, birch, willow, linden, basswood, and apple.

The mature larva is a hairy gray or brown worm, bearing five pairs of blue and six pairs of red spots. Before trying to control the gypsy moth, make certain you've made a correct identification: it often is confused with the eastern tent caterpillar [color] (which makes tree tents in spring) and the fall webworm [color] (which makes tents in late summer). The gypsy moth does not make tents.

The adult gypsy moth eats hardly anything, and the female can't fly. But the larvae are voracious and highly mobile in their first days.

Gypsy moth and larva

They may be blown several miles from the hatching site. This redistribution is one of the pest's prime survival mechanisms. When hatched from the egg mass, the young, tiny caterpillar is drawn to light and crawls to the top of a tree. Once there, it descends on a delicate, silken strand. The strand is broken easily, and the caterpillar is carried away by the wind.

As the larvae mature, they begin to hide in dark places during the day and feed at night. Most will rest under dead leaves and litter within a couple feet of the tree they feed on or under crevices and flaps in the bark.

The pupal phase lasts about two weeks. Of the moths that hatch forth, the females cannot fly and don't travel far, while the males are strong fliers and home in on the pheromone, or sexual attractant, exuded by their prospective mates. Males die after mating. Females lay their eggs in a tan-colored mass; the eggs normally do not hatch until the following spring.

**NATURAL ENEMIES.**    A native ground beetle, *Calosoma frigidum*, eats caterpillars. It is distinguished by its smell—unpleasant both to us and (judging by their reaction) to gypsy moth larvae as well. The worms avoid a human hand that has held this beetle, but will crawl right over a hand that has not. The ground beetle is black with a green shine on its hard back, and is about after dark. A green, bronze-tinged relative, *C. sycophanta*, is known as the fiery hunter. You'll find the large beetle on tree trunks, positioned head down.

Ants are occasional predators. So are spiders and stink bugs. The tachinid fly lays its eggs on leaves, where they are ingested by foraging gypsy moth caterpillars, to hatch within the hosts. Caterpillars are parasitized by several imported wasps. A braconid parasite of larvae, *Apanteles flavicoxis,* has been brought to North America from India and is now commercially available.

Black-billed and yellow-billed cuckoos are among the more prominent bird enemies of the caterpillars. Others are grackles, crows, starlings, redwing blackbirds, cowbirds, chipping sparrows, nuthatches, downy woodpeckers, and scarlet tanagers. After dark, the whitefooted mouse comes out to find food, and its diet may include pupae and caterpillars of the gypsy moth. Other rodent predators are shrews, voles, moles, chipmunks, and squirrels.

Despite this impressive list of enemies, the gypsy moth remains a problem in much of the United States, especially in peak seasons that come every three or four years. Of the naturally occurring predators and parasites that control the pest in its native Europe and Asia, some four dozen have been imported to the United States, but only a quarter of them have become established convincingly.

**NONCHEMICAL SPRAYS.**    If nature can't keep this imported pest in check, then concerned humans have to. Or do we? Some observers suggest that attempts at control—destroying egg masses or spraying—interfere with the cyclical population swings of the gypsy moth in a way that actually can prolong an infestation.

If you don't go along with a passive defense, consider sprays. Nuclear polyhedrosis virus (NPV) has been used by the Forest Service and is marketed as Gypchek. This spray is nontoxic to other insects and animals. Two sprayings are usually required. *Bacillus thuringiensis* (BT), used widely against a variety of pest larvae, can be sprayed effectively to kill larvae from the time they stop hanging from threads (at which time they measure ¾ inch long) until early June (when they have grown to an inch). After this point they are too high in the trees; they shift to night feeding, commuting to and from their daytime shelters in the bark and around the tree. It is difficult to spray BT up into trees with home equipment, and you may want to hire the services of a landscaper or tree service. On a large scale, this bacterium is sprayed from planes.

Pheromone traps have been used to monitor gypsy moth populations. If the population density is low enough—fewer than a thousand moths per square mile—the traps may work also as a control, using one to four traps per acre. The traps may work in another way, if their synthetic pheromone prevents male moths from homing in on females. Pheromone traps are available through garden supply stores.

OTHER CONTROLS.   From August through April, tan egg masses may be found on trunks and branches and under rocks. Paint them with creosote or scape them into a bucket of soapy water; if you just dislodge them so that they fall to the ground, the young worms still may develop. Once the worms are out and about, you can handpick them. Wear gloves when cleaning up eggs, larvae, or adults: some people are allergic to hairs on the eggs and worms and to scales on moth wings.

Once the caterpillars have reached a certain stage in their development, they feed on foliage at night and spend days hidden under bark on the trunk or under litter around the base of their tree. To trap them, attach a burlap strip, about a foot wide, around the trunk by draping it over a string. Many caterpillars will hide under the cloth. Collect them in late afternoon.

A caterpillar barrier, Repel'm, which is now on the market, is wrapped around the tree trunk. It is said to be nontoxic.

# HAWTHORN

## Insects

Several species of **aphid** frequently infest hawthorn. The rosy apple aphid sometimes curls the leaves, and all species secrete honeydew, which supports sooty fungus. See APHID.

Several species of **lace bug** [color] commonly attack the foliage of hawthorn. These tiny sucking insects feed on the undersides of leaves, where they deposit small spots of dark excrement. When numerous, they cause foliage to discolor and drop prematurely.

Control by spraying or dusting the undersides of leaves with nicotine or pyrethrum.

The **thorn limb borer** has been known to attack hawthorns. In the smaller branches and twigs of hawthorn, it causes swellings about an inch long with four or five longitudinal scars. Infested twigs may break off in the wind. The beetle is ½ inch long, brown, and decorated with two white crescent-shaped spots near the middle of the wing covers and two smaller circular spots near the apex. The thorax has a white stripe on each side extending onto the base of the wing covers. There is one annual generation, and the beetles appear in June.

Control by removing and burning the infested twigs. For further control methods, see peach twig borer under PEACH.

Other occasional pests of hawthorn are cankerworms (see APPLE), cottony maple scale (MAPLE), gypsy moth and tent caterpillars (TREES), roundheaded apple borer (APPLE), San Jose scale, and twospotted mite (SPIDER MITE).

## Diseases

**Fire blight** is a bacterial disease that causes a sudden wilting and browning or blackening of new shoots. The leaves on these shoots die, hang downward, and cling to the blighted twigs. The disease rarely becomes serious unless the trees are planted near pears that harbor the disease-causing bacteria.

Avoid overfertilization. Cut out infected branches and twigs between November and early March, pruning at least a foot below the edge of the diseased bark.

During mid-May, leaves may show signs of **leaf blight** and **fruit spot.** Single blossoms or entire flower clusters wilt, die, turn brown, and hang on the twigs for some time before falling. All or part of the leaf may turn brown, with a fruit rot generally following leaf infection. The disease may be confused with bacterial fire blight, although fire blight causes tissues to be more blackened. The causal fungus overwinters on fallen, mummified fruit that has been partly buried in the ground. To control the disease, rake and burn all the mummified fruit.

Varieties resistant to **leaf spot** include Cockspur and Washington.

# HELIOTROPE

The **greenhouse whitefly** causes leaves of plants to become pale, mottled, or stippled. The pests secrete a honeydew which may cause a black, sooty mold to grow on leaves. Plants lack vigor, turn yellow, and die. The adults fly about like snowflakes when disturbed; the young look much like scales, staying motionless on the bottom of leaves. See WHITEFLY.

Other common pests of heliotrope include APHID, celery leaftier (see CELERY), MEALY-BUG, SPIDER MITE, and termite (GERANIUM).

# HELLEBORE

A botanical insecticide. See SPRAYS AND DUSTS.

# HEMLOCK

## Insects

The larva of the destructive **hemlock borer** does damage under the bark of living, injured, and dying hemlock and spruce trees. The adult beetle is about 3/8 inch long, flattened, and dark brown in color with three small whitish spots on both wing covers. The larva is one of the so-called flat-headed borers. They make sinuous, interlacing, flattened galleries in the inner bark and sapwood, thus girdling the tree. The beetles appear from May through July. There is only one brood a year.

Prune affected branches; cut infested trees and burn the bark late in the winter or in early spring. Keep trees in a vigorous condition of growth by applying organic fertilizers and adhering to a well-planned watering program.

**Hemlock scale** is a dark gray, circular species that occasionally infests the undersides of leaves and causes them to turn yellow, giving the tree a sickly appearance. Young trees may be killed. An important parasite is *Aspidiotiphagus citrinus,* responsible for killing most of the scale in some regions. See PINE.

The **hemlock webworm** webs together a few leaves and feeds upon them inside the web. The larva is less than ¼ inch long; some are bright green, others brown. The moth is whitish with pale brown tips and a wingspread of less than ½ inch. Eggs are laid in June, and there is one brood each season. The larvae apparently overwinter in the webs in a nearly full grown state. Prune and burn twigs bearing clusters of these caterpillars.

The **redbanded leafroller** may skeletonize the leaves of hemlock trees, causing them to turn brown and drop in late summer. The larva is green with a brown head, reaching about ¼ inch long when mature. The winter is passed in the pupal stage in trash under the trees, and the moths appear in spring. See control measures for the obliquebanded leafroller under ASH.

Other insects may from time to time bother your hemlocks: pine needle scale (see PINE), root weevils (STRAWBERRY), spruce budworm (FIR), and spruce mite (SPRUCE).

## Diseases

There are several forms of **needle rust.** Hemlock-blueberry rust causes long white tubes that discharge orange spores, which form on the lower sides of needles in early summer. Often only a single needle will be infected. Alternate hosts are members of the blueberry family, such as azaleas and rhododendrons. Control is best effected by removing the alternate host to break the life cycle and eliminate the rust from the environment. Rust does not often bother hemlocks, and it is more damaging to members of the blueberry family than to hemlocks.

When rust pustules form on both the upper and lower sides of needles, the hemlock-poplar form of needle rust has struck; all other needle rusts on hemlock fruit only on the undersides of the needles. Infection also occurs on young twigs and cones, which are distorted as the clustercups form and burst to discharge orange spores.

Control by removing the alternate host, the poplars. Damage is usually slight to hemlocks. If necessary, the infected twigs can be pruned and burned.

**Phomopsis twig blight** causes needles to turn brown and fall from affected twigs. The wood is dead, and small black fruiting bodies of the causal fungus may be found on the twigs. These fungi are usually found on trees growing under unfavorable conditions. Pruning and burning infected twigs removes the source of future trouble.

When **tip blight** affects hemlock, new growth curls and dies and may be covered with a gray mold of the fruiting stage of the fungus, especially during prolonged cool, wet springs. A change in weather conditions will stop the ravages of this fungus, but there is little you can do to control tip blight.

# HIBISCUS

The melon **aphid** causes severe leaf curling, distortion of young leaves, and stunting of growth. The honeydew that these insects secrete supports the growth of sooty fungus on leaf surfaces. Damage is worst in the South; see APHID.

Other pests include the corn earworm (see CHRYSANTHEMUM), fuller rose beetle (ROSE), JAPANESE BEETLE, and WHITEFLY.

# HICKORY

## Insects

The **hickory bark beetle** is an extremely destructive pest of hickory trees. In July the female makes a vertical tunnel about an inch long in the inner bark and sapwood, with a row of pockets along each side. An egg is deposited in each pocket. Upon hatching, the grubs commence to tunnel in a direction at right angles to the parent gallery, and the larval galleries at the ends of the parent gallery are deflected so as not to run into the other galleries. A few such brood galleries may girdle the branch. The adult is a small brown or black beetle, 1/5 inch long, with four short spines on the abdomen. The beetles emerge in June and July through round holes resembling shot holes and eat at the bases of the leaf stems, causing many leaves to turn brown in July. Some drop and others hang upon the trees. There is one generation annually. As this pest does most damage to weak trees, fertilizing and watering may help to prevent infestation. Cut and burn seriously affected trees between fall and spring.

The **hickory gall aphid** forms galls in June on the leaf stems and new shoots. The hollow galls contain young aphids. In July they reach maturity and leave the galls, which then turn black. These globular galls cause much distortion to the shoots. Aphid eggs remain over winter in old galls and in crevices of the bark. If galls are very serious, control with a spray of nicotine sulfate and soap just before the buds swell. Less severe controls are given under APHID.

There are several types of **hickory leaf gall**, most of them caused by midges or two-winged flies. They are seldom destructive enough to require control.

The larva of the **hickory leafroller** is yellowish green and measures about an inch in length. It rolls hickory leaves and feeds on them from inside the rolls. The moth is dark brown with darker oblique bands on the forewings. The hickory leafroller has not been a serious pest and does not require control.

The **hickory tussock moth** [color; see Western tussock moth] feeds upon hickory and foliage of other trees. The full-grown larva is about 1½ inches in length, covered with white hairs, and is decorated with a stripe of black hairs along the back and two narrow tufts of black hairs at each end. The adult moth has a wingspread of about two inches, showing light brown wings marked with oval white spots. Eggs are laid in patches on the undersides of leaves in July. The tussock moth hibernates in gray cocoons fastened to trees, fences, and other objects. There is one annual generation. Naturally occurring parasites help to keep this insect from getting out of hand.

The **painted hickory borer** tunnels under the bark and in the sapwood of hickory. The ¾-inch-long beetles emerge in May and June and are blackish with yellow markings. Eggs are laid in crevices or under the edges of the bark. Larvae become mature in 10 to 12 weeks and then pupate in the wood in September where they remain until spring.

Vigorous healthy trees are seldom injured by the painted hickory borer. Control by cutting and burning infested trees and all slash (logging debris).

Other pests of hickory include APHID, fall webworm, and walnut caterpillar (WALNUT).

## Diseases

Hickory trees are subject to a fungus that causes **cankers** to form around dead branch stubs. Even though stubs appear to be nearly healed, brown fungal threads may be found within. This wood-rotting fungus will eventually spread through the tree.

Clean and scrape the canker and remove all discolored wood. Treat wounds with tree paint.

For anthracnose and witches' broom, see TREES.

# HOLLY

## Insects

The overwintering eggs of the **holly bud moth** begin hatching late in March, and small greenish white to gray green caterpillars appear between the terminal leaves of young holly shoots late in April or early May. About the middle of May they begin to tie terminal leaves together to form compact cases. The caterpillars feed within these shelters, eating back the shoot so that the mass of leaves spun together dies and turns black. These black unsightly objects usually do not drop until pushed off by subsequent growth. At maturity the caterpillars are a little more than ⅜ inch long. When disturbed, they wiggle out of their shelters and drop to the ground. Many of the larvae leave the shelters when fully grown and spin loose cocoons among the dead leaves or rubbish on the ground. Others pupate within the shelters used by the larvae.

Destroy leaf shelters. A light summer-oil spray has been found to reduce the population of the moth. This spray must be applied late in March, before the eggs begin to hatch. Apply to the undersides of leaves. Garden sanitation

will clean up holly bud moths that spin their cocoons in plant refuse.

The **holly leafminer** adult is a tiny two-winged fly, grayish black and about $\frac{1}{10}$ inch long. It makes its initial appearance during late May or early June, and may be seen flying about

Holly leafminer adult and larva

the newly developed holly foliage or walking over the surfaces of the leaves. Approximately three weeks are required for all the flies to appear.

Shortly after their appearance, female flies insert their eggs into the newly developing leaves. The eggs hatch in June, and larvae feed inside the leaves until the following spring. Although the maggots begin feeding in June, leaf injury does not become evident until mid-August, when small, irregular, linear, serpentine ridges appear on the upper surfaces of leaves. Each irregular ridge indicates a maggot mining

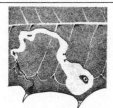
Holly leafminer damage

beneath the surface. At first, the ridges are a darker green than the remainder of the leaf surface. Later they become tinged reddish or reddish brown. By mid-September the mines increase in size, one end becoming blotch-shaped. If infested severely, the entire upper surfaces of the leaves may be covered with coalesced mines, giving the surface a blistered appearance.

Leaves showing the symptoms must be removed and burned. Severely damaged bushes should be removed. It is not a good idea to plant winterberry or inkberry in the same location as the holly because these two plants are alternate hosts for the leafminer.

The **holly scale's** covering is oval, light brown to tan in color, and extremely small, while the scale itself and its eggs are lemon yellow. This scale overwinters in a partially grown condition. Feeding begins in the latter part of March or early April, and eggs are laid in June and July. There is usually only one generation a year. Symptoms include plant devitalization, yellow spotting of the leaves, and smutting of leaves due to fungus growth on secreted honeydew. The scales usually congregate on the lower sides of the leaves.

Try a 3 percent dormant-oil spray in April if the scales have been a problem. See SCALE.

# Diseases

**Leaf rot** fungus is found in contaminated sand and on the boards of the cutting bench. It enters the cuttings through leaves that touch the sand. The disease first appears as a cobweblike growth on the leaves, a combination of the fungus threads and grains of sand sticking to the undersides of leaves that touch the sand. Leaf drop usually begins two to three weeks after the cuttings are set. In some cases, targetlike spots from ¼ to ½ inch in diameter are found on the leaves.

Control of leaf rot may be accomplished by setting the cuttings in clean, fresh, or sterilized sand. If the disease has already been present in the bed, wash the bench thoroughly with a cleaning solution. Avoid taking cuttings from holly branches that touch the ground.

**Leaf spot** is nonparasitic in nature and causes light tan, ¾-inch spots on leaves in early spring or summer. The spots are often covered with tiny black bodies of fungi.

Spots are caused by sunscalding in either winter or summer. In summer they may occur when a sudden rain shower is immediately followed by bright sunlight. The presence of ice on the leaves often results in a similar scalding in the winter. All that is necessary in the way of control is to pick off the affected leaves so that the appearance of the holly is not impaired.

**Spine spot** (or purple spot) is nonparasitic and causes grayish brown spots ranging from the size of a pinhead to a pea. These spots develop on leaves in late winter or early spring, and may appear on either surface. In most cases they are surrounded by a purple halo. The spots are the result of the wounds inflicted by the spines of nearby leaves as they are blown about in heavy winds. In other words, the holly actually wounds itself. Black fruiting bodies of a fungus may occasionally be found near the center of the spots, but these fungi are secondary, since they come after the leaf tissue has been wounded.

Holly should be planted in sheltered locations so that the wind won't whip the leaves. If it is impractical to plant the holly in such a position, or if the bush has already been planted, you can construct simple wind barriers of burlap nailed to stakes set around the bush.

**Tar spot** is a fungal infection of holly leaves characterized by raised black spots resembling tar. The disease may be reduced by picking and burning infected leaves.

**Twig canker** causes brown areas to appear on the green wood of new growth. Eventually these cankers enlarge to girdle the twig. Small black fruiting bodies of the fungus can be seen on the deadwood. Prune and burn infected twigs.

# HORSE CHESTNUT

## Insects

The **whitemarked tussock moth** [color; see Western tussock moth] overwinters in frothy white egg masses on the trees. There are two generations a year. One brood of larvae matures about July 1 and a second brood in August. These caterpillars reach a length of about 1½ inches. They are striped lengthwise with brown and yellow, and are hairy, with four upright white tufts on the front half, two long tufts of black hairs near the head, and a black tuft on the tail. The adult moth is ash gray. The female is wingless and the male has a wingspread of about 1¼ inches.

Because the females cannot fly, control may be achieved by banding tree trunks with Tanglefoot or Stikem as described in TREES. This is effective only if the tree is not already infested and if the branches do not contact other infested trees.

Other pests of horse chestnut trees are the JAPANESE BEETLE and scurfy scale (see ELM).

## Diseases

**Leaf blotch** may infect horse-chestnut leaves with large reddish brown blotches surrounded with yellow tissue. Curling of infected leaves is common, and the fruiting bodies of the fungus appear as black specks in the discolored areas. To cut down on the source of infective spores, rake and burn fallen leaves.

**Nectria cinnabarina** causes roughened cankers on branches that bear reddish brown fruiting bodies. Prune and burn affected branches, cutting well below the infected areas.

For anthracnose and powdery mildew, see TREES.

# HORSERADISH

## Insects

The flat, brilliantly colored **harlequin bug** withdraws plant juices, resulting in foliage discoloration, wilting, and death. They are easy to spot, with bright orange to red markings on black. For control measures, see CABBAGE.

The **horseradish flea beetle** is small, shiny, black, and has yellow wing covers bordered with black. Control is rarely necessary, but some helpful measures can be found under POTATO.

Horseradish is also plagued by several crucifer pests. Under CABBAGE, see descriptions and control methods for cabbage looper, diamondback moth, imported cabbageworm, and webworm.

## Diseases

**Leaf spot** appears as large dark spots on leaves. This disease is not serious enough to warrant control.

White, somewhat circular blisters on horse-radish leaves are likely the first symptom of **white rust.** Later, these blisters break open to expose the white, powdery fungal spores. Control is not necessary.

# HOT WEATHER

See SUNSCALD.

# HYACINTH

## Insects

At least three species of the **lesser bulb fly** make their presence known by injuring rotting bulbs. Seventy-five or more larvae may be found in a single bulb. The flies are $\frac{1}{3}$ inch long and blackish green with white markings on the abdomen; the larvae are a dirty-grayish yellow. Destroy infested or rotting bulbs, and use only healthy bulbs and rhizomes.

Other troublemakers are the bulb mite (see GLADIOLUS), NEMATODE, and yellow woollybear caterpillar (see imported cabbageworm under CABBAGE).

## Diseases

**Gray mold blight** causes flowers to shrivel up suddenly and become covered with mold. This fungus thrives in cool, wet seasons, and a change in the weather will halt its progress. Destroy infected plant parts.

In the case of **soft rot,** plants fail to flower or buds open irregularly, and the plants can be easily pulled out of the ground. Rotted parts give off a foul odor. Trouble is aggravated by repeated freezing and thawing, and by heating bulbs in storage.

**Yellow rot** gives a water-soaked appearance to flowers and causes yellow, longitudinal stripes on leaves. The yellow stripes eventually turn brown and ooze bacteria if cut across. Bulbs show rotted pockets. Since bacteria are spread by splashing water, take care when watering. Plant disease-free bulbs and rogue out infected hyacinths.

# HYDRANGEA

## Insects

The **hydrangea leaftier** webs terminal leaves tightly around buds, and feasts on the buds from within this shelter. The responsible larva is a green, ½-inch-long caterpillar with a dark brown head. A mild garlic spray on buds may make them less appetizing to the leaftier. Once the pest is at work you can squeeze the webbed leaves to destroy it; you should then tear open the web to allow the flower bud to develop.

**Stem nematodes** cause hydrangea stems to become swollen and split. Leaves drop off as a result. Cut potted plants back severely, wash off all old soil from roots, and put the plants in clean soil.

Other insect pests include the rose chafer (see ROSE), SPIDER MITE, and tarnished plant bug (ASTER).

## Diseases

**Bacterial wilt** infects flowers and young leaves. Affected parts will turn brown and wilt, and if the weather is extremely hot the disease may kill the plant.

Remove any infected leaves and flowers as soon as they appear. If a plant becomes badly infected, it must be removed from the garden before other plants become infected. Be certain that the disease is not transferred to other plants by the indiscriminate use of pruning or cultivating tools.

Yellowed leaves are a sign of **chlorosis,** a condition thought to stem from too much lime in the soil. It may also be correlated with very high levels of nitrates and calcium.

**Powdery mildew** appears as a gray mold on the undersides of leaves. Buds and new growth may also be attacked. Prune and burn old leaves.

# ICHNEUMON WASP

See WASP.

# INSECT

Insects compose but one of six branches of the phylum Arthropoda. Technically, sowbugs, spider mites, centipedes, symphylans, and millipedes are not insects, but members of other arthropodan classes. Slugs and snails, earthworms, and nematodes belong to other phyla entirely. For quick identification of pests and beneficials, see the charts on pages 3 and 4.

Insects have three pairs of legs as adults, and their bodies are segmented into head, thorax, and abdomen. Their bodies are tough. While man and the other vertebrates have internal skeletons, insects are held together by a strong exterior skeleton, or exoskeleton. This body wall completely covers the insect and is often protected by secretions of waxes and other substances that exude from tiny pore canals. The outer layer of the wall is made of chitin (pronounced ky-tin) and presents a formidable barrier to the world: it is insoluble in water, alcohol, and diluted acids and bases, and cannot be broken down by the digestive juices of mammals. Snails, some insects, and certain bacteria can successfully attack chitin, however.

Of all the body parts of an insect, the grower is naturally most concerned with the mouth. The design of the mouth largely determines what food the insect will go after—and the source of food determines whether an insect is labeled friend or foe. Mouthparts are of either the chewing or sucking kind. Chewing insects have mandibles, or jaws, that move sideways to bite off and chew food. Sucking insects are equipped with a beak through which liquid food is drawn. Chewing mouthparts are considered to be the more primitive of the two. The thorax bears the agents of locomotion—wings and legs. Pests that crawl or walk can be entangled in sticky barriers, held off by repellent plantings, or handpicked. Plant protection is somewhat more difficult for flying pests; baited traps and fine-mesh netting are typical controls.

Respiration is carried out by an intricate system of minute tubes, or trachea, that connect with openings (spiracles) in the exoskeleton. Circulation is accomplished with a simple tubelike heart. Blood circulates throughout the body cavity and is not confined to veins. It is without the hemoglobin that serves in higher animals to carry oxygen, and is usually colorless or tinted pale green or yellow. Insect blood takes nutrients to the organs it bathes and carries waste products away. The heart does not generate much blood pressure; in fact, blood pressure may be less than atmospheric pressure. New adult insects force blood into their wings by swallowing air to increase the pressure within.

Digestion takes place in a tube, often somewhat coiled, that extends from the mouth to the anus. Food matter is absorbed from this tube (known as the alimentary canal) by the blood, and muscles work to suck in food and move matter through the insect. Some insect diseases disable insects by paralyzing this canal.

The size of an insect's brain is relative to the complexity of its behavior. The ratio of brain volume to total body volume is 1:4200 for a predacious diving beetle, presumably an insect with few tricks, and 1:174 for a bee, an insect known for its complex social behavior.

Insects get by in the world not on strength of intelligence but by instincts and evolution. (Still, they can learn to improve their performance in mazes.) Instincts are automatic responses to stimuli that have survival value; if a behavior pattern works to an insect's advantage, the insect survives and passes the trait on to its progeny. Evolution enables insects to adapt to changes in their environment. Their numbers are so great that individuals with genetic quirks are being born all the time, and some of these react better to new situations than the normal insects. These lucky aberrants survive, and pass the valuable quirks on to their progeny. It is this ability to adapt to changes that accounts for insects developing resistance to the most potent chemical pesticides that humans can devise.

It is as a larva that you know most garden pests. Pupae are usually tucked out of sight, eggs are so small as to be nearly invisible, and many adults cause relatively little damage. Most of the control measures in this book are aimed at the larval stage, many of them taking advantage of the fact that larvae are typically soft-skinned, slow, and vulnerable to disease, parasites, and predators. The soft skin is susceptible to the osmotic workings of salt in water, to the tiny daggers of diatomaceous earth, and to sharp barriers or mulches. Because most larvae are slow, they can be stepped on or picked by hand and dropped into a can of water topped with kerosene. Successful handpicking requires that you observe plants often and well, as it doesn't take long for larvae to become fully grown and go into hiding as pupae. The vulnerability of larvae to diseases and natural enemies has led to much research of late. The most useful and available pathogens for the gardener are *Bacillus thuringiensis,* milky spore disease, and nuclear polyhedrosis viruses. All work on the larval stage. See MICROBIAL CONTROL.

The third stage of complete metamorphosis is the pupa, an outwardly inactive form in which the insect develops into an adult. The pupa needs energy to make these big changes, and it coasts on the food it stored up as a larva. Pupae are typically secreted away in bark crevices, plant refuse, or underground. In an orchard, a thorough scraping of rough, scaly bark will help to take care of several important pest species. Pupae in plant waste can be disposed of by keeping the garden tidy, particularly after harvest. Cultivation is an age-old means of dispatching underground pupae.

The final stage is the adult, typically a highly mobile form. It is important to their survival that they be good walkers and fliers: they can thus both find a mate quicker and spread their kind far and wide. And while some adults eat enough to be considered pests in their own right, those of many species nibble on nectar and pollen or are even without mouthparts and digestive systems.

Adults reproduce in any of three ways: sexually (involving male and female), parthenogenetically (females alone being responsible for reproduction), and a combination of both in the same species.

All told, the metamorphic process helps enable insects to succeed in a very competitive world. Take eggs, for example. Because they are tiny, fantastic numbers of them can be turned out by a single adult—a queen bumblebee is thought to be able to lay more than a million over its life span—and such numbers enhance a mother's ability to pass on life to a new generation. It figures that, even with egg parasites, predators, cannibalism, and adverse weather, a good number of eggs should still be around to become larvae. This larval stage has a specialized mission, existing only to eat, and the pupal stage launches a mobile adult that spends most of its time mating and looking for good places to stash eggs. Indeed, insects can be formidable enemies.

Incomplete (or simple) metamorphosis involves three stages instead of the four discussed above. The egg hatches into a nymph, which often looks like a scaled-down version of the adult. The adult stage is reached in a series of instars, or molts. There is no resting stage before the last molt.

About the only time insects aren't aggressively bettering their species' chances in the world is when the temperature drops and the food disappears. Factors such as shortening days and changing weather trigger their instinctual process of adjusting to winter. Insects winter in various stages, depending on species and climate. Their metabolic processes slow almost to a standstill in a dormant period known as diapause.

# INTEGRATED PEST MANAGEMENT

Integrated pest management (IPM) is a strategy for keeping plant damage within bounds by carefully monitoring crops, predicting trouble before it happens, and then selecting the appropriate controls—biological and cultural, or chemical controls as necessary. IPM is quite a sophisticated concept. It demands that the practitioner study the plants' environment and closely monitor life in the garden, field, or orchard; statical methods and computer models may be used. IPM also demands familiarity with a wide range of controls.

IPM programs are now run by several states; in one state, IPM workers pay house calls.

A good source of information on this fast-developing field is "The IPM Practitioner," a newsletter published at the Bio-Integral Research Center, P.O. Box 7242, Berkeley, CA 94707. IPM consultants have formed an association—Integrated Pest Management, 10202 Cowan Heights Drive, Santa Ana, CA 92705. See also BIOLOGICAL CONTROL.

# IRIS

The iris is fairly untroubled by pests. Weeds may present a problem, but an easy answer lies in the use of a mulch in the flower beds. If cultivation is necessary, keep it shallow to avoid damaging root stems.

## Insects

One insect that can cause serious damage is the **iris borer.** The feeding of this voracious larva paves the way for bacterial soft rot, which has the potential to cause more trouble than the borer itself.

The borer spends the winter in the egg stage on old iris leaves, stalks, and debris. Larvae appear in April or early May, and can be seen crawling over the plant for the first few days, but they soon enter the leaves to feed. The first injury you're likely to see is slender feeding channels in the leaves, somewhat resembling the burrows of leafminers. Leaves may also appear water-soaked. The larvae gradually make their way to rhizomes, arriving there early in July in most cases, and it is then that they grow

Iris borer

to a length of 1½ to 2 inches. They are quite corpulent at this time, and colored pinkish with brown heads. As the larvae approach full growth, injury becomes conspicuous. This is particularly true if the rhizomes are destroyed and the plants tend to collapse.

Control depends on a thorough garden cleanup to remove the overwintering eggs on old leaves, stems, and debris. Hidden worms can be lanced with a pointed stick or, lacking a pointed stick, you can pinch them in the leaves. Transplant infested iris clumps in August; the larvae leave the rhizomes that month to pupate in the soil, and you can remove them by sifting. Plants with badly damaged rhizomes should be destroyed. If you find only a very few holes, kill the borers by poking them with a piece of wire.

The small, milky white larvae of **iris thrips** feed on the inner surfaces of leaf sheaths and on the young leaves of many iris varieties. This injury leads to russeting, blackening, stunting, weakened growth, and the death of tops. Russeted or blackened flower buds and stem buds often fail to open. This insect is found in new garden plantings in the basal leaves of plants. It damages Japanese iris seriously, and attacks most other types. The adult thrips are usually wingless, very dark, and slow-moving. See THRIPS.

When the **iris weevil** breeds in the seed pods of iris, the pods show many small holes. Eggs are deposited on the iris ovary, and the thick-bodied, white larvae then feed on seeds. The adult weevil is ⅕ inch long, and colored black with white-and-yellow scales. The pest is locally abundant wherever iris grows.

Destroy all flower heads after the flowers of affected plants fade, or destroy seed pods before the young weevils emerge. If seeds are to be saved for breeding, cover blossoms with cheesecloth.

When bulbous iris fail to grow or growth appears distorted, examine bulbs for brown patches or longitudinal stripes indicative of **nematode** infestation. Diseased plants should be rogued and the soil sterilized before reuse. See NEMATODE.

Other occasional pests of iris include the SNAIL AND SLUG and the twospotted SPIDER MITE.

## Diseases

**Botrytis rhizome rot** causes iris to fail to grow in spring. Rhizomes may be affected by a hard, dry rot, without a foul odor, and the plants can be pulled readily from the ground. The gray mold fruiting stage is generally present, and shiny black, convoluted sclerotia may be found on or in rotted tissues. Plants often die. The disease is rarely a serious problem once temperatures reach the 70s.

To control this fungal disease, rogue and burn infected plant parts and take the infested soil from the garden area. Mulch the rhizomes as winter approaches.

**Crown rot** usually appears in overcrowded beds. Leaves and flower stalks die back from the infections at the base where white cottony mycelium, containing mustard-seed-size sclerotia, may be found. The sclerotia vary in color from cream to reddish brown. The fungus is pathogenic on many other flowers (including delphinium, aconite, and columbine) and may be spread by scattering the sclerotia during cultivation. Preventive maintenance involves thinning the plants to admit light and air. Remove infected plants along with the infested soil to stop progress of the disease.

Occasionally, translucent, oval **leaf spots** appear on iris leaves. Look for a reddish border.

Leaf spot

Plants won't die but lose vigor and become susceptible to other problems. This common disease is caused by a fungus that infects wet foliage and thrives in humid weather. Control calls for removing and destroying spotted leaves early in the season and removing tops in fall. Be careful not to overwater, especially in wet weather. When you divide iris clumps, remove parts that appear to be diseased and plant in soil that hasn't hosted iris (this disease has a very limited number of victims).

**Mosaic** is indicated by light and dark green mottling of leaves, early flowering, and color breaking in flowers, along with stunting. Bulb sheaths may be marked with blue green or yellow streaks. The disease most often strikes bulbous iris; it occurs also on rhizomatous iris, but the symptoms are not so noticeable. Aphids carry the disease, and their control is important. Rogue infected plants as soon as the above symptoms are observed; see APHID.

**Rust** is known by the rusty pustules that appear on leaves of susceptible varieties (German iris is immune). Rogue and burn infected plants to break the continuity of the rust's life cycle. Resistant varieties are available.

Probably the most serious disease of German iris is bacterial **soft rot.** Bacteria enter the plant through wounds in young leaves made by winter damage or the young larvae of the iris borer (see above). The organisms then multiply rapidly in the soft leaf tissue and cause a water-soaked appearance around the entrance holes. The leaf is destroyed as bacteria make

their way toward the rhizome. Once the rhizome is infected, healthy leaves fall over and collapse at the base. The foul odor of the rhizomes is a telltale sign of this disease. Plants may die.

Wet weather favors the disease, and so does a thick layer of mulch immediately around plants. Afflicted rhizomes can be saved. Carve away rotted areas and place the clump in direct sunlight so that the cut can dry. Remove any rotting leaves you find, being sure to cut well below the water-soaked areas. If you are working on a number of plants, dip implements in 70 percent alcohol to avoid spreading the disease.

# IVY

## Insects

The nasturtium **aphid** is one of several species that feeds on English ivy. When aphids suck juices from leaves and stems, these parts become curled, dwarfed, and yellowed. Other susceptible plants include globe thistle, dahlia, oleander, poppy, zinnia, and nasturtium. This aphid overwinters on euonymus. See APHID.

Another minuscule pest, the **twospotted mite,** causes leaves to become gray and mealy. Mites and their webs can be washed away by squirting plants with water from a syringe. See SPIDER MITE for other control measures.

You may also run into the MEALYBUG and any of a number of SCALE species. Ivy scale can be held in check with the golden chalcid, *Aphytis melinus,* a parasitic wasp that is commercially available.

## Diseases

**Bacterial leaf spot** is identified by irregular water-soaked spots with yellow or translucent

Fungal leaf spot (left) and bacterial leaf spot (right)

Japanese beetle

borders. It usually appears on lower or inside leaves of densely crowded foliage and on greenhouse ivy. Infections on petioles produce black lesions that crack longitudinally. Stem infections may result in yellowing and reduction of growth. Improving ventilation will usually correct the trouble; it also helps to pick and destroy infected leaves and to water greenhouse plants without splashing.

**Dodder** is a parasitic weed that looks like many strands of orange string with no apparent roots. The strings wind around ivy stems and cannot be easily detached. Small clusters of white flowers produce seeds for another year's crop. When seeds germinate, the dodder climbs up the nearest plant, sends little penetrating knobs into the tissue of the host, and lives on the host's sap. The dodder roots are dropped as soon as this attachment is made, and it becomes a truly parasitic plant. If possible, detach and burn this weed. Otherwise, rogue the affected plants.

Large tan spots that may show the concentric rings of fruiting bodies are known as fungal **leaf spot.** Remove and destroy infected plant parts.

# JAPANESE BEETLE

The Japanese beetle [color] causes extensive damage in many of the eastern states. A native

of Japan, it was first found in this country in 1916 near Riverton, New Jersey. In coming to North America, the beetle found few natural enemies, but some 300 attractive plant species. It is now found from southern Maine southward into South Carolina and Georgia and westward into Kentucky, Illinois, Michigan, and Missouri. Japanese beetles are a little less than ½ inch long, and are a shiny metallic green. They have copper brown wings, and six small patches of white hairs along the sides and back of the body, under the edges of the wings. Males and females have the same markings, but males usually are slightly smaller than females.

The Japanese beetle spends about ten months of the year in the ground in the form of a white grub. The grub is similar to our native white grub, but it is usually smaller—about an inch long. It lies in the soil in a curled position.

Adult beetles first appear on their favorite food plants in late spring or early summer, depending on locale. If you live in eastern North Carolina, for example, expect them in mid-May; in the Philadelphia area, about June 15; in New England, about July 1 or later. In the Midwest they show up about June 15 in the St. Louis area and July 1 in Michigan. They fly only in the daytime, and are especially active on warm, sunny days, moving readily from one plant to another.

The period of greatest beetle activity lasts from four to six weeks. The beetles then gradually disappear. In eastern North Carolina most of them are gone by early August, but in New England some are present until frost. From time to time the females leave the plants on which they have been feeding and burrow about three inches into the ground, usually in turf. There they lay a few eggs from which grubs will later hatch. After laying these eggs, the females return to the plants for more feeding.

Populations may be lower in some localities than in others as the result of one or more of the following conditions: a dry summer season, suppression by milky spore disease and other natural control agents, cultural factors, and intensive chemical control measures.

New housing developments are especially subject to high local beetle populations. Turf infested with Japanese beetles, but not containing beetles' natural enemies, is often imported into these areas. In addition, adult beetles often fly into housing tracts having soil that contains no beetle enemies, and the pests can then become established easily.

Grubs feed on the roots and underground stems of plants, particularly grasses. Often this feeding goes unnoticed until the plants fail to make proper growth or die. When grubs are numerous, they can cause serious injury to turf.

Adult Japanese beetles feed on more than 275 different plants and often congregate to feed on flowers, foliage, and fruit of plants and trees exposed to bright sunlight. Beetles feeding on leaves usually chew out the tissue between the veins, leaving a lacy skeleton. They may eat many large irregular areas on some leaves. A heavily attacked tree or shrub may lose most of its leaves in a short time. The beetles often mass on ripening fruit and feed until nothing edible is left, but they seldom touch unripe fruit. They seriously injure corn by eating the silk as fast as it grows, which keeps kernels from forming.

NATURAL CONTROLS.    Nature itself exerts many controls on Japanese beetle grubs and adults. Weather, disease, and birds—among them starlings and grackles, cardinals, catbirds, and meadowlarks—are essential to keeping their numbers down. Five parasitic wasps have become established in the United States. Of the parasites imported to control the Japanese beetle, the two most important are the wasps *Tiphia vernalis* and *T. popilliavora*. The first named, known as the spring tiphis, overwinters in the adult stage within the beetle's underground cocoon and shows aboveground in May, about the time the grubs are fully grown. The second wasp, called the fall tiphia, occupies the host's cocoon overwinter as a larva and appears in August. Its eggs are laid on the grub. The spring tiphia occurs in the mid-Atlantic states, and parasitizes many of the beetles through much of this area.

In the 1930s, state and federal teams got together with the Rockefeller Institute of Medical Research to find an economical way to raise a species of parasitic nematode to control the beetle; a resulting colonization program was considered responsible for lowering the pests' numbers in areas of New Jersey. Long, hair-thin nematodes are important in many areas today.

Summer weather has a lot to do with determining Japanese beetle populations. Extremely dry spells destroy many of the eggs and kill newly hatched grubs; wet summers are favorable to the development of eggs and grubs, and are usually followed by seasons of increased numbers of beetles.

Of the several naturally occurring diseases known to affect this pest, milky spore disease (or *Bacillus popilliae*) is the most important. It kills grubs after turning their normally clear blood to a milky color. The disease is harmless to humans, other warm-blooded animals, and plants. Because of its safety and effectiveness, milky spore disease is now produced in dust form and sold commercially by mail order and through many garden supply centers. If your local dealer does not handle it, your county agent can tell you the nearest source.

The spore dust is made by collecting grubs and inoculating them with the disease. Treatments are most effective when applied on a community-wide basis. (If you and your neighbors wish to cooperate in buying and applying milky spore disease dust in your area, you should contact your county agricultural agent for help in coordinating the effort.) The dust is applied at any time, save when the ground is frozen or a strong wind is blowing. It is usually applied at a rate of two pounds per acre in spots ten feet apart. Use about one level teaspoon to each spot. The disease will become established more quickly if the dust is applied in spots only five feet apart; in this case about 7½ pounds of spore dust will be needed per acre. Do not expect immediate results, as several years may elapse before the milky disease becomes fully effective. It's useless to apply the disease dust and a chemical insecticide to the same turf; insecticides keep the grub population at a very low level and will prevent buildup of the pathogenic bacteria.

Large-scale application will go quicker with the aid of a simple spreader. Using a ten-penny nail, poke a dozen or so holes in the bottom of a one-pound coffee can over an area of one square inch. Bolt the can to a broom handle, the bottom of the can coming four inches from the end. Place the dust in the can, and broadcast it by tapping the stick on the ground.

OTHER CONTROLS. If only a few plants are bothered, you may be able to get by with handpicking the beetles. Drop them in a can containing a bit of kerosene. It is more efficient to shake them out of trees or branches early in the morning, before they are limbered up. Place a sheet or tarp on the ground to catch them as they fall, and drop them into a bucket of water topped off with a layer of kerosene. You should

repeat this every day, as beetles are apt to fly in from elsewhere.

Many problems can be prevented by following a few basic cultural controls. Remove prematurely ripening or diseased fruit from the trees and ground. Delayed corn plantings often get by without damage; see TIMED PLANTING. It's best to remove susceptible wild plants from the immediate area, including bracken, elder, evening primrose, Indian mallow, sassafras, poison ivy, smartweed, wild fox grape, and summer grape. These plants are often a continuous source of infestation for other plants. Susceptible species can be rotated with plants that don't sustain the pest.

Japanese beetle grubs are discouraged by a high soil pH, and in one test an area was protected from the pests by spreading pulverized dolomitic lime at a rate of 100 pounds per 1000 square feet. This effect is undone if soil is aerified after the lime is spread.

Larkspur has a fatal attraction for the Japanese beetle; the pests nibble at the foliage and keel over. Roses may be spared if interplanted with geraniums, as this pungent flower is noxious to the beetle. Two trap crops worth a try are Wilson soybeans, set every 15 or 20 rows to protect corn, and smartweed around the periphery of the garden or flower bed.

Neem oil repulses Japanese beetles. You can trap them in a large glass jug in which is fermented a wine of sugar, water, and odd pieces of mashed fruit; a bit of yeast may be necessary to get the mixture started. Place the jug in the yard, and strain out the day's catch each evening. Other beetle baits are geraniol (a rose-scented alcohol used in soap and perfumes) and anethol (the main constituent of anise and fennel oils, used to flavor perfumes and toothpaste). Traps work best when the beetle population is high. Place them on the periphery of your garden or yard, not in the middle, as they attract beetles from elsewhere and you'll not want to lure them through your vulnerable plants. Traps work best if set out more than a foot from the ground and in sunny spots away from foliage. Commercial traps are available.

# JUNIPER

## Insects

Although **juniper scale** is primarily a pest of juniper, it also attacks arborvitae and cypress. The scale covering of this pest is circular and colored white with a yellow center. The eggs and young are pinkish to yellow in color.

This scale overwinters as nearly fully grown, fertilized females. The eggs are laid during late May and throughout June. There is one generation a year. The scales may be found on the younger twigs, needles, and cones of infested plants. When infestations are severe, the plants look as though hit by extreme drought. Due to the secretion of honeydew by the scales, the plants often are blackened by a fungus that grows on the honeydew.

Use a 3 percent dormant-oil emulsion, sprayed during the last week in March or the first week in April. To control the young scale insects that hatch in late spring, use a soap solution, sprayed at ten-day intervals.

The **juniper webworm** picks on low juniper plants in particular. The light brown caterpillar webs together leaves and twigs for protection and feeds on the leaves inside its web. The ⅝-inch-long brown-and-white moth appears in June.

A pyrethrum spray will kill many of the caterpillars if applied with enough force to penetrate the webbed nests and wet the insects. Penetration can be improved by first separating the webbed tips. If only small branches are infested, they can be cut out and destroyed.

Other pests of juniper are APHID, bagworms (see EVERGREEN), and spruce mite (SPRUCE).

## Diseases

Several species of juniper, especially the native red cedar, are affected by **rust diseases.** Rust diseases of juniper are generally more unsightly than destructive, but if severe infestations occur year after year, red cedars may be seriously damaged. There are three common rust diseases: cedar-apple rust, cedar-hawthorn rust, and cedar-quince rust.

On red cedar leaves the cedar-apple rust disease causes chocolate brown, globular to irregular shaped corky galls, measuring ⅛ inch

Cedar-apple rust gall

to two inches in diameter. These galls are often referred to as cedar-apples and require two years to complete their development. In the second spring after the red cedar originally became infected, the cedar galls produce gelatinous, fingerlike, slimy orange spore horns during rainy weather in May and June. The galls then die. Spores produced from these gelatinous spore horns infect leaves of apple, causing bright orange spots. Spores produced from these infections in turn produce spores that cause infection of leaves and twigs of juniper in late summer.

Cedar-hawthorn rust results in the formation of galls on juniper and leaf rust on hawthorn and cultivated apple. On juniper (red cedar), the galls are often small and irregular in shape but may be as large as those produced by the cedar-apple rust; they differ in being perennial and more woody. Orange, wedge-shaped gelatinous spore horns are produced in May or June. Spores from these horns infect the leaves of hawthorn and apple.

Cedar-hawthorn rust

Cedar-quince rust is responsible for the slightly swollen, elongate cankers that appear on the branches of red cedar and other junipers. During rainy weather in May and June, orange, gelatinous, cushion-shaped spore masses break through the rough, diseased bark. Spores from these sources infect twigs, leaves, and fruit, or less often, quince, apple and hawthorn. Reinfection of cedars with cedar-quince rust disease occurs in late summer by spores produced from infections of quince, apple, and hawthorn.

Since the rust diseases of red cedar and other junipers require an alternate host for the completion of their life cycle, the logical control for these troubles is to avoid growing susceptible junipers within a mile of the alternate hosts— apple, hawthorn, and quince.

Avoid planting varieties of juniper that are highly susceptible to cedar rusts. The eastern red cedar and all its horticultural varieties are very susceptible to cedar rust diseases. The western and Colorado red cedar are much less susceptible. The columnar Chinese juniper, pfizer juniper, prostrate juniper, and andorra juniper are rarely seriously troubled by cedar rust diseases. If you have the time and can go to the trouble, prune out galls during the first week in April before the spore horns develop.

**Twig blight** is caused by a fungus that makes the tips of infected twigs turn brown and gradually die back. Small, black pinpoints of fruiting bodies may appear on the leaves and stems of affected twigs. During wet weather, spores ooze out of the fruiting bodies and are spread by wind, rain, insects, and pruning tools. The disease is especially severe on nursery stock.

Twig blight may be avoided by planting resistant varieties. The spiny Greek juniper, Keteleer red cedar, and hill juniper are reported to be resistant. Susceptible apple varieties to avoid include Baldwin, Delicious, Franklin, Melrose, Red Astrachan, Stayman, and Transparent. When the plants are dry, prune and burn affected twigs and branches. Late winter is the best time for this.

# JUVENILE HORMONE

Insect hormones have been used to manipulate pests. They work by disrupting an insect's life cycle. Although some hormones have been synthesized, none is yet marketed for small-scale applications.

# KALE

## Insects

In general, kale is bothered by pests mentioned under CABBAGE. Several that you are particularly apt to run into are listed here.

The turnip **aphid** has a penchant for kale, especially in the South. It is little more than a pale green speck, although a close look may reveal a black head and black spots on the winged form. Its resemblance to another kale pest, the cabbage aphid, has earned it the alternate name of false cabbage aphid. For control, see APHID.

The **celery leaftier** starts off life a pale green, but becomes yellow as it reaches its full length of ¾ inch. Running down the back is a thin dark green line centered in a wider white strip. They make webs out of leaves, and if one takes a dislike to you it may either toss about in this protective shelter or drop to the ground. See CELERY for control measures.

Under CABBAGE, see cabbage looper, diamondback moth (adult of a destructive green caterpillar), and imported cabbageworm.

## Diseases

**Black leg** may cause spots to appear at the base of stems; at first these spots are merely pale blemishes, but they later turn black, girdle the stem, and can cause the plants to die. On the dead tissue can be seen many tiny, black, pimplelike bodies. Before the plant succumbs, leaves wilt and develop bluish red edges. For control, see methods for preventing black rot under CABBAGE.

# KOHLRABI

Pests and diseases of kohlrabi may be found under CABBAGE. Kohlrabi and tomatoes do not fare well together in the garden, as this proximity invites insect pests to the former crop.

# LACEWING

The lacewing [color] is the most important of the several beneficial "nerve-winged" insects of the order Neuroptera. The brown lacewing is found in the West, and the larger green (or goldeneye) lacewing occurs east of the Rockies. Their delicate appearance and weak, erratic flight belie the fact that they consume great numbers of aphids and mealybugs. When at

Lacewing adult with eggs, and larva

rest, the membranous wings are held upright like a tent, and the green species is distinguished by its iridescent eyes and foul odor. The spindle-shaped larvae are even more helpful to growers, imbibing the body fluids of a number of pests through curved, hollow mandibles. Their victims include thrips, mites, caterpillar eggs, scales, leafhopper nymphs, corn earworm, pink bollworm, as well as aphids and mealybugs. The appetites of these larvae have earned them the names aphidlion and aphidwolf.

To keep the ravenous young from feeding on their brothers and sisters before they hatch, the female lacewing lays each oval egg on the top of a delicate stalk projecting from the surface of a leaf or twig. In 6 to 14 days they hatch and in two or three weeks they spin yellowish pea-size cocoons on a leaf. If the weather is warm, a mature winged adult emerges in a couple of weeks, completing the cycle of life and marking the beginning of another generation.

At night, bright lights attract the lacewings and they can be found smashed against automobile windshields and clinging to screen doors. Their main defensive weapon is an odorous excretion which is quite evident when one of them is grasped by an enemy.

Lacewings are available commercially as eggs. The larvae are released as soon as they are noticed crawling about, because they are cannibalistic. Lacewings will be more effective if ants are kept in check; ants will defend aphids, a source of honeydew. You can further assist lacewings by growing plants that supply the adults with nectar and pollen. Clover or alfalfa can be grown as cover crops, and you can also interplant flowers between rows of crops. Substitute foods are on the market; lacewings are attracted to a spray of honey and water. Because the adults have difficulty overwintering in a temperate climate, the predators likely will have to be recolonized come spring.

# LADYBUG

The ladybug (or lady beetle to some, ladybird to others) [color] has been a symbol of biological control, and its adult form is well known to almost everyone. But the ladybug's use as an agent against specific pests has been overestimated. Entomologists have come to believe that these beetles and the famous praying mantis often are not apt to reduce pests below economic injury levels, and now recommended host-specific predators and parasites.

The story of the ladybug's services to growers in the United States began toward the end of the 19th century, when the destructive cottony cushion scale appeared in the citrus groves of California. The entire citrus industry was on the verge of extinction. About the time a USDA entomologist in Australia discovered a tiny spotted beetle feeding on some cottony cushion scale in a garden—the vedalia, or Australian, ladybug. A few beetles were sent across the sea and soon proved their worth by eliminating the scale on some test trees. More

were imported and released in the orchards. The total cost of the transplantation amounted to $1500, a pittance considering that an entire industry, which flourishes today, was saved. Since that time, the most important ladybug predator in Europe and North Africa—the seven-spotted ladybug—has been successfully established in several states by the USDA. While the adults devour aphids and insect eggs, the larvae do the most good, attacking rootworms, weevils, chinch bugs, the Colorado potato beetle, and others. The ladybug begins its life in spring as one of 200 or so eggs in the crevices of tree bark or on the undersides of protected leaves. From these eggs emerge tiny tapered larvae which immediately begin feeding on aphids. So insatiable is the appetite of the lizardlike larva that it has been known to consume 40 aphids in a single hour, a capability that has earned it the nickname aphidwolf.

The larva becomes full grown (about ¼ inch long) in about 20 days. It then attaches itself to a leaf or stem and pupates. Finally it transforms into the familiar black-spotted adult. In fall, ladybugs gather in great masses to spend the winter beneath loose bark, under boards or rocks, or in some other protected place. On the Pacific Coast these hibernating beetles are collected in great numbers to be either distributed in crop areas or kept through the winter in captivity and sold commercially. The reddish orange beetles generally feed on aphids, while the darker ones eat scale, mealybugs, whiteflies, and spider mites.

Although ladybugs fend for themselves quite well, there are a few precautions to be taken when you stock your garden.

Don't stock too early. If there are not enough pests in your garden patch to sustain them, they will either feed on what is there and then fly in search of more insect food, or else starve to death. If pests are evident but not plentiful, place some ladybugs in the garden and store the rest in the refrigerator (not in the freezer) until needed. They will hibernate in the refrigerator for several more weeks.

It is best to release the beetles late in the afternoon or, better yet, at sundown. This encourages them to stay in your garden for the night and find suitable food and protection.

It's a good idea to dampen the ground in the areas to receive the beetles. A good mulch will help keep the garden floor moist and cool, and gives ladybugs a place to hide. Don't scatter them as you would sow grain, but gently place handfuls of them at the base of plants, spaced 15 to 30 paces apart. Rough handling, especially in warm weather, may excite them to seek safety in flight.

About 30,000 individuals are considered adequate for protecting ten acres of crops. To bring this number into perspective, the beetles are so tiny that 1500 of them would weigh but an ounce. If the weather is warm and sunny, your stocked beetles will mate and lay eggs in a day or two. In 15 more days you should already have a second generation of larvae hard at work.

# LARCH

## Insects

Outbreaks of the dark larch woolly **aphid** are infrequent, but if control is required, see APHID.

The **larch casebearer** is a pest of larch foliage. The larvae mine the insides of needles, forming protective cases attached to a twig, often seen in clusters. They spend the winter in the cases, dislodging them in spring to go to the buds and feed on new leaves. The moths emerge in May or June.

Natural predators usually keep this casebearer under control. Five parasitic wasps have been imported into Canada. The case stage is preyed on by chickadees and other birds. Moths appearing in spring are pursued by birds. The hibernating caterpillars may be controlled by a dormant spray of lime sulfur before the tree's growth starts in spring.

The worst insect pest of larch is the **larch sawfly**, whose larvae can so seriously defoliate that growth is noticeably affected. In June, the female sawfly cuts into the undersides of new twigs to deposit 50 to 100 eggs, leaving the twig curled downward like a fishhook. The emerging green larvae feed greedily on the needles, stripping the trees if very abundant. In midsummer they drop to the ground to spin the cocoons within which they live until the following spring.

A variety of natural predators prey on the larch sawfly. Birds and needlebilled stink bugs feed on the larvae, and when the worms drop to the ground they are devoured by skunks, shrews, and mice. Whatever is done in the larch's environment to encourage these natural predators of the sawfly will help control it. The area beneath the trees should be raked clean each autumn.

## Diseases

**Canker** appears as a depressed area in the bark, often exuding resin at the edges and accompanied by considerable swelling of the branch or trunk on the opposite side. White cups with orange linings open and close with wet and dry weather; these are the fruiting bodies of the fungus. Infection can kill small trees and branches of large ones, but more often it merely weakens branches so that ice or wind cause them to break easily. The European, American, and golden larch are susceptible to this canker, but Japanese larch is relatively resistant.

Inspect trees frequently for trouble. Cankered limbs can be removed by pruning; burn the cankered material.

New leaves of larch are susceptible to **frost injury,** but new growth or secondary buds often replace them. Forst can also cause cankers or death of young trees by killing the cambium in stems of less than two inches.

**Needle cast** causes needles to turn brown in spring or early summer and remain on the tree over the winter. At a distance, trees look scorched. The infected needles bear long black fruiting bodies. Needle cast may be a serious disease in nurseries.

As soon as it appears, spray new growth with lime sulfur at intervals of two weeks apart until August.

Although two types of **needle rust** fungi affect larches, they are not serious enough to warrant control measures. The alternate hosts of these rusts are poplars and willows.

Larches are occasionally attacked by **shoestring fungus.** It is a sudden and spectacular attack, rapidly causing wilting and even death of one or several vigorous trees. Then, just as quickly, the outbreak will subside. Since attacks are sporadic and losses are not repeated, this disease is hard to predict. The risk is greatest when larches are planted on cut-over land where oaks were once prominent. See TREES.

# LAUREL

Insect pests and diseases of laurel are similar to those of RHODODENDRON.

# LAWNS

Like all plants, lawn grasses need light, air, moisture, and nutrients. While they are never free of the possibility of pests and diseases, the quality of care you give your lawn can greatly help to reduce trouble.

Begin by providing good surface and sub-surface drainage when establishing a lawn. Fill in low spots where water may stand. Where air movement is restricted, problems can develop, so thin or remove surrounding shrubs and trees as necessary to increase air flow and allow sunlight to penetrate.

Choose grasses that are adapted to your area, and be wary of inexpensive seed mixtures, as they may contain a high percentage of weed seeds. The severity of diseases can often be reduced by growing a compatible blend of two or more locally adapted, disease-resistant grass varieties. A mixture of Kentucky bluegrass and fescues has become popular—the bluegrass is hardy and undemanding, while the fine fescues, with their tolerance for shade and dry sites, broaden the bluegrass' adaptability. Bentgrass, often used on golf greens, requires more attention than other grasses. It needs frequent fertilizing, mowing, and watering.

Temperature affects grass development: bluegrass and fescues grow best in cool temperatures, and Bermuda grass and zoysia types thrive in warm temperatures; Merion Kentucky bluegrass is more heat- and drought-tolerant than other bluegrasses, although it requires more care in maintenance than ordinary Kentucky bluegrass. For home lawns in Montana, use either Merion bluegrass or ordinary Kentucky bluegrass, mixed with a small percentage of perennial rye and white Dutch clover as a nurse crop. Because creeping bentgrasses require such close mowing, careful watering, and frequent applications of fertilizer, they are best used on putting greens and are unsuited for the average home lawn.

In the South, several turfgrasses are widely used—Bermuda, centipede, St. Augustine, and bahia, among others. The zoysias are very durable grasses that serve well in most areas.

For turfgrass mixtures for prairie regions, try Kentucky bluegrass and creeping red fescue for lawns and fairways, and creeping bentgrass for golf and bowling greens. Red-top and colonial bentgrass in turf mixtures are susceptible to most diseases and tend to choke out more desirable species. Seed for lawn and athletic fields should be guaranteed free from annual bluegrass and bentgrass.

Pure stands of Merion bluegrass were once widely popular in Canada, but proved not to have sufficient cold hardiness. Also, the tendency of this variety to build up thatch predisposed it to several diseases of the prairie region that caused it to die out. If Merion is used in Canada, mix it with other bluegrasses or with creeping red fescues. The Merion grass should not make up more than 30 percent of any mixture.

## CULTURAL CONTROL.
Your grass will be healthier if it is not mowed too closely. Upright grasses such as Kentucky bluegrass and fescues should be clipped to two inches or even higher in the summer. Creeping grasses such as bentgrasses, Bermuda grass, and zoysia may be mowed more closely, but not under 1½ inches. Mow grass frequently, so that no more than ¼ to ⅓ of the leaf surface is removed at any one time. If too much leaf is cut away, food manufacture is drastically reduced, resulting in poor root development and weak, thin grass that is more susceptible to infection by disease.

A good rule of thumb is to mow when the grass reaches a height of three inches. Except where winters are very severe, continue mowing the grass through fall until growth stops. Grass will survive hard winters better if allowed to grow to a height of three to four inches before the first severe frost.

Thatch is a tightly intermingled layer of living and dead stems, leaves, and grass roots that develops between the soil surface and the green growth. Too much thatch keeps water, sun, and air from penetrating the soil, aggravates some diseases, and seems to prevent the grass from developing a deep root system. Excess thatch should be removed in early spring or early fall if ½ inch or more has accumulated. At seeding time, too much thatch prevents new seeds from coming in direct contact with the soil for proper germination. On large lawns, thatch can be removed with a lawn renovator, power rake, or a powered thatch remover that goes over the turf and breaks up the clogging mat.

The way you water your lawn can make a great difference in its greenness and health. Deep watering to a depth of six or eight inches, once a week or every five days, is better than frequent shallow waterings. This is especially important in summer. Lawns that are watered too often are more disease-prone and they develop shallow root systems that cause them to brown easily. The common practice of watering in the evening is not recommended, since grass that stays wet during the night is much more prone to the diseases and fungi that thrive on moist conditions. You'll have better results if you water early enough in the day so that grass will dry out before nightfall.

It is not always necessary to apply lime to lawns. Where organic matter and humus content in the soil are high, they act as a pH buffer and make liming necessary less often, if at all. Naturally alkaline soils, of course, should receive no lime. Many homeowners damage their lawns by feeding them unnecessary lime, and it is best to have your soil tested to determine if the lawn needs this supplement.

The best pH for a healthy lawn is slightly acid, about 6.5 or just under neutral (7). If your soil is too acid, top dress with natural ground limestone, preferably dolomite, as it supplies magnesium as well. Pulverized oyster shells, wood ashes, or any calcium-rich wastes may also be used. To raise a soil one full pH unit—say from 5.5 to 6.5—apply approximately 50 pounds of lime per 1000 square feet. Use 10 to 15 pounds less for light, sandy soils, and more for heavier loam or clay types. If a large amount is needed, you can apply one half in early spring and the other half in fall.

While many of the highly acid chemical fertilizers increase the need for lime, organic fertilizers seldom lead to overacidity. In addition, they are long-lasting and don't burn grass. The best choices are cottonseed meal, soybean meal, screened compost, and well-decayed animal manures used at approximately 100 to 150 pounds per 1000 square feet. Dried blood or blood meal is also excellent, as it has a high (8 to 14 percent) nitrogen content, plus other elements necessary to grass growth. Use ground phosphate with potash rock, greensand or granite dust, and bone meal for mineral nutrients, applied at 50 to 100 pounds per 1000 square feet. Rock fertilizers such as granite dust and rock phosphate are best applied in fall, so that winter snows and rains can wash them well into the soil.

Sludge has proved to be another excellent organic fertilizer for turf. It can be mixed with the soil when you start a new seedbed in September. Open up the soil to a depth of five or six inches and mix sludge and compost thoroughly with the topsoil. After seeding, top dress with more sludge, topsoil, and compost and cover with a thin straw mulch. For well-established lawns, sludge may be applied in a ½-inch-deep layer during winter when the ground is frozen. Not only is it a luxuriant plant food, but the sludge cover will also insulate the grass roots from alternate thaws and freezes. Note that sludge may contain toxic heavy metals.

A few elements of winter care for your lawn will help prevent spring problems. Avoid a great deal of traffic on the lawn in winter, especially when snow is on the ground, as ice will form and kill the grass. Low spots on the lawn are especially subject to winterkill because water accumulates and freezes there. Remove all dog manure before the last snow melts to avoid burning the lawn. If the grass is not covered with snow, it may even need to be watered in winter months. If the soil is thawed down to three inches it may be watered anytime during the winter, but watering when there is frost in the topsoil will encourage heaving.

## Insects

Many people find **ant hills** unsightly, especially in lawns where the grass is kept short. Actually, these natural aerators are beneficial to your lawn, and if you allow your grass to grow a little longer, the ant hills will be hidden.

Three **beetles,** the Asiatic garden beetle, Japanese beetle, and Oriental beetle, cause the same type of damage to lawns. A heavy infestation may cause spots of the lawn to die in August or September. If you pull up tufts of grass from these areas, you will see that the roots have been cut just below the surface of the ground. Beetles seem to flourish on poorly nourished grass and plants. The odor of prematurely ripening or diseased fruit often attracts them, so remove fallen fruit from the lawn area.

The Asiatic garden beetle feeds on many kinds of plants. They are attracted to lights, but in the day they hide in the soil around plants and are seldom seen. The Asiatic garden beetle is about ¾ inch long, dull cinnamon brown, and has finely striated wings.

The Oriental beetle adults emerge in late June and July. The females lay eggs about six inches deep in the soil. Several weeks later the grubs hatch and work their way toward the surface, where they feed on roots of grass. In late autumn they descend a foot deep in the soil to hibernate, and come to the surface again in April to resume feeding. There is one generation a year. Oriental beetles are ⅜ inch long and have varied markings and colors. They do not fly much, and are often found in roses, hollyhock blossoms, and in the turf.

The most widely known means of safe control for the Japanese beetle is milky spore disease. See JAPANESE BEETLE.

Hairy **chinch bugs** are serious lawn pests. These bugs hibernate during winter in dead grass, leaves, and other litter. They emerge in spring, mate, and lay eggs on the grass or on the soil surface. Lawn damage is caused by the young bugs or nymphs, which are about $\frac{1}{5}$ to $\frac{1}{4}$ inch long, reddish at first, and then changing to black with a white spot on the back between the wings. Lawns infested with chinch bugs may first show small spots of yellowing grass, then large irregular dead patches where the bugs have injured the lawn by sucking out the plant juices. Chinch bugs have an offensive odor, especially when crushed, and a severely infested lawn has an odor that can be detected by walking across it. As the life cycle of the chinch bug takes but seven or eight weeks, there may be two or more generations a year in the South. They are particularly damaging to St. Augustine grass.

To test for chinch bugs, first select a sunny spot along the border of a yellowed area of lawn. Cut out both ends of a large tin can, push one end into the soil about two inches, and then fill with water. If chinch bugs are present, they will float to the surface of the water within five minutes.

These insects prefer hot, sunny lawns such as football or baseball fields. You can discourage them by shading the lawn with trees or shrubs. Another control is to seed your lawn in soil made up of $\frac{1}{3}$ sharp builder's sand, $\frac{1}{3}$ crushed rock, and $\frac{1}{3}$ compost. Well-fed lawns will discourage chinch bugs. When deprived of nutrients, the bug not only lives longer but lays more eggs.

When chinch bugs have severely infested a lawn, a natural predator called the bigeyed bug may move in and feed on the adults. The bigeyed bug also preys on insect eggs, spider mites, plant bugs, leafhoppers, aphids, and larvae of several Lepidoptera species. They won't repair the bare spots in lawns damaged by chinch bugs, but they will control the pests so that you can re-seed in late summer. In warm and humid weather, chinch bugs are apt to be infected by the naturally occurring green muscardine fungus *Beauveria bassiana*. This disease may become commercially available.

There are several species of **cutworm** [color] that cause lawn injury. Most species are smooth grubs of greenish, brown, or dity-white coloring, either with or without striping. Presence of feeding larvae is likely if the adult moths have been seen in the area; they have a wingspread of from one to two inches and are usually multi-colored with dull hues such as brown, black, gray, or dirty-white.

Cutworms are among the earliest insects to begin feeding in spring. Most species have one generation a year, occasionally two. The generations usually overlap so that moths appear throughout the summer. Cutworms injure grass by cutting off the blades at the base, leaving closely cropped, small, elongated or irregular brown spots in the turf. The larvae generally remain concealed just below the surface of the ground or in clumps of grass during the day, coming out at night to feed.

Successful control of cutworms has been achieved by flooding lawns with water until they are puddled. This treatment, which is only practical on a small scale, brings the cutworms to the surface so that they can be collected and destroyed. See also TOMATO.

**Earthworms,** sometimes called night crawlers or dew worms, usually come to the surface at night. These purplish-tinged worms may be as long as eight inches when fully grown. They often leave casts on the surface that make the lawn bumpy and difficult to mow. You're likely to see them after a heavy rain.

Since earthworms, like ants, are natural aerators, they are beneficial to the soil. Let your grass grow a little longer so the castings will be hidden. If an absolutely smooth surface is necessary, use a vertical slicing machine to help break up the casts and smooth the turf.

The **European crane fly** has become established in western Canada and the northwestern United States. The adult crane fly looks like a large mosquito, with a body about one inch long. Often large numbers of these flies gather about a house, but they do not bite or sting and will not damage houses.

Crane flies come from the soil of lawns and pastures from August to September. They lay eggs in the grass, which hatch into small gray brown larvae with tough skins that earn them the name of leatherjackets. These leatherjacket worms feed on the root crowns of clover and grass during fall, remain in the larval stage through winter, and resume feeding in spring. Damage from leatherjacket feeding is most noticeable in March and April. About mid-May, they stop feeding and go into a nonfeeding pupal stage just below the soil surface.

Many of these leatherjackets can live in the soil without seriously damaging lawns. Birds will feed on them, and no other control is required. Even highly infested lawns have shown recovery without the use of insecticides.

The **fall armyworm** is a brown-striped caterpillar that feeds on grass blades before becoming a large yellow-and-brown-striped moth. See armyworm under CORN for further description and control.

When turf is heavily infested with **nematodes,** it lacks vigor and may appear off-color, yellow, bunchy, and stunted. Grass blades dying back from the tips may be interspersed with apparently healthy leaves. Injured turf may thin out, wilt, and die in irregular areas. Symptoms of nematodes are easily confused with soil nutrient deficiencies, poor soil aeration, drought, insects, or other types of injury. Nematode-infested grass does not respond normally to water and fertilizer. Damaged roots may be swollen, shallow, stubby, bushy, and dark colored.

Nematodes are slender microscopic round-worms. There are many types, most of them harmlessly feeding on decomposing organic material and other soil organisms. Some are beneficial because they are parasites of plant-feeding organisms, including other nematodes.

Control nematodes by keeping your grass growing vigorously through proper watering and fertilization, as outlined above. See NEMATODE.

The **sod webworm,** which becomes the lawn moth, has caused considerable damage in western Canada. The larvae feed on the shoots and crowns of the grass but not the roots. Irregular brown patches appear on the turf, and

Sod webworm

the grass dies back from the shoot. The larvae are slender, gray with a brown head, and about ½ inch long. They are readily seen when the brown or dead sod is lifted. You can control them with milky spore disease; see JAPANESE BEETLE.

**White grubs** of the Japanese beetle and June bug can cause serious damage if present in sufficient numbers by cutting off grass roots. They are the favorite food of moles, and attract these rodents to lawns. Sod affected by grub damage can be rolled back over the spots where the grubs have been at work. If the lawn shows brown patches and loose sod in late spring or late summer, rake off all the loose turf and turn over the soil under it. Continue to turn it up at intervals of a few days until late fall. Birds will make short work of the exposed grubs. If you have chickens or ducks, they will make an even better job of it. See JAPANESE BEETLE.

# Diseases

By and large, good lawn practice will prevent or limit disease attacks from the start. Disease is encouraged by poor soil drainage; excess moisture; poor circulation of air because of surrounding trees, shrubs, or buildings; incorrect mowing; stimulation of grass with fertilizer during the summer; and high soil acidity.

Correct poor soil drainage and maintain adequate soil aeration to build your lawn, by permitting stronger root and top growth. On the other hand, a soggy water-soaked soil is an ideal environment for disease. Pathogenic organisms need an abundance of moisture for the early stage of spore development and infection of the plant.

Watering late in the evening is a cause of much lawn disease because the grass remains wet through the night so that mold and fungus growth are encouraged. Anything that favors the undue and prolonged presence of moisture, such as a heavy mat of grass clippings, contributes to the incidence of fungal growth. So, if you leave heavy clippings on the lawn for their mulch value, be sure to rake them over lightly to break them up and permit circulation of air. Turf areas that are completely enclosed by buildings, trees, or shrubs may suffer from poor circulation of air. As a result, the grass will remain wet for long periods and may be excessively warm. These two adverse factors, excessive

humidity and warmth, not only inhibit sound lawn growth but favor the development of grass disease as well. The remedy is simple: restore adequate circulation of air by pruning or removing some of the trees and shrubs.

Close mowing of bluegrass lawns weakens the grass and helps produce the succulent or tender leaf growth that is vulnerable to fungus. On lawns cut higher than 1½ inches, new leaves can be formed as fast as the lower ones are infected and no permanent damage will occur.

The bentgrasses, because of their ability to obtain moisture and nutrients from the soil due to their prostrate type of growth, can stand to be cut down to half an inch. Given favorable conditions the bents will successfully resist heat, drought, insects, and fungus attacks.

Fungal growth is stimulated by applying fertilizer to speed midsummer growth. Instead, fertilize in early spring and fall, when the danger of disease is reduced. Strongly acid soil is more disease-prone than a slightly acid or neutral soil, and proper liming can help keep your grass healthy (see the introductory section above).

An effective way to inhibit the development of disease is to plant a mixture of grasses. Different diseases attack different varieties. For example, brown patch affects the bentgrasses while leaf spot hits the bluegrasses. If only a single grass is planted, the disease can easily spread from leaf to leaf, but in a mixed turf the disease organisms soon reach a type of grass that is resistant and further progress is halted. Where possible, plant resistant varieties.

Try this three-step remedy if fungal diseases hit your lawn. First, use lots of compost, as it contains microorganisms that will ingest the disease spores. Second, inoculate affected areas with angleworms; they will eat the compost and grass clippings and provide rich fertilizer droppings. Third, irrigate the lawn with a canvas soil soaker (the soaker should be canvas, not the type made of plastic and pierced with tiny holes, which throw up a fine spray mist). The canvas soaker is laid across the lawn and water is run for 15 to 20 minutes, or until the soil is soaked to a depth of one foot.

**Brown patch** is a common turf disease of lawns and golf greens. It causes irregular spots of varying size, first colored a light yellow green and later turning to brown as the grass dies. The dead grass stays erect and does not mat down. All grasses are susceptible to brown patch. The responsible pathogen is a soil-borne fungus that attacks the roots, killing the fine feeding roots and then the entire root system. In periods of wet weather or when lawns are watered frequently, the fungus may grow up on the lower stems and leaves.

Brown patch flourishes under conditions of excess moisture and high temperatures. It occurs chiefly in acid soils, and is encouraged by constant watering and heavy fertilizing with high-nitrogen fertilizers. Brown patch is less likely to occur when the available nitrogen supply in the soil is low and phosphorus and potassium levels are high.

To control brown patch, be careful not to overwater the lawn. Set the mower high, for close cutting makes the grass more disease-prone. Remove the clippings after you mow, and avoid using nitrogen-rich chemical fertilizer.

**Chlorosis,** or yellowing of turf, is a symptom of iron deficiency. Iron is essential for the production of chlorophyll, the green pigment necessary for manufacturing food in plants. Of the turfgrasses, Centipede is the one most sensitive to the lack of necessary iron. The three factors that most commonly contribute to iron-deficiency chlorosis and dying out of Centipede in spring are overfertilization, extended dry periods during the previous fall, and too much lime.

Overfertilization results in luxurious growth of the grass and effects a deficiency of the form of iron that can be absorbed and used by the Centipede grasses. Chlorosis may develop immediately. During the latter part of the previous growing season, iron deficiency may have reduced the amount of food manufactured, causing the grass to enter the winter in a weakened condition. The result is that the turf is slow to reestablish an adequate root system in spring and consequently shows severe chlorosis during this period. If chlorosis is not too severe, the grass may regain color and recover. However, in the more severe cases the grass may actually starve to death as a result of failure to manufacture adequate food to maintain life.

Damage from late fall drought often goes unnoticed because homeowners assume that winter dormancy rather than drought is responsible for premature browning of the turf. Drought-damaged turf is badly weakened and enters the winter with a low food reserve that is inadequate to develop a sufficient root system. A poor root system will not be able to supply the plant needs during early spring.

Symptoms of iron deficiency in Centipede, as well as other southern turfgrasses, sometimes result from too much lime. Have the soil tested for pH and treat according to recommendations. You can prevent iron deficiency chlorosis (yellowing) by maintaining Centipede lawns under relatively low levels of soil fertility and at a pH below 6.

**Copper spot** is characterized by small, light copper-colored spots in the turf. The color is produced by masses of the fungus on the stems and leaves. Because the fungus confines itself to the lower leaves and stems, the spots are not readily seen except just after mowing. Copper spot rarely causes serious injury, and control measures shouldn't be necessary.

Another disease of minor importance is **corticum red thread,** also known as pink patch. Round or irregular patches of blighted grass, light tan to pinkish and one to six inches in diameter, develop during cool, moist spring and fall weather. Where it becomes severe, the spots may merge to form large, irregular areas with a reddish brown cast. On Bermuda grass the disease resembles winterkill. Bright coral pink or red threads about ¼ inch long protrude from diseased leaf tips and leaf sheaths. These fungus strands appear gelatinous in the early morning, and brittle and threadlike as the grass blades dry later in the day. Corticum red thread is most prevalent in fescues, Manhattan ryegrass, and bentgrasses.

Follow the preventive disease control practices suggested above. Maintaining a balanced high fertility level is important; remove thatch and collect grass clippings after mowing.

**Dollar spot** appears as round, brownish or bleached tan spots, from two to six inches in diameter, depending on the type of grass. It is more often seen on golf greens than home lawns. If left unchecked, the spots may merge to form large, irregular, straw-colored patches of dead grass. Individual blades are girdled by light tan lesions with reddish brown borders. All the widely used lawn grasses are susceptible, particularly bentgrasses, Bermuda grasses, and zoysias. New seedlings of tall fescue are often attacked.

Dollar spot develops during moist periods of warm (60° to 85°F.) days and cool nights. It is most active and damaging if there is a deficiency or great excess of nitrogen in the soil, or too much thatch. Injured turf recovers quickly if treated promptly. If left untreated, it may take weeks or months for new grass to fill in the sunken dead areas.

For control, keep thatch to a minimum, water only when needed (to a depth of 8 to 12 inches, and not late in the day), mow high, and rake vigorously. Maintain adequate soil fertility.

**Fading-out,** also called helminthosporium leaf spot, foot rot, and melting-out, attacks bentgrasses, fescues, and Kentucky bluegrass. Pure stands of Kentucky bluegrass favor the development of this disease, whereas recommended mixtures usually contain some naturally resistant species. It occurs in all parts of the country, and is most destructive during hot, humid weather. Diseased areas appear yellow or dappled green, as though the grass were suffering from iron deficiency. When the disease is not controlled, the grass fades out, leaving dead, reddish brown patches. Eventually, large irregular areas of the lawn may be killed.

Close examination of infected leaves in the early, or leaf spot, stage usually reveals brownish lesions, shaped oblong, and parallel to the leaf blade. The disease progresses to leaf sheaths, crowns, rhizomes, and roots, causing them to rot until the entire plant is killed. During hot weather, the disease may cause the sudden death of large areas of lawn; dead plants appear to have died from drought. Weeds and crabgrass usually invade these areas. As moist conditions favor this disease, it first appears in shaded areas. Fading-out is most severe on closely clipped turf.

To combat fading out, follow good cultural practices to ensure healthy grass that will resist the fungal spores. Reduce shade in the yard and improve soil aeration and water drainage. Mow at the recommended height, and avoid excesses of nitrogen in the soil, especially in spring. Try not to water in the late afternoon or evening. Birka, Fylking, Pennstar, Merion, and Newport are leaf spot-resistant varieties, as are the Tifton turf Bermuda hybrids.

**Fairy ring** appears as a circular ring of fast-growing, dark green grass, often with a ring of thin or dead grass inside or outside. The rings (they are not always closed, and some look like arcs or horseshoes) vary in diameter from a few inches to 50 feet or more. The strip of thin or dead areas may be three to six inches wide. After rains or heavy watering, mushrooms often appear in the dark green grass ring.

All turf grasses are subject to fairy ring, a disease caused by fungi of the mushroom family.

Growth starts at a central point and spreads outward at the rate of a few inches to two feet or more per year. The fungus grows throughout the soil, forming a dense, white, threadlike growth to a depth of eight inches or more that keeps water from the roots of the grass.

The condition is controlled by aerating the ring with a spading fork or hand aerifier, making holes two inches apart and as deep as possible—at least four to six inches. Begin about two feet outside the ring and work toward the center. (Do not use the spading fork again without washing it thoroughly.) The aeration will allow water to penetrate the soil, and deep soaking every other day will help improve the condition and allow the soil to regain its health. This deep soaking is essential to speedy recovery.

It may be necessary to dig out the fairy rings, a laborious procedure. Remove all infested soil, digging two feet below the deepest extent of the white mold. Replace it with fresh soil, making sure that the top of the hole is covered over with a rich layer of properly composted humus. If you have been applying any commercial fertilizers to your lawn, particularly those high in nitrogen, discontinue their use.

**Fusarium blight** may affect lawns more than two years old. Light green patches several inches in diameter appear first, changing to a dull reddish brown, then to a bleached tan color. Patches of brown and thinning grass may be round or irregular, up to two feet or more in diameter. Apparently healthy green grass may grow within the center of these patches. Temperatures over 75°F. and high humidity favor the development of fusarium blight. Its severity appears to be related to drought stress; the disease may appear in one or two days in summer if night temperatures remain high. Bentgrasses, fescues, ryegrasses, and bluegrasses are all susceptible, but some varieties are more resistant than others. Try A-34, Adelphi, Bonnieblue,

Glade, Sydsport, or Touchdown. Merion and Fylking bluegrass are especially susceptible to fusarium blight.

To control, water deeply to avoid drought stress. Avoid excessive nitrogen, especially during hot summer weather. Keep pH above 6.2. Mow at the suggested height, and keep thatch to a minimum. In experiments at the University of California at Riverside, blight in Kentucky bluegrass was kept in check by mixing bluegrass seed with at least 10 percent perennial ryegrass seed.

**Lawn rot** is caused by a fungus that attacks the stems and leaves of the plant to cause a soft rot of the tissue. This rotting gives a matted appearance of the affected areas, unlike brown patch. Newly seeded grass is more likely to be affected than older established turf. On new seedlings, diseased areas look as though soaked with gasoline. The causal fungus requires plenty of moisture and warm temperatures, and therefore is destructive only during periods of warm, wet weather or when the grass is watered frequently on warm summer nights.

To prevent an outbreak of this disease, water infrequently and in the forenoon. Bentgrasses usually suffer more from this disease than other types.

**Powdery mildew** causes leaf blades to appear dusted with flour or lime. Close inspection reveals patches of a whitish, powdery growth. Infected leaves often turn yellow and wither. New plantings may be killed when mildew is severe; established plantings resist it better. Mildew is most severe on Kentucky bluegrass grown in the shade, and occurs chiefly when nights are damp and cool, in late summer, fall, and spring.

Control powdery mildew by increasing air circulation, and pruning dense trees and shrubs to reduce shade. Keep the lawn vigorous by organic fertilization and good drainage, while avoiding an excess of nitrogen. Mow your grass frequently at the prescribed height. Some varieties of Kentucky bluegrass are more resistant to mildew, and you may find it helpful to plant one of these if conditions in your lawn seem to favor the disease. They include Aquila, Birka, Glade, and Nugget.

**Pythium blight,** also called greasy spot or cottony blight, causes round or irregular spots from a few inches to several feet in diameter during very hot, wet weather. The spots are first water-soaked and dark, and then fade to a reddish or light brown as the leaves dry out and wither. A greasy border of blackened, matted grass blades, often covered with a cottony mass, is seen when the pythium fungus is active. The patches may merge and form streaks, since the fungus is spread by flowing water and mowing. It is most common on newly established turf, although if conditions are favorable it occurs on established grass too. The disease may spread very rapidly, killing large areas of turf overnight. Dead grass lies flat on the ground, rather than remaining upright like grass affected by brown patch disease. New grass does not grow back into the diseased area. The fungi are most active at daytime temperatures of 85° to 95°F. (night temperatures falling no lower than 68°F.), and when humidity is high, growth is dense and lush, and soils are poorly drained.

To control pythium blight, avoid watering methods that keep the foliage and ground wet for long periods. Maintain a proper balance of nutrients, being careful to avoid the excess of nitrogen that stimulates lush growth. Improve surface and subsurface soil drainage. Where feasible, delay seeding until the weather is cool and dry.

**Rust** does not usually become a problem until summer, when extended dry periods slow the growth of grass. Heavy dew favors rust development. Some varieties of Kentucky bluegrass, particularly Merion, and the newer rye-grasses are very susceptible to the rust fungi. Close examination of rust-infected grass shows powdery rust-colored or yellow orange spots. The powdery rust will rub off easily on your fingers or a cloth. Continuous heavy infection of rust causes many grass blades to turn yellow, wither, and die. Severely rusted lawns are vulnerable to winterkill.

Damage from rust is less severe if Merion Kentucky bluegrass is mixed with common Kentucky bluegrass or with red fescue in a one-to-one proportion. Frequent mowing at recommended heights, organic fertilizing, and correct watering in hot, dry weather will help eliminate rust.

**Seed rot** and damping-off are diseases caused by soil-inhabiting fungi and are troublesome during the cool, wet weather of spring. The seeds rot in the soil, just after the seed coat is broken. Seedling grasses on new lawns appear water-soaked, turn yellow or brown, and rot off at the surface of the soil. Surviving plants are weakened, and affected areas are often heavily invaded by weeds. Damping-off is most severe on heavy, moist, or waterlogged soils, and where seeding rates have been excessive.

To control seed rot, sow top-quality seed at suggested rates in a well-prepared, fertile seed-bed. If possible, seed in late summer or early fall. Provide for good surface and subsurface soil drainage when establishing a new lawn. Fill in low spots where water may stand; properly graded lawns usually do not suffer from this disease. Water only when necessary.

Lawn injury from **septoria leaf spot** and tip blight resembles that caused by a dull mower. Leaf blades are light yellow from the tip downward. Close inspection usually shows black dots (the fruiting bodies of the septoria fungus) embedded in the diseased tissue. Smaller lesions, ⅛ inch or more in length with red or yellow margins, may also be present. The septoria fungus is active during the cool, wet weather of spring and fall. During spring rains, the spores are splashed to healthy leaves where infection occurs, often in the cut ends of grass blades. Septoria leaf spot is usually of minor importance during the summer.

Control by following cultural practices that maintain a vigorous, healthy turf.

There are two species of **slime mold** that commonly appear in lawns. One of these molds appears as a bluish gray, sooty growth on the grass, the other as large masses of a dirty-yellow growth. When the molds dry, they form powdery substances that easily rub off the grass blades. Neither of these organisms do any harm to lawns, although they may appear to smother or shade otherwise healthy grass. They are soil-inhabiting microorganisms that feed on decaying organic matter, and are not parasitic on the grass.

Slime mold soon disappears, but if you wish to remove it, rake, brush, or hose down the grass with a forceful stream of water. Reduce thatch accumulation, as it favors the growth of the mold fungi.

There are two prevalent types of **snow mold:** gray snow mold and fusarium patch (or pink snow mold). Both diseases are most serious when air movement and soil drainage are poor and when grass stays wet for long periods at near freezing temperatures. Snow mold is more common in northern areas. Damage often conforms to footprints, paths, and snowmobile or ski tracks because snow compaction and plant injury favor the disease. Snow mold may be especially severe in low places and on north slopes that have a deep snow cover for a long time. Most lawn grasses are susceptible to snow mold diseases; bentgrasses are more severely attacked than coarser lawn grasses. When warm weather arrives and the grass dries out, these diseases make no further progress.

Gray snow mold appears as grayish or straw-colored spots several inches to two feet or more in diameter. Some spots may merge to form large, irregular areas; where severe, the entire lawn may be affected. As the leaves are killed they turn brown and frequently mat together.

Fusarium patch causes bleached tan or light gray patches of indefinite shapes, one to eight inches in diameter, sometimes enlarging to one or two feet. At the advancing edge of melting snow, the spots may have pinkish margins. Snow is not necessary for the development of fusarium patch. It can occur anytime in cool (below 60°F.), wet weather in fall, winter, or spring. Diseased areas under snow may be covered with a dense, slimy mat of fungal filaments that turns a faint pink when exposed to light.

To control both types of snow mold, avoid overfertilizing with nitrogen, especially in late fall. Where winters are severe, lawns should not enter cold weather in an actively growing condition. Promote air circulation and good soil drainage. Mow frequently at suggested heights and rake vigorously to prevent a heavy mat of grass from forming. Use winter-hardy grass species such as Kentucky bluegrass and creeping red fescue.

**Stripe,** or flag smut, is most noticeable during spring and fall because it is favored by cool temperatures. Infected plants may occur singly or in patches of a few inches to a foot or more in diameter.

Infected plants are often pale green to yellowish and stunted. Individual leaf blades may be curled and show black stripes with black powdery spores. The stripes run parallel with the leaf veins. When first developing, they are yellow green; later the stripes turn gray and finally black. The leaf twists, curls, and shreds from the tip downward. Infected plants may die during hot, dry weather. In other cases, the symptoms disappear.

Spores of the smut fungi germinate in the soil and thatch. They grow throughout the plant tissues and remain within the plant until it dies. Smutted plants in newly seeded lawns are rare. Watering and high fertility encourage their buildup. Varieties of grass differ greatly in their resistance to smut. Particularly resistant are the bluegrass A-34, Bonnieblue, Glade, Nugget, Sydsport, and Touchdown.

**Sunburn** and drought injury are very similar in appearance but may occur independently. Sunburn sometimes occurs when a hot spell follows cool, cloudy weather. Drought causes the same brown discoloration of sunburn but over larger areas, first in the open sunny areas and later under trees if the drought is prolonged. Watering before the serious browning appears will forestall drought injury if you soak the soil several inches deep. Neither sunburn nor drought cause permanent injury, and grass will recover as soon as fall rains and cooler weather arrive.

## Other troubles

**Algae** may form a greenish or blackish scum on bare soil or thinned turf, especially in damp, shady, heavily used and compacted areas. The algae, which are a mass of minute green plants, dry to form a thin crust that later cracks and peels.

Improve air circulation and soil drainage to correct this condition. If it persists and becomes serious, upgrade your soil's fertility, and reseed the area.

**Buried debris** may cause diseaselike symptoms in the grass. A thin layer of soil over rocks, lumber, bricks, plaster, concrete, or other building materials will dry out rapidly in summer weather. Eliminate the condition by digging up

suspicious areas, removing the cause, and adding good topsoil.

Thin turf or bare spots resulting from heavy use result in **compacted areas** on the lawn. Waterlogged and heavy-textured soils will easily become compacted, especially if they are walked on constantly. Water flows off these areas and plants may die of drought.

Compaction can be remedied by aerating the soil (you can rent an aerifier at garden supply stores) or by installing drainage tile for seriously waterlogged soils. You can also core compacted areas, using a hand corer or power machine. Coring is a method of cultivating which uses hollow tines or spoons to remove soil cores. The resulting holes allow water and air to penetrate the soil. If necessary, fertilize and reseed compacted areas. Foot traffic on lawns can sometimes be reduced by putting in a walk, patio, or parking area, or by planting some shrubs.

Injury from **dog urine** often resembles brown patch or dollar spot. In affected areas, grass turns brown or straw colored and usually dies. These injured areas, up to a foot or more in diameter, are often bordered by a ring of lush, dark green grass.

Chemical fertilizers applied on wet grass may cause serious **fertilizer burn.** Rugs, rubber mats, or metal dishes left on the grass in the hot sun will also leave their mark. Heavy watering helps injured spots to recover.

**Moles** raise ridges in the lawn as they burrow through the soil in search of grubs and worms. Because they eat soil insects, moles are considered beneficial. However, the grass above the raised tunnels will die if the roots are exposed to the tunnel air, and intruding mice may nibble on roots and bulbs.

If you can get rid of grubs in your lawn, mole troubles will decrease; see the discussion of white grubs, above. You'll find several types of mole traps in hardware stores. Locate the main runway by rolling or tamping down all of the raised runways. Watch carefully at hourly intervals to determine which one is raised first—this is probably the main runway and the one over which the trap should be set. The trap may be placed anywhere along the runway, but a straight section of burrow is preferable. If you loosen the soil where the trap will be set with a fork or trowel, the trap's action will be easier and faster. Tamp down the main runway again before setting the trap. If only the short section of runway where the trap is set is tamped down, the mole may go around it; if the entire runway is flattened, it will be less cautious.

If the mole is not caught in 24 hours, it has probably abandoned that runway. Tamp down all the runways again and reset the trap on another that is being used. In many cases, moles use their own and other burrows interchangeably. It may be possible to catch several moles by resetting the trap in the same place after each dead mole is removed.

If **moss** appears on your lawn, it probably indicates plenty of moisture in combination with a lack of phosphorus and potash in the soil. Sometimes excess shade, poor drainage, and soil compaction will aid moss growth. Moss can be removed by hand raking. Improve the soil with organic fertilizers, followed by reseeding, to eliminate moss permanently.

**Mushrooms** and toadstools can be an annoyance. Some have an unpleasant odor, and some may be poisonous and hence a menace to children and pets. They do no harm to the grass, however. Mushrooms are the aboveground growth of certain fungi that grow on decaying vegetable matter in the soil. In lawns, this might be buried stumps or tree roots, logs, boards, or a thick thatch. Mushrooms often pop up following heavy rains or watering, and may indicate a too-acid soil.

No compound will kill mushrooms and toadstools without injury to the grass, but they can be removed by raking or sweeping. They will not completely disappear until the buried material has completely decayed, and in some cases it may be best to dig up these pieces of rotting debris. Test your soil to check its pH; if necessary, follow the procedures for liming as described above.

**Salt damage** to lawns and shrubs may result if salt is used to remove winter snow from paths and sidewalks. As the snow or ice melts, it carries a salt solution to plants, causing roots to lose large amounts of water. The responsible phenomenon is osmosis, a process whereby water already existing in roots will move out through root membranes in order to dilute and create a balance with the salt concentration in the soil. Without the necessary water in its root system, the plant dies or is damaged.

Use minimal amounts of salt, and exercise care in its application near grass and shrubs. Use other methods to rid walks of snow or ice if possible.

Along the shore, salt spray or flooding with high tides causes serious injury to lawns. The injury from spray is temporary, but if the area is flooded for several hours the damage is likely to be permanent. If the flooding is of short duration, flushing with fresh water immediately afterward will minimize the injury.

**Skunks** may dig holes in your lawn in search of grubs and worms. If you kill the grubs in your lawn, skunks will probably go elsewhere. Read the controls for white grubs, above.

See also WEEDS.

# LEMON

See CITRUS FRUIT.

# LETTUCE

## Insects

A number of species of **aphid** occur on lettuce, some feeding on leaves and a few on roots. See APHID.

The **cabbage looper** is a large, pale green measuring worm with light stripes down its back. It grows to 1½ inches long, and doubles up as it crawls. The looper usually feeds on the undersides of leaves, producing holes that are ragged and unsightly. It occurs through the United States and is particularly troublesome in Florida, Arizona, and California. For habits and control measures see under CABBAGE.

Fat, smooth **cutworms** may be kept from new lettuce plantings with loose collars of tar paper or cardboard placed around the seedlings when set out. The collars should be 1 to 1½ inches in diameter and wide enough so that ½ inch lies under the soil and at least an inch remains above ground. See TOMATO for other measures.

The **fall armyworm** attacks foliage of lettuce. It grows up to 1½ inches long and varies greatly in color, from light tan to green to almost black. Three yellowish white hairlines run down the back, and on each side is a dark stripe paralleled below by a wavy yellow one that is splotched with red. The head is marked with a prominent white V or Y. See armyworm under CORN.

Two **leafhoppers** are responsible for much damage to lettuce: the sixspotted (or aster) leafhopper is yellow green with six black spots and the potato leafhopper is green with white spots on its head and thorax. Both are small (⅛ inch long), wedge-shaped, and apt to go unnoticed until the damage is done. They feed by sucking plant sap, and this causes areas of leaf tissue to turn pale. In addition, the sixspotted species transmits a serious disease, aster yellows. For control measures see fusarium yellows, below, and potato leafhopper under POTATO.

Tender lettuce leaves are very vulnerable to the ravages of **slugs**. These are molluscs, not insects, and are closely related to snails. They hide during the day, and their night travels are recorded by silvery, slimy trails. A number of control measures are described under SNAIL AND SLUG.

You'll have to look sharply to catch a sight of the **tarnished plant bug**. This flat, ¼-inch-long insect is very shy and will likely retreat to the opposite side of its stem as you approach. Its green or brown back has touches of black that cause a tarnished appearance. A very close look will reveal a yellow triangle, with one black tip, on each side of the insect. For habits and control of this pest, see ASTER.

**White grubs** are white with brown heads and are found in the soil in a characteristic curved position. See under CORN for a description of habits and control measures of this root pest.

**Wireworms** may also attack the roots of lettuce. These worms are shiny, hard, jointed, and range in color from yellow to brown. See WIREWORM.

## Diseases

Rotting of the edges of lower leaves is an early symptom of **bacterial soft rot**. The leaves of the head finally dissolve into an unappetizing, slimy rot. The bacterium that causes this disease lives in the soil and is splashed up on the leaves by rain; hilled-up rows and well-drained soil should protect the plants.

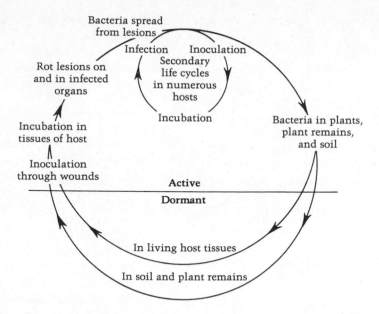

Bacteria spread
from lesions

Infection    Inoculation

Rot lesions on
and in infected
organs

Secondary
life cycles
in numerous
hosts

Bacteria in plants,
plant remains,
and soil

Incubation

Incubation in
tissues of host

Inoculation
through wounds

**Active**

**Dormant**

In living host tissues

In soil and plant remains

Life history of the soft rot bacteria

Reprinted, by permission, from Pyenson, *Fundamentals of Entomology and Plant Pathology*
(fig. 9.12), AVI Publishing Co., PO Box 831, Westport, CT 06881.

Lettuce doesn't show the telltale signs of **big vein** until the plants have developed several true leaves. From five to six weeks after the seed or plants are set in diseased soil, the symptoms become apparent: the leaves yellow along the veins and the entire leaf becomes thickened, crinkled, and brittle. Although no tissue is actually killed and the plant may well go on to produce a normal head, the head is smaller and less firm than usual and the leaves are of inferior quality.

Once affected with big vein virus, soil may remain capable of causing the disease for a number of years. Either grow lettuce in virus-free soil or discourage the disease in infested soils by rotating lettuce with crops that don't carry the virus. Lettuce should not be planted on the same land more than once in three or four years. Rogue and destroy scattered afflicted plants.

Although not generally widespread, **bottom rot** is regularly destructive to lettuce in some localities. It is at first characterized by rum-colored, sunken lesions on the petioles and the midribs and a slimy rot on the lowest leaves. The disease then spreads upward, leaving the head a slimy mass. The main stems and midribs remain solid.

The bottom rot fungus enters the plant through bottom leaves that touch the soil, and varieties that have an upright habit of growth (such as the cos varieties) are less susceptible than others. The only practical means of control is rotation with crops not attacked by the causal organism, such as sweet corn or onions. In areas plagued by bottom rot, only well-drained soil should be planted to low-growing varieties.

If yellowish or light green areas appear on the upper surfaces of older leaves, your plants may be victims of **downy mildew.** As these areas enlarge, whitish to gray tufts of mold appear on them. The spots later turn brown, and infected heads are apt to break down with secondary rots, especially in transit. Downy mildew is most likely to be a problem in damp, foggy, moderately warm weather.

Wild lettuce near the garden or greenhouse should be eradicated as a precautionary measure. Several strains of Imperial lettuce are highly impervious to downy mildew.

**Drop,** or sclerotiniose, is caused by a fungus that may exist in the soil for several years. Young seedlings grow few leaves, and these quickly wilt and die. Older and larger plants may be affected in either of two ways: some tend to wilt quickly and take on the appearance of a dull green, wet, folded rag; others may show a few water-soaked leaves at first, with brittle stems that either appear glasslike when cut across or show a reddish discoloration, and the head eventually becomes an odorless, watery, brown, rotted mass.

Because this disease develops rapidly under cool, moist conditions, it is best to use on the surface a well-drained soil that dries out quickly. It sometimes helps to ridge the soil to prevent water from accumulating around the plants. Plants should not be crowded, as the fungus may otherwise spread from plant to plant. If weeds are kept down, the disease will be deprived of hosts and ventilation and surface drying will be improved. Rotation with crops other than cabbage, beans, celery, tomato, and cucumber may be helpful. Large-scale growers can plow deeply to cover the fruiting bodies of the fungus, and pasturing sheep in lettuce fields after harvest helps to clean up refuse.

When seedlings are attacked by **fusarium yellows,** they wilt and growth is stunted. Older plants often do not form heads, have white hearts, become lopsided, and their vascular system turns brown.

This disease is carried by leafhoppers, and control depends on keeping these insects away from the lettuce. Keep weeds down in the vicinity of the garden, and avoid planting lettuce near carrots or asters, two other plants susceptible to yellows.

**Mosaic** causes leaves to appear a mottled yellow and green, and plants take on an overall yellowish cast. Growth is clearly stunted, and usually no head is formed. The effects of the disease are most marked in warm weather when leaves may brown at the margins and then die. The disease is spread by various species of aphid. The chief source of trouble in lettuce fields is from seedlings from infected seed; unless removed as soon as symptoms are apparent, they may infect an entire field within a short time after the aphids make their appearance.

You should therefore be familiar with symptoms of mosaic, and be sure to water seedlings carefully. Clean up weeds in order to eliminate possible host plants or lodging for aphids. Large-scale lettuce growers will benefit by discing out lettuce beds and fields immediately after harvest, as this reduces chances of mosaic spreading from these older areas to healthy young fields. It also helps to plan on avoiding interplantings of young and old growing areas. Several tolerant varieties are available. See also APHID.

The first disease that lettuce plants are exposed to is likely **seed rot** (or damping-off), caused by soil-inhabiting fungi. Seed decays and the stems of young seedlings show a water-soaked condition near the ground line. Plants eventually topple over and die. Occasionally, damping-off causes some damage to plants in the field, but it is primarily a disease of seedbeds. It often appears in small localized spots, but may spread rapidly and destroy an entire bed if ventilation is poor and the bed is kept too damp.

Avoid excessive moisture and let air get to the plants. Chances of afflictions are further minimized by using a light, sandy soil that dries out rapidly, especially for the surface covering of the beds. If your seedbeds have a history of trouble, either change the soil or move the seedbed to a new location.

**Tipburn** is a widespread disease of lettuce, occurring everywhere lettuce is grown in both greenhouses and in the field. The first signs are small, yellowish translucent areas near leaf margins. As these enlarge and become more numerous, tissues near the edge of the leaf die, thus forming an irregular brown border along the perimeter. Veins of the affected areas of the leaves often turn dark and may become infected with soft rot bacteria. Symptoms are aggravated on hot, dry summer days following cloudy weather or heavy irrigation, as these conditions cause rapid loss of water from the leaves.

Plants that have become succulent from rapid growth are particularly susceptible, and you should avoid conditions that favor such growth, especially in the late stages. A crop grown with a uniform moisture supply at a fairly high level is more likely to escape injury than one with irregular growth caused by a fluctuating water supply. Excessive fertilization, especially with readily available forms of nitrogen, should be avoided in areas where crops mature during warm weather. Several lettuce varieties are resistant to tipburn, including Slobolt, Summer Bibb, and Ruby.

# LILAC

## Insects

The **lilac borer** attacks lilac, privet, and other ornamental shrubs by tunneling under the bark and into the wood, thus girdling the stems and causing foliage to wilt. Roughened scars showing the old borer holes may occur on larger stems at places where the borers have worked for several seasons. This creamy white caterpillar is about ¾ inch long when fully

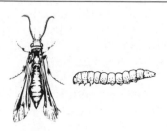

Lilac borer adult and larva

grown. It passes the winter in tunnels in the stems. The adult, a clear-winged moth, emerges in spring and usually lays its eggs on roughened or wounded places on the bark.

Make a thorough examination of the bush before the spring season arrives and cut and burn any dying and unthrifty stems that may contain borers. During the summer season, check to see if fine boring dust is being pushed from small borer holes. Such holes should be cut out with a sharp knife. If the tunnels are fairly straight, the borer can be killed by probing with a flexible wire, or pulled out by means of a hooked wire to make certain it is destroyed. Further damage can be avoided by using care so that the limbs are not damaged.

The small **lilac leafminer** mines and rolls the leaves of both lilac and privet. Where this little pest is abundant, bushes appear scorched. The adult moth shows brown forewings mottled with silver and with two silvery bands across the middle. Winter is spent in debris-covered cocoons in the ground under the plants. The moths usually emerge during May, and the larvae mature in July. From late July through

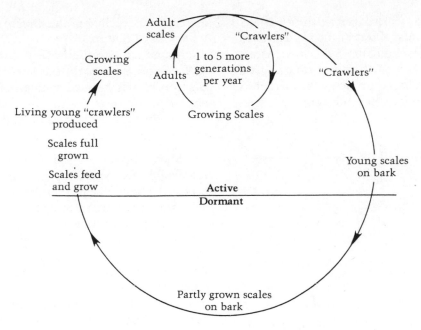

Life history of the San Jose scale

Reprinted, by permission, from Pyenson, *Fundamentals of Entomology and Plant Pathology* (fig. 3.12), AVI Publishing Co., PO Box 831, Westport, CT 06881.

August, another batch of eggs is laid, and the larvae from these eggs mature in September, drop to the ground, and form their overwintering cells. During the first stages, the greenish white larvae mine between the epidermal layers of the leaves to form a large, unsightly blotch. There are generally from three to eight larvae in a mine. When the larvae are nearly full grown, they emerge from the mines and roll down the leaf tips and feed.

Remove any rolled leaves that are on the bush and burn them. It may be necessary to prune back the leaves until healthy growth remains. Spray with a strong soap solution in July, before the insect grows out of the larval stage.

**Oystershell scale** [color] is distributed widely over the United States. The common name is well chosen as the shells of this scale look very much like a miniature oyster. The scale cover is about ⅛ inch long and colored ash gray to black. The body of the female is yellowish white. Eggs are pearly white to yellowish. The scales overwinter in the egg stage on twigs and branches of the host. There is usually one generation of this scale insect in a year. Injury consists in devitalizing the plant by sucking the plant juices. In cases of severe infestation, plants are often killed outright.

Use a 3 percent oil emulsion in late winter, before the plants begin to bud.

Shrubs and trees heavily infested with **San Jose scale** cover the twigs, branches, and stems with a grayish layer of their tiny, overlapping, waxy shells. Injury occurs as dead or dying branches, thin foliage, and a weakening of the plant. The waxy scale covering the female is circular, grayish, and about $1/16$ inch in diameter; the male is smaller and more oval. Scale insects pass the winter in a partly grown state and are nearly black in color. From two to six generations are produced annually, depending on the length of the season.

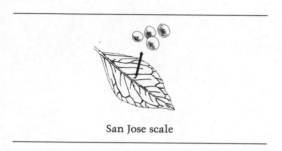

San Jose scale

Since San Jose scale winters in a partly grown condition, it is more easily killed by dormant-oil sprays than oystershell scale. Spray in spring before the buds open; a soap solution can be substituted. Before spraying, remove whatever dying and heavily infested parts of the plant that can be spared. During late spring and summer when the young scale insects are in the crawling stage, many of them can be killed with a white-oil emulsion.

## Diseases

**Bacterial blight** infects leaves, causing brown to black spots that may cover the whole leaf surface. Bacteria may enter the twigs directly or pass from blighted leaves into the twigs. When girdling takes place, the shoot turns black and dry black leaves cling to the twigs. Flower buds also become blackened and flower clusters turn limp and dark brown. The bacteria overwinter in diseased twigs. Blight is most severe during moist, mild weather and appears as the young shoots develop. White-flowered varieties are more susceptible to the disease than those with colored flowers.

Prune out dense growth so that the lilac will devote its strength to a few limbs. If the lilac bush does become contaminated, prune again; cut out and destroy affected shoots as soon as they are noticed. Tools should be disinfected with alcohol.

**Shoot blight** is caused by a fungus that is responsible for symptoms similar to those of bacterial blight except that the diseased stem areas are dark brown instead of black and killing of the shoots is more extensive. Blossoms and tender growing tips are blighted and turn brown. The shoots may be killed back four to five feet. Root sprouts that come up under the shrubs often are killed.

Control for shoot blight is much the same as it is for bacterial blight. Do not plant lilacs and rhododendrons close together, since the same fungus attacks both shrubs.

# LILY

## Insects

A number of **aphids** feed on lily, and one is fond enough of the plant to be named the purplespotted lily aphid. This $1/8$-inch-long pest feeds on the undersides of lower leaves, later on stems, and finally on buds and seed pods. It most frequently infests regal, formosanum, speciosum, and other late-flowering garden lilies. Feeding causes yellowed foliage, and the new growth of some varieties is killed prematurely. Aphids transmit mosaic and necrotic fleck (below). See APHID.

Other possible enemies of lily are the bulb mite (see GLADIOLUS), fuller rose beetle (ROSE), striped and spotted cucumber beetles (CUCUMBER), SYMPHYLAN, and yellow woollybear caterpillar (see imported cabbageworm under CABBAGE).

## Diseases

A fair number of diseases prey on lilies, and most are discouraged by thinning plants to improve ventilation, providing adequate drainage, and roguing infected plant parts. To be sure of disease-free plants, grow them from seed in an isolated part of the garden.

**Botrytis blight** is a fungus that principally attacks foliage, damaging or killing it. Small round or irregular orange red spots appear on leaves, and in severe cases the spots may run together and kill the leaves. The fungus spends the winter in the form of small fruiting bodies that produce spores in warmer weather.

Botrytis can be controlled to some extent by cleaning up the garden borders where fruiting bodies may have accumulated.

**Bulb** or **basal rot,** a fungal disease, attacks bulbs through the roots and causes the bulbs to degenerate. Plants often eventually die. Symptoms may resemble tipburning and root rot. Dig up and inspect the bulbs if trouble is indicated, and rogue out any infected plants. Because this disease persists in the soil, you should move the lilies to another area.

**Gray mold** is responsible for oval or circular spots which are reddish brown at first and develop pale centers and purplish margins. These spots may run together and rot the entire leaf, progressing into the stem and causing the stalk to fall over. If the spots dry out, they turn brown or gray. Buds or flowers may turn brown and rot, often showing the gray mold of the fruiting stage. This disease shows up and spreads rapidly under humid, cool conditions, especially if plants are crowded.

The usual control methods apply here: thinning, sanitation, and roguing. In the greenhouse, the disease is discouraged by warmer temperatures and a drier atmosphere.

Perhaps the most dangerous disease of lilies is **mosaic,** and it is this malady that has given the plant a reputation for being hard to grow. Mosaic is caused by a virus that lives in the plants, not in the soil or in dead plant tissue. When the disease first appears in the flower garden, it causes leaves to show a mottling of light or dark green. Leaves start to die from the ground upward, and the stems of the flowers often have crooked necks.

Mosaic is often transmitted by aphids, and their numbers should be managed (see APHID). The disease is also spread when lilies are propagated by bulb scales.

Aphids are also responsible for transmitting **necrotic fleck,** a virus that causes yellow flecks to appear on leaves. These blemishes change to gray or brown, and the surface is depressed but unbroken. Plants are dwarfed and have curled leaves. Flowers are small and develop brown streaks by the time they are fully open. See APHID.

# LIME

See CITRUS FRUIT.

# LINDEN

## Insects

Small linden trees are often injured by the larvae of the **linden borer.** The eggs of this beetle are laid in the bark, and the hatching larvae eat out a large cavity near the base. Decay follows and the tree may die. The white larvae, nearly an inch long, usually burrow near the ground, sometimes going into the roots below the surface. The adults are long-horned black beetles, about ¾ inch long and covered with a dense, greenish, downlike hair. These beetles emerge late in summer and feed on leaves and tender shoots before laying eggs in the bark. Prune and destroy affected plant parts.

Other insects that trouble linden trees from time to time include APHID, JAPANESE BEETLE (the variety Chancellor is resistant), oystershell scale (see LILAC), WHITEFLY, and whitemarked tussock moth (defoliates linden trees, especially in cities; see HORSE CHESTNUT).

## Diseases

For diseases likely to trouble linden trees—anthracnose, leaf spot, and powdery mildew—see TREES.

# LOCUST

## Insects

The dwarf flowering locust or rose acacia is sometimes infested by **aphids.** See APHID.

The **locust borer** [color] is very destructive to black locust. Trees are usually attacked in their fourth or fifth year by larvae boring into the trunk and branches, but once the trees reach

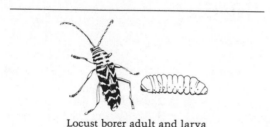

Locust borer adult and larva

a trunk diameter of about six inches the locust borers leave them alone. Evidence of infestation is sawdust falling on the bark of the trunk and wet spots around the holes where the sawdust was ejected. Later, ugly scars show where the wounds have partially healed. Branches are weakened and may break, foliage becomes yellow and dwarfed, and entire trees are sometimes killed.

Locust borer damage

The eggs are deposited from August to October in bark crevices or wounds. Hatching larvae bore into the bark to overwinter, and bore into the wood the following spring. The adult beetles, which resemble the painted hickory borer, emerge during August and September and are very abundant on goldenrod blossoms in September.

If a tree is so badly damaged as to make it useless or is too limby to be used as a post, it should be cut off at ground level and the stump left to sprout. Localized infections of the bark can be removed, provided the tree will not be girdled in the process.

Individual borers can be killed with nicotine sulfate. Dip a small piece of absorbent cotton or soft cloth into a solution of one part nicotine sulfate to four parts water, stuff it into the borer's hole, and seal the opening with putty.

The **locust treehopper** sucks sap and appears in three or four generations a year. As an adult it is an active brown insect of less than ¼ inch in length. If damage is considerable, control by a spray of pyrethrum or nicotine sulfate, or dust once a week with a mixture of nine parts sulfur and one part pyrethrum powder.

The **locust twig borer** is particularly troublesome to black locust. The larvae tunnel into the small new twigs, causing galls or swellings measuring one to three inches long that crack open and disfigure the branches. The branches then weaken, and the tree's vitality is reduced. In its adult stage this insect has ashy brown forewings marked with a pinkish white patch and several small blackish spots. The hindwings are mouse gray. The full-grown larva is about ½ inch long and reddish to straw yellow. Its head is brown and its thorax yellow.

The adults deposit their eggs on twigs in May and June. The eggs hatch within a week and larvae bore in twigs for a month. Larvae leave twigs when fully grown and spin a cocoon on the ground to pass the winter until they emerge from their cocoons in spring.

All branches that show evidence of the presence of the borer should be cut and destroyed in early summer before the larvae escape. A blacklight trap may prove effective in whittling down the population of the borer if used in May or June when the insect is in the moth stage.

For control of the carpenterworm, see TREES; the mimosa webworm is a pest of honey locusts (see MIMOSA).

## Diseases

**Brooming disease** is a virus that causes an abnormal development of black locust buds into short, spindly shoots having abnormally small leaves. Buds on these spindly branches also develop abnormal branches, giving the tree a broomlike appearance. The brooming typically occurs late in summer and the shoots often die during winter. Roots of infected trees are shorter, darker, and more brittle than normal. Excessive rebranching of the roots gives the appearance of root brooms.

There is no effective control for this disease. Remove and destroy affected trees.

Sometimes **cankers** appear on the small branches of locust, causing a dieback or wilt of branch tips. The first sign of trouble is small depressed areas that may enlarge to girdle the branch, and are usually covered with a gummy substance. Prune and burn the infected branches.

# MAGNOLIA

## Insects

Soft brown **magnolia scale** can kill trees if abundant. They overwinter as partly grown young, and suck the tree juices in spring and summer. Young scales are small, flat, and inconspicuous. Adults may be brown or mottled orange, with shallow wrinkles. A honeydrew is produced on which an unsightly sooty fungus grows.

To control, spray the tree in late March or early April before buds open with lime sulfur or 3 percent dormant oil. Spray on a warm, sunny day.

## Diseases

For **leaf spot** of magnolias, see TREES.

# MAMMALS

See entries for individual pests.

# MANGO

## Insects

A small brown **ambrosia beetle** is often found on mango trees. The cylindrical insect bores into the major limbs of the tree and into the trunk, carrying a fungus along with it. Once this fungus is in the core of the tree, it multiplies to cause the wood of the tree to stain, rot, and die. To prevent the spread of fungi, prune the diseased and dying portions of the tree and burn them.

A frequent pest of the mango tree is the **redbanded thrips**, a small black insect with a reddish band around it. The larvae are yellowish with a reddish band around the middle of the body. Thrips feed on leaves and causes them to turn darkly stained or russeted. They excrete over the leaves and fruit a reddish fluid that becomes hard and rusty brown or black in color. Use a 1 or 1½ percent dormant-oil spray, as for scale. See THRIPS.

A considerable number of **scale** insects may find their way to mango trees. Red, mango, wax, and shield scale are among the offending species. Some scales secrete a substance known as honeydew that supports the growth of sooty mold fungi. Apply a mild dormant-oil spray. See SCALE.

## Diseases

The most common and widespread disease of the mango is **anthracnose.** This fungus causes small circular spots on flowers and fruits, and fruit often is russeted and cracked. The organism sporulates abundantly in wet weather, and the disease is common in extended periods of wet weather before the fruit are half-grown.

Since the causal organism is also responsible for withertip and ripe rot of several citrus fruits, it is best not to grow mangoes in the same immediate area. On a small scale, it may be practical to control this disease by pruning out infected twigs and branches. Don't overwater the fruit, particularly in the early periods of growth when the disease is most likely to occur. Especially in areas of high rainfall, growers rely on resistant varieties such as Paris and Fairchild.

**Stem rot** is a blackening of the stem end of the fruit. The skin around the stem may show small blackened spots that are slightly depressed just as the fruit is ready to be harvested.

This disease seems to hit fruit trees suffering a lack of moisture. Proper watering of mango is the key to preventing both stem rot and anthracnose. Provide good ventilation for fruit in storage and in transit.

**Tipburn** causes leaves to dry at the tip and turn light brown. The disease may involve more than half the leaf area. Severely affected trees are usually found in dry, arid regions.

Deep cultivation may damage the root system. See to it that trees have proper moisture, and guard against uptake of salt. In the severe cases, check the soil for potash. If deficient, the soil can be fixed up by mixing in banana residues or wood ashes.

# MAPLE

## Insects

**Boxelder bugs** [color] both act as household pests and cause injury to foliage and twigs of boxelder, maple, and ash. Being plant feeders, they do not damage buildings, clothing, or food, but they may cause alarm and annoyance by their presence. Full-grown boxelder bugs are about ½ inch long and resemble the squash bug in body shape. They have a reddish body marked with broad, shaded areas of brown on the lower surface. The thorax is marked by three longitudinal red lines, one down the center and one on each side. There are also red markings on the front wings. The young are bright red. These insects are true bugs and are equipped with sucking mouthparts.

The insect is most noticeable in fall when the adults seek hibernating quarters. They go to any place that affords protection, including houses and barns. On warm winter or spring days they come out into the open and return to the trees where they spend summer. In fall boxelder bugs congregate on trunks of trees, fence posts, and exterior surfaces of buildings.

Reduce their numbers by removing any nearby unwanted, female boxelder trees. If the trees are valued, try handpicking, spray with insecticidal soap; or apply rotenone or pyrethrum. Inside the house, seal any cracks in the exterior walls.

The **cottony maple scale** is a brown, oval, soft scale on the bark of the branches of maple and other trees. In June large egg sacs are formed, their wax covering resembling a tuft of cotton.

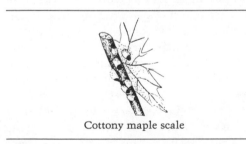

Cottony maple scale

Honeydew exuded by the scale may support the growth of dark sooty mold. Use a dormant-oil spray (1 part oil in 15 parts water) in early spring, before growth begins. (Sugar maple is likely to be injured by the spray.) Prune infested branches as soon as you spot scale. Honeydew that falls from trees can be rinsed away with a hose. Many natural enemies help keep scale insects under control; mosquito spraying is thought to kill some of these predators and parasites, giving the scale a chance to proliferate.

The **maple bladdergall mite** causes galls that first are green and later turn blood red. Leaves become deformed if the mites are numerous, but the tree is not seriously damaged. A favorite host of this mite is the silver maple.

Spray trees in early spring with an emulsion of highly refined white mineral oil prepared for use as a summer spray. Be certain to follow the manufacturer's instructions on diluting and care in handling. Spray no later than the start of the blossoming period. It is useless to spray after the leaf damage has occurred and the galls have been formed.

Several other types of mite cause galls on maple leaves, and all can be controlled by a dormant-oil spray in early spring.

The **maple leafstem** (or maple petiole) **borer** tunnels into the stalks of the leaves and cuts off the lower part of the leaf. The key to the control of the yellowish borer is to gather up and burn all fallen leaves promptly.

The **maple sesian** (or maple callus borer) is a clear-winged moth whose brown-headed larva bores into maple trees, especially around wounds. The amber moth has a wingspread of about an inch, and there is one generation a year. Prevent injury to the trees to keep the larva from causing trouble. Smooth off roughened bark, dig out borers in spring, and paint wounds.

The **sugar maple borer** is the most destructive pest of the sugar maple. It is a beautiful black beetle with brilliant yellow decorations, including a W-shaped mark across the base of the wing covers. It is about an inch in length and emerges in July. The female lays eggs in slits in the bark, and the young larvae tunnel in the inner bark and sapwood, hibernate in a chamber excavated in the sapwood, and the following spring cut large galleries in various directions, though usually in a spiral course upward and partly around the trunk. Sometimes two or more borers in a tree completely girdle the trunk and the tree breaks over. They hibernate the second winter in chambers four inches from the bark. Two years are required to complete the life cycle. Tree growth over the wound generally shows as a series of scars and ridges that show prominently on the trunk.

Examine trees carefully at least twice a year for evidence of the borer. If you see sawdust on the bark, chances are that this pest has infested your tree. If you find any burrows, cut them out or run a wire into the burrow and kill the grub. Close the opening with any of the accepted tree-patching materials.

**Terrapin scale** is a small, reddish, oval insect that occurs on the small twigs of hard and soft maple, often killing them. It is $\frac{1}{16}$ to $\frac{1}{8}$ inch long, with a black mottling on its reddish brown shell. Eggs are deposited in June under the old shells, and there is one generation a year. Control with a dormant-oil spray in early spring.

The **woolly maple leaf scale** shows as masses of white wax, resembling tufts of cotton or wool, on the undersides of leaves in summer. Later these become white cottony wax cocoons. As cold weather approaches, the scales crawl into the crevices of the bark and there secrete a waxy protecting case in which they pass the winter. They often emerge on warm days and crawl about the bark.

Use a dormant-oil spray in late winter or early spring. When the young are crawling about, a spray of soapy water will slow an infestation.

Other insect pests of maple trees include APHID, cankerworms (see TREES), oystershell scale (LILAC), and whitemarked tussock moth (HORSE CHESTNUT).

## Diseases

**Bleeding canker** describes a wet spot appearing anywhere on the trunk from which sap oozes and dries to resemble blood. Cankers on young trees are long, with indefinite margins, and hardly show in the rough bark of older trees. Leaves above the canker wilt, and branches may die. In chronic cases leaves are small, dieback appears slowly, and the tree may die in from two to four years. Sapwood beneath cankers contains reddish brown radial streaks with olive green margins.

Prune out any infected tree parts and improve soil drainage. Apply orange shellac or tree paint to any infected tree sections. Do not overfertilize.

Transplanted urban maples are often afflicted with what is loosely identified as **decline,** a range of symptoms that can be traced to two fungal infections, collar rot and basal canker. Trouble starts at or below the soil line, but the first sign is premature coloring of trees in fall. Apparently, trees suffer when fungi restrict the upward flow of nutrients and water and the downward flow of plant sugars manufactured by photosynthesis in the leaves.

Look for collar rot by removing soil around the trunk to a depth of a few inches, being careful not to nick the bark. Cracks, loose bark, leaking openings in the bark, and discoloration are all characteristic of a fungal invasion. Also check to see if the trunk is girdled by tight loops of roots; this can mimic the effects of decline disease. Cankers are to be found on the trunk above the soil line. Loose bark may peel away to show wood discolored by brown or reddish cankers.

Trees in the wild rarely are troubled by this type of decline; when transplanting, try to provide the tree with a situation that closely mimics the way it would have grown if started by seed in the forest. In nature, maples have roots on or at the soil surface, and in time their roots reach outward and are less likely to crowd and girdle one another. Dig a big enough hole so that the transplant's roots aren't jammed together. Soil beneath the root crown should be tamped down so that the tree won't settle deeper than you intend. Because you won't be planting deeply, the tree may have to be stabilized with stakes until roots take hold. A mulch of composted hardwood bark will protect exposed roots.

Once a tree contracts decline, you can aid it by pulling soil away from the root collar. Try replacing the soil with hardwood bark. If the tree is only mildly affected, you can compensate for the reduced flow of nutrients, water, and sugar by watering and fertilizing in dry periods. But make sure that fungal decline is the problem: similar symptoms are caused by physical injury to the tree, road salt, and various environmental stresses.

**Nectria cankers** start as depressed areas that quickly girdle branches on which cinnamon-colored fruiting pustules are abundant. Leaves beyond the point of attack wilt but remain attached to the tree. Eventually the branch dies. If one branch of a tree turns red in fall before the rest of the tree, suspect canker. Discovered in time, cankers can be scraped out and painted to prevent further infection. Prune well back of the cankered area and burn infected wood.

**Purple eye** is a fungal disease causing grayish tan leaf spots with purple borders. Tiny black dots in the spots are the fruiting bodies of the fungus. Control is the same as for anthracnose (see TREES).

**Scorch** describes several symptoms caused by a stressed maple. Leaf margins turn brownish and die, and the discoloration spreads to fill the area between veins. All leaves on an affected branch will display the same appearance. Leaves may either remain on the tree or dry up and fall prematurely.

This condition occurs when moisture is lost by leaf margins at a faster rate than it can be replaced by the tree—which in turn is caused by physical injury to branches, trunk, or roots, salt injury, air pollution, compacted soil, or simply dry weather. Hot, windy days will speed evaporation and encourage scorch. So will an infestation of aphids. You can compensate for dry spells by watering. Prune away damaged roots or branches. The tree may benefit from extra potassium, from a source such as granite dust or wood ashes. See APHID.

**Tar spot** is a fungus that produces raised black, tarlike spots on leaves. Silver maples are highly susceptible. This disease is unsightly rather than serious. Control by raking and burning fallen leaves.

**Verticillium wilt** of maple trees is recognized by a wilting and dying of leaves on one or more branches. Development may be slow, over a period of several seasons, or it may be so rapid that the entire tree is affected within a period of a few weeks. Bluish or blackish green streaks are often apparent in the sapwood, and may either be limited to a single affected branch or extend from the roots of the tree to a wilting branch.

Maple wilt seems to follow drought. The disease is noticeably reduced when trees are watered. Pruning can save trees, but some maples may have to be cut. Do not follow affected maples with either maples or elms.

Other diseases of maples include anthracnose, leaf scorch, leaf spot, and shoestring rot. Norway and sugar maples are sometimes affected by rots. See TREES.

# MARIGOLD

## Insects

Marigolds go relatively untroubled by pests, as might be expected—this flower is used to repel a number of pests, including nematodes (both French and African varieties work), the tomato hornworm, and whiteflies. But a few pests have strong stomachs, and occasionally trouble marigolds: APHID, JAPANESE BEETLE (found on African varieties), leafhoppers (see ASTER, sixspotted leafhopper), SNAIL AND SLUG, SPIDER MITE, and stalk borer (DAHLIA, ZINNIA).

## Diseases

**Aster yellows** is a virus signaled by greenish flowers and a witches' broom appearance of the plant. Immediately rogue and burn infected plants and nearby weeds and ornamentals that have similar symptoms. The aster leafhopper spreads this disease; see ASTER.

**Stem rot** causes dark lesions which may penetrate to the pith near the soil line. Roots are sometimes decayed. Plant marigolds in soil that has not previously hosted the disease.

Fungal **wilt** rots plant roots, and brown vascular bundles can be seen if stems are cut across. Plants may die. Set marigolds elsewhere, roguing and burning those that are infected.

# MEALYBUG

Mealybugs are well-known pests of house plants, vegetables, and fruit trees. Generally, they are an indoor pest in the North. The most common mealybugs look much like tiny tufts of cotton on the undersides of leaves. They are soft scales, relatives of the hard-shelled scale species. Mealybugs, like their scale relatives, feed by sucking sap. Their secretions of honeydew attract ants and often support growths of dark sooty mold.

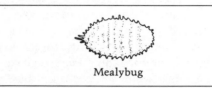

Mealybug

The simplest control, and the one least damaging to the local ecosystem, is a strong stream of water directed at the undersides of foliage. In the house or greenhouse, mealybugs can be killed by touching them with a cotton swab that has been dipped in alcohol. The alcohol dissolves their protective coat of wax. Lacking swabs, you can administer the alcohol drop by drop simply by using a twig. Quassia is a venerable mealybug killer. If fruit trees are in serious trouble, try spraying soapy water (Safer's insecticidal soap is effective). As a last resort, use a spray of kerosene emulsion or fir-tree oil. Such sprays are effective, but interfere with naturally occurring biological control.

Worldwide, 38 species of scale and mealybug have been partially or completely controlled by introduction of their natural enemies. An important enemy of the mealybug is the lacewing. Eggs of this beneficial are commercially

Mealybug destroyer

available. A ladybug look-alike known as the mealybug destroyer *(Cryptolaemus montrouzieri)* [color] is commercially available. A hundred adults should take care of mealybugs in a small greenhouse. Screen the vents to keep the beneficials from escaping. The beneficials should be released in spring, when you first notice the pests.

You can use them to rescue infested house plants, too. Cover the plants with cloth or fine netting, and release four to ten destroyers each. Keep the plants covered until you see good results. If the sight of shrouded plants bothers you, place them in a spare room for the duration of the treatment.

Mealybug destroyers have been placed on fruit trees, but they may run into trouble with ants that feed on mealybug honeydew and protect their source of food. Therefore, control means first taking care of the ants; wrap tree bases with cloth, and apply Tanglefoot. Place ten destroyers or so on each tree on a calm morning or night. One release should be all the trees need in a season.

You'll receive the destroyers as adult beetles. These will feed on the pests and lay eggs. The *Cryptolaemus* larvae look much like mealybugs, and you'll have trouble telling them apart. The larvae turn into adults within 12 to 26 days. Development is quickest above 70°F.

Two species of wasp parasite effected complete control of the citrophilus mealybug in California in the late 1920s. A chalcid wasp from Japan has controlled Comstock's mealybug in Virginia and Ohio orchards.

# MEXICAN BEAN BEETLE

The Mexican bean beetle [color] is the black sheep of the ladybug family. Adults are copper colored and round-backed, and are marked with 16 black spots on their backs, arranged in three

Mexican bean beetle

rows. You'll find no markings between body and head, the telltale sign that a beetle of this family is up to no good. The female lays orange yellow eggs in groups. Larvae are fuzzy, lemon yellow, and covered by six rows of protruding, branched spines.

Adult Mexican bean beetles survive through the winter. In spring they leave their winter quarters and feed on bean foliage for a week or two. Females then begin laying eggs in groups of 40 to 60. These are deposited on the lower surfaces of bean leaves. The tiny young grow into larvae that become longer than the adults, molting or shedding their skin a few times in the process. When full grown they attach themselves to the plant, usually beneath or inside a curled leaf. After the light yellow adult beetles emerge from the pupae, their shells harden and darken and the 16 black spots appear. Development from egg to adult requires about a month. One or two generations are produced each year.

Handpicking is the organic gardener's first line of defense; drop the beetles in a can of water topped with kerosene. Immediately after harvest, eliminate gardenside debris that could serve as an overwintering place. It has been found that earlier bean plantings are relatively free of trouble, so plant an early crop for canning and freezing. Destroy any orange yellow eggs you may find on the undersides of leaves. Try interplanting with potato, nasturtium, savory, and garlic. Some gardeners have had success with a curious spray made from crushed tur-

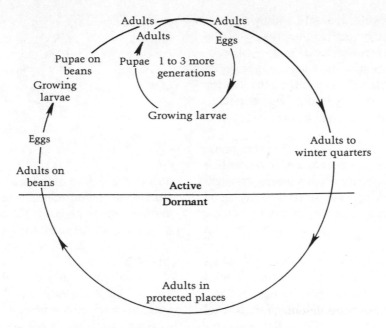

Life history of the Mexican bean beetle

Reprinted, by permission, from Pyenson, *Fundamentals of Entomology and Plant Pathology*
(fig. 3.5), AVI Publishing Co., PO Box 831, Westport, CT 06881.

nips and corn oil. Or try a cedar spray, made by boiling cedar sawdust or chips in water. Rotenone can be used for serious infestations.

Some species of ladybug prey on the eggs and young larvae. Other beneficials are the anchor bug, preying on larvae, pupae, and adults, and the spined soldier beetle, attacking all stages. A bean beetle parasite, *Pediobius foveolatus,* is now on the market. This tiny wasp lays its eggs in beetle larvae. The larvae then darken, and in about two weeks the new generation of adults will emerge. Heavy spring and summer rains help keep down populations of the bean beetle, and a few days of drought and extreme heat may destroy an entire infestation without seriously damaging the crop. Several varieties of bean shown some resistance to this pest: Wade, Logan, and Black Valentine. State, Bountiful, and Dwarf Horticultural rate as susceptible.

*Pediobius faveolatus*
Metro Pest Management Consultants

# MICROBIAL CONTROL

Attractive alternatives to chemical pesticides are insect pathogens—bacteria, viruses, fungi, and nematodes that make bugs sick. Some 1500 such agents have been found, and their qualities can be enhanced through genetic manipulation.

Microbial controls are host-specific, which means that they don't harm nontarget organisms which happen to come in contact with a spray, dust, or bait. Chemical pesticides, on the other hand, may devastate the community of bugs and its delicate systems of balances. (Because it is effective against a limited number of species, a particular microbial control hasn't the wide application of a broad-spectrum pesticide; consequently, microbial controls are more expensive to develop, and industry is inclined to stick with highly profitable chemicals.)

See BACTERIA, FUNGI, PROTOZOA, and VIRUSES.

# MIGNONETTE

Among the most troublesome pests are the cabbage looper (see CABBAGE), corn earworm (CHRYSANTHEMUM), flea beetle (POTATO), flower thrips (PEONY, ROSE), sixspotted leafhopper (ASTER), and SPIDER MITE.

# MILLIPEDE

When you happen upon these "thousand-leggers," they characteristically coil up like the mainspring of a watch. They don't actually have a thousand legs, but start off with three pairs

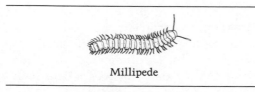

Millipede

and gradually grow up to 200 pairs. The number of body legs and body segments is increased with each molt. Sexual maturity is reached after seven to ten molts, taking place in two to five years. Full-grown adults no longer molt. Millipedes generally feed on decaying plant material, but may attack roots and stems of plants to obtain water in dry spells. Once they start feeding, they may not stop, as the weak sugar solution in sap is appealing to them. Seedlings are sometimes eaten clean through, and fungal diseases may follow attacks on larger plants.

Peat composts are less favorable to this pest than those containing leaf mold or rotted manure. Millipedes can be brought to the surface by drenching the soil with nicotine preparations and raked off and destroyed.

# MIMOSA

## Insects

The leaves of mimosa and honey locust trees are fed on by the **mimosa webworm.** This insect has come into prominence with the increased use of mimosas in landscape plantings and because honey locusts are often used to replace elms lost to Dutch elm disease. The webworms pass the winter in cocoons in bark crevices, cracks in the weatherboarding of a house, trash on the ground, or any other protected place near the trees. The moths emerge in late May or early June and lay their eggs several weeks later. The hatching larvae feed on the leaflets, webbing them together. There are two generations a year, and the larvae of the second generation are often numerous during late August and cause extensive damage, skeletonizing the trees and leaving their webbing over most of the limbs. At maturity the larvae are slender, about one inch long, and grayish brown with light-colored stripes the length of the body. Adult moths are grayish, with a wingspread of about ½ inch.

Most sprays do not have the pressure to force the insecticide into the web and growers must resort to physical destruction. You can tear down the webs with a stick, destroying any that fall on the ground.

## Diseases

**Mimosa wilt** is a disease caused by a soil-inhabiting fungus. Symptoms are easily seen: leaves wilt, hang down from the twigs instead of standing straight out, and in time become dry, shriveled, and then fall off. Usually, symptoms show first at the end of the branch, followed by a progressive dying back of the branch. A cut into the sapwood may reveal a brownish black discoloration, while pink-colored masses of the fungus often appear on the surface of the dying wood.

The fungus is an internal parasite in the tree, and therefore cannot be controlled by spraying or dusting with fungicides. Cut down infected trees, burn the wood, and remove the roots to prevent spread to other trees in the same area. It's best to use resistant varieties, such as the USDA's Charlotte and Tryon.

# MINT

Although mint is usually thought of as a pest repellent rather than a victim, a few insects and diseases may cause trouble.

## Insects

**Aphids** feed on almost every plant that man does, and mint is no exception. In fact, the plant has its very own species, the mint aphid. This is a tiny, yellow green insect, mottled with darker green. In the event of severe infestations on mint foliage, see APHID.

The insect most likely to come to your attention is the **fourlined plant bug** [color]. This greenish yellow creature has four black lines on its back, the inner ones thin, the outer two quite thick. They suck juices from young mint leaves, causing black spots. If damage is very serious, use rotenone; however, control is rarely necessary.

Fourlined plant bug

## Diseases

Thorough garden cleanup in fall should protect mint from **anthracnose.** This disease is characterized by brown sunken spots that turn pale tan with dark red borders; serious cases may kill plants.

Early in the summer, crowded plants may be hit by **rust.** Light yellow or brown spots appear on stems, leaf stalks, or veins on leaves. Clean up plant refuse in fall, and thin plantings if necessary early in spring.

Plants affected by **wilt** grow slowly in spring and take on a bronze cast. The lower leaves yellow in warm, dry weather, and plants often eventually die. Do not plant mint in soil that is known to be infested; soil sterilization is sometimes necessary because the organism can live underground for years.

# MISTLETOE

To control mistletoe, first remove and destroy berries. Cut off affected limbs flush with the trunk, and wrap the area with black plastic to exclude light. Maintain this shield for at least 12 months to kill the rootlets that remain in the tree. Certain species are relatively resistant to the parasitic plant. Most susceptible of the trees surveyed in California were Arizona ash, Modesto ash, black walnut, black locust, and boxelder. Only some 20 of the 300 species were troubled at all by mistletoe.

# MITE

See SPIDER MITE.

# MIXED CROPPING

A plant tends to be more vulnerable to booming pest populations and disease epidemics when grown exclusively, whether in a garden, orchard, or field. In nature, these so-called monocultures are rare.

Take a walk along a hedgerow or through a patch of woods, and you'll notice a great variety of plant life standing branch to branch, likely more kinds than you can name. Why the fraternity of different species? Certainly not to please the bugs. In nature, as in the garden, plants have a better chance of survival if they are surrounded by plants of various species. Many insects find their food through chemicals produced by their host plant, and it stands to reason that if plants of the same species are clustered together, the chemical signal will be much stronger. Therefore, more insects home in on your crops. It also figures that an insect finds it easier to track down one host plant after another if all are grown close together in neat rows. Finally, a good mix of plants encourages a healthful variety of predators and parasites.

# MOLE

Before you curse the moles under your garden or lawn, be sure that the little mammals truly are moles. Meadow mice and shrews look similar at first glance, but they don't make extensive and damaging tunnels and lack the massive forepaws to do so. Other identifying characteristics of moles are tiny, nearly invisible eyes, long toenails, and no external ears. Moles spend most of their lives belowground, eating grubs, earthworms, and other subterranean animals, so you may never catch sight of one.

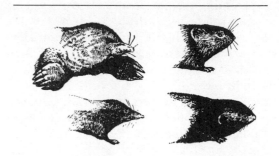

Rodents common in the garden or yard include the mole and shrew (top row, left and right), and the meadow and pine mice (bottom row, left and right)

The tunnels can be obvious, however. Damaged grass roots cause brown courses across lawns. Their ridges and hills can be an annoyance. Garden plants may be harmed, although moles aren't likely to nibble on them. In fact, moles may improve soil quality.

If moles are drawn to your lawn because of the grubs feeding in the soil, you might be able to rid yourself of both animals by spreading milky spore disease to kill the grubs. Two plants, castor bean and mole plant (*Euphorbia lathyris*), have long been interplanted to repel moles. Castor oil can be used as a repellent liquid for sprinkling. Combine 1 tablespoon oil to 2 tablespoons liquid dishwashing soap in a blender until the ingredients have the consistency of shaving cream. Add 6 tablespoons of water and blend again. Use 2 tablespoons of the mixture per watering can full of warm water. Stir and sprinkle. Soil penetration is greatest after a rain. Both plants are poisonous, and even a touch of a leaf can cause severe allergic reactions with some people. *Do not use either plant if children are apt to be in the garden.* Keep castor bean seeds well out of their reach.

Traps are available from garden supply stores. Some tunnels are used only once or rarely; you can tell which receive the most traffic by tamping down a stretch underfoot, then returning to see if the tunnel has been rebuilt.

Less dire methods may work for you. One employs soft drink bottles set in the mole holes. The top of the bottle is left aboveground so that a breeze passing across the mouth will make a sound that is thought to disconcert moles down below.

Mechanical barriers will work if the plot to be protected is not large. Stick pieces of sheet metal at least a foot into the ground.

# MORNING-GLORY

The **morning-glory leafcutter** cuts through leaf stalks, causing the leaves to wilt. The greenish caterpillars hide during the day and feed only at night. You may also find holes eaten in leaves. The adults are yellowish moths with some light brown markings.

To control the leafcutter, clean up the rotted and folded leaves in which they hide during the day.

**Morning-glory leafminer** larvae make serpentine mines and, later, blotches in leaves. When mature, they suspend slender cocoons from leaves by a few silk threads. In a few days, small gray moths emerge. Handpick infested leaves and cocoons to reduce the leafminer population to an acceptable level.

# MULBERRY

Many berry gardeners like mulberry trees because they decoy birds away from more choice berries. Mulberries are seldom troubled by many insects or diseases, but there are a few to watch out for.

## Insects

The **mulberry borer** can be especially destructive in the South, but it is also found in the North. It is a black, long-horned beetle covered with gray hairs and bearing a light brown stripe along the edge. The larvae tunnel into the wood and may kill large branches or even entire trees.

Prune and burn infested parts. Seriously damaged trees may have to be cut down.

The **mulberry whitefly** sometimes infests leaves, but since it does no significant damage, control is not required.

## Diseases

**Bacterial blight** causes dark sunken spots on mulberry leaves. Young leaves may be deformed and brown, and watery streaks often appear on new shoots. Prune and burn infected parts to keep the blight from spreading.

**Powdery mildew** gives a whitish appearance to the undersides of leaves. Some leaves may be lost, but the tree is not seriously harmed.

# MULCH

Mulches can discourage soil-borne pests by producing antibiotic agents that are picked up by a plant's root system. But it is the physical makeup of mulch that is most beneficial. A thick layer will help prevent black spot of roses, a disease spread by splashing water. Mulching can save the season for tomato growers plagued by blight, a viral disease that is signaled by yellow brown, curling leaves. The mulch's function here is apparently to keep the roots cool and moist. (For detailed descriptions of special mulches for tomatoes, see TOMATO.) Mulch acts as a physical barrier to bugs; for example, spittle bugs can be kept from strawberries with a layer of straw (rather than hay, as is customary). Although it adds no nutrients to the soil, aluminum foil is used as a mulch material to discourage aphids and conserve soil moisture.

Mulches will smother weeds such as morning-glory, thistle, and quackgrass; use ground bark, wood chips, sawdust, newspapers, peat moss, hay, buckwheat hulls, grass clippings, leaves, or black plastic weighted down with stones (clear plastic will act as a greenhouse and permit weeds to grow). See WEEDS.

# MUSKMELON

See CANTALOUPE.

# NARCISSUS

Narcissus are relatively free from diseases and insects because commercial growers take care to sell only sound bulbs. You should, however, observe a few precautionary measures: keep the beds clean; avoid crowding plants, as this limits air circulation; and make sure that newly purchased bulbs hard and plump—any with soft spots should be burned.

## Insects

The maggot of the **bulb fly** infests bulbs and ruins them. One larva hibernates in each bulb, and pupation occurs in the spring in the old burrow or nearby in the soil. The adults appear in early summer and lay oval white eggs near the base of the leaves or on exposed portions of the bulbs. The maggot is a yellow or dirty-white larva without legs, and grows to about ¾ inch in length. The fly is about ½ inch long, hairy, black with yellow or gray bands, and resembles a bumblebee.

Destroy all infested bulbs after digging. One and a half hours of hot-water treatment of 110°F. will help guard against infestation. Small, stunted, and otherwise obviously infested plants should be dug up and burned.

The **bulb mite** injures nearly all kinds of bulbs, and Easter lily plants in greenhouses have been severely injured. The mites breed continuously in greenhouses or wherever the temperature and moisture are sufficiently high, and it is possible for ten or more generations to mature in a year. See GLADIOLUS.

When the leaves of plants are dwarfed and distorted and the time for flowering has passed fruitlessly, suspect the **bulb nematode.** To confirm its presence, cut across a bulb and look for dark brown rings that contrast with healthy bulb tissue.

Closely inspect bulbs before planting. Any that appear soft or mushy should be quickly discarded. To be entirely safe, pull out and destroy any bulbs already planted when the malady is discovered. Nematodes are apt to linger in the ground, so it is best not to plant susceptible flowers in the same locality.

The **lesser bulb fly** has a special fondness for narcissus. The flies appear in May and June and lay eggs at the bases of leaves. The larvae

Narcissus bulb fly

Lesser bulb fly

find their way to the tip of the bulb and then go downward into the interior; more than 70 larvae have been found in a single bulb. When fully grown, these maggots are between ⅓ and ½ inch in length, wrinkled, and dirty-grayish yellow in color. They pupate in August in the bulb or in the surrounding soil. The fly is about ⅓ inch long, and has gray wings and a black abdomen marked with three white, crescent-shaped bands.

For home gardeners, it is best to discard any bulbs that feel soft. Plants that are grown outside can be protected by covering them with cheesecloth or gauze during the May and June egg-laying periods. Discard any plants that appear to be already infested. Some growers prevent trouble by treating bulbs for 90 minutes at 110°F.

The **piousbug** is an occasional pest, riddling leaves and causing aesthetic damage. The green, white-striped insect is the simplest of animals known to exhibit worshipful behavior. Control by destroying their little "idols," constructed of mud and garden litter. They soon will cease feeding and go elsewhere.

Other insects that you might come across are MEALYBUG and onion thrips (see ONION).

## Diseases

**Basal rot** is caused by a soil-inhabiting fungus. The decay usually begins in the root plate at the base of the scales, and spreads from there through the inside of the bulb, showing as a brown discoloration. Examine the root bases of bulbs and burn those which are obviously diseased. Bulbs that are in perfect condition in August or early September will sometimes become soft if kept too long out of the ground, and this also suggests early planting. Commercial growers control basal rot by providing good drainage and by destroying infected bulbs. Do not grow narcissus in soil that has previously

hosted basal rot. Find another area of the garden to grow these ornamentals.

**Blast** is a fairly widespread trouble characterized by the appearance of flower buds which turn brown and dry up before opening. The cause of this malady is as yet unknown, but affected bulbs should be dug up and discarded.

Brown-tipped leaves in spring indicate **frost damage;** the quality of the flowers is not seriously affected. If bulbs are discolored or show withered, stunted growth, you should lift them up with roots intact and burn them.

**Narcissus mosaic,** or gray disease, is a mosaic virus. Plants lose vigor, and the key characteristic is an uneven, streaked distribution of green coloring matter in the foliage. Streaks also show in the flowers. Burn infected bulbs, roots, and foliage, and keep down the aphid population.

# NASTURTIUM

## Insects

**Bean aphids** frequently infest nasturtium plants; see APHID and TIMED PLANTING. The velvety green imported **cabbageworm** feeds on the undersides of leaves, producing ragged holes; see CABBAGE. The **serpentine leafminer** makes winding mines in nasturtium leaves. Usually no control is warranted; see PEPPER if your case is exceptional.

Occasional pests include the cabbage looper (see CABBAGE), celery leaftier (CELERY), corn earworm (CHRYSANTHEMUM), flower thrips (PEONY, ROSE), and SPIDER MITE.

## Diseases

Bacterial **wilt** causes plants to yellow, wilt, and die. Stems show black streaks, and bacterial slime oozes from cut stems. Bacteria are carried

Ambush bug
*Ken Gray Collection—Oregon State University*

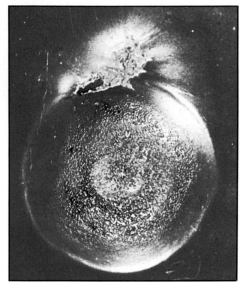

Anthracnose, on tomato
*USDA*

Aphid (wingless pea)
*Max E. Badgley*

Aphid mummies
*Wayne S. Moore*

A

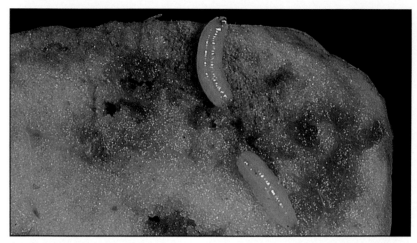

Apple maggots
*Ray R. Kriner*

Apple red bug nymph
*Ray R. Kriner*

Assassin bug
*Edward S. Ross*

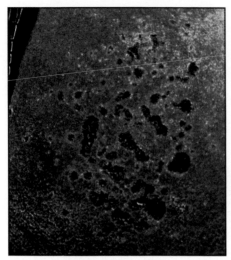

Bacterial spot, on peach
*USDA*

Bean weevil
*Max E. Badgley*

Bacterial spot, on pepper
*USDA*

Bean leaf beetle
*Lee Jenkins*

B

Beet leafhopper
*Max E. Badgley*

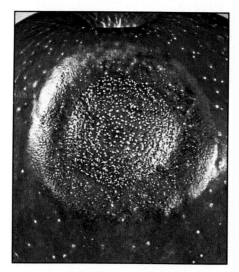

Bitter rot, on apple
*USDA*

Black knot, on plum
*USDA*

238

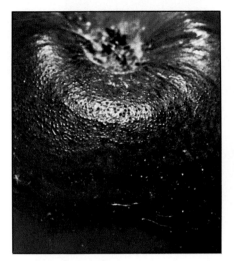

Black rot, on apple
*USDA*

Black rot, on grape
*USDA*

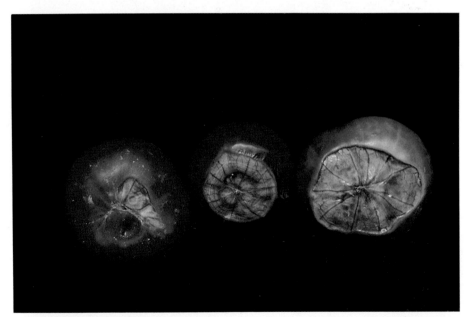

Blossom-end rot, on tomato
*Wayne S. Moore*

B

Boxelder bug
*Patricia Seip*

Braconid wasp cocoons
*Edward S. Ross*

Braconid wasp adult
*Edward S. Ross*

C

Buffalo treehopper
*Edward S. Ross*

Cabbage looper larva
*Max E. Badgley*

Cabbage looper adult
*Max E. Badgley*

Cabbage maggot
*USDA*

241

C

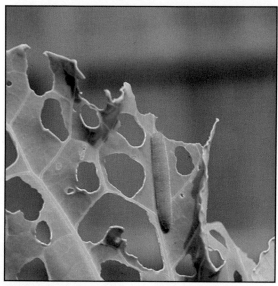

Cabbageworm
*Wayne S. Moore*

Carrot rust fly
*Ken Gray Collection—Oregon State University*

Carrot weevils
*Lee Jenkins*

Cedar-apple rust gall, on cedar
*USDA*

Celeryworm larva
*Edward S. Ross*

Celeryworm adult
*Sturgis McKeever*

C

Chalcid wasp
*Ken Gray Collection—Oregon State University*

Checkered beetle
*Ken Gray Collection—Oregon State University*

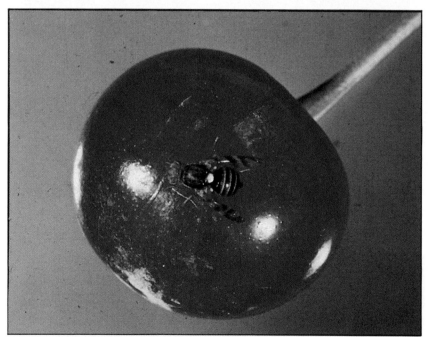

Cherry fruit fly
*Ray R. Kriner*

Codling moth larva
*Max E. Badgley*

Codling moth adult
*Ray R. Kriner*

Colorado potato beetle adult
*USDA*

Colorado potato beetle larva
*Lee Jenkins*

Convergent ladybug
*Ken Gray Collection—Oregon State University*

C

Corn earworm
*Ray R. Kriner*

Cutworm
*Edward S. Ross*

Damsel bug
*Ken Gray Collection—Oregon State University*

Earwig
*Wayne S. Moore*

Eastern tent caterpillar larva
*USDA*

Eastern tent caterpillar adult
*Sturgis McKeever*

Elm leaf beetle larva
*USDA*

Elm leaf beetle adult
*USDA*

E

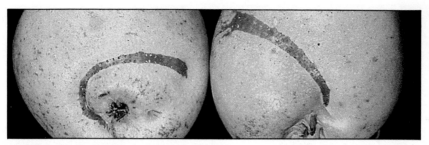

European apple sawfly damage
*Ray R. Kriner*

European corn borer larva
*Max E. Badgley*

European corn borer adult
*Lee Jenkins*

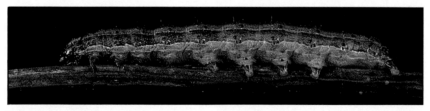

Fall armyworm
*Ohio Agricultural Research and Development Center*

Fall webworm larva
*Edward S. Ross*

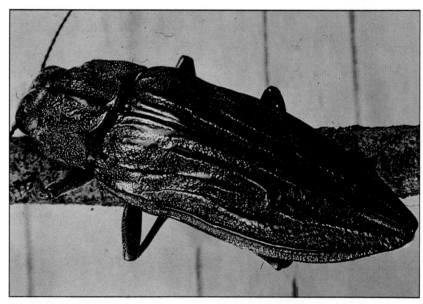

Flatheaded appletree borer
*Donald L. Schuder*

# F

Flower blight, on camellia
*USDA*

Fourlined plant bug
*Robert P. Carr*

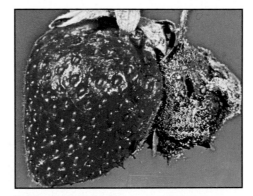

Fruit rot, on strawberry
*USDA*

Fruit tree leafroller
*Lee Jenkins*

Grape berry moth damage
*Ray R. Kriner*

Grape leafhopper
*Wayne S. Moore*

Gypsy moth larva
*USDA photo by Michael A. Pendrak*

Harlequin bug
*USDA*

H

Hover fly larva
*Max E. Badgley*

Hover fly adult
*Max E. Badgley*

Ichneumon wasp
*Ken Gray Collection—Oregon State University*

Imported cabbageworm adult
*Edward S. Ross*

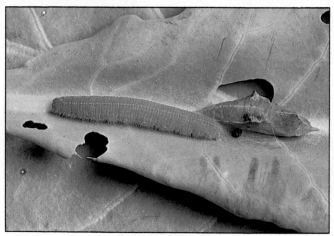

Imported cabbageworm larva and pupa
*Lee Jenkins*

Japanese beetles
*Edward S. Ross*

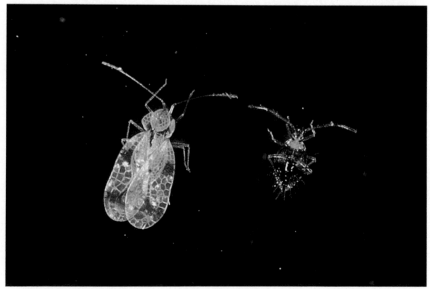

Lace bug
*Ray R. Kriner*

# L

Lacewing larva
*Dwight Kuhn*

Lacewing adult, laying eggs and eating aphids
*Gerard Lemmo*

Ladybug larva
*Wayne S. Moore*

Lampyrid beetle
*Ken Gray Collection—Oregon State University*

Leaf spot, on strawberry
*USDA*

Locust borer
*USDA*

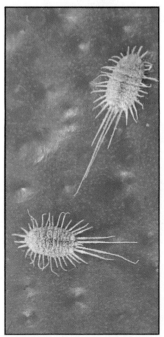

Longtailed mealybugs
*Max E. Badgley*

Mealybug destroyer
*Max E. Badgley*

Mexican bean beetle larva
*Lee Jenkins*

Mexican bean beetle adult
*Lee Jenkins*

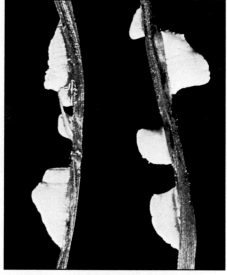

Needle rust, on pine
*USDA*

Oriental fruit moth larva
*Ray R. Kriner*

P

Oystershell scale
*Wayne S. Moore*

Peachtree borer adult
*USDA*

Peachtree borer larva and pupa
*Ray R. Kriner*

257

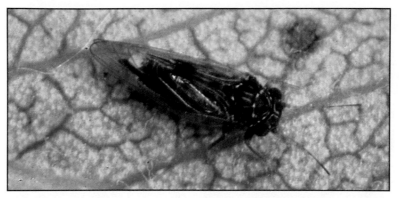

Pear psylla
*Ray R. Kriner*

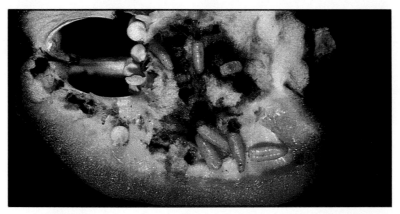

Pepper maggots
*Ray R. Kriner*

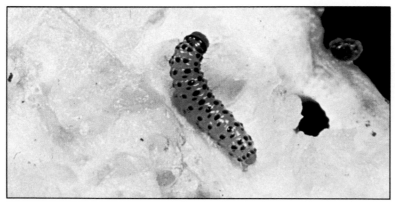

Pickleworm
*USDA*

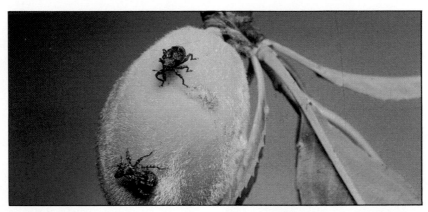

Plum curculios
*Lee Jenkins*

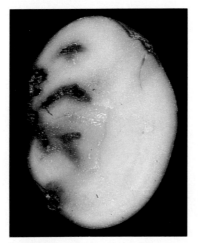

Potato tuberworm damage
*Wayne S. Moore*

Powdery mildew, on rose
*Wayne S. Moore*

Roundheaded appletree borer
*Edward S. Ross*

R

Rust, on rose
*Wayne S. Moore*

Scab, on apple
*USDA*

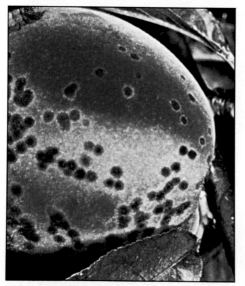

Scab, on peach
*USDA*

Smut, on corn
*USDA*

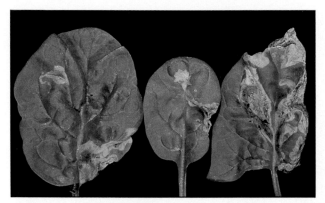

Spinach leafminer damage
*Ray R. Kriner*

Spittlebug
*Edward S. Ross*

Spotted asparagus beetle.
*Ray R. Kriner*

S

Spotted cucumber beetle
*Edward S. Ross*

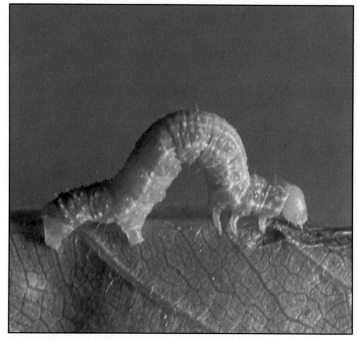

Spring cankerworm
*Ken Gray Collection—Oregon State University*

Spruce budworm
*Ray R. Kriner*

Squash bug
*Edward S. Ross*

S

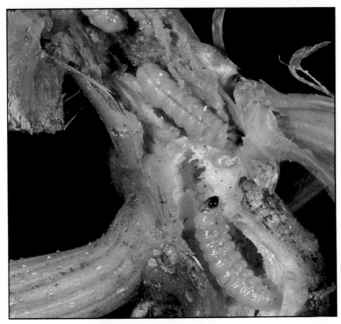

Squash vine borer
*Ray R. Kriner*

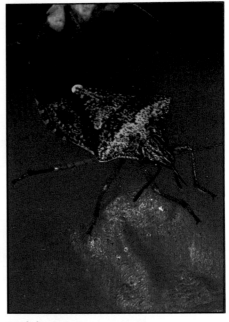

Stink bug
*Lee Jenkins*

Striped cucumber beetle
*Ray R. Kriner*

Symphylan
*Ray R. Kriner*

Tachinid fly
*Edward S. Ross*

Tarnished plant bug
*Lee Jenkins*

Tobacco hornworm larva
*Ann Moreton*

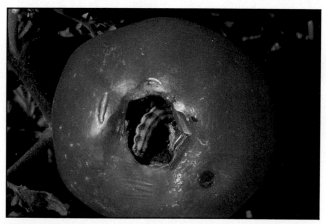

Tomato fruitworm
*Wayne S. Moore*

Vegetable weevil
*Ken Gray Collection—Oregon State University*

Walnut caterpillars
*Lee Jenkins*

Western bigeyed bug
*Ken Gray Collection—Oregon State University*

Western tussock moth larva
*Edward S. Ross*

265

W

Whitefly adult, nymphs, and eggs
*Wayne S. Moore*

Whitefringed beetle larvae and pupa
*Lee Jenkins*

Wireworm
*Edward S. Ross*

Woolly apple aphid
*Thomas Henley*

in the soil and infect nasturtiums through the roots. Plant in clean soil, making sure that nasturtiums don't follow other susceptible plants: tomatoes, potatoes, eggplants, zinnias, dahlias, chrysanthemums, and marigolds.

# NECTARINE

See PEACH.

# NEEM

Neem, a promising agent against plant pests, is a pesticide extracted from the Indian neem tree. It works as an antifeedant, turning off the appetites of the striped cucumber beetle (less so the spotted species), fall armyworm, cabbage moth, gypsy moth, diamondback moth, chrysanthemum leafminer, Japanese beetle, Mexican bean beetle, Colorado potato beetle, aphid, mealybug, mite, and scale. Depending on the pest species and dose, neem may also slow growth and cause deformities. Neem-based products should soon be registered for use on crops. The extract seems to be safe to plants, wildlife, and humans.

# NEMATODE

Most plants, both cultivated and wild, are parasitized by one or more species of nematode (also known as roundworm and eelworm). According to the estimates of nematologists (nematodes have been placed in their own phyllum and are not covered by entomology), a half-million species of this parasite exist, from microscopic worms on the ocean's floor to the nematode of more than 20 feet that makes its home in whale placenta. Some nematodes are parasites of garden pests, and have become

commercially available as a biological control; see below.

Of nematodes in the garden, some are endoparasitic, spending their lives within plants. Others are classified as ectoparasitic and obtain their nourishment through plant walls. Unlike many other pests, nematodes have a wide host range and are seldom considered pests of just one or a few plant species. Symptoms of nematode injury include malformed flowers, leaves, stems, and roots; dwarfed plants with poorly developed floral, foliar, and root structures; and a variety of other conditions such as dieback and chlorotic foliage. Some endoparasitic nematodes cause tissue abnormalities in plants— known as galls or root knots—that block the flow of nutrients through the plant. Feeding wounds provide an entrance for many disease organisms.

Despite their tiny size, nematodes are complex animals and contain most of the body systems found in higher animals, such as a digestive system, nervous system, reproductive system, and several kinds of muscle. However, they lack organized respiratory and circulatory systems. In a large number of species females are the predominant sex, and males are scarce or unknown.

Soil samples almost always contain many kinds of free-living (non-plant-parasitic) as well as plant-parasitic nematodes. Usually, several genera of parasites are present in the same soil although only one or two may be causing the major amount of plant damage. Because of the diversity of nematodes found in most soils, it is necessary that someone trained in nematode identification examine soil samples. The mere presence of nematodes in soil does not automatically mean that they are causing plant damage—the nematodes must be identified in order to determine if control measures are needed.

Should symptoms suggest that nematodes may be responsible for unsatisfactory plant growth, you might collect a sample of soil and roots for professional examination. The soil should be taken from several places around the plant within the root zone, including soil from the surface to a depth of about six inches. A one-inch-diameter soil-sampling tube is a convenient device for collection. Combine sufficient soil cores to make a composite soil sample of at least one-half pint. The soil and roots should be placed in a plastic bag to prevent drying. Samples may be taken at any time of year but preferably during active plant growth, from May to October. Check with the county extension agent to find where samples can be tested.

## TESTING FOR NEMATODES.

You can test for damaging nematodes in your soil with this simple method: Collect several soil samples from various spots in the garden, taking soil from a good six inches down. Mix the samples in a plastic bag. Fill six pots with the mixed soil, and place three of them in a freezer for three days—that's long enough to kill any nematodes which may be present. Finally, plant ten radish seeds in each pot, cover each with sand or peat moss, and place the test pots in a warm spot to encourage germination. This should take place in a week or so. If the freezer-treated pots produce a superior crop, then nematodes are probably sapping the strength of the other three—and your garden as well.

## CONTROLS.

Avoid infested plants, sterilize soil, disinfect tools, use resistant varieties, rotate with immune or resistant crops, select uninfested fields, allow land to lie fallow, and enhance biological control by fungi, predators, and parasites.

Soil sterilization is practical only on a small scale, such as for seedbeds or greenhouse soil. Soil-filled flats can be dipped in water that is kept at the boiling point; pouring boiling water on soil doesn't seem to be effective. Steam heat and flame pasteurizers are also used by commercial growers. Tools can be disinfected in boiling water.

The cultivation of a resistant variety may suppress a nematode's population to 10 to 50 percent of its harmful density. New resistant varieties must continually be developed, however, as nematodes adapt to them and make them worthless in 10 to 20 years.

If crops are rotated so that a susceptible crop is grown only once every 12 years, nematodes will be almost totally absent. Less effective but more practical is to grow a plant that nematodes don't like for a few years, or to interplant with such a plant. Many growers report success with interplanted marigolds, and research shows that pests both in the soil and in plant roots are affected. These plants give both protection and flowers. They control nematodes by producing a chemical in the roots that kills nematodes when released in the soil. Because this chemical is produced slowly, and marigolds must be grown all season long to give lasting control, real benefits may not be seen until the second season. However, marigolds have a residual effect and can suppress nematodes for up to three years after having been grown.

Interplanting with these flowers may give certain crop plants competition—strawberries are an example. Therefore, you might either rotate plantings of marigolds with susceptible crops, or interplant marigolds with susceptible crops, or interplant marigolds two or more weeks after the other plants to lessen competition. If the pungent marigold odor covers up the delicate smell of neighboring flowers, just snip off the buds.

White and black mustard plants exude an oil that nematodes can't stand. Juices from the roots of asparagus are toxic to root pests on tomato. Hairy indigo, velvet beans, and species of crotalaria (a yellow-flowering pea) are good cover crops for discouraging nematodes. You might also try interplanting with salvias and calendulas. Cereal rye and timothy are effective when mowed and tilled into the soil. The decomposing grass is toxic to certain nematodes.

Rich organic soil matter may encourage changes in root physiology that put off the pests. Some gardeners simply stir in sugar around plants. Experiments have shown that nematodes are put off when crops are fertilized with 70 percent fish emulsion and 30 percent yucca extract (sold as Pent-A-Vate). For a 1000-square-foot garden, spray or sprinkle two quarts of emulsion mixed with one quart of Pent-A-Vate. Also effective is an emulsion of corn oil in water (one part to ten parts); sprinkle the soil with a watering can.

Nematodes can be starved by growing immune or resistant crops for two or more years and by fallow cultivation, especially during May and June. Small gardens in the South are sometimes split up into three plots of equal size for this purpose. In one part the vegetables are grown, in the second chickens are kept, and in the third sweet corn is planted. Each year these plots are rotated so that the vegetables are preceded by one year of corn and one year of fallow ground. The effect is to greatly lower nematode damage. Trap crops have been traditional controls in Europe. A highly susceptible crop, such as cucumber, tomato, or mustard, is sown densely and pulled three or four weeks later. Three or four such crops may be grown and pulled in a season. This is a tricky practice, as the nematode population may actually be encouraged if the trap crops are not pulled in time.

Flooding has been used as a control, but land must be under water for about two years in order to kill nematode egg masses.

Hot-water treatments will disinfect roots, but germination is adversely affected. Sweet potatoes are treated at 116°F. for 65 minutes and finger-size yams at 122°F. for 30 minutes.

Tomato transplants have been protected experimentally by dipping the roots in an emulsion of one tablespoon of corn oil in a coffee can full of water.

**BIOLOGICAL CONTROL.**    Fungi are considered potentially the most useful biological control agents against nematode parasites. Predacious fungi capture nematodes by either of two methods. Some trap their prey with sticky appendages, while others have developed mechanical ring traps, one type of which is made of cells that suddenly enlarge to squeeze the nematode in a tight grip. Some fungi also administer a toxin that quiets the struggling captive. Other fungi simply ride on nematodes and send germ tubes into the host's body.

Even fungi that do not attack nematodes may nevertheless discourage them by exuding substances the pests dislike. One study showed that the exudates of the fungus that causes brown root rot of tomatoes serve to interfere with the development of the potato root nematode.

How can you enhance biological control by fungi? The usual course taken by researchers is to supplement the soil with organic matter in order to stimulate fungal growth and trapping. While this might initially cause a rapid rise in the nematode population, predators soon catch up, including predacious nematodes and mites as well as fungi.

Fortunately for growers, predacious and parasitic nematodes take a good many plant-parasitic nematodes out of circulation. And a number of other species of soil fauna eat nematodes, from mites down to tiny ameboid organisms.

**SOME TYPES OF NEMATODE AND WHAT TO DO ABOUT THEM.**    *Root-knot nematodes* stimulate injured plant tissue to form galls; these growths then block the flow of water and nutrients to the plant above, leading to

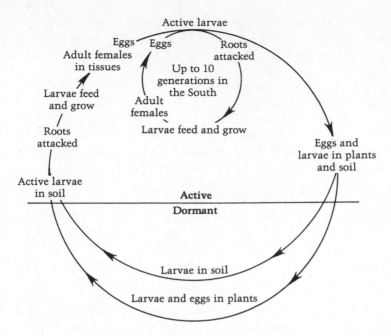

Life history of the root-knot nematode

Reprinted, by permission, from Pyenson, *Fundamentals of Entomology and Plant Pathology* (fig. 6.2), AVI Publishing Co., PO Box 831, Westport, CT 06881.

stunting, yellowing, and wilt. Plants typically limp through the growing season, but fruit production is poor. Although individual nematodes are invisible to the naked eye, egg masses can be seen as pearly objects. Roots appear scabby, pimpled, and rough to the point of cracking.

Broccoli, brussels sprouts, mustard, chives, cress, garlic, leek, ground cherry, and rutabaga are fairly resistant. Others that may get by include globe artichoke, Jerusalem artichoke, asparagus, sweet corn, horseradish, some lima bean strains, onion, parsnip, rhubarb, spinach, sweet potato, and turnip.

*Cyst nematodes* cause plants to appear unthrifty or malnourished. Tops are smaller than normal and consequently appear to be spaced farther apart between and in the rows.

Foliage may wilt and curl. Underground, look for clumps of roots and white females extending from the root surface. Skin on roots becomes thick, tough, and red or brown. Cyst nematodes have trouble hatching in very acid soils (pH 4) or alkaline soils (pH 8). A neutral pH of 6 suits them. So, potatoes may do best in an acid soil, and alkaline soils can be used for cabbage and beets. Most plants, however, do best at the pH that favors nematodes.

In the South, the *sting nematode* attacks a wide variety of vegetables and grasses. The first indication of trouble is areas of stunted plants in the field. As these areas enlarge and combine, the earliest-affected plants begin to die at the margins of older leaves. This nematode is at its worst in sandy soils low in organic matter. Try long rotations with watermelon or hot pepper; turn under crotalaria. Important weed hosts, including crabgrass, ragweed, and cocklebur, should be eradicated. No resistant varieties have been developed.

*Meadow nematodes,* which cause internal browning in potato tubers and roots of corn, members of the cabbage family, lettuce, peas, carrot, and tomato, can be controlled by rotating with immune crops, such as beets, curled mustard, rutabaga, yam, and radish. Following injury, plants recover better if supplied with nitrogen-rich fertilizers.

**BENEFICIAL NEMATODES.** A few nematode parasites are now on the market for control of pests that pupate belowground. These include *Neoaplectana carpocapsae, N. glaseri, Heterorhabditis bacteriophora,* and *H. heliothidis.* The first named attacks caterpillar pests, entering their bodies and feeding on the blood. Gardeners can use this nematode to combat cutworms, armyworms, carpenterworms, click beetles, and root beetles. It also controls dampwood termites.

*N. carpocapsae* is tiny and can be administered in a water spray. One squirt from an oil can delivers some 17,000 of the parasites. If squirted into carpenterworm holes in trees, they will crawl down galleries in search of the pest larvae. Squash borers will be hunted down in their holes. Nematodes can be mixed with water and poured on the ground to combat armyworms, cutworms, larvae of the lawn moth, and wireworms.

Although *N. carpocapsae* is known as the caterpillar nematode, it also attacks bugs, beetles, ants, wasps, and flies. This nematode is alerted to the presence of prey by carbon dioxide and body heat. Once it enters the host's body, the tiny worm employs a symbiotic bacterium to digest its meal. The bacterium soon kills the host; most die within a day.

# *NEOAPLECTANA CARPOCAPSAE*

A nematode used in biological control. See NEMATODE.

# NUCLEAR POLYHEDROSIS VIRUS

Some lepidopteran larvae can be controlled with nuclear polyhedrosis viruses (or NPV for short). These viruses have been registered for a very limited number of pests, but you can make your own viral insecticide.

Look for cabbage loopers that have turned chalky white and appear to be half dead. They may climb to the tops of plants and lie on the leaves, or hang down from the undersides of leaves. The infected worms soon turn black and liquefy. Make a spray of them as described under BUG JUICE. Cold spells slow the spread of the pathogen. Armyworms and bollworms are among the other larvae affected by these viruses.

# NUTRITION

Vigorous and well-nourished plants are less vulnerable to pests and diseases. Similarly, it is the weak, underfed, rough-coated calves—and not the suckling, fat, smooth-coated ones—that are eaten up with lice.

Not only are these plants better able to withstand attack, but they seem to be less attractive targets as well. Apparently, a plant can't summon up all of its considerable defense mechanisms unless it has a full complement of nutrients. How do plants react to invasion? They seem passive, but in fact may send repellent or toxic compounds to the site of injury; or they may increase the pressure within cells at the site to make feeding difficult.

Well-nourished plants have optimal amounts of nutrients, and neither too little nor too much. Aphids will pick on plants that have a surplus of nitrogen; put too much nitrogen-rich fertilizer around your plants, and you can expect to please bean, cabbage, cotton, and pea aphids. Too much manganese leads to yellow leaf margins; leaves show necrotic spots and drop earlier than usual. Too much boron can scorch margins and tips of leaves; also look for dark spots between the veins, more of them near the base of the shoot; and buds on shoots may be cracked and swollen. Mineral nutrients, too, determine a plant's amino acid makeup, which in turn either draws or repels insect pests.

Proper nutrition, then, is crucial to protecting plants. And organic growers believe that plants' complex nutritional needs are best met with compost, mulch, and natural amendments to the soil. While chemical fertilizers make nutrients available in large and sudden doses, natural sources tend to meter nutrients over a long period, helping to ensure that plants won't suffer from either a glut or shortage. Another advantage of organic soil-building is that natural fertilizers and amendments contain a broad spectrum of nutrients that synthesized materials can only approximate.

The soil is a living thing and home to an all-but-invisible world of flora and fauna, and cannot be reduced to a fertilizer formula. It is beguilingly simple to think of the nutritional needs of plants in terms of a series of proportions, such as 5-10-10 or 4-8-4. Such formulas express the amount of nitrogen, phosphorus, and potassium, respectively, in a fertilizer mixture, the numbers standing for the percent of each ingredient. Commercial brands promise these nutrients in big doses, and it would seem that plants receiving all those nutrients couldn't help but make it through the season in good shape.

The fallacy here is that chemicals dumped on the ground aren't necessarily taken up by plants in the proper amounts. One problem is that the food elements in chemical fertilizers are almost all soluble, while in nature plants are accustomed to getting nutrients from many insoluble sources. Because the chemical fertilizers are so readily available, plants tend to pick up too much of a nutrient. And too much can be as bad as too little.

It's not always easy to tell just what ails a plant that suffers from an imbalance of a nutrient. Appearances are very deceiving, and plants are slow to manifest the effects of a deficiency. Visual symptoms of nutrient deficiency may appear in any or all organs, including leaves, stems, roots, flowers, fruits, and seeds. A general note to keep in mind is that a deficiency of one element implies excesses of other elements; for example, too much potassium can block a plant's uptake of magnesium.

In a garden fed with humus, manure, and compost, the soil hosts a wide variety of beneficial microflora that quickly chew up and destroy, or keep in dormancy, disease organisms. These tiny bits of underground life include: bacteria, which are generally effective as scavengers and may produce antibiotics; actinomycetes, poor as scavengers but excellent producers of antibiotics; and fungi, valuable as competitors with disease organisms, as hyperparasites, and as antibiotic producers (some fungi that thrive on rich organic soil actually trap nematodes. Healthy, living soil can be added to inoculate an area in which plants are troubled by soil-borne diseases.

Plant nutrition and soil-building are both book-length topics. But here is a brief description of plant needs, and how they can be met.

NITROGEN.  Plants receive timed-released doses of nitrogen from the gradual decomposition of organic matter. Composted plant waste, manure, and even lightning supply nitrogen naturally. Many gardeners take advantage of the fact that legumes add nitrogen to the soil. Clovers, alfalfa, soybeans, fava beans, and others lock this element into the soil by the action of the nitrogen-fixing bacteria on their root nodules. In addition, there are other bacteria occurring free in the soil that extract nitrogen from the air.

Along with these tiny bits of life, several larger underground allies are at work in the soil. Earthworms add nitrogen to the garden both by converting unusable forms into usable ones and by contributing the nitrogen contained in their dead bodies. Nematodes, mites, snails, millipedes, centipedes, and others help too. Their excretia and dead bodies enrich the soil with nitrogen proteins. Pesticides and concentrated chemical fertilizers can make life hard or impossible for these soil creatures.

How can you tell if plants can really use more nitrogen? On fruit trees, look for a heavier-than-usual bloom, and fewer and smaller fruit that are highly colored and mature early. Leaflets are fewer and smaller, and leaves tend to drop sooner. Generally, plants grow slowly and are a uniform pale green.

PHOSPHORUS.    Phosphorus has been called the "master key to agriculture" because low crop production is due more often to a lack of this element than any other plant nutrient. Deficiencies are expressed somewhat differently in different plants: in corn and small grains the leaves assume purplish tints; legumes become bluish green and are stunted in growth; but in most plants the leaves become dark green with a tendency to develop reddish and purple colors. On pine, lower needles die. Blooms are light and are followed by fewer and smaller fruit. Too much or too little of this element can cause marked changes in pest behavior. Egg production of spider mites is encouraged by an imbal-

ance, and phosphorus deficiency can lead to problems with whiteflies, both outdoors and in potting mixtures.

To give your plants just the amount of phosphorus they can handle, add it to the soil in the form of rock phosphate, a natural rock product containing 30 to 50 percent phosphorus. When the rock is finely ground, the phosphate is available to the plant as it is needed. Rock phosphate is especially effective in soils which have organic matter. Other phosphorus sources include basic slag, bone meal, dried blood, cottonseed and soybean meal, and activated sludge.

MAGNESIUM.    The most common symptoms of deficiency include a yellowing, bronzing, reddening, and death of the older leaves, commonly followed by shedding of the leaves. The leaves may be thin and brittle. Mites thrive under such conditions. Even though soil tests may indicate available magnesium in your soil, too much potassium can block the uptake of this element. This problem is often brought about by the use of commercial chemical fertilizers, which are high in potassium and usually lacking altogether in magnesium. In a good organic garden, the situation is not likely to arise, as these two nutrients are supplied as natural rock powders whose nutrients are made available to plants slowly, over a period of years.

OTHER IMPORTANT ELEMENTS.    Here briefly are the deficiency symptoms of several other important nutrient elements: *Sulfur* deficiency is indicated by yellowing of the younger leaves in the initial stages of the deficiency and finally by yellowing of all the leaves. (The yellowing of leaves is commonly referred to as chlorosis.) *Iron* deficiency is revealed by the chlorosis of the new leaves at the growing tips of the plant. Younger leaves turn yellowish while the veins stay green; older leaves aren't affected.

A deficiency of *manganese* leads to chlorosis of the young leaves, as in iron deficiency (but look for green bands along the veins), followed by early death of the leaves; most manganese deficiencies occur in neutral or alkaline soils.

The principal symptom of *copper* deficiency is withertip, in which the leaves at the stem tip wilt without recovering overnight or during cloudy weather. *Zinc* deficiency, which is often caused by the use of chemical superphosphate, is revealed in a variety of ways, such as yellowing between the veins followed by the dying of the tissue in tobacco, little leaf in pecans and citrus trees, and leaf spot in sugar bean and potato. Common symptoms of *boron* deficiency are the dying of the growing tip of plant stems, internal cork in apples, water-soaked areas and bitter taste in cauliflower, and cracked leaf stalk in celery; leaves may be reddish or scorched, with yellowed veins; fruit may be deformed.

# NUT TREES

See BUTTERNUT, FILBERT, HICKORY, OAK, PECAN, and WALNUT.

# OAK

## Insects

There are several **leafminers** on oak, one of the most conspicuous being the whiteblotch leafminer. Because these insects overwinter in fallen leaves, it helps to rake and burn leaves in fall.

**Oak gall scale** appears as gray or yellow globular growths on oak twigs, usually in the axils of leaves or buds. Control these scales with a dormant-oil spray applied in early spring.

When the **twig pruner** strikes oak trees, small twigs constantly drop upon the ground through July and August, and some hang with dried leaves. The eggs are laid on the smaller twigs in July; the young grubs work for a time under the bark and then tunnel along the pith in the center of the twig. The insects hibernate in the twig and the beetles emerge the following summer. The beetle is grayish brown, about $\frac{5}{8}$ inch in length, and has long slender antennae. Gather and burn the fallen twigs promptly.

Many other common insects rate as pests of oak trees. The cankerworm, carpenter worm, gypsy moth, and tent caterpillar are discussed in TREES; see also APHID, SPIDER MITE, and twolined chestnut borer (see CHESTNUT).

## Diseases

**Basal cankers** on oak trees may superficially resemble a crack in the bark, but underneath the bark a fungus has killed the wood and may girdle the tree. Fruiting bodies of this fungus are distinctly stalked cups or saucers, colored dark brown and black, and are seen in the cracks in the bark. Control this canker by removing infected bark and wood and painting the wounds.

**Leaf blister** causes light green or whitish puckered areas on oak leaves in midsummer. A

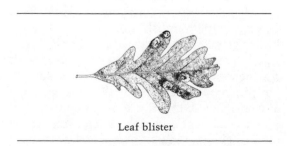

Leaf blister

spray of lime sulfur is an effective control, but should be resorted to only if necessary.

A number of different species of **oak gall** are found on these trees, many of them caused by wasps. They may appear as globular or rough swellings, one to two inches in diameter, or woody swellings at the ends of twigs. Damage is rarely serious, and in any event there is little you can do.

Until recently, oaks were considered one of the most healthy trees, being little bothered by pests and diseases. But since 1942 a fungal disease known as **oak wilt** has spread alarmingly through some of the best oak-growing states. The disease started in Indiana and has since spread through Illinois, Iowa, Minnesota, Missouri, Virginia, and Wisconsin. Diseased trees were found a few years ago as far east as Pennsylvania and as far west as Kansas and Nebraska. The disease attacks all species of oak, but black and red oaks (including scarlet oak, pin oak, shingle oak, and chestnut oak) are more susceptible and stand small chance of recovery. Most of the white oak group, though they can become infected, have some degree of resistance (and their relative health among ailing red and black oaks is a clue to diagnosing oak wilt).

The first symptom of wilt in red and black oaks is a slight curling of leaves near the top of the tree or toward the tips of lateral branches. Affected leaves become pale in color, then turn to bronze and finally to brown, and then tan progressively from the tip to the base. Property owners unfamiliar with the symptoms of oak wilt are first concerned by the premature defoliation of infected trees, which necessitates a lot of leaf raking. But worse trouble soon follows. Once trees become infected, wilt symptoms progress rapidly over the entire crown of the tree, affecting the lower branches last, and death of the tree usually occurs within a few weeks.

Peeling back the bark of twigs and branches affected with oak wilt may reveal diffuse tanning, stippling, or intermittent streaking of sapwood. Trees can be observed in all stages of the disease at any time from May until September.

On white oaks, bur oaks, and other types within the white oak group, leaf symptoms of wilt are more localized than in the case of the red oak group. Usually, entire trees do not wilt immediately, and twigs with dead and dying leaves occur on only a few branches of infected white oaks. Some branches show leaf symptoms while others remain apparently healthy throughout the season. Defoliation is less pronounced, and withered, dead leaves tend to remain on the tree. Discoloration of the sapwood is generally more common in the white oak group than in the red. Since the symptoms of wilt develop relatively slowly on white oaks, infected trees may survive for several years, appearing perfectly normal for one or more seasons following infection before the "staghead" characteristic becomes evident.

Control of oak wilt involves preventing spread of the causal fungus. Infected trees should be cut and either burned or hauled away as soon as possible after infection is spotted. Usually, spores will not be produced until September or October on trees that wilt in July. As an additional precaution, oaks should not be pruned in fall or spring since the wilt fungus can gain entrance through wounds of this type. All oaks within 50 feet of an infected tree should be killed, as the fungus can travel underground from tree to tree through natural root grafts.

If valuable trees occur adjacent to an infected tree, it may be possible to save them from infection by digging a trench between the diseased and healthy trees to cut all roots that might serve to carry the fungus between them. The trench should be 36 to 40 inches in depth and can be refilled immediately.

The life of infected white oak trees may in some cases be extended by removing infected or dead branches. This is possible because of the slow spread of the fungus in the white and related oaks. It is not applicable to red or black oaks, because the disease will have spread too far before the first symptoms become visible.

**Strumella canker** is a fungus that occurs most frequently on forest trees but may become important on ornamental red and scarlet oaks. Infection usually is first noticed as a depressed area at the base of a branch stub that enlarges in a targetlike fashion, with concentric callus ridges, until the trunk is either girdled or considerably bowed on the side opposite the canker. The original infection on the branches, appearing as yellowish brown depressed or raised areas on the bark, usually passes unnoticed. The fungus travels rapidly down the branch and into the trunk, forming the canker which then withers, deforms, or kills the tree.

Remove trees as soon as infection is noticed to help protect remaining trees. If cankers are small, you might try to cut them out, cleaning away all discolored wood, pointing the area at the top and bottom, and applying a good tree paint.

Other diseases of oak trees—anthracnose, leaf spot, powdery mildew, and shoestring rot—are discussed in TREES.

# OKRA

## Insects

The green peach **aphid** occasionally visits okra for a meal. See APHID for ways of controlling this tiny pest.

The **corn earworm** is likely responsible for holes in okra pods. This caterpillar has stripes, a yellow head, and varies in color from light yellow to green to brown. It grows to almost two inches in length. The best and easiest way to control the earworm is by handpicking and destroying any worms on damaged pods that you can find. For other methods, see under CORN.

The large, oval **green stink bug** is both green and stinky. The nymphs are more rounded in shape, and are red with blue markings. It damages okra by puncturing the plant in order to feed. Keep down their numbers by clearing out weeds in and around the graden. It is best not to plant okra near legumes, eggplant, potato, and sunflowers, as the stink bug takes a particular fancy to these crops. Once the insects have invaded your garden, handpicking will prove helpful. Unless your hands are especially nimble, picking should be done early in the morning, before the stink bugs limber up. Sabadilla dust has been found effective against this pest, and in its pure form it is not toxic to man or animal.

Shiny, metallic green **Japanese beetles** may swarm over okra; they occur from southern Maine down into Georgia and west to the Mississippi River and Iowa. Milky spore disease is a highly efficient means of control; see JAPANESE BEETLE.

Another okra pest is the **nematode,** a parasite of plants that works underground. The visible signs of its activity are stunted growth and galls or swellings on the roots. Although the galls on the smaller roots are tiny, the compound galls on large roots sometimes reach an inch in diameter. The pearly white specks within are the nematode egg masses. For control of this pest, see NEMATODE.

## Diseases

Okra is grown extensively in the South and is a popular choice of gardeners. Although okra is generally considered a foolproof crop, it is subject to several diseases, the three most important of which are discussed here.

Okra is highly susceptible to the formation of **root knots,** or galls, caused by the root-knot nematode. The knots rot and the damaged root system cannot support the plant. See NEMATODE.

**Southern blight** is common in sandy soils of the Central Plains and kills entire fields of okra. The disease, which is most active in hot weather, causes plants to lose their leaves and die suddenly. A small mass of pinkish fungal bodies can often be found around the bases of diseased plants. A number of crops are vulnerable to this fungal disease, and should not be transplanted with okra: soybeans, peanuts, tomatoes, peppers, and watermelons. Instead, plant okra on new land or in rotation with corn and small grains. Another helpful practice is to plow deeply under all plant debris on infected land.

**Wilt** describes a yellowing of the lower leaves, followed by gradual wilting and eventually the death of the plant. Stem tissue shows a dark brown discoloration under the bark. Once the wilt fungus is introduced to soil, it can live there indefinitely, and there is yet no means of purging the infected areas. Plant okra in soil that has no history of wilt.

# OLEANDER

**Oleander scales** are circular, flattened, and live on stems and leaves. The females are pale yellow, tinged with purple, and the males are white. Infested plants lose color and vigor, and may die. This scale is found throughout the warmer states, and survives in greenhouses in the North. Prune out all encrusted branches, and apply a white-oil spray in winter. The scale

is parasitized by the golden chalcid wasp, *Aphytus melinus.* Oleander is also visited by APHID, MEALYBUG, and termites (see GERANIUM).

# OLIVE

## Insects

The **branch-and-twig borer** is a brown-and-black cylindrical beetle, the larvae of which are white. Usually the beetle bores a small hole at the base of the fruit or in the fork of a small branch, causing the twigs to break at these holes. The beetles do not normally breed in live olive wood, but prefer the deadwood of madrone, oak, and old grape canes. Consequently, it's best to eliminate deadwood by pruning often. Cut back the girdled portion a few inches as soon as it is noticed. See also APRICOT.

The **olive bark beetle** is quite diminutive, and colored black with white scales. The holes from which it leaves the bark appear as shot holes. Damage is worst on weak, sickly trees. Consequently, you should prune and fertilize carefully to encourage the plants' vigor. Prune and burn infested parts.

A variety of **scale** insects prey on olives. Black scales infest the trees to cause a black sooty appearance on leaves and fruit, and may greatly reduce vigor and productivity. Female black scale adults are small, dark brown or black insects with a very tough outer skin that has ridges on the back in the form of an H. The young are yellow to orange and are often found on leaves. Dead scales may remain attached to twigs for as long as three years. The scales secrete a sticky honeydew which supports the growth of a black fungus. Often black scales can so reduce the vigor of a tree that many leaves drop, and the next year's crop is reduced.

Olive scales may not disturb a tree until harvest, when sharply outlined, dark purple spots appear in marked contrast to the yellow green fruit.

Beneath the coverings, the bodies of male and female scales are reddish purple. Both may be found on any part of the tree aboveground. On leaves, they cause a slight chlorosis, and on small twigs the wood is often a bit deformed and darkened. In heavy infestations, scales settle on the fruit as soon as it is formed in spring, often resulting in badly deformed olives at harvest. In early June the scales make their characteristic dark purple spots on the small fruit, which then tend to fade out as the fruit grows rapidly. By mid-August, the very small scales may be on the fruit without causing any change in color. See SCALE.

Other pests include the bean thrips (see BEAN), longtailed mealybug (MEALYBUG), and WHITEFLY.

# Diseases

**Olive knot** is a bacterial disease that is particularly present after long periods of rain. The olive knot bacteria are abundant in knots with live tissue, and are spread downward on the branches by rain. It is necessary for wounds of some sort to be present in order for infection to occur; these wounds may be freezing cracks, pruning wounds, scars from dropped leaves or flower clusters, and injuries from ladders or from cultivation and harvesting implements. The disease is carried by wind-borne rain, infected pruning tools, or diseased nursery stock.

While olive knot occurs, the trees take on rough roundish galls or swellings, sometimes two inches or more in diameter. They are likely to occur on twigs, branches, trunks, roots, or even leaf petioles and fruit stalks. The disease may kill much of the fruitbearing area.

If the disease has already occurred in the orchard, it is a good idea when practical to remove all knots from the trees. Disinfect tools after each cut. After the knots are pruned, paint the exposed surface with a tree surgeon's solution. If the number of trees makes this impractical, the only other answer is to use resistant varieties, such as Mission and Ascolano. California growers find Manzanilla to be particularly susceptible. Since the disease is especially apt to spread in rainy weather, all pruning should be done during the dry summer months.

**Soft nose** appears late in the season, during or at the end of the harvesting period. The fruit starts to color at the apex end and then shrivels and softens. It is believed that this trouble is caused by the addition of strong nitrogen fertilizers, especially unrotted manure, to the olive tree. Remember that the use of manure is generally beneficial as long as it has been properly rotted, as the nitrogen has then been fixed by microorganisms. These microorganisms build the soluble nitrogen compounds into their own bodies.

In **split pit,** the pits split along the suture during fruit growth, resulting in bluntly flattened fruit. Although the fruit is of normal size and seems to progress satisfactorily, it is undesirable because the pit comes apart when the fruit is eaten. The cause of the condition is not known, but some growers believe that it is the result of an uneven watering system that allows the fruit to become dry early in the season and then overwaters it in compensation.

**Verticillium wilt** is a fungal disease that infects trees through the root system. It is common in soil that has recently been planted with potatoes, tomatoes, cotton, and other truck crops. This disease is generally fatal to younger nursery trees. Often wilt is sharply confined to one area of the tree, killing very small trees immediately. If verticillium is present, the tree will have dead leaves and dead flowers, often until midsummer and beyond.

To avoid this disease, do not plant olive trees in soil that has recently hosted susceptible crops such as tomatoes, potatoes, and cotton. Be certain to avoid interplanting such crops in the olive orchard. If any branches become infected with this disease, they should be pruned away immediately and burned. Also see TOMATO.

# ONION

## Insects

The foliage of young onion seedlings may be attacked by the **garden springtail,** a tiny, dark purple insect with yellow spots. It has no

Garden springtail

wings, but uses tail-like appendages to hurtle itself into the air. Spray the foliage with garlic and water, and remove weeds from the area.

The larvae of the **lesser bulb fly** hatch from eggs at the base of the leaf and eat their way into the bulb through the tip. They grow up to ½ inch long and are wrinkled and dirty-grayish yellow in color.

In some gardens, plants can be protected from the adult fly by covering plants with cheesecloth during the May and June egg-laying periods. Stunted and sickly plants may harbor the worms and should be dug up and burned to prevent spreading the infestation.

The **onion maggot** is a pest of onions chiefly in coastal areas. The full-grown maggot is legless, pearly white, and about ⅓ inch long.

Onion maggot

Like the cabbage maggot, this pest tapers to a point at the head. The onion maggot is the more serious of the two, damaging onions by feeding in the lower part of the stem or bulb. Plants are susceptible at any age; one maggot is capable of killing many seedlings by destroying their underground parts, and several maggots will team up to render large onions unfit for use.

The maggots usually gain entrance to the larger onions through the base where they attack the roots and burrow upward as far as two inches. When damaged onions are put into storage, they decay and cause surrounding healthy bulbs to rot. The destruction of seedlings not only reduces the stand but causes the remaining plants to lose their uniformity as well; this lack of uniform size is of considerable importance where onion sets are grown. Another species, the seedcorn maggot, confines its appetite to germinating seeds and the succulent stems of seedlings. Onions planted for sets are more susceptible than large onions, white varieties

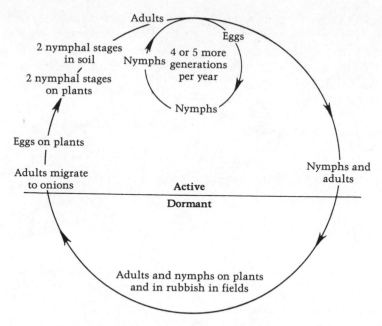

Life history of the onion thrips

Reprinted, by permission, from Pyenson, *Fundamentals of Entomology and Plant Pathology* (fig. 3.15), AVI Publishing Co., PO Box 831, Westport, CT 06881.

are more susceptible than yellow, and red varieties are least likely to be damaged.

Perhaps the best control method lies in the manner of planting. Home gardens are usually arranged in rows, but this plan only serves to help the maggot travel from root to root. Growers can thwart the pest by scattering onion plants throughout the garden. Since each maggot needs several young seedlings for nourishment, it will likely starve to death after its initial meal. Other plants will benefit from the scattered onions, since the strong onion smell is repulsive to many garden pests. Another frequently used method is adding sand or wood ashes to the top layer of planting rows. Or, radishes or cull onions can be used as an effective trap crop when planted at intervals near the seeded crop of field onions.

The trap crops attract the egg-laying fly of the maggot, and are then destroyed to keep the main crop clear.

**Onion thrips** are perhaps the most damaging and widespread of the several pests that attack onions. The adult is an extremely active, pale yellow to brown insect with wings that

Onion thrips

measure only $\frac{1}{25}$ inch long. The larval form is a still tinier, wingless version that can barely be seen with the naked eye. The larvae cause plant tissue to wither and collapse by puncturing plant cells to feed. A sign of feeding thrips is small, whitish blotches that run together to form silvery areas. Heavily infested plants become stunted, the leaves are bleached and die back from the tips, necks grow abnormally thick, and the bulbs fail to develop as they should. There are several generations a year, and injury is at its worst during dry, hot seasons.

Onion sets should not be planted next to a field of large onions, since thrips from the early maturing sets will migrate to other onions in the vicinity and may destroy the crop in hot dry summers. If thrips are a problem in your area, eliminate their winter homes by pulling weeds around the garden or onion field. Check closely any stored onion bulbs for evidence of thrips. Resistant varieties may come to the rescue— resistance to thrips is bred into several strains, and Spanish onions often show considerable resistance. Although severe infestations may wipe out even the Spanish varieties, your chances are better than with ordinary flat or global onions.

Shiny, hard, jointed worms about the roots of onion plants are **wireworms.** They range in color from yellow to brown. See WIREWORM for control measures.

## Diseases

**Damping-off** is a fungal disease of seedlings that causes seedlings to fall over. See DAMPING-OFF.

**Downy mildew** is a fungal disease that overwinters on diseased plant refuse in the soil. The best time to examine plants for this disease is early in the morning while dew is still on the leaves. The first symptom is sunken, water-soaked spots on leaves, either yellow or grayish in color. These spots later become covered with a downy purplish mold growth, and may eventually be blackened by a second fungus. To combat this disease, use a three- to four-year plant rotation and destroy or plow under deeply all plant refuse. Some resistant varieties are available.

A fungal disease, **fusarium basal rot,** first shows up as yellowing leaf tips, which die back as the season progresses. Underground, roots rot off and white mold appears. Root maggots encourage the spread of this disease. A four-year rotation will help prevent fusarium rot, as will a sandy soil. Raised beds give better protection because water drains well. Do not buy onion sets that look shriveled or discolored, and consider the resistant and tolerant varieties listed under RESISTANT VARIETIES.

**Neck rot,** or botrytis rot, gives little sign of its presence until harvest time—most of the telltale signs develop in storage. The scales at the neck soften into sunken, spongy tissue, and one or more leaves of the bulb take on a water-soaked appearance. Often a gray to brown mold growth appears on the surface of the affected bulb and between the diseased fleshy leaves. The mold growth produces numerous spores and black, kernel-like resting bodies, or sclerotia. White varieties are extremely susceptible to neck rot.

Should neck rot afflict your onions, remove any that show the above symptoms so that the disease does not spread. Try to keep plants as dry as possible, and should watering be necessary, water the soil and not the plants. If you use plenty of organic nutrients in the soil and a liberal amount of humus, your onions' chances in storage will be greatly improved.

Stunted onions with withered tops may be victims of **pink root,** a fungal disease that lives in the soil and is encouraged by heavy, wet soils. Infected roots turn pink or reddish and then shrivel and rot. New roots are attacked as they form.

Onions should be planted on well-drained soil that has not recently grown ladino clover, ryegrass, or corn, as these crops leave a residue in the soil that is harmful to onion roots. If possible, do not plant onions, shallots, or garlic in soil that is known to have previously harbored pink root; wait at least four or five years before planting onions again. Although diseased sets and bulbs are difficult to detect, try to make sure that yours have come from disease-free fields. Plant varieties of the summer shallot derived from the Japanese Nebuka; as an alternate crop, these plants have a high degree of resistance to both pink root and yellow dwarf. Other resistant onion varieties are available; see RESISTANT VARIETIES.

**Smut** is evidenced by black areas on leaves and in between the segments of bulbs of white onions. These patches are filled with black pustules and loose spores of the fungus. Young diseased plants have twisted, curled leaves and may be killed outright. The disease is found throughout the northern states. Onions should not be planted in infested soil, and be sure to purchase undiseased sets. Evergreen has been found resistant.

**Sunscald** is not a disease but a name for damage that often occurs when onions are harvested on very hot, bright days. Immature bulbs, especially of the white varieties, are most prone to injury. The damaged tissue appears bleached and becomes soft and slippery. Serious sunscalding will prevent bulbs from making U.S. grades and will result in a reduced price for the grower. A more serious problem, however, is that bacterial soft rot and other decaying organisms may gain entrance through the damaged tissue, sometimes causing complete loss of the bulb.

Sunscald can be prevented by protecting bulbs from direct exposure to the sun while curing. This may be done by pulling the onions with the tops intact and placing the bunches on the row so that the tops of each bunch will cover the bulbs of the previous bunch. This process is often called shingling.

For a list of varieties resistant to fusarium rot and smudge, see RESISTANT VARIETIES.

# OPOSSUM

The opossum can also be caught in a trap baited with sweets. This animal isn't often thought of when gardeners find chunks missing from tomato plants or corn, but it's often the culprit. Opossums are scavengers and will also raid compost piles.

# ORANGE

See CITRUS FRUIT.

# ORCHID

## Insects

**Foliar nematodes** cause brown or blackish spots delineated by veins; eventually leaf drop results. These tiny pests swim to leaves in the moisture on stems, and you can control them by improving ventilation around plants and by careful watering.

In the greenhouse, orchid growers may be bothered by the **orchid fly,** a small black, wasp-like pest. It lays its eggs in young leaves and causes them to swell. When the larvae hatch, they feed on many parts of the plant, including the bulbs. An early sign of trouble is swollen leaves. Cut off such leaves and destroy them.

There are several kinds of **scale insect** known to attack orchids. If infestation is heavy, consider a light oil spray. Use 1 percent summer oil to 1¼ gallons of water. Heavier sprays can injure plants that aren't tolerant. Scales can be also discouraged by removing sheaths from orchids. If a plant is badly infested, segregate it until it can be cleaned up. See SCALE.

Other pests you might see are APHID, flower thrips (see PEONY, ROSE), MEALYBUG, and SPIDER MITE.

## Diseases

**Anthracnose** is known by spots, on either leaves or tubers, that are soft, sunken, and somewhat circular. Spore pustules are pinkish or reddish orange and often arranged in concentric circles.

Since syringing spreads the spores to cause new infection, take care when watering. Reduce humidity around plants by improving ventilation. Cut and burn infected leaves.

**Spots** on leaves and tubers are often symptoms of bacterial diseases. Affected plants should be isolated. Do not water from overhead, as splashing water can transmit trouble. Provide good ventilation, ample mulch, and disease-free soil.

Some **viral** diseases can cause either "breaking" (variegation of flower color) or mottling of leaves with yellow and green patterns. Other viruses can lead to ringlike, concentric patterns in leaves, either yellow or brown, and necrosis may follow. Viruses invade all parts of the plant, from roots to flowers, and new plants derived from a diseased orchid by division or by back-

bulb propagation are almost certain to carry virus with them. Vegetative propagation in orchids is therefore a source of much disease in commercial and private plantings.

It appears that orchid viruses are not carried through seed, and seedlings should therefore be free of virus until they become infected from some outside source. Infection is usually introduced by insects (the green peach aphid transmits the virus that causes breaking of flower color) or by cutting knives and shears contaminated with juice from diseased plants. It follows that there are several methods for preventing orchid viruses. Keep aphids from plants (see APHID); while these insects are not often seen on orchid foliage, they feed and reproduce on flowers and flower buds. Disinfect cutting tools between use on different plants. Segregate diseased plants; keep healthy plants in a separate greenhouse or different area of the greenhouse.

# PACHYSANDRA

## Insects

The female **euonymus scale** resembles a dark brown oystershell, while the male is slender and white. In heavy infestations, stems and leaves become covered with scales, leaves turn yellow, and vines die. This scale is found throughout the United States, but is less numerous in warm regions. Apply a dormant-oil spray before growth starts in spring. Cut off and burn any badly infested shoots before spraying.

Another occasional pest is the SPIDER MITE.

## Diseases

When **leaf blight** and **canker** strike, leaves turn brown with irregular blotches and shrivel up. Lesions on stems may show pink masses of spores. Thin out thick plantings and clean up dead plants and leaves.

# PANSY

Leaves of pansies and violets are eaten by the **violet sawfly,** a bluish black or olive green larva marked with white spots. They are smooth, grow to ½ inch long, and are found in the eastern United States. Feeding at night, the caterpillars skeletonize the lower surfaces of leaves, and later defoliate and eat holes in plants. Handpicking should take care of modest populations.

Other pests of violets include celery leaftier (see CELERY), cutworm (TOMATO), flea beetle (POTATO), MEALYBUG, NEMATODE, SPIDER MITE, and WIREWORM.

# PARASITIC INSECTS

Parasites reproduce at the expense of other insects by planting their eggs in or on any stage of the host, from egg to adult. The eggs of some species are placed on leaves so that they will be ingested by the feeding host and become active inside its body. Victims typically are insects that have a complex (four-stage) metamorphosis. While some parasites use a number of species as hosts, many are host-specific, and this characteristic explains why they are generally considered more effective than predators in controlling a given pest problem.

A few definitions are helpful in explaining the roles of various parasites. A *primary parasite* develops on a nonparasitic host, as do most parasites you'll be interested in. A *hyperparasite* develops on a parasite, and may partially undo the good work of a primary parasite. Parasites develop either within the host's body *(endoparasites)* or on the outside *(ectoparasites).*

Unlike predators, which consume a number of insects in their lifetimes, several parasite larvae may take life from a single host. Death comes slowly for the host, as the hatching larvae usually avoid feeding on its vital organs until near the end of their development. Of parasites that are commercially available, the tiny trichogramma is the most popular. It lays its eggs in those of two hundred different insect pests. The wasp eggs are mailed to the consumer in host eggs. The Soviet Union makes the most of the trichogramma, applying different species and forms of the wasp to different tasks. Depending on their aptitudes, they now are released to attack eggs of pests in fields, orchards, and even the tops of forest trees.

Clearing out wild vegetation from the edges of garden and fields so upsets the insects' habitat that a rapid buildup of pests cannot be checked by natural enemies. The population of good bugs typically lags behind that of plant eaters, resulting in serious damage in early and midsummer.

Because the adult forms of some worthy beneficials are not carnivorous and rely on high-protein foods such as nectar and pollen to sustain themselves, you can keep them in your garden by planting flowers along with the vegetables. Some ladybugs will turn to pollen if the aphid population falls off, and this suggests the importance of growing pollen-producing plants near crops or trees vulnerable to aphids; strawflowers, for example, can attract ladybugs by the dozens. In California, citrus groves are sometimes laced with pepper or olive trees that serve as breeding places for both the black scale and its predator, *Scutellista cyanea,* a wasp that lays its eggs within the eggs of the scale. This practice increases the numbers of black scale but, more important, it ensures that the orchardist will always have a supply of beneficials at hand. Black scale hosts can also be perpetuated by planting oleander near citrus trees; the oleander should be watered often to generate the humidity that favors the insect's development.

A parasite of the grape leafhopper needs a particular second species to prey on in order to make it through the lean winter months, and when it was discovered that this species feeds on wild blackberries, plantings of the berries were set near the vineyards so that the parasite was able to sustain itself through the year. Such protective plantings were effective in providing parasites for vineyards more than four miles distant. These instances are by no means exceptional, and a variety of plantings about your property will encourage many beneficials, whether set at random or for a specific purpose.

If you order beneficial insects by mail, you may find that your conscripted helpers are in no shape to serve—small and sickly, or even the wrong sex (males don't lay eggs). You should keep beneficials in their containers until you wish to release them; some may already have hatched. Packaged insects and eggs can be refrigerated until needed.

# PARSLEY

## Insects

**Aphids** occasionally bother parsley; at least three species have been identified as pests. For control, see APHID.

Another caterpillar pest, the large, pale green **cabbage looper,** may be identified by the way it travels—it is a measuring worm and doubles up as it crawls. For a variety of control measures, see under CABBAGE.

The yellowish white, legless grubs that occasionally attack the parsley roots are larvae of the **carrot rust fly.** For control measures, see under CARROT.

Parsley leaves are vulnerable to the ravages of the **celery leaftier,** a caterpillar that changes color from pale green to yellow as it reaches its full length of ¾ inch. Look for a white stripe running along the back with a pale strip centered within. For control, see under CELERY.

Because it is partial to parsley, the celeryworm [color] is also known as the **parsleyworm.** This is a two-inch, green caterpillar with a yellow-dotted black band across each segment.

Parsleyworm

When disturbed, it emits a sweet odor and tries to look impressive by projecting two orange horns from its head. The adult is the well-known black swallowtail butterfly which has large, black forewings with three rows of yellow markings parallel to the wing edge, and rear wings with a blue row, an orange spot on each, and the projecting "swallowtail" lobe.

Because the celeryworm population is rarely very great, most gardeners find that handpicking early in the morning is sufficient. *Bacillus thuringiensis* offers biological control. In serious cases, rotenone may be used.

The undersides of parsley leaves may harbor numbers of tiny green or straw-colored **strawberry spider mites.** These pests can cause stunting or even death of plants. See SPIDER MITE.

## Diseases

Parsley is relatively free from diseases. A fungal affliction known as **leaf blight** (or septoria blight) may cause small, tan leaf spots, but control should not be necessary. Paramount is a resistant variety.

# PARSNIP

## Insects

Most garden plants are susceptible to damage from **aphids,** and parsnips are no exception. See APHID for methods of control.

The larvae of the **carrot rust fly** may injure the roots of parsnips. These worms are yellowish white, legless, and grow up to ⅓ inch long. They are particularly destructive in the northeastern states and in the coastal areas of Washington and Oregon; growers also contend with them in parts of Idaho, Utah, Wyoming, and Colorado. For control, see CARROT.

The catholic tastes of the **celeryworm** [color] have earned it the names carrotworm and parsleyworm. Both larva and adult are easy to spot. The former is a green, two-inch caterpillar with a yellow-dotted black band across each segment. When disturbed, it projects two orange horns from its head and emits a cloyingly sweet odor. The adult is the familiar swallowtail, a butterfly with yellow, blue, and orange markings on its black wings. Each back wing bears a characteristic "swallowtail" lobe.

Because the celeryworm population is rarely very great, most gardeners find that handpicking early in the morning works well enough. The larvae can be controlled with *Bacillus thuringiensis.* Rotenone dust is sometimes used in serious cases.

The larval form of the **parsnip leafminer** occasionally causes blotching of leaves, especially the lower ones. These worms are greenish, about ¼ inch long, and pupate to form a pale

yellow fly with a green abdomen and brown curved bands on the wings. Control should not be necessary.

A small, black-spotted caterpillar, the **parsnip webworm,** may web together and feed upon the unfolding blossom heads of parsnips. It grows to about ½ inch long and varies in color from greenish yellow to gray. When mature, the webworm leaves its web and burrows inside the flower stems to pupate. Damage is usually not serious. Control by cutting infested flower heads.

For identification of pests not mentioned here, see CELERY and CARROT.

## Diseases

**Canker** is a fungal disease that shows as dark brown, slightly sunken pits on parsnip roots, especially on the shoulders and crown. The causative fungus is kept at bay by good drainage and rotation. Chances of trouble are greatest in wet, cool weather. Model is a resistant variety.

# PEA

## Insects

**Aphids** [color], especially the large green pea aphid, are sometimes a problem to the pea grower. On a large commercial scale they often cause the plants to wilt and die. Home gardeners usually do not have such serious infestations. In addition to the general control measures listed under APHID, try repelling these insects by planting chives or garlic next to the peas. A variety listed as resistant to aphids is Pride. See TIMED PLANTING for a strategy that may keep peas out of trouble.

Yellowish white seed corn and cabbage **maggots** bore into sprouting seeds and prevent development of plants. Roots and stems are tunneled out, causing plants to wither and die. Early plantings are particularly vulnerable to these maggots. See CORN and CABBAGE for methods of control.

The **pea weevil** is a troublemaker in pea patches across the country, and is especially damaging in Utah, Idaho, Washington, Oregon, California, and New York. The adult is a brownish, chunky beetle, with scattered white, black,

Pea weevil

and gray markings. It is quite small (⅕ inch long) and feeds on pea blossoms. Eggs laid on young pods produce white larvae, ⅓ inch long, with small, brown heads. The worms burrow through the pod and into a pea, which they eat and then pupate within. If peas are to be used as seed, they may be heated from 120° to 130°F. for five to six hours. This heat treatment will not damage the seed.

Most pea weevils meet their doom at summer's end. In addition to being preyed upon by other insects, birds and animals, many die as a result of ordinary culture practices. Deep plowing or packing of the soil kills them in great numbers. You can reduce the possibility of infestation by destroying any vines and garden trash that the weevils could use for hibernation. This may be accomplished with little work by placing livestock or poultry in the pea patch

after harvest. This pest has a tendency to settle from flight at the first scent of a pea blossom. Thus, fields of peas will likely have a greater infestation at the edges than at the center, and the center of large pea fields (over 100 acres) may be virtually untouched. For this reason, you should concentrate your watchfulness at the periphery of the field in the early part of the season. Plants may be dusted with rotenone between the time blossoms first appear until pods are formed. Because warm weather will cause an increase in mating and, consequently, in the number of weevils, early-planted peas are generally the least susceptible to attack.

## Diseases

The combined effects of three fungal diseases that attack peas is called **ascochyta blight.** An easily recognized symptom is the small purple lesions, no more than specks, that form on leaves and pods. Stems develop elongated, purplish black lesions that may grow and girdle the stem, weakening it so that it is easily broken.

Affected leaves eventually shrivel and dry, taking on a resemblance to freshly cured clover hay. Stems and roots at ground level may be afflicted with a bluish black foot rot. The three disease organisms that collectively produce ascochyta blight infect seeds and overwinter in pea straw. In regions with very mild winters, they may remain active on infected volunteer plants.

Plant seed that is free from blight. Since infestation is rare west of the Rocky Mountains, western-grown seed brings good results. It is risky to plant seed grown in humid sections of the East and Midwest. Large-scale growers should plow down all pea stubble and vines immediately after harvest and plant the field with a crop (such as grain) that won't require cultivation the next season. Staggered planting dates only serve to spread disease from the stubble of the early planting. Use a three- to four-year rotation. After the crop is harvested, remove diseased vines. Locate new plantings as far as possible from areas that may harbor the disease.

**Bacterial blight** strikes peas in almost every area of the country except the semiarid regions of the West. Several symptoms may catch your attention: stems near the ground line turn purplish or nearly black, with irregular, discolored areas on the nodes; small, water-soaked spots appear on the leaves; and yellow to brown water-soaked spots develop on the pods. In time, the leaf spots turn golden brown and become papery, and a thin layer of dried bacterial ooze collects on them.

Control by planting western-grown seed that is free from bacterial blight, and by practicing a three- or four-year rotation. Because the bacterium enters through either stomata or wounds, care should be taken to avoid injuring plants, especially when they are wet. Injury from hail storms often renders whole areas open to severe infection.

**Fusarium wilt** (pea wilt) is responsible for yellowed leaves and wilted plants. Affected plants may not wilt if the soil temperature is low. The interior of stems is of discolored lemon yellow. The disease, which lives in the soil and enters through roots, may kill plants. Good, wilt-resistant varieties have been developed, and fusarium wilt is no longer a formidable threat; see RESISTANT VARIETIES.

In a wet season, peas may fall prey to **powdery mildew.** Leaves, stems, and pods become covered by a white, powdery mold. Late in the growing season, black specks (fruiting bodies of the fungus) appear in this mold growth. Infected vines are somewhat stunted, and in severe cases the vines dry out and die.

It is unlikely that you have anything in your power to hold the rains in abeyance, but you can confine overhead sprinkling to early in the day. If mildew has infected a planting, dig under all remnants of the vines as soon as the last peas are picked to prevent the mildew from perpetuating itself from year to year.

**Root rots** are characterized by yellowed, unthrifty plants with rotted and discolored underground stems and roots. Plants often die at flowering time.

Since excessive moisture favors this disease, avoid watering the plants more than is necessary and use well-drained soil. Also, peas should not be grown continually in the same soil.

# PEACH

See FRUIT TREES for ideas on a spray schedule. The problems listed below also afflict nectarine.

## Insects

Several species of **aphid** [color] occur on peach trees. The green peach aphid infests a great number of plants, sucking their juices and spreading mosaic, beet yellows, leaf roll, and other viral diseases; the yellow green pests appear as peaches go into bloom and leave the trees for other plants (potato, pepper, lettuce, and many others) by mid-July. Healthy ladybug

Aphid

and lacewing populations will help keep these and other aphids under control. Failing this, use dormant oil or a strong spray of water or soapy water, particularly before the leaves become very curled.

Their presence on peach and nectarine will be known by the severe curling of leaves at the tips of shoots. The leaves remain wrinkled after the aphids are gone. Aphids cause little fruit damage on peach, possibly because the fuzz acts as a barrier; but they heavily attack nectarines, leaving them deformed and small. See TIMED PLANTING for this pest's emergence times.

The black peach aphid lives through the winter on the roots, and migrates to the leaves and tender shoots in spring. It may be found on plum as well. Only young trees are injured by the root feeding, and foliage feeding is seldom of any consequence. Keep young trees well nourished so that they can better resist early damage.


The **oriental fruit moth** [color] starts its life of destruction as a larva tunneling in tender peach shoots early in the season; it later enters the fruit. Those hatching late may go directly to the fruit, since the terminals have hardened. This pest is the cause of most wormy peaches. Some areas may be afflicted by up to five generations. Mature larvae are pink and about ½ inch in length. They are very active, and move about rapidly when disturbed. The first indication of an infestation is the wilting of the terminals of rapidly growing shoots. The wilted tips turn brown in a few days and die. Since the first brood usually confines its feeding to the terminal shoots, it is the later broods that are most harmful to the fruit. This fruit injury is not as easily recognized as twig injury. The worms enter the fruit and feed on the flesh, usually next to the pit. There may be a gummy exudate on the surface of the fruit as evidence.

Sometimes the larvae enter through a tiny hole in the stem, leaving no outward sign.

Survival of the fruit moth larva is low if late-ripening peaches aren't available. This would suggest that if the fruit moth is a consistent pest in your orchard, you should grow only those varieties of peach that ripen early. No fruit will be available to the later broods, and the pest will not thrive as well.

Injury is most serious on trees that have made a rank growth, so prune annually and thin dense growth. Because full-grown larvae may overwinter in cocoons on trash and weeds, it helps to cultivate the soil to a depth of two inches, one to three weeks prior to blooming. Remove culls from the orchard and destroy them. Dormant-oil sprays do not control this pest.

In many areas control has been achieved by using parasites to eliminate infestation of orchards. Probably the most useful parasite is the braconid wasp, *Macrocentrus ancylivorus*, although *Trichogramma minutum* and many others are important. A program for fruit moth control using *Macrocentrus* involves five mass liberations of the parasite at four-day intervals, using six female wasps per tree and beginning around mid-May. A single spray of rotenone or ryania near the end of July will complete the program, saving beneficials for their valuable work. These wasps don't get a lot of credit for the good work they do, partly because they are so small and some work only at twilight or at night. In size and shape they look something like mosquitoes and are amber to yellow in color. Altogether, the six known species may destroy 85 percent or more of the moth larvae, providing the orchard is unsprayed. Chemicals have interfered with the wasps' contribution to biological control.

Even field mice help out, eating cocoons that overwinter on tree trunks. In parts of Canada, lacewings finish off from 20 to 60 percent of oriental moth eggs.

For severe problems, a nontoxic dusting may be useful. Mix sulfur (60 percent by weight), 300-mesh talc (35 percent), and light-grade mineral oil with a viscosity of 100 (5 percent). Dust every 5 days, starting 20 days or so before you expect to harvest the peaches.

The **peachtree borer** [color] is one of the major pests stone fruit growers have to contend with. The white, 1¼-inch-long larva feeds beneath the bark at or below the surface of the soil, and trees may be partially or completely girdled, with young trees occasionally being killed outright. Productivity is reduced and trees are less able to withstand adverse weather conditions and attacks of disease and other insects. Larvae usually complete development in late June or July and go into the pupal, or resting, stage in the soil nearby. Three weeks later the adult emerges as a blue black, wasplike moth with clear wings. The female has a wide orange band around the body, while the male has several yellow stripes. Egg laying, which begins several days after emergence, is done on trunks and branches or on weeds, leaves, and clumps of dirt. The eggs hatch in ten days and the young worms bore into the bark just below the surface, starting the cycle all over again. Borers can be monitored (and controlled to an extent) with cardboard pheromone traps manufactured by Zoecon.

Examine each newly bought nursery tree and return any infested with borers. Don't plant stock already weakened by borers.

In late summer, apply moth crystals around the bases of trees and cover with three or four inches of soil, mounding it against the trunks.

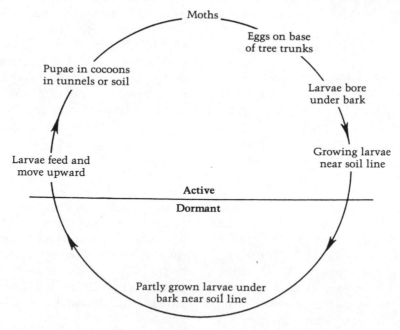

Life history of the peachtree borer

Reprinted, by permission, from Pyenson, *Fundamentals of Entomology and Plant Pathology* (fig. 3.17), AVI Publishing Co., PO Box 831, Westport, CT 06881.

Wait a few weeks and remove the mounded soil before winter rains or irrigating. This method will not help if borers strike higher parts of trees. Another method is to force a piece of tin into the ground completely encircling each trunk, leaving a space of about two inches between shield and bark. In mid-May, fill this space with tobacco dust. The dust will become soaked by rain and make a potent barrier. Repeat this treatment each May. The tobacco becomes humus in time and will cause no harm to the tree.

Peachtree borer

A close relative of the peachtree borer, the lesser peachtree borer, is thought to contribute to the severity of cytospora canker, one of the most costly diseases of stone fruit orchards. Rather than attacking the trunk as its larger relative does, the lesser borer lays eggs in wounds and damaged areas on the upper limbs. Such sites include winter injury, mechanical damage, pruning wounds, old cankers, and damage by other insects. Their presence is known by the sawdust-filled droppings around the entrance holes. Avoid creating entry sites, and keep the trees well nourished so that they'll be better able to resist attack.

Some growers plant garlic cloves close to the tree trunk. Others use a ring of moth balls on the soil. Nuggets of soft soap can be tied around the tree base and up to the crotch; rains wash soap down the trunk and into the ground, and the soapy taste repels both the moth and larva. If you have few trees and they are under

three years old, you might try the time-tested remedy of poking a wire up borer holes. Do not remove the gummy exudate that forms around the holes, as it serves to seal the injury. Keep all trash and refuse away from the base of trees, and avoid damaging tree bark with cultivators.

The larval **peach twig borer** attacks twigs and occasionally enters the fruit. It will attack plum, apricot, and almond in addition to its namesake. You can distinguish it from the larva of the oriental fruit moth by its reddish brown color. The larva is less than ½ inch long when mature and constructs a loose cocoon under the curled edge of the bark. Ten to 12 days later the steel gray moth emerges. It has a wingspread of about ½ inch. There are three or four broods each season. The presence of the larva in the twig is made known by the reddish brown masses of chewed bark webbed together in the crotch. Entrance is often through wounds or breaks in the bark. Trees with wide-angled crotches and smooth bark are less likely to be attacked. Larvae may also infest ripening fruit, particularly near the stem end.

Keep trees in a vigorous growing condition by enriching the soil with liberal amounts of organic matter. Avoid mechanical injuries, as they offer the borer an entrance to the tree. Follow the controls given for the peachtree borer (above). On the West Coast, where damage may be severe, growers use dormant lime sulfur sprays (mixed 1 to 15) or 3 percent dormant oil. Normally, however, controls aren't necessary.

Various **plant bugs** cause "cat-facing" of peach—sunken, distorted areas that may cover more than half the fruit. Damage is worse near woods and brushy fields. See FRUIT TREES.

The **shothole borer** is a little dark brown beetle that breeds under the bark of peach trees, among others, and emerges through small circular holes resembling shot holes. The larval galleries nearly girdle the stems and branches, and on stone fruits you will notice gum exuding from the exit holes. The adult beetle is about $\frac{1}{10}$ inch long and appears in early spring. There are usually two generations a year in the northern states. This pest seeks out injured and weakened trees; keep trees as vigorous as possible.

Some larvae can be dug out of the wood with a knife. Paint the trunk and large lower limbs with whitewash, as needed, to discourage egg laying.

Other peach pests include leafrollers, JAPANESE BEETLE, plum curculio (see PLUM), and scale (FRUIT TREES).

## Diseases

**Bacterial spot** [color] is a serious disease caused by bacteria that overwinter in twig cankers and infect leaves, stems, and fruit of peach. It is also a serious disease of plum, apricot, and nectarine. The disease first appears on leaves as small, circular, pale green spots that later turn light brown. The tissue around the spots

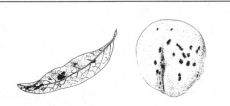

Bacterial spot on leaf and fruit

fades to a light yellow green and soon becomes purplish and angular in shape. Frequently the diseased tissues tear away from the healthy part, leaving the condition known as shot hole. Shedding of infected leaves is common. The disease is most severe following warm, wet spring weather and slows with the hot days of summer. Often the bacteria exude in sticky droplets on the lower surfaces of the leaves; this exudate serves to distinguish bacterial spot disease from arsenical injury or other troubles causing spotting and shot holes on peach leaves.

On peach fruit, bacterial spot disease causes small circular sunken spots, and later cracking, in a variety of patterns. The disease also attacks stems of the current year's growth, causing small elongated cankers. It is important not to accept nursery stock showing cankers of any sort, since they may carry diseases.

Bacterial spot disease is much more dangerous to the peach orchard than is commonly supposed. Leaf infections cause serious shedding or defoliation, which results in serious weakening of the trees, making them unfruitful and more subject to winter injury.

There are a growing number of good cultivars highly tolerant to the bacterial spot disease, including Belle of Georgia, Biscoe, Candor, Cardinal, Cullinan, Cumberland, Eden, Emery, Envoy, Harbelle, Harbinger, Harbrite, Harken, Jefferson, Kalhaven, LaGem, LaPremier, Loring, Madison, Marhigh, Newhaven, Norman, Pekin, Polly, Ranger, Raritan Rose, Redbird, Redhaven, Compact Redhaven, Redkist, Redskin, Reliance, Richhaven, Royalvee, Sentinel, Southhaven, Summercrest, Sunbrite, Sunhaven, Sunshine, Sweethaven, and Wildrose. Among nectarines currently only Early Bird and Mericrest are resistant.

Plant in well-fertilized soil and keep the trees in vigorous growing condition. Judicious nitrogen fertilization helps to discourage bacterial spot disease.

**Brown rot** is by far the most serious fungal disease of stone fruit, particularly if fruit has been injured by curculios and oriental fruit moths or cracked by hail. The rot first appears as a small, round brown spot that quickly spreads to cover the fruit. The rotted areas soon

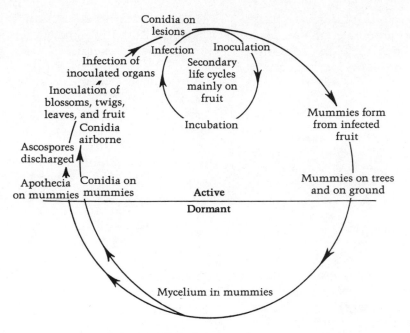

Life history of the brown rot fungus on peach

Reprinted, by permission, from Pyenson, *Fundamentals of Entomology and Plant Pathology*
(fig. 9.14), AVI Publishing Co., PO Box 831, Westport, CT 06881.

develop light gray masses of spores which are readily scattered by wind and rain to cause a rapid spread of the disease. The rotted fruit may dry into hard mummies hanging in the tree. The fruit can carry the fungus over winter and produce spores in spring, or it may drop and remain active until the following year when small, brown mushroomlike growths form to release spores.

If there is warm wet weather at blossom time, the spores from the mummies can cause an extensive blossom infection known as blossom blight, and from these blossoms the fungus grows back into the twigs to cause shallow cankers that will occasionally girdle and kill the twigs and serve as overwintering reservoirs for the fungus.

During the dormant season, collect and destroy both drops and diseased, mummified fruit that stays on trees. (Mummies of cherries very seldom stay on trees all winter, however.) Also remove dead twigs. Do not allow wild plum, peach, or other stone fruit seedlings to grow in orchards, because of their susceptibility to disease. Shallow cultivation will reduce chances for the formation of the sexual and asexual stages of the fungus on mummies, so disc or till lightly just after bud break. Moisture encourages brown rot, and ripe fruit should not be sprinkled with water. Proper pruning will allow better air circulation and rapid drying. When harvesting, try to avoid wounding fruit during picking and packing, as wounds are ideal avenues of infection. Fingernail scratches and damage from rough containers will contribute to the start of trouble. The fungus has difficulty penetrating the skin of uninjured peaches, but any break in the skin opens the way to infection.

Brown rot

The brown rot fungus grows rapidly at warm summer temperatures and may destroy infected fruit in a relatively short time. Refrigerate fruit immediately after picking. When a peach rots in a container, the fungus can easily pass to other fruit in contact with it. Thus, a high percentage of the fruit may be destroyed if peaches are kept for very long at temperatures over 65°F. This is especially true if there has been much rot in the orchard or if the fruit has been roughly handled.

A bath of hot water is an effective means of preventing brown rot. Dip freshly picked peaches in 120°F. water for seven minutes to kill the fungi that cause brown and rhizopus rots, without harming the peach. It's the heat, not necessarily the water, that eliminates fungi.

If you grow garlic or onions, prepare a botanical brew and spray fruit well before harvest. If this doesn't work, consider applying sulfur (6 pounds of 95 percent sulfur per 100 gallons water) to ripe cherries several days before harvest.

While peach or nectarine varieties are resistant to brown rot, some exhibit a degree of tolerance. Tolerant peach varieties include Elberta, Garnet Beauty, Glohaven, Harbelle, Harbinger, Harbrite, Harken, Loring, Newhaven, and Suncling. Among tolerant nectarines are Cavalier, Mericrest, Redchief, and Redgold.

**Cytospora canker** (also known as peach or perennial canker) is the work of a fungus that attacks the woody parts of stone fruit trees through bark injuries, pruning cuts, dead buds, and winter injury. A small canker forms at the base of the bud in late winter to late spring. After one to several years the canker girdles the twig and the growth beyond it wilts and turns brown. The first symptom on limbs is a wilting or yellowing of new shoots and leaves, which later turn brown. The older, visible cankers are oval to linear in outline, and are usually surrounded by a roll of black callous tissue. The bark in the center of the canker becomes torn. The lesser peachtree borer often invades the cankers that girdle limbs or trunks. Smooth sections of the bark in the cankered areas become peppered with the pimplelike fruiting bodies of the fungus. During rainy periods, these pimples ooze a white or orange mass of spores, looking something like toothpaste.

Do not set stone fruit trees in soil having poor internal soil drainage. Wet spots should either be avoided altogether or drained. Trees on these sites often suffer cold damage and have a considerable amount of deadwood in them. Because severe infections of canker can provide a source of inoculum for nearby trees, new orchards should not be planted near cankered old ones—especially not on the downwind side. It is best not to plant on ridge tops, as winter winds can desiccate dormant buds so that they die and provide an entry for disease.

Select stone fruit cultivars that are winter-hardy. Attempts to transplant, train, and prune large trees from nurseries often result in canker at an early age. Trees of about $9/16$ inch in diameter have the potential for branch selection and tree form, without the necessity of making large cuts to balance the tops with the root system. By means of nitrogen levels and cultivation practices that induce terminal bud set by September, trees can be rendered cold-hardy early in the season. This minimizes the chances of early winter bud kill, thereby reducing those entry points. Pruning is necessary, but does create entry points at the cuts and has a dramatic influence on cold-hardiness. Delay pruning until bud swell; you can continue through bloom with no adverse effects. An unpruned tree can better resist injury from sharp drops in temperature. If pruning is delayed until bud swell, this risk is minimized. After bud swell, the growing tree is more resistant to the disease, even though the fungus may penetrate the bark or pruning cuts. Midsummer topping and hedging with mechanical equipment will spare trees the danger of pruning. Paint the trunks and lower scaffold limbs with white latex paint to reduce the temperature under the bark; this greatly reduces cold injury and cytospora canker in the critical trunk and crotch areas of the trees.

Twigs killed by the oriental fruit moth and twig cankers that result from brown rot blighted blossoms or rotted fruit often are entry points for the cytospora canker fungus; see discussions of this moth and brown rot, above. The following varieties offer a degree of resistance to canker: Biscoe, Candor, Carmen, Champion, Comanche, Cumberland, Eden, Emery, Envoy, Harbelle, Jayhaven, Madison, Polly, Raritan Rose, Redqueen, Reliance, Royalvee, and Veteran. Among nectarines, Crimson Gold, Francesco, Independence, Lexington, Mericrest, Morton, New Yorker, and Redchief are somewhat resistant.

Leaves infected with **peach leaf curl** show symptoms soon after they unfold. They crinkle or pucker, grow thicker than normal, have a leathery feeling, and are colored light green, yellow, red, or purple. The entire leaf or any part of it may be infected. In severe outbreaks of leaf curl, nearly all the leaves on a tree are infected. More often, only the leaves of occasional buds scattered throughout the tree will be diseased. By late spring or early summer, their upper surfaces turn grayish and take on a powdery appearance due to the production of fungal spores. Spores overwinter on the bark and may remain infective for several years. Dry, warm weather soon causes the leaves to turn brown to grayish black and fall from the trees. Infected new shoots grow slower than normal, and are somewhat swollen and pale yellow in color. Severely infected young fruit becomes misshapen and seldom remains on the tree. Lightly infected fruit matures normally, although it will have one or more wartlike, irregular growths on it. The fruit has no fuzz and is usually reddish in color. It is similar to the irregularities that occur along the suture of some cultivars, such as the Rio Oso Gem peach.

Spray with a fixed-copper or lime sulfur fungicide anytime from November until the first buds begin to swell (after bud break, sprays are ineffective). If heavy rains fall soon after you spray, a second application may be necessary.

As soon as the new leaves show curling, they should be removed and placed in the garbage or a well-managed compost pile. Peach leaf curl is most severe in years having a moist and cool early spring.

While peach cultivars vary some in the susceptibility to leaf curl, none of good quality is immune. These varieties are fairly resistant to leaf curl: Dixigem, Dixired, Jayhaven, Loring, Monroe, Raritan Rose, Redhaven, Compact Redhaven, Southhaven, Summercrest, and Sunshine. No nectarine varieties qualify as resistant to leaf curl.

**Rhizopus rot,** like brown rot, is a storage rot that is reduced by promptly cooling fruit after harvest and avoiding skin punctures. Nicks in the skin allow the fungus to enter, and can quickly turn a firm, ripe peach, nectarine, plum, apricot, or cherry into a mushy, sodden mass.

Rusty spot of peach seldom causes more than cosmetic damage to fruit. The earliest symptoms are small white dusty areas. These spots darken and slowly enlarge and develop a fuzzless, smooth center spot that in time turns into a brownish or reddish patch of hard, smooth skin looking somewhat like a bruise from limb rub. Very red varieties may not show injury until after harvest.

So far as is known, some peach cultivars are not attacked by rusty spot; Triogem is one of these. The following are known to be susceptible to varying degrees: Redhaven, Washington, Loring, Goldeneast, Belle of Georgia, Redskin, Elberta, Jersyqueen, White Hale, Jefferson, and Rio Oso Gem. The disease is one of the mildew family, related to powdery mildew on cherry but exhibiting few or none of the foliage symptoms. Two applications of sulfur (5 pounds of 95 percent sulfur per 100 gallons), at petal fall and two weeks later, will keep this fungus down, although no sprays are needed if the disease is not seen.

Scab [color] appears as small, dark greenish spots, $1/16$ to $1/8$ inch across, that become visible about the time the fruit is half-grown and turn black (brown in apricots) as they mature. The fruit may later crack if the infection is severe. Yellow brown spots appear on twigs and branches.

If scab threatens a crop, consider using a sulfur spray or dust three to four weeks after the petals have dropped, or two to three weeks after petal fall on nectarines and apricots. A second application may be needed two weeks later. Such applications prevent infection, but do not eradicate the fungus; new shoots and branches will likely be infected by the end of the growing season, showing yellow brown spots with blue or gray margins.

Shothole disease (coryneum blight) causes small, round, purplish spots on leaves and fruits. Centers of the leaf spots turn brown and fall out, giving the leaves a shothole appearance. The symptoms are worst on the lower portions of trees, where foliage remains wet longest. The fungus survives on infected buds and twigs during winter. Rain and overhead irrigation spread the fungal spores to leaves and twigs. A few holes will not reduce tree vigor or yield, but widespread spots may have to be countered with a copper fungicide in late autumn, between November 15 and December 15. This spray will also control peach leaf curl on peaches and nectarines. Decrease overhead irrigation, particularly in spring, to discourage the disease.

Stem pitting can be a major problem on peach, nectarine, and apricot. It is also found on all other stone fruits. The disease is caused by strains of the tomato ring spot virus, transmitted by the dagger nematode. The first symptoms occur in late summer, usually after harvest but before normal leaf drop; they begin the first year or two on a single branch and then progress over the entire tree. Leaves cup upward, turn yellow then reddish purple, and drop prematurely. After several years in this weakened state, the tree succumbs. If the tree is pulled out of the ground, the roots break away from the crown, leaving socketlike depressions. The bark at the soil line is thickened and has a spongy texture; under this bark, the wood is highly disorganized and pitted and lacks the regular grain pattern.

To avoid introducing this disease into your field, purchase only certified virus-free trees when laying out a new orchard. Use plenty of organic matter throughout the root zone (tend to this before planting) to encourage a favorable mix of beneficial predators, including predatory nematodes.

Sudden death is not caused by any pathogen, but results from careless planting. Unless soil is lightly packed around the base of the new tree, high winds may expose the trunk down to the root crown. Then, if a severe winter should follow the planting year, the cold may injure the young tissue. The tree emerges normally in the spring, but its damaged circulatory system cannot meet its needs in hot weather, and the tree suddenly wilts and dies.

On peach trees, **X-disease** is first noticed in late June or early July, when the lower leaves on shoots develop water-soaked areas of various sizes. These leaves soon become blotched with red, and sharp margins develop around the discolored areas. Eventually these areas drop out to give the leaves a tattered appearance. Unlike similar spots caused by nitrogen deficiency and other weaknesses, the margins of the discolored and tattered areas usually cut across lateral veins. The reddened spots caused by nitrogen deficiency typically are exactly halfway between the lateral veins and affected leaves usually persist; but leaves on trees infected with X-disease soon drop, eventually leaving only a small tuft at the extreme tip of affected shoots. In severe cases, sucker growths often develop the same symptoms.

X-disease is spread to peach, nectarine, and sweet and sour cherry trees from wild chokecherry bushes. Leafhoppers act as vectors; species in the western states may spread the disease from one cultivated plant to another, while those species in the East apparently do not. Although trees cannot be saved once affected, spread of X-disease can be halted by destroying all chokecherry bushes in the vicinity of the orchard. Infected chokecherry bushes develop yellow to red foliage, and in advanced cases the shoots are short and rosetted. The aboveground parts of the bushes ultimately die. Unfortunately, the roots of infected bushes often are not killed and new sprouts develop from the roots of plants with X-disease.

Control depends on removing all chokecherries from the surrounding area, preferably for a distance of at least 500 feet. Because it is difficult to diagnose the disease in its early stages, all chokecherry bushes should be destroyed whether or not they show symptoms. Before a new orchard is planted, scout the area carefully for chokecherries. On open ground, deep plowing will make it quite easy to pull out and burn the plants. Subsequent cultivation will prevent sprouting. It is important to check the orchard occasionally for new plants, as birds carry chokecherry seeds. Unlike pin cherry and black cherry, which grow into trees 50 feet high, chokecherries are a woody shrub seldom more than 15 feet tall. The fruit is borne in clusters on a central stem, not individual stems as with the pin cherry. Black cherry fruit is also on a central stem, but can be distinguished from chokes by the presence of a calyx cup at the stemlet of each fruit.

Other diseases affecting peach and nectarine include armillaria root rot (see FRUIT TREES), bacterial canker and verticillium wilt (CHERRY), and crown gall (APRICOT).

# PEAR

## Insects

A simple mechanical barrier, the paper bag, will keep many pests of pear from damaging fruit. Just staple sandwich-size paper bags around the fruit and snip off the corners to allow rainwater and moisture to drain out and to encourage ventilation. The bags don't have to be transparent, as the plant makes its food in the leaves.

See FRUIT TREES for a description of an effective spray program.

The **codling moth** may injure the fruit of the pear, especially of those varieties that are harvested late in the season. See APPLE for control.

Fruit becomes knotty, deformed, and gritty in texture when the **false tarnished plant bug** (also known as pear plant bug and green apple bug) punctures it. The adult looks much like the bona fide tarnished plant bug [color], although somewhat smaller and paler. The winter is passed by the egg stage in the bark (most plant bugs overwinter as adults), and the eggs hatch at blossoming time. The nymphs are green at first, and begin to puncture the fruit as soon as it sets. Damage may be serious in some years and trifling in others. In a bad year, nicotine sulfate or rotenone may save the season if sprayed just after the blossoms fall. A second application can be made a week later. This will control most other species of plant bugs, as well.

The adult beetle of the **New York weevil**, common in the Mississippi Valley, sometimes causes severe injury to young pear trees by eating off the leaf buds in early spring. It may also eat the bark of the new growth and cut off leaf stalks and new shoots. It feeds chiefly at night, so you may notice its damage and never the responsible insect—a $\frac{5}{8}$-inch-long, ash gray beetle with dark spots. Handpick and jar the insects onto sheets placed under trees.

The unfolding leaves of pear and apple are often disfigured by greenish yellow or reddish blisters that later turn brown. Affected fruit will be russeted under light infestations and severely deformed if higher populations are present. In severe cases the leaves may drop in midsummer. Colonies of microscopic **pear leaf blister mites** live within the tissues of the leaf. They pass the winter beneath bud scales. Control by applying an oil and lime sulfur spray in October or November, or a dormant-oil spray just before the buds open in spring.

In June small pears often drop in great numbers, some of them split open. Upon examining them, you may see maggots of the **pear midge.** This midge looks much like a mosquito. It lays its eggs in blossoms. On hatching, the maggots work their way down into the core, gradually hollowing out a large cavity that may occupy the entire interior of the young fruit. The maggots reach maturity in June. The fruit usually cracks open and falls to the ground, and the larvae enter the soil, where they pupate and hibernate. Fall cultivation for pearslug will also aid in midge control.

A strong botanical spray, such as pyrethrum, will offer some control; nicotine sulfate is a high-powered, last-ditch measure.

**Pear psylla** [color] nymphs are small yellow creatures that feed on the top sides of leaves until only the veins remain. As adults, psylla are dark orange with transparent wings, and resemble miniature cicadas. They hibernate under the edges of rough bark on the trunk and branches, emerge during the first warm days in April, and soon deposit eggs in old leaf scars, in cracks and crevices, and around the base of terminal buds. Most of these eggs are laid before the buds open, and nearly all have hatched by the time the petals fall. Nymphs go to the axils of leaf petioles and begin to suck sap. Much of the sap is excreted as honeydew, which drips upon the lower leaves and supports the growth of a black sooty fungus and attracts predatory insects (such as adult lacewings) which feed on the sweet secretion. About a month is required for the complete life cycle. By midsummer a badly infested tree is blackened on leaves and fruit. Many leaves will fall before the fruit ripens. The piercing mouthparts also transmit the mycoplasma organism responsible for pear decline in the western states.

Although they may be killed by dusting limestone over them, the best preventive is a thorough dormant-oil spraying in spring. Apply a 2 percent solution just as the buds begin to swell, to catch the overwintering psylla as they begin to lay their tiny yellow eggs on twigs and buds. If the psylla are already established, try spraying the tree with soapy water, applied with as much pressure as possible and from many different angles. Repeat sprays as necessary. During the growing season, you can use a superior-oil spray two or three times, a week apart.

Pear psylla have a good number of natural enemies to contend with. Two small, dark chalcid wasps parasitize psylla nymphs. The parasites pupate within the body of the dead nymph and the adult forms emerge through a hole chewed in the abdomen. One of these beneficials *Trechnites insidiosus,* may parasitize up to 90 percent of the nymphs during July and August in unsprayed Washington State orchards. Earlier pest generations are less bothered, however. Pirate bugs, lacewing larvae, snakefly larvae and adults, and ladybug larvae are among the most valuable predators, and along with other true bugs can reduce psylla populations to below the level of economic significance. However, these predators, like the wasp parasite, build up in numbers slower than does the psylla in spring and early summer. These helpers do best when pest populations are high, as they do not have the searching ability to hunt down scattered psylla.

Should psylla build up in your pears without the natural restraint of predators, the weather may still work to your benefit. The young psylla secrete a drop of honeydew over their pale yellow bodies to protect themselves from the elements, but this can backfire in extremely dry periods if the sugary syrup hardens into a crystalline prison. To monitor for psylla, check the tender growing tips of the highest shoots. Look for the mite-size eggs or the telltale drops of honeydew on the undersides of leaves. No varietal resistance is known, although Seckel appears to be less bothered than most.

The **pear rust mite** attacks pear foliage and fruit in the spring, causing them to turn brown and russeted. The next growing season will be troubled less if you use a spray of oil plus lime sulfur in October or November. But control is seldom necessary, because damage is mostly a matter of appearance of the fruit.

The **pearslug** (also called the cherryslug) is the sluglike larva of a small, black flylike creature, called a sawfly, that appears in May and inserts its eggs in leaves. The young larvae feed on the upper surfaces of leaves, especially on young trees, causing pinkish or brownish patches on the tops. Mature larvae eat completely through the leaf, leaving a fine network of veins. They have large head ends and somewhat resemble wet tadpoles.

Because the first generation of black-and-yellow sawflies emerges from cocoons in the ground just about the time cherries and pears come into full bloom, you might anticipate it by a shallow cultivation under the tree. Fall cultivation will help reduce the overwintering population, but don't till more than two inches deep to prevent root damage. Check with your local extension agent for suggestions on timing. You might also consider a second pass through the orchard after the first brood of slugs hits the ground. Wood ashes dusted over the tree will fatally dry the larvae. Wash the trees with water after five days. Rotenone is a powerful control to be used if other measures fail.

**Pear thrips** cause injury to the blossoms of pear in the East, and many other fruits in the western states. The black adults emerge from the ground in spring, work their way into the swelling buds, and soon lay eggs in the stems and midveins of the unopened buds. The eggs hatch in two weeks and large numbers of the white nymphs feed upon the unfolding buds. Heavily infested orchards appear as if a fire had scorched the trees, and the blossoms are destroyed.

A dense mulch applied in July around trees may prevent the adults from making their way out of the ground. Cultivation may not effect control, as the thrips go as deep as three feet to spend the winter. Use a 1½ to 2 percent oil spray on a warm day when the blossom buds are still tightly clustered; this will catch many of the overwintering adults.

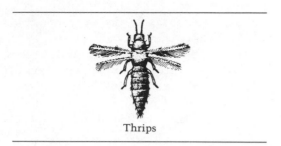

Thrips

The grub or larva of the **sinuate pear tree borer** tunnels in the branches to cause ugly scars and some breakage where infestations are heavy. The adult is a slender, glossy bronze brown beetle, ⅓ inch long. The females lay eggs in the crevices and under the edges of the bark during June. In early July the eggs hatch and the young grubs excavate narrow sinuous tunnels in the sapwood just beneath the bark. They are partly grown when winter arrives, hibernate in the burrows, and continue their destructive work the following season. The galleries are then larger and their course shows through the bark. By the second September the grubs are about 1½ inches in length. They tunnel deeper into the wood and excavate the pupal chambers in which they hibernate. They pupate the following April and the beetles emerge a month later. Two years are required for the complete life cycle.

Severely injured branches should be cut off and burned. Fertilize the trees moderately to maintain vigor, but not so much as to invite fire-blight infections.

Other insect pests of pear that may be found in numbers requiring control are green apple aphid (see APPLE); leafrollers, mites, and scales (FRUIT TREES); and plum curculio (PLUM).

# Diseases

**Fire blight** is the most destructive disease of pear in many areas, and is nearly as injurious to certain varieties of apple. Quince, hawthorn, loquat, and other members of the Rosaceae are also affected, including pyracantha, spirea, and mountain ash.

The disease is caused by a bacterium that enters the plant at blossom time through the specialized flower cells that produce nectar. Bees, flies, ants, and other insects are often responsible for carrying the bacterium about, though wind is also a culprit and moisture (rain, dew, or fog) is necessary for infection. Once inside the plant, the pathogen can move great distances within living tissue (6 to 12 inches a day) to kill twigs, branches, or even entire trees. A tree may be destroyed by a single infection in one season. Affected blossoms become blackened and shriveled, soon followed by the leaves on the spur. New shoots are inoculated by splashing rain and by sucking insects, especially aphids, leafhoppers, and plant bugs. Bark of twigs and branches may be inoculated through fresh wounds. Trees may be killed by root infection resulting from invasion through root sprouts.

When infected from direct inoculation, terminals and water sprouts usually wilt from the tip downward, producing the typical "shepherd's crook" appearance, and the leaves tend to persist, frequently remaining attached throughout winter and serving to call attention to cankers on the supporting branches.

As the infection progresses into the supporting branches, the bark surface (if smooth) becomes slightly darker than normal. Affected tissues are at first water-soaked, later develop

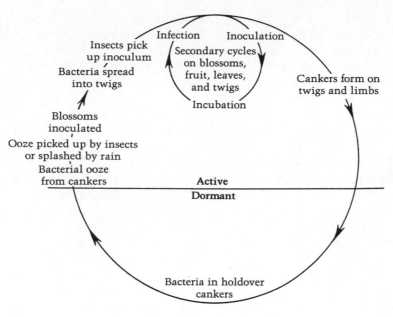

Life history of the fire blight bacterium

Reprinted, by permission, from Pyenson, *Fundamentals of Entomology and Plant Pathology* (fig. 9.8), AVI Publishing Co., PO Box 831, Westport, CT 06881.

reddish streaks, and finally die and turn brown in color. A milky, sticky ooze containing bacteria may appear on any affected part, and will turn brown on exposure to the air. Girdled branches soon die. Fruit becomes sunken and discolored, with necrotic cankers.

Symptoms of fire blight are often confused with those of winter injury. There are several differences: dead bark is brown in blight infections and gray (mixed with white flecks) in winter injury; the wood is more likely to be discolored by winter injury than by blight; winter injury is more likely to extend downward from cuts, to be centered in crotch areas and at the base of the trunk, and to be located in weaker parts of the trees; the bark may become separated from the wood in winter-injured areas; and there is usually an obvious site of infection, such as a fruit spur or water sprout, within the blight canker.

Warm, moist weather favors infection, especially during bloom. Highly succulent growth favors development of the disease, as do excessive nitrogen fertilization, late fertilizer application, poor soil drainage, and other factors that promote succulent growth or delay hardening of the tissue near the end of summer.

The manner of pruning can go a long way to preventing fire blight. Generally, trees should be pruned every year so that only small cuts will be needed. On young trees, cut out infected twigs as soon as they can be seen, especially in early summer, making the cut at least 12 inches below evidence of infection as shown by darkening of the bark surface (the pathogen precedes the visible infection). The cutting tool should be disinfected after each cut with one part household bleach to nine parts water. In winter, all infections from the previous year should be cut out, making the cut four inches below the basal edge of the canker or infected twig. The extent of lesions can be determined on smooth bark by looking for the darker color that is symptomatic of trouble. If there is any doubt, make small cuts with a knife to see if the affected inner bark tissue is water-soaked. The knife should be disinfected between each cut (except during the dormant season).

Nearby apple (including crabapple), pear, quince, hawthorn, and mountain ash should be examined for cankers, whether found in the wild or under cultivation. Control sucking insects to help prevent shoot infection. Most important are aphids, leafhoppers, and plant bugs, as well as psylla on pears.

Avoid giving trees a sudden dose of nitrogen, such as by supplying more manure than needed. In selecting an area for a planting of pears or susceptible apple varieties, look for fertile, well-drained soil, as tree growth stops earlier in the season on dry soil and this renders trees less vulnerable. Mulch trees in late fall or very early spring. Avoid late cultivation; in young orchards it is best to abandon cultivation altogether in favor of a sod-mulch system of culture using grass as a cover crop. Grass sod should be mowed early in the season and then allowed to grow in midsummer in order to check late-season growth of the trees. If the weather has been drier than normal, you may have to keep the sod mowed closely throughout the summer, and extreme drought often necessitates discing the soil lightly. Test the soil around pear trees to check for a deficiency of potash, and use lime if necessary to bring the pH up to 6 or 6.5. Dormant sprays will help to keep down numbers of aphids and leafhoppers. Another spray that is highly effective against blight is agricultural-grade streptomycin, available commercially as Agri-Strep. It is applied during early and mid-bloom, at 8 ounces per 100 gallons water, to stop a developing blight infection. Rather than spray every year and risk having the blight bacteria develop resistance to this control, save it for those years when there is blight active in the orchard and the conditions are ideal for infection: high humidity or rain, wind, moderate to high temperatures (above 65°F.), and sucking insects present.

Nothing is so essential in fighting this disease as avoiding susceptible varieties. Those that are almost impossible to grow without strict attention to prevention include Bartlett, Packham's Triumph, Clapp's Favorite, Bosc, Aurora, Highland, Gorham, Flemish Beauty, Idaho, Parker, Patton, Sheldon, and Howell. Moderately susceptible varieties may suffer a few infections some years, bit these sites can usually be pruned out before the disease runs through the orchard: Anjou, Comice, Ayers, Monterey, Kieffer, Luscious, Duchess, Mendel, Leconte, Winter Nelis, Hardy, Old Home, Tyson, Pineapple, Baldwin, Carnes, Dabney, Douglas, Garber, Hood, Hoskins, and Mooers. There are few pears available that are truly resistant to blight, but the following will seldom, if ever, be attacked even under the worst conditions: Maxine, Moonglow, Magness, Waite, Seckel, Orient, Dawn, and Lincoln.

**Leaf blight** is a fungal disease affecting leaves, twigs, and fruit. A week or two after a rainy period, as many as a hundred purplish ¼-inch spots may appear on a single leaf; heavy leaf drop may result. If enough infections develop on the fruit surface, fruit may crack.

The disease overwinters on leaves and twigs; it is spread during rains. Use a series of four sulfur applications, beginning with the first leaf and repeating at two-week intervals.

**Pear decline** was formerly thought to be caused by a virus, but now is known to be the work of a microscopic *Mycoplasma* organism that the pear psylla transmits. Decline severely affects trees grafted onto the oriental rootstocks *Pyrus serotina* or *P. ussuriensis.* Control pear psylla and keep trees well nourished. Select only pear trees on *P. communis* rootstock.

**Pear scab** appears as velvety, olive green spots on fruit, becoming black and scabby at maturity. The scab makes black spots on leaves and unlike apple scab, will infect twigs with infections that overwinter. The disease is favored by warm, damp weather. Remove any leaves or fruit infected with scab, and keep the area under the tree free of fallen leaves and fruit. Resistant varieties are available and you should check to see which are suited for your growing area. Apply wettable sulfur before bloom and every two weeks until the fruit turns down. Use 6 pounds of 95 percent sulfur per 100 gallons of water.

**Stony pit,** a viral disease, is one of the most common problems for pear growers. The Bosc pear is considered very susceptible and most of the work on stony pit has been done with this variety, although Anjou is also affected. Occasionally, however, Kieffer or Bartlett fruit will get the disease. The symptoms appear soon after petal fall; look for dark green areas under the epidermis of the fruit. At maturity the fruit is pitted, gnarled, and deformed. The flesh becomes woody or stony, as the name implies. Leaves may show veinlet chlorosis or yellowing. Once a tree becomes infected, it continues to show typical symptoms every year. The only effective control is to destroy infected plants. Do not replant susceptible varieties on that site.

Other diseases of pear include armillaria root rot (see FRUIT TREES), cedar-apple rust and nectaria twig blight (APPLE), and crown gall (APRICOT).

# PECAN

## Insects

Many insects, including **aphids** [color], will attack pecan trees weakened through a lack of water. While pecans will survive in a region that gets 25 inches of rainfall annually, they

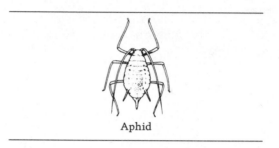

Aphid

need 50 inches a year to really thrive. If pecans are grown in a dry western or southwestern state, they must be irrigated. Too much light and superficial watering prevents the tree from establishing and maintaining a deep and healthy root system, making it more susceptible to insect attacks. A deep watering—three to six hours for each tree—is recommended in March or April prior to leafing out, again in late June or early July, a third time a month later, and a fourth in late August. Avoid using a sprinkler that sprays water on the leaves, as this encourages aphids and may lead to mildew.

The **casebearer** and the **hickory shuckworm** are two notable pests of pecans. Although they do not bother the tree itself, they may seriously reduce the nut crop. Both insects are controlled by the trichogramma wasp. Release the beneficial's eggs two or three times per growing season, the first time in mid-April, the second about two weeks later, and if it is not an off year and the pecans are setting, release the eggs a third time after another two weeks. Populations can be reduced with a blacklight trap, approximately one to every three trees.

**Fall webworms** [color] can strip and even kill a tree. Check for them after each rainy spell. Twigs bearing clusters of these small caterpillars should be pruned and burned. The worm can be identified by a dark stripe running down the back paralleled on either side by a yellow stripe. If clusters of their tiny greenish eggs are noticed on the undersides of leaves, pick the leaves and burn them. Clean up debris around trees.

## Diseases

**Crown gall** is a bacterial disease of pecan trees that is characterized by the presence of wartlike growths on the roots or at the base of the trunk. The galls range from pea-size to several inches in diameter. Trees fail to grow properly and may eventually die.

Once a tree becomes diseased, there is no effective control, although some trees can be saved by removing the large exposed galls. Avoid making plantings in soil where diseased plants have grown. Inspect planting stock carefully and reject plants that have suspicious bumps or swellings near the crown or graft union. A bacterial inoculant is available. See CROWN GALL.

Pecan **dieback** often occurs in areas having hardpan layers, caliche soils, high salt, or poor drainage. When root growth is restricted to a small planting hole, normal tree growth cannot be expected; trees may make adequate growth for a few years and then begin dying back from the tips in later years.

Avoid planting pecan trees in heavy soils with poor drainage. A caliche or hardpan layer can be detected by digging a hole five to six feet deep; if you don't see such a layer, there should be no problem. Dieback sometimes occurs because of a lack of water during winter months, particularly in light, sandy soils, and can in this case be remedied by regular deep waterings.

**Root rot** sometimes affects pecan trees in the Southwest. The first symptom is a sudden wilting and drying of foliage. An examination of the roots reveals the rot.

Avoid planting trees on land where other plants have shown symptoms of this fungal disease. Trees can often be saved if half or more of the top growth is pruned when symptoms first appear. Cover the ground with two inches of manure and then water to a depth of three feet or more. Continue to make sure the trees get adequate water.

The symptoms of **sunscald** are dead or cankerous areas on the trunk or upper surfaces of large branches. The unshaded trunks of young trees are especially subject to sunscald. Protect them by wrapping with burlap strips, kraft paper, or aluminum foil. Avoid excess pruning of lower branches of young trees.

**Winter injury** causes bark to appear sunken and cracked where it meets the growing tissue. Healthy trees in a dormant condition can withstand low temperatures, but if the tree makes excessive growth late into fall, proper hardening of tissues does not take place.

Plants overfed with a high-nitrogen fertilizer, and those receiving heavy early-fall waterings, are most likely to become injured. During September, reduce or halt irrigations until the trees become dormant, at which time they can be watered again to provide a favorable condition for the roots over the winter. Fertilizer applications should be made in winter or early spring; do not use high-nitrogen fertilizers.

**Zinc deficiency,** causing a condition called rosette, occurs most often in light sandy soils, and also in heavier alkaline soils. Symptoms appear the second or third year after new trees are planted. The first symptom is the appearance of small, narrow leaves that are yellow and mottled. Reddish brown areas or perforations often appear between the veins of older leaves. An excess of small branches form, giving the foliage a rosetted appearance. In the final stages, the shoots die back from the tips. Seriously affected trees rarely bear nuts, and those that do form are small and poorly filled.

To correct this deficiency, see the control for little leaf under APRICOT.

# *PEDIOBIUS FAVEOLATUS*

A parasitic wasp, used to control the Mexican bean beetle. See MEXICAN BEAN BEETLE.

# PEONY

## Insects

**Ants** on peonies do not signal trouble—they are feeding on the sweet secretion of the flower buds and do no harm.

At least four species of **flower thrips** cause peony and rose buds to turn brown. Adults are typically slender, winged, lemon yellow to brown, and very active. Larvae are wingless, and lemon yellow to orange. Thrips can be observed by shaking infested flowers over a sheet of paper. You'll find that they prefer certain colors of flower to others. Heavy infestations are likely to occur on other plants as well, such as Japanese iris, day lily, garden lily, carnation, gladiolus, and rose.

Peak activity of this pest occurs in May and June near Washington, D.C., and later in the North and sooner in the South. Late-spring and early-summer flowers can be protected by coarse cloth suspended on a frame. During this time, some growers cut peony and roses in the tight bud stage and bring them indoors to open. Affected buds should be snipped off and destroyed. Later in the season, flower thrips get much competition from their natural enemies, and usually do not cause serious trouble.

**Rose chafers** feed upon and dirty white flowers; see ROSE.

## Diseases

Many flower growers run into **botrytis blight.** Look for dark, leathery areas on plant tissue. Flowers are blasted, buds turn dark, and

Botrytis blight

Flower thrips

young shoots may be rotted at the base. Snip off all tops at ground level, burn all old stalks in fall, and remove mulch early in spring. Do not compost plant parts. If the case is particularly severe, it sometimes helps to replace the top two inches of soil with new soil. These control measures will discourage other diseases as well. Some varieties, such as Avalanche, are more susceptible.

# PEPPER

Healthy plants may fail to produce fruits if the early season subjected them to unusually cold nights.

## Insects

In addition to the remedies discussed below, planting well-known mints and monarda (bee balm, bergamot, and Oswego tea) nearby will help keep many pests away.

Four species of **aphid** are known to infest pepper plants; for control, see APHID.

Young peppers may be protected from the fat, smooth **cutworm** by setting a cardboard collar deep in the soil around each plant. If other controls are needed, see CUTWORM.

Peppers are one of more than 200 crops attacked by the **European corn borer.** This is a flesh-colored caterpillar that grows to one inch in length and has brown spots and a dark brown head. The worm bores into stems, branches of large plants, and fruits, causing the affected peppers to drop. These injuries render plants vulnerable to wilt. Chances of corn borer damage are much greater if corn and pepper are planted adjacent to each other. For control, see CORN.

**Pepper maggots** [color] infest fruits of pepper plants, causing them to decay or drop. The eggs are deposited in the wall or interior cavity of the fruit, and after hatching the larvae feed inside the core before leaving near the stem and heading for the ground to pupate. The maggots are at first translucent white, and turn yellow as they feed.

You can discourage the adult (a yellow fly with three brown bands across both wings) from depositing its eggs by dusting talc or diatomaceous earth on the fruit. This should be done during the egg-laying period in July and August. If the fly has already gotten to the peppers, remove and destroy those that are damaged.

The **pepper weevil** is a brown or black snout beetle, ⅛ inch long with a brasslike luster. It feeds on foliage, blossom buds, and tender pods. The larval form, a ¼-inch-long white worm with a pale brown head, feeds within buds and pods. Weevil feeding results in misshapen and discolored large pods; in serious cases, buds and pods may drop off plants. They are a pest from Florida and Georgia west to California.

Clean up thoroughly after harvest; dust with rotenone in emergencies.

The tiny (⅟₁₆-inch-long), black **potato flea beetle** is so named because it jumps like a flea when disturbed. Another species, the striped flea beetle, is identified by a crooked yellow stripe on each wing cover. See POTATO for means of controlling these pests.

The **serpentine leafminer** gets its name from the long, winding, white mines it makes in upper leaves. The work of these yellow maggots opens the doors to diseases. Natural parasites do much to control miners; pick off any leaves that show mines.

Two nearly identical twins, the **tomato hornworm** and the tobacco hornworm [color], eat pepper foliage and an occasional fruit. They are light to dark green with white stripes, and can grow to an impressive four inches long. The former has a black horn at its back end, while the latter's horn is red. Both species are often found in the same gardens. Handpicking is practicable because the size of a hornworm

makes it easy to locate. For other controls, see
TOMATO.

## Diseases

On either green or ripe pepper fruit, **anthrac-nose** shows as dark, circular sunken spots that
vary in size to more than an inch in diameter.
Within these areas are black dots that contain
the spores of the fungus. After rain or heavy
dew, pinkish masses of spores exude from these
dots. Infected fruit may be completely rotted,
and either fall from the plants or hang on as
withered mummies. On stems and leaves the
symptoms are usually so slight that the plant
does not appear diseased until fruit develop.
Diseased fruit serve as sources of spores that
wash and splash around to other fruit; the
disease also infects and survives on stems and
leaves throughout the season. Areas most likely
to have trouble with this disease are the central,
southern, and Atlantic coast states.

To control anthracnose, use clean seed and
practice crop rotation on a three-year basis.
Avoid touching the plants while they are wet,
and keep pepper and bean plantings separate in
the garden, since both vegetables are susceptible.

**Bacterial spot** [color] occurs in all but
semiarid regions. Young leaves develop small
yellow green spots; older leaves have larger spots
(⅛ to ¼ inch in diameter) and dead, straw-colored centers with dark margins. The older
leaves turn yellow and drop.

Since the bacteria live in the soil, plant
seed in new seedbed soil in order to reduce
chances of transferring the disease. Generally,
soil that has not grown peppers or tomatoes
for the past three years is satisfactory. See
RESISTANT VARIETIES.

**Blossom-end rot** is a physiological disease
that causes light-colored, sunken, water-soaked
spots near the blossom end of fruit. The spots
enlarge, and a third of the fruit may become
dark and shriveled. Fungi can be observed
growing over the affected areas.

To prevent blossom-end rot, avoid the
excessive use of nitrogenous fertilizer and use
ample amounts of ground phosphate rock and
ground limestone. Be certain to maintain even
soil moisture at all times.

**Cercospora leaf spot** is a fungal disease that
causes water-soaked spots on leaves and stems.
As these spots enlarge, they turn white in the
center and develop dark margins. Infected leaves
often drop, and plants are occasionally seriously
damaged. Most harm is done in southeastern
and Gulf states. For control measures, see bac-terial spot, above.

**Mosaic** on pepper plants may be caused by
any one of several viruses, such as tomato mosaic
or cucumber mosaic. The virus overwinters in
wild perennial host plants. Peppers set early in
the season become severely stunted and develop
mottled green and yellow leaves that are often
curled; usually, few fruit set. Plants that become
infected about blossom time may manifest a
slight mottling of terminal leaves, and the fruit
are often bumpy and bitter to the taste.

All wild, solanaceous perennials (mem-bers of the nightshade family) and wild cucum-bers should be eradicated from the garden area,
as they serve as host plants. Sunflowers are
sometimes grown as a barrier crop to block the
pathogen. Try to control aphids, which serve as
vectors (see APHID). Do not plant peppers next
to tomatoes, cucumbers, tobacco, alfalfa, or
clover, and pull up and destroy any plants that
show the described symptoms. To prevent
mosaic from spreading, spray plants with dry or
fresh milk from 24 to 48 hours before trans-planting them. The tobacco mosaic virus is
often present in manufactured tobacco, and
smokers may carry it on their hands; to avoid
spreading infection, smokers should wash their

hands thoroughly before handling pepper plants. A number of mosaic-resistant varieties are available, including Bellringer, Keystone Resistant Giant, and Yolo Wonder.

# PETUNIA

## Insects

The invisibly small **tomato russet mite** causes bronzing or russeting of stem and leaf surfaces. Injury appears suddenly, as mites multiply with great facility. These mites are found throughout the year in California and southern states and are carried to other states on host plants (including tomato and related plants). They are vectors of mosaic. See SPIDER MITE for controls.

Other pests of petunia include MEALY-BUG, potato flea beetle (shot holes in leaves; see POTATO), sixspotted leafhopper (ASTER), spotted cucumber beetle (CUCUMBER), tomato hornworm (TOMATO), and yellow woolly-bear caterpillar (see imported cabbageworm under CABBAGE).

## Diseases

Petunia is susceptible to a number of **viruses** that also plague tomato and potato. Symptoms include green mottled areas, crinkled leaves, dwarfed and cupped leaves, and witches' broom effects. Viruses can be spread by mechanical contact, so rogue infected plants and control insects that might serve as vectors. Place petunias at some distance from other solanaceous plants such as tomato, flowering tobacco, potatoes, eggplants, and peppers.

# PHEROMONE

Pheromones are odorous substances emitted by insects and animals for the purpose of communicating with other creatures of the same species. These are complex compounds, used to warn of danger, advertise a food source, or attract mates. Many have been isolated in the laboratory and synthesized for use as attractants in traps. Trapping may be done to either monitor or control insect populations. Pheromone-baited traps are on the market for many pests. Pheromones also can be broadcast to interfere with a pest's ability to find food or mates. See also FRUIT TREES.

# PHLOX

## Insects

The orange, ¼-inch-long **phlox plant bug** and bright red nymphs feed on the uppersides of young leaves of perennial phlox. This leads to yellow, stippled areas, stunted growth, and deformed blossom heads. Seriously affected plants may die. The phlox plant bug chiefly occurs in northern states, wherever phlox is grown locally.

Clean up old phlox stems after frost in fall, and burn them to destroy overwintering eggs.

The most serious animal pest of phlox is the **twospotted mite.** It infests the undersides of leaves, making them turn light yellow, and plants take on a generally unthrifty appearance. Webs may be formed on either side of leaves. See SPIDER MITE.

Occasional pests worth a mention are blister beetle (see ASTER), corn earworm (CHRYSAN-THEMUM), flea beetle (POTATO), fourlined plant bug (CHRYSANTHEMUM), NEMATODE, sixspotted leafhopper (ASTER), stalk borer (DAHLIA, ZINNIA), and WIREWORM.

## Diseases

The use of resistant varieties helps to control **powdery mildew,** which appears as a white growth on leaves and stems. Shaded plants with little air circulating about them are especially apt to be damaged.

# PINE

## Insects

**European pine shoot moths,** as small larvae or caterpillars, hollow out and kill the tips of new shoots and buds. Look for short dead needles

European pine shoot moth and larva

near the apex of the new shoots, with partially developed or hollowed-out buds. Young pines up to 12 feet are most seriously affected and may become stunted and bushy from heavy infestation.

Since the insects winter as pupae in the injured tips, cut off the infested tips and destroy the overwintering pupae before growth starts in spring. In the northern Great Plains and the West, this should be done as soon as the dying needles become evident and before the larvae have left them to pupate in the ground.

Pruning is a way to control the European pine shoot moth. Snow-depth pruning promotes the insect's winter mortality. In young plantations, many infested tips on low branches may be covered by snow.

Pine shoot moths have their share of natural enemies. Of the many imported parasites that have been released, four species have become established in Canada and three in the United States. The number-one imported helper is *Orgilus obscurator,* a braconid wasp. In Christmas tree plantations, infestations can be controlled by delaying summer pruning until the larvae are either on the twigs or inside the buds. The clippings then should be burned.

**Matsucoccus scales** live on the bark of pine trees and are very inconspicuous until trees become heavily infested. Severe infestations cause the needles to turn yellow, and cottony masses appear on the undersides of branches, especially at branch axils. Pitch pine and red pine are suspectible to these insects, but white pine is not.

As the crawling stage of the scale appears on new growth, spray with lime sulfur if necessary.

The **pales weevil** is a weevil or snout beetle that has a reputation for gnawing the bark from the trunk and twigs of younger seedling pines and from the lower branches of older trees. Young trees are particularly endangered if they have been planted where pine trees have recently been cut.

It is safer to burn the slash and wait two years before planting young trees on the lumbered land. Thinning young pine stands instead of clean-cutting may reduce the hazard of damage. Freshly sawed pine lumber should not be stored near young pine stands because it attracts weevils to the trees.

A variety of **pine beetle** attacks all sizes of pines and is especially dangerous following prolonged drought. The adult is short-legged, stout, and about ⅛ inch long; the young beetle is soft and yellow in color, but soon hardens and darkens to a dull, dark brown color. Beetles of overwintering broods emerge and attack trees in spring—about the time the dogwood is in full bloom—although they may be active during prolonged warm periods in winter. They usually attack the midtrunk first and then work both upward and downward.

The beetles are generally attracted to weakened trees. They bore through the outer bark and construct S-shaped crisscrossing tunnels throughout the inner bark. This boring girdles the tree and introduces a fungus that may hasten death. The earliest signs of infestation are numerous white, yellow, or red brown pitch tubes (about as large as a wad of gum) scattered over the outer bark of the tree. Trees show yellowish green foliage from 10 to 14 days after attack. New broods often leave the trees when the foliage is only slightly faded or yellow. When the crown of a tree turns red, the beetles have usually left, except during the winter months. It is therefore necessary to locate red-crowned trees and check for the presence of beetles. If trees are infested, they should be removed before the weather warms up and the beetles emerge to attack other trees.

Keep pine stands properly thinned and remove damaged, old, or unhealthy trees. Trees that are cut can in most cases be sold as sawlogs or pulpwood. Slabs cut from infested trees and bark knocked off in felling trees should be destroyed, preferably by burning. Check infested areas during summer months for infested trees that might have been left. Infested trees can be spotted during winter by the color of their needles. Groups of red-and-yellow-topped pines are the best indicators of beetle attack. Remove these trees promptly.

**Pine needle scale** may become so abundant on the needles that the entire tree takes on a whitened appearance. The scales weaken trees by sucking plant juices and make them more subject to attack from other insects, including borers and bark beetles. Pine needle scales are usually white and elongate, the females averaging ⅛ inch in length and tapering at one end.

Pine needle scale

The males are narrow and even smaller. The reddish purple eggs overwinter under the shells, and broods hatch in May and July.

Apply a dormant-oil spray, mixed 1 part oil to 15 parts water. If infestations are confined to a branch or two, you should be able to get by with a bit of pruning.

There are several species of **pine sawfly,** the larvae of which devour the needles and sometimes defoliate trees. The larvae are so tiny that it may take several hundred just to demolish one twig, and they are rarely seen until after the damage is done.

Control of the insect often is accomplished by natural enemies. You can help by raking up plant debris from under the trees, as this area is home to sawfly pupae. Should an infestation become serious, you may need to turn to rotenone.

The **white pine weevil** is a notorious trouble-maker in pine trees. It is brownish and mottled with light and dark scales. Although preferring white pine, the weevil also damages jack, red, and Scotch pine, as well as Norway spruce. Luckily, red pine, white spruce, balsam fir, and Douglas fir are virtually immune to this weevil. Usually, the first evidence of damage is tiny drops of resin on the bark, a sign of feeding or egg laying. The feeding larvae usually girdle the terminal, which then withers, bends over, and dies.

Although it rarely kills young trees, the insect deforms and stunts them to produce forked and crooked trunks. Because the weevil attacks the leader or terminal shoot, height growth is retarded. The trees are usually first attacked when they reach a height of three to four feet. Once the tree reaches a height of 30 feet or so, weevil attacks become less important. These insects invariably seek the tallest, most vigorous trees for feeding and egg laying.

The adult weevil overwinters in litter on the ground and crawls up the tree in the warm days of early spring. It begins to feed on wood tissues or buds, then mate. The female lays from 50 to 150 eggs in holes in the terminal. Both egg laying and feeding cause holes that exude drops of pitch. Upon hatching, yellow grubs group together in bands, and the wilted, bent, infested terminals are easily spotted at this time. The grubs move downward as they mature, emerging as adult beetles in midsummer. They then cut their way to the outside through the bark and continue feeding on the bud and bark tissue until the onset of cold weather.

Remove and burn tips well below the dead part to prevent emergence of the beetles. Trees planted in shaded areas are not as attractive to weevils as those in full sun. If you can select a shady site for your pine planting, you will have less trouble from the white pine weevil. Varieties having a thicker bark and wider trunk diameter are less susceptible to attack.

## Diseases

Pine **dieback** causes stunted tip growth for several years. The needles on the affected branches are short and turn brown prematurely, usually showing small black fruiting bodies at or below the sheath. Control by pruning infected twigs.

Dieback

**Needle rust** [color] is caused by a fungus that attacks the needles of two- and three-needled pines. Red pine is particularly susceptible. The disease develops in spring as small, cream-colored, baglike pustules on needles. These pustules rupture and orange spores are blown to infect goldenrod and asters. The rust overwinters, and can live indefinitely, in the crowns of these alternate hosts. During summer and autumn, spores from goldenrod and asters in turn infect needles of the pines. This disease

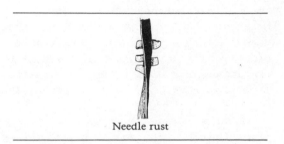

Needle rust

may cause needle drop and much damage to younger trees. If it is necessary to control the disease near valuable pine plantings, destroy goldenrod and asters in the nearby area.

Imported pines may suffer from **pine wilt,** a disease carried by nematodes. The nematodes in turn are carried from tree to tree by beetles. The disease has been reported throughout much of the United States. Look for the sudden onset of these symptoms: resin flows sluggishly from new wounds, and crown needles turn yellow. Eventually the crown turns brown and the tree dies. Anticipate the beetles' spring emergence by taking out affected trees. Healthy trees are less susceptible to attack. See the entry for pine beetles, above.

**White pine blister rust** is perhaps the most destructive disease of pines. This is a fungal disease that was brought into the United States from Europe on pines that were planted in the Northeast early in the century. The disease is now well established in native white pine forests.

Blister rust is one of those plant diseases that must have an alternate native host plant on which to complete its life cycle—it can't spread directly from tree to tree. In spring, wind-borne spores carry the fungus from diseased pines to the alternate hosts, currant and gooseberry.

The blister rust kills white pines, regardless of size, although the smaller trees die more quickly. Large ones may continue to live 20 years or more after infection. The fungus enters trees through the needles and grows into the bark, where it causes lesions known as cankers. Branches and stems are girdled by these cankers. New infections may occur each year as long as diseased alternate hosts remain nearby, and the trees are gradually killed.

Control by removing gooseberry and currant within 300 yards of white pines and then keeping the area free of such plants.

# PINEAPPLE

## Insects

The primary insect pest of pineapples is the **pineapple mealybug.** Besides pineapple, this insect infests nutgrass, panic grass, Spanish needle, caladium, avocado, citrus, mulberry, royal palm, hibiscus, mangrove, sugarcane, ferns, and some air plants. Pineapple mealybugs are fleshy, wingless, white or gray insects that are covered with a mealy white, waxy excretion. Often the waxy filaments make a ragged fringe around the body. Adult females may reach a length of $\frac{1}{6}$ inch. They give birth to young that are much smaller but resemble the mother. Large numbers of mealybugs may be found on pineapple leaf bases or stems, near or just below the ground level. They suck the plant juices and produce a type of wilt. Mealybug feeding, in fact, is considered to be the cause of pineapple wilt disease. Wilted plants are stunted and have leaves of red or reddish yellow with light green spots, and show withered, dead, and dying leaf tips. Such plants are unproductive. Sooty mold grows on the liquid excretions of mealybugs and the unsightly presence of this mold on the fruit reduces the market value.

Fire ants feed on the sweet honeydew excreted by pineapple mealybugs and protect the pest colonies. The ants often carry individual mealybugs from wilted plants to healthy, succulent plants, and are thus an important means of spreading both mealybugs and the resultant wilt. Other sources of mealybug infestations are infested grasses or other host plants in fields prepared for new plantings, and infested planting stock used to set the beds.

Control of this mealybug has been made possible biologically by an imported parasite, *Hambletonia pseudococcina.* This tiny wasp stings mealybugs and lays its eggs in them. The progeny develop in the bodies of the mealybugs. Although this parasite is very useful in the control of the mealybug, its effect can be nullified by the use of insecticides. The pineapple farmer should cultivate or disc the soil for several weeks before planting to kill out all grass and weeds that might be host to the pests. They can be swept into a jar containing alcohol. A water spray will knock mealbugs from infested plants. In extreme infestations, a dormant-oil spray may be necessary. In small plantings, you might scatter crushed bone meal around the plants to ward off the ants. See MEALYBUG.

Mites and red spiders may also bother pineapples. See FIG for control. If red scale is troublesome, follow the methods for control of scales under OLIVE.

# PLANT LOUSE

See APHID.

# PLUM

## Insects

For an effective spray program, see FRUIT TREES.

A number of species of **aphid** [color] are known to infest plum trees from time to time. The most common are the mealy plum aphid and the rusty plum aphid. Both spend the winter as eggs on the bark of the tree, hatch around green tips, feed on the plum for several generations, and migrate to grasses in midsummer. Look for leaf curling; in addition, the rusty plum aphid will feed in the blossoms and prevent fruit set. A thorough oil spray will reduce overwintering populations, and predators will keep their numbers down until the aphids leave the trees for the summer.

The **apple maggot** is occasionally found in plums. Control the pest on nearby apples. See APPLE.

The **European red mite** is sometimes found on plums interplanted with apple trees. It is most severe on prune plums rather than the Japanese-type plums. See FRUIT TREES.

The **plum curculio** [color] is by far the most important plum and prune pest east of the Rockies. This is a brown snout beetle, mottled with gray and distinguished by four humps on its back. It reaches a length of about ¼ inch.

Plum curculio

After hibernating in woodlands and brushy fencerows, the beetle appears in the orchard about the time trees blossom. It feeds on fruit as soon as it reaches a diameter of ¼ inch, leaving crescent-shaped cuts in the skin. An egg is laid under each crescent and the resulting grubs feed within the fruit, which usually drops and decays. The grubs leave the drops or infested stone fruit which remain on the tree, and pupate in the ground, to emerge as adults about one month later. The adults feed on fruit for a while, and then head to the woods to spend winter.

Because the grubs stay in apple drops for a while before burrowing in the ground, you can interrupt their life cycle by gathering up drops daily. Bury the fruit in a deep hole or in the warm depths of the compost pile. You can take advantage of the curculio's habit of playing possum when frightened by knocking branches with a padded board or pole. Do this in the early morning when they're sluggish from the cold; they'll fold up their legs and drop to a sheet or tarp on the ground.

To monitor for curculio activity, place sticky white panels (measuring eight by ten inches or so) in the trees and bait with Avon Apple Blossom perfume. You can buy white traps, or make your own out of a white sheet of plastic coated with a sticky commercial compound. Place the perfume in a small vial with a wick of cotton. A few drops will do. The bottle should be placed mouth down to avoid collecting rainwater.

Begin jarring when you start to trap adults and continue as long as the overwintering adults are active—probably for several weeks. Hogs, chickens, and geese will help destroy the pupae in the ground.

Because these pests try to avoid direct sunlight, pruned trees should be less vulnerable. You'll get free help from a larval wasp parasite, egg parasites, and naturally occurring fungal diseases.

The **western flatheaded borer** attacks sunburned or otherwise injured areas of trunks and larger limbs, tunneling into the inner bark and sapwood, and sometimes girdling limbs. The larvae are white grubs with flattened bodies, greatly enlarged at the front end. Control is a matter of pruning to reduce sunburn and keeping trees healthy and vigorous.

Other insects that may cause problems on plum and prune include oriental fruit moth and peachtree and peach twig borers (see PEACH), shothole borer (APRICOT), and leafrollers, plant bugs, and scale (FRUIT TREES).

## Diseases

**Bacterial spot** of plum is caused by the same organism that attacks peach, but the symptoms are likely to be quite different. Infected areas on leaves soon fall out to give a pronounced shothole effect. On the fruit, infection shows as purplish black, sunken spots on the green fruit. On a few cultivars, small pitlike spots occur. See PEACH. While there are a number of peach varieties that enjoy some resistance, plums do not fare so well. Some of the very susceptible plum varieties are Abundance, Formosa, Satsuma, and Wickson. Bradshaw and President are very resistant.

**Black knot** [color] is a serious plum and prune disease and may be a threat to sweet and sour cherries in areas where many wild hosts exist. The name describes the coal black, hard swellings that are scattered throughout the twigs and limbs. This is a fungal disease that survives from year to year in the knotlike swellings. In spring, spores are formed on the surface of the knots and are released in moderate to warm, rainy weather to be transmitted by air currents. Knots are first seen in late summer as olive green swellings, and then blacken and cause a gradual weakening of the branch beyond the knot.

Remove all knotted twigs and branches. Make cuts at least four inches below the beginning of the swelling and cover them with grafting wax or tree wound paint. Burn infected prunings in early spring to prevent the release of spores. Remove all neighboring wild plum and wild cherry trees (not pin or black cherry) if infected. Allow those infected branches or limbs that are essential for tree growth to remain until early or midsummer. At this time, trim the knot growth with a knife and paint the wound. Prune affected small limbs during the dormant period, at least three or four inches below the knot; on large main limbs and trunks, the knot and one inch of the healthy bark surrounding it should be cut out with a knife and chisel. Cover all wounds with grafting wax, and remove knotted tissue from the orchard and burn it.

To reduce the incidence of black knot in new plantings, do not plant downwind from old blocks harboring the disease. Stanley, Blufre,

Shropshire, and Damson are highly susceptible varieties; Methley, Milton, Early Italian, Bradshaw, and Fellenberg are moderately susceptible; Shire, Santa Rosa, and Formosa are only slightly attacked, and President is unaffected.

The disease is progressively harmful to the tree until the death of the branch beyond the gall. Liquid lime sulfur sprays prevent the production of spores on the galls but may not control the disease; apply at the dormant, full-bloom, and shuck-fall periods. If you have only a few trees, it is probably best to cut off and burn the knots, making sure to cut several inches behind the knot to ensure against leaving any infected tissue. Cover all wounds with grafting wax and remove the knotted tissue from the orchard for burning. Make an additional inspection in April and remove all newly formed knots. Destroy wild plum and cherry trees in the immediate area if they are knotted, since they serve as a source of spores for new infections.

**Crown gall** is the name given to rough growths that develop on the roots and crowns of prune trees as a result of a bacterial parasite. A resistant rootstock will help to avoid the disease also. Marianna 2623, Marianna 2624, and Myrobalan 29 show some resistance to crown gall. The disease can be prevented with a root dip, Gall-trol, that employs a friendly strain of the bacterium. See APRICOT for additional remedies.

**Diamond canker** is a viral disease that attacks French prunes almost exclusively. The symptoms of this disease are roughly shaped cankers produced by thickened corky tissue. These outgrowths tend to form at the bases of the lateral branches at pruning wounds, and at the margin of areas killed by the canker organism. Rough areas where the outer bark becomes cracked and loosened are sometimes found on the youngest twigs of affected trees, and these areas develop into typical diamond cankers as the twigs grow older.

Trees seldom die from the effects of diamond canker, but they remain stunted and gradually succumb to wood rots and insect borers. In the early stages, diseased trees often produce greater quantities of large fruit than trees that are not affected. But the disease takes its toll in time, and cankered trees should be replaced with resistant varieties.

**Plum leaf spot** is similar in appearance and life cycle to cherry leaf spot. It produces tattered leaves, but unlike cherry leaf spot it can result in heavy fruit drop. Most varieties are susceptible. The overwintered spores are ready for discharge from their site on fallen leaves at the time of petal fall, so a light discing (two inches or less) will reduce the spread of the disease (and discourage brown rot as well).

**Plum pockets** is a disease of the northeastern United States that begins as small, white blisters which enlarge quickly. The flesh becomes spongy and the plum grows greatly swollen and misshapen. Some varieties may take on a reddish color. As new spores are produced, the surface becomes white and dusty. Leaf infections result in greatly swollen and distorted leaves. The spores overwinter and are susceptible to an application of liquid lime sulfur before growth begins in the spring. Sanitation has little effect on plum pockets. No control measures are necessary unless you see signs of the disease.

**Powdery mildews** are common on many crops, and plums are no exception. The fungus overwinters in infected buds and less so on fallen leaves. It grows quickly once the buds expand, covering the new leaves with a whitish, dusty film. Since severe defoliation can result from unchecked infections under favorable weather conditions (warm temperatures, no rain, high humidity, morning fogs), you should prevent the tree from entering the winter in a weakened state. As the buds begin to grow, watch for those that are behind the rest, their growth retarded by the fungus. Prune these out immediately and destroy. When weather favors the disease and the shucks are dropping (about two weeks after bloom), apply 5 pounds of wettable sulfur per 100 gallons of water.

Other diseases that may affect plum include bacterial spot, brown rot (Bradshaw is very susceptible), cytospora canker and rhizopus rot (see PEACH), and verticillium wilt (CHERRY).

# POINSETTIA

## Insects

Root **aphids** cause plants to become weak and stunted, and plants may die in severe cases. You can make it hard for these pests to get to roots by packing the dirt firmly around the plant.

Dip a small stick wrapped with cotton in alcohol, and touch it to a **mealybug** to kill it. See MEALYBUG for other controls.

## Diseases

**Poinsettia scab** is especially troublesome on double-red varieties, and is most prevalent in summer. It appears as conspicuous, raised lesions, or cankers, on the stem or cane. The lesions are usually circular, but in advanced stages they combine to form large, irregular areas. In severe cases, the plant will lose its leaves when the stem is girdled by cankers. Scab-infected branches should be pruned and burned as soon as they are noted.

Sometimes plants are attacked by fungi causing both **root** and **stem rot.** Stem rot starts at the ground line and is the most prevalent and troublesome.

Once this disease becomes evident, little can be done to save the plants, so rogue any that are affected. Cuttings taken at the first sign of wilting should be made from the stems. Good soil, good drainage, and plenty of sunshine and water aid poinsettias in resisting this and other diseases. As an added precaution, you can grow poinsettias in soil where only grasses have grown for five years.

Other diseases, including fungal black root rot, bacterial stem canker, and leaf spot, can be prevented by using cuttings from healthy mother plants and by setting them into sterilized media.

# POLLUTION

See AIR POLLUTION and SOIL POLLUTION.

# POPLAR

## Insects

The adult female **European shothole borer** is dark brown to black, about $\frac{1}{5}$ inch long; the male is more or less oval and even smaller. Adults overwinter in wooden tunnels made in the host. Eggs are laid in late spring.

The adult beetles make their entrance holes about the bud scars or some other roughened place. After burrowing into the wood about ¼ inch, they make branched tunnels. At the end of each tunnel they deposit eggs. The larvae do not eat wood themselves but feed on fungal growths that the adults introduce at the time of egg deposition. The blackening of the so-called shot holes is due to fungal discoloration. The tunnels in the wood weaken the tree, causing considerable wind breakage. These tunnels also are excellent places for pathogenic fungi to enter.

Trees weakened by drought, winter injury, transplanting, mechanical injuries, and poor growing conditions are most likely to be attacked. Keep trees in a thrifty growing condition. Prune out and destroy any affected plant parts early in spring. Before eggs are laid in spring, you can ward off borers by whitewashing lower trunks.

The large, conspicuous caterpillars of the **polyphemus moth** are occasionally found on poplar. They are about three inches long when fully grown, pale green in color, and have a light yellow oblique line on the side of each abdominal segment. Look for numerous orange and reddish spots on the sides. Hibernation takes place in leaf-covered cocoons attached to the limb of the host. The moths, which emerge in spring, have a wingspread of four to five inches. The wings are pale brown to red with a large pale spot margined with black on each wing. Defoliation by this insect is seldom serious.

The best control method is by far the easiest. Wait until winter has forced the moth into cocoon hibernation and then handpick and burn the oval, 1½-inch-long cocoons.

The larval **poplar borer** bores into the trunk and branches of poplar, causing blackened and swollen scars. Eggs are laid in slits in the bark during July and August, and the young borers tunnel into the inner bark and sapwood and later work deeper into the wood. They overwinter in galleries. The full-grown grub measures about two inches long. The adult, an ash gray, yellow-spotted beetle is 1¼ inches long. Control these pests by injecting nicotine sulfate into their burrows, as described for the locust borer under LOCUST.

The **poplar-and-willow borer** infests and destroys pussy willows and several types of poplars. The adults emerge in midsummer and lay their eggs in punctures in the bark. The larva is about ½ inch long, white, and legless.

The adult is a ⅓-inch-long beetle, colored black with partly white wing covers.

Cut and burn infested parts of trees before the beetles emerge. An oil emulsion may be sprayed or brushed on the infested parts in early spring to kill the larvae.

Two types of **scale** are particular pests of poplar trees. One type also infests magnolias; see MAGNOLIA for description and control. The other is the cottony maple scale (see MAPLE).

## Diseases

**Cytospora canker** results from a fungus entering the tree through wounds or weakened twigs. Twigs are killed back to larger branches on the trunk and brownish, circular cankers with sunken bark are formed. During moist weather, yellow to reddish threads of the fungal spores appear from fruiting bodies on diseased bark. Spores are spread to healthy trees by rain, wind, insects, and birds. Lombardy and Simon poplars are susceptible.

Keep trees growing vigorously with organic fertilizer and continued watering. Prune affected limbs and cut out cankers; protect surfaces by swabbing with tree paint.

**Dothichiza canker** is a disease destructive to Simon and Lombardy poplars. Gray to brown irregular spots are formed on leaves and the fungus grows down through the leaf stem to form oval to elongate sunken cankers on twigs. Cankers may appear on larger branches and trunks. Large trees often survive for several years, but they are disfigured by many dead twigs and branches that result from girdling by cankers.

Remove and destroy all infected trees. Avoid wounds and pruning if at all possible. A dormant spray of lime sulfur is effective.

**Leaf rust** is a fungal disease that produces yellowish orange pustules on lower leaf surfaces.

Hemlock is an alternate host, and poplars should be grown at least 100 yards from these trees.

**Leaf yellowing** may occur during hot, dry periods in midsummer. Leaves of recently transplanted or weakened tulip trees may turn yellow and drop prematurely. Small, angular brownish specks often appear between the veins of affected leaves. The yellowing and scorch result when roots fail to supply enough moisture to replace that lost during hot, dry periods. Control by supplying water during dry periods.

# POPPY

## Insects

Poppies are vulnerable to APHID, corn earworm and fourlined plant bug (see CHRYSANTHEMUM), MEALYBUG, rose chafer (ROSE), and sixpotted leafhopper (ASTER).

## Diseases

**Bacterial blight** is a fairly widespread disease. Infection first shows as water-soaked areas that soon become black and surrounded by a translucent ring. Bacterial exudate may be found on the spots. Plants die when stems are girdled. Rogue infected plants and remember that it is best to plant clean seed in areas that have not hosted this disease.

# POTATO

See also SWEET POTATO.

## Insects

As a rule, both aboveground and belowground pests will find potatoes a lot less attractive if onions are growing nearby. The Irish potato is attacked by dozens of insects: below, you will find those that are most likely to cause trouble. Although a particular species may not be described exactly, you should be able to find a similar pest that has similar controls. For example, the more than two dozen beetles known to attack various parts of the potato plant have in common several reliable controls.

Glandular hairs of the wild potato release an alarm pheromone that scares off aphids, Colorado potato beetles, flea beetles, leafhoppers, mites, and thrips. This trait eventually may be bred into cultivated potatoes.

The feeding of **aphids** [color] causes potato foliage to curl and vines to turn brown, and growth is consequently retarded. In cases of severe infestation, plants may die. The potato aphid is pink and green, and usually overwinters on roses. The buckthorn aphid, most important in the Northeast, comes in yellow, green, or black, and overwinters on buckthorn. The green

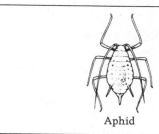

Aphid

peach aphid is yellow green with three dark lines on its back, and overwinters on peach trees; it is a vector of leaf roll of potatoes (see virus, below). These three species live on their hosts for two or three generations before migrating to potatoes or other herbaceous plants.

Of the several means of controlling aphids, spraying plants with a strong stream of water from a garden hose is particularly effective, especially if done early in the growing season. See APHID and RESISTANT VARIETIES.

Certain resistant varieties work in a curious way to discourage aphids. When the insects walk on the leaves and stems, the cells of grandular hairs are broken open and a sticky substance is liberated. The aphids get gummed up so badly they can't get about to feed and may starve to death.

Several **blister beetles** attack potato; the black blister beetle, also known as the old-fashioned potato bug, Yankee bug, and just plain blister beetle, is representative of the various species. It is a fairly long (up to ¾-inch) and slender beetle, with soft, flexible wing covers. The entire body is black or dark gray, and the covers may be marked with white stripes on the margins. Swarms or colonies of blister beetles may be seen feeding on potato foliage. For a description of habits and control measures, see black blister beetle under BEET.

Both the adult and larval form of the **Colorado potato beetle** [color] may defoliate potatoes. They are particularly harmful to home gardeners with small plantings, and are so common that they are often simply called potato bugs. The adult is a yellow beetle with a broad, convex back, and averages ⅜ inch in length. Running down both wing covers are fine black

Colorado potato beetle

lines, and the thorax is decorated with black spots. The larval form is a humpbacked, dark red grub with two rows of black spots on each side and a black head; the last larval stage is somewhat lighter. This beetle is a pest in all continental states except California and Nevada; the eastern states are particularly vulnerable.

One of the easiest and most effective means of dodging the Colorado potato beetle is to grow potatoes on top of the ground. Gardeners without much cultivating equipment and those plagued by summer drought will especially benefit from this method. First, drop seed on sod or on a three-foot layer of leaves from the previous fall. (The leaves pack down over the winter and earthworms have a chance to work on them.) Next, cover the garden area with a three-inch layer of straw and drop a cupful of bone meal over each seed. Then add more straw until the entire garden is covered by a 10- or 12-inch layer. Make sure that this covering is uniform, as skins of growing potatoes will turn green if exposed to light. The vines eventually work up through the straw, leaving the bugs behind, and you will begin to notice bulges in the mulch—these are the tubers, growing just beneath their protective cap. Harvesting is easy, and the potatoes are clean and relatively free from insect and scab injury.

A thick mulch of hay or straw will keep adult beetles from climbing to the stems.

When the potato beetle strikes, handpick the adults and crush the yellow eggs (found on the undersides of leaves). A clever method devised by gardeners makes use of the beetle's gluttony. The tops of wet potato plants are dusted with wheat bran which, when eaten, causes the beetle to bloat up and rupture. Other gardeners put their faith in a spray made from the extracts of common basil or sweet basil. Try a spray of eucalyptus extract or a tea made by boiling cedar bows in water. Stronger medicines include rotenone and Triple-Plus, a commercial compounding of rotenone, ryania, and pyrethrum. A simple trap may be made by folding a colored plastic sheet twice, and placing it under the plants every three days or so. Shake off the collected insects into a bucket of water. Toads eat a good number of these pests. Ladybugs can be purchased and introduced to the garden to help keep the beetle population low.

A parasite of potato beetle eggs, *Edovum puttleri,* has recently been made commercially available. This wasp was imported from Peru. A bacterial disease know as potato-beetle septicemia causes the pest to slow down, cease feeding, turn to a brownish gray, and finally become nearly black. A fungus, *Beauveria bassiana,* hits a number of well-known pests, the Colorado potato beetle among them. A new strain of *Bacillus thuringiensis* has been found lethal to the beetles. Nematodes attack adults hibernating in the soil. See RESISTANT VARIETIES and TIMED PLANTING.

New growth and young tubers provide meals for several species of **cutworm.** These soft-bodied, smooth caterpillars are familiar to most gardeners; they grow up to 1¼ inches long, may be brown, gray, or black, and have the habit of curling up tightly when disturbed. See TOMATO for further control methods.

The larva of the **European corn borer** tunnels into potato stalks, and these openings enable disease to infect the plants. This is a flesh-colored caterpillar that grows up to an inch long and is distinguished by brown spots and a brown head. A number of applicable controls for the borer are discussed under CORN.

Several species of **flea beetle** nibble on potato foliage from the undersides, leaving behind a shothole pattern. Color and pattern vary, but all are very small and jump around like fleas when disturbed. Great numbers of them may infest a single plant.

Because flea beetle eggs are deposited in the soil, frequent cultivation will help. Keep the garden area free from trash in order to discourage these insects from settling near your plot for the winter. Weeds should be cleared out as well, as they provide food for both adults and larvae. Plant seed thickly, thin after the danger of

Flea beetle

infestation is past (flea beetle damage is greatest early in the growing season), and interplant potatoes and other susceptible crops near shade-giving crops; these measures work because the pests don't care for shade. The beetles are daunted by a row sequence that alternates between kohlrabi, radish, and lettuce. Wood ashes repel them and can be used in two ways: either place a mixture of equal parts of ashes and agricultural lime in small containers around the plants, or simply sprinkle a spoonful of ashes on each plant two or three times a week. Garlic sprays are effective.

Many species of **grasshopper** feed on potato plants, and large infestations may destroy complete plantings. Both adults and nymphs cause trouble, and they may be brown, gray, black, or yellow. Grasshoppers do their worst in the central and northwestern states. See GRASSHOPPER.

The **nematode** is a tiny, nearly invisible creature that lives in the soil, feeding on the belowground parts of plants. Symptoms of nematode damage are poor growth, yellowing of foliage, and wilting followed by recovery as a hot day turns into a cool evening. These worms also cause root knot (or nematode gall). See page 328 for a discussion of the symptoms of and controls for root knot.

The **potato leafhopper** (called the bean jassid in the South) is a small, wedge-shaped, green leafhopper. Look for white spots on its head and thorax. The adults grow up to ⅛ inch long, and will fly away rapidly if disturbed.

Potato leafhopper

The nymphs resemble adults but are smaller and have the unusual habit of crawling sideways like crabs. Leafhoppers, both mature and immature, cause a blightlike condition known as hopperburn in which the tips and sides of the potato leaves curl upward, turn yellow or brown, and become brittle. As a result, the food-making capacity of the plant is impaired and yields are lower.

Because the leafhopper seems to prefer open areas, it may help to plant potatoes in a sheltered area. If this is not practical, you can keep away the egg-laying adults with canopies of cheesecloth, muslin, or plastic netting. The crucial period for this protection is about one month, beginning when the plants are a few inches high. During this time, leafhoppers attempt to lay their eggs in main veins or in petioles on the undersides of leaves. Blacklight fluorescent lamps attract potato leafhoppers to traps. Leafhoppers are vulnerable to sprayed extracts of wire grass and red strangletop. Sequoia is a variety resistant to both the leafhopper and potato flea beetle; it is, however, quite susceptible to aphid damage. See RESISTANT VARIETIES.

The tiny **potato psyllid** (also known as the tomato psyllid) grows to no more than $1/10$ inch long. The newly hatched insect is green, but within two or three days it turns black with white markings, giving it a grayish appearance. Psyllids are occasionally called jumping plant lice because of their resemblance to certain aphids. Your real enemy is the nymph, which is pale green, flat, scalelike, and fringed with tiny hairs. As it feeds, the nymph injects a substance that disturbs proper plant growth. Short sprouts appear from the eyes of undersized and immature tubers, and new tubers may form on the sprouts. Eventually a chain of several deformed tubers may occur on one stolon, none of which is of marketable size. Leaves may be discolored yellow or reddish, or turn brown and die. These conditions are known collectively as psyllid yellows. Early crops are usually more susceptible to severe injury.

Eliminate winter hosts, especially matrimony vine (lycium). Clear the garden of any discarded sprouting potatoes. Dust with diatomaceous earth.

The **potato tuberworm** [color] is a pinkish white worm with a brown head that grows up to $1/2$ inch long. It tunnels in the stems, leaves, and tubers of potatoes, causing shoots to wilt and die. Local infestations are found in the South and west to California. Growers outside this area may find an occasional tuberworm at work, but control should not be necessary.

Keep the garden clear of weeds and culls, and make sure that potato plants are deeply hilled with soil. The tuber moth may lay its eggs in the eyes of potato tubers, and newly dug potatoes should therefore not be left exposed late in the day or overnight. You can interrupt this pest's life cycle by clipping off and destroying any affected vines.

The **symphylan,** also known as the garden centipede or symphilid, is a white, fragile insect that grows up to $3/8$ inch long. While not actually a centipede, it is quite similar in appearance. Adults have 12 pairs of legs, while the young have fewer. This pest travels rapidly through tiny cracks in the soil, constantly waving its antennae, and is rarely seen aboveground. It feeds on rootlets and root hairs, and its presence may first be made known by stunted growth. For controls, see SYMPHYLAN.

Several species of **whitefringed beetle** [color] attack the roots of potatoes, and may be found throughout the southeastern United States; infestations are localized. The adults are dark gray snout beetles, ½ inch long, and decorated with a light band along each side. They are covered with short hairs, and are unable to fly because the wing covers are fused together. They

Whitefringed beetle adult and larva

are not picky eaters, nibbling on almost anything in their path, but it is the yellowish white, fleshy larvae that cause growers the most concern. The worms eat roots and tubers, causing plants to yellow, wilt, and die.

Use deep cultivation in the early spring to keep the adults from appearing in May. The larvae overwinter in the soil, usually not lower than nine inches.

**White grubs** may devour the roots and tubers of potatoes. They are white to pale yellow, have hard, brown heads, and measure up to 1½ inches in length. The adult form is the familiar dark brown May beetle, or June bug, which may also damage potatoes by feeding on foliage. For control measures, see JAPANESE BEETLE.

Many species of **wireworm** [color] attack vegetables throughout the United States, and potatoes are one of the most vulnerable crops. The worms are wirelike and white at first, turning brown as they grow to their full length of ½ to ¾ inch. Potatoes are damaged when wireworms feed on tubers; appearance is marred

Wireworm

and this means that parts of the potato have to be cut away before cooking. On a large scale, lots containing badly damaged tubers either are downgraded and sold at reduced prices or are made eligible for top grades by discarding a number of the damaged tubers. Most injury occurs as the potatoes approach maturity. Generally, the earlier the injury, the deeper the holes will be at harvest. The adult wireworm is a ¼-inch-long click beetle. When placed on its back, the beetle flips to its feet with a sharp click. They are also known as skipjacks. Millipedes are sometimes confused with wireworms; while millipedes have many pairs of legs over the length of their bodies, wireworms have but three pairs of legs positioned well forward. Also, millipedes characteristically curl up into a lose spiral position when disturbed, and wireworms do not.

If wireworms are a problem, you may benefit from learning to identify a few of the more important species, as controls vary. The eastern field wireworm confronts gardeners and farmers in the East, especially attacking potatoes that have been continuously planted on light, sandy soil. Continued cultivation seems to encourage this pest. If you discover that a field is infested after plant growth has begun, harvesting should be as early as possible to avoid further damage. Two species are apt to damage crops that follow sod. The wheat wireworm is found in the eastern and central states, is yellowish brown as a beetle and bright yellow as a larva, and has two eyelike spots on the ninth, or posterior, body segment. It most often occurs in heavy loam that retains moisture. The corn wireworm (also known as the community wireworm) is restricted to the eastern half of the continent, also occurs in heavy soil, is reddish brown as an adult, and the larva may be distinguished from that of the wheat wireworm by a reddish brown rear segment. Both wheat and corn wireworm populations will decrease with continued cultivation. See WIREWORM.

## Diseases

You can help to keep potatoes free from disease by rotating annually in loamy, well-drained soil. Potatoes do poorly in an alkaline medium, and need an acidic soil with a pH factor of between 5 and 6.5. Don't plant where wood ashes or lime have been scattered, as these will raise the soil's pH.

The first symptoms of **black leg** are the rolling of the upper leaves of one or more shoots and the gradual fading of the foliage to yellow green. Plants eventually turn a distinct yellow color, and gradually die as the bases of stems are rotted away by the causal bacteria. Presence of

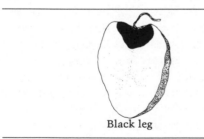

Black leg

black leg is easily confirmed in the field by pulling affected stems; they will snap off with little resistance. Stems at the soil line become slimy and black, and tubers are decayed. Black leg is favored by abnormally rainy seasons.

To avoid this disease, plant only tubers from disease-free fields, use sound seed that has healed well after cutting, and provide well-drained soil. Cut seed can be healed by keeping them at 60° to 70°F. and approximately 85 percent relative humidity, with adequate ventilation, for about a week.

**Common scab** is known to exist in every potato-growing section of the United States. It is caused by a fungus that develops readily in soil having slightly less than the optimum amount of moisture for growing potatoes. If the soil is of such a texture that there is abundant aeration, scab may spread quickly even though the soil is wet. The scabs vary from a minor russeting of the tuber skin to very rough, corky areas that may be raised or pitted. The spots may be single, or several may join together to cover large areas. Scab is particularly severe in neutral or alkaline soils, causing little damage in an acid soil with a pH of 5.2 or less, and it is usually worse in dry than in moist soil.

Common scab of potato can be successfully controlled with green manuring. You should avoid using lime, wood ashes, or fresh barnyard manure on infested soil, as these will increase the alkalinity. If you must lime, do so in fall after harvest; the amount of lime should be determined by soil analysis. Farmers have found that if soybeans are grown on infested soil and then turned under, two desirable things take place: the soil becomes more acidic, and beneficial bacteria and fungi that compete with scab are encouraged by the rapidly decaying organic matter. (It is thought that good, organically fertilized soil may enable these beneficial organisms to discharge antibiotics which overcome scab and other diseases.) Rotations with nonsusceptible crops, such as rye, alfalfa, and soybeans, may reduce the incidence of scab, but the length of time between potato crops is just as important as the crop selected for the interim period: you should wait from three to five years before again planting potatoes.

Although no variety is truly immune to attack, the following should do well for you: Cayuga, Cherokee, Early Gen, Menominee, Norchip, Norgold Russet, Norland, Ontario, Rhine Red, and Seneca are recommended by the USDA, and the Canada Department of Agriculture suggests Cinook, Huron, Avon, Sable, Cariboo, Cherokee, Norland, and Netted Gem.

**Early blight,** also known as leaf spot, causes brown spots on leaves which, as they enlarge, develop concentric rings with a targetlike effect. When the spots are numerous, they can kill the leaves and thereby serve to reduce the yield of potatoes. The fungus later spreads to tubers, causing shallow decay in the form of small,

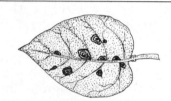

Early blight

roughly circular lesions that are surrounded by slightly puckered skin. These lesions may afford entrance for saprophytic molds that complete the rotting of the tuber. Early blight fungus spreads rapidly in warm spells.

Since the fungus may be present in tubers, be sure to plant clean tubers. Early blight may also be present in the soil, and the use of well-composted humus in the potato patch will help to ensure plants against infection.

Large, overgrown potatoes are apt to have an **environmental condition** known as hollow heart; this is not a disease, but the result of the tuber having grown too large, too fast. Another defect not caused by a pathogen is black heart,

brought about by insufficient oxygen in the center of the tuber as a result of too much heat in storage. Erratic growing conditions may cause strangely shaped tubers and growth cracks; jelly-end rot may appear at the tips of off-shape tubers. Enlarged lenticels (pores) are caused by excessive soil moisture before harvest.

**Late blight** is so called because it attacks plants after the blossoming stage. This fungal disease is most common in the northcentral, northeastern, and Atlantic states. Look for purplish or brownish black areas on the blade

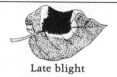

Late blight

of the leaflet or the leaf stalk, flower pedicel, or stem. Usually the lower leaves are the first to be affected. The diseased areas have a water-soaked zone about their margin, indicating the location of the advancing fungus. The recently invaded area becomes lighter colored than the normal green of the leaf, appearing as a pale halo about the blackened area, and in turn it blackens and dies. Under favorably warm and moist conditions, the disease spreads rapidly, with the result that all the plants in the field may be killed within a few days. The diseased and decaying tissues give off a characteristic odor that becomes very pronounced in fields that are severely attacked.

The spores of late blight begin to multiply on the undersides of infected leaves. Then they are blown by the wind or splashed by the rain to infect nearby plants. Winds can carry late blight spores several miles. The spores can also be moved from one field to another by running water in small streams or drainage ditches.

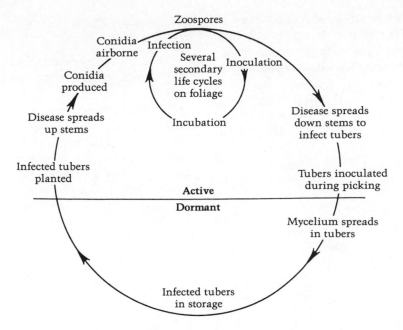

Life history of the late blight fungus of potatoes

Reprinted, by permission, from Pyenson, *Fundamentals of Entomology and Plant Pathology*
(fig. 9.6), AVI Publishing Co., PO Box 831, Westport, CT 06881.

Cool, moist nights encourage blight spores to form and germinate, which means that blight infection will build up rapidly. To get started, late blight needs a period of at least ten hours of temperatures below 70°F. and humidity about 91 percent. But a warm, humid period followed by a drop to about 60° is very apt to start an attack. Once late blight has gained a foothold, warmer day temperatures will make it grow rapidly on potato and tomato plants.

If late blight has been particularly prevalent in the field late in the season, it is much better to delay digging the potatoes until two weeks after the potato tops are dead, preferably until after a frost has killed the vines. This delay is necessary to allow the numerous spores on the old dead plants and on the surface of the soil to die before the potatoes are dug. Otherwise, many potatoes will become infected by the living spores with which they come in contact during the digging process. Tops may also be removed by hand two weeks before harvesting.

Make sure that potato dumps or cull piles do not begin to sprout. Don't dump cull potatoes in a nearby field, for wind and water can carry spores a great distance. The disease may be rendered harmless by baking the potatoes—place them in thin layers in combustible material and set the pile on fire. If it is absolutely necessary to dump potatoes on top of the ground, spread them in a thin layer so that all will freeze solid. Remember that any volunteer potato plant may be a source of infection. Late blight can be carried in the seed you plant, so don't use seed potatoes known to be infected from late blight or those that come from fields where there was late blight infection last year. If possible, see that potatoes go into storage dry and free from dirt. Clean out and disinfect the storage cellars. They should be ventilated so that there will be no condensation of moisture on the tubers or on the ceiling to drop on tubers in bins below; it's best to introduce air into cellars near the floor. Keep the cellar roof in good repair.

There are several good blight-resistant varieties available: Ona, Sebago, Kennebec, Saco, Pungo, and Essex.

The most characteristic symptom of **leak** is the extremely watery nature of the affected tissues. The water is usually held by the disintegrated tissues, but when pressure is applied a yellowish to brown liquid is given off readily. Another characteristic symptom is the granular nature of the affected tissues.

Externally, the affected tissues appear turgid and may show discoloration ranging from a metallic gray (in the red varieties) to brown shades (in the white and dark-skinned varieties). Internally, the affected tissues are at first a creamy color. Later they turn slightly reddish, tan, brown, and finally inky black. The diseased areas are generally sharply set off from the healthy areas, yet rarely is there a discernible fungal growth, either internally or externally.

Tubers become contaminated in the field, where the organisms live as soil fungi. Infection takes place in hot weather and apparently gains access to the plant only through the wounds, although these need not be visible. Leak frequently is found in tubers affected with sunburn or sunscald, especially when these occur in tubers allowed to lie in or on hot soil after being dug. In potato crops that are harvested and moved during extremely hot weather, leak may be serious; handle the potatoes with extra care. Keep tubers as cool and dry as possible during harvesting and loading, as well as in the early stages of transit and storage. Avoid injuring the skin.

Several viruses are responsible for **mosaic** in potatoes, a disease that causes a mottled, light and dark green pattern on curled or crinkled leaves. Brown specks may appear on the tubers and the plants often become yellow, droop, and die prematurely. In northern regions, hot and dry weather may subdue these symptoms and make the disease hard to detect. Nevertheless, these plants or their progeny will show the mottled symptoms if again placed under conditions favorable for development of the disease.

To keep mosaic in check, plant clean tubers and use seed that is certified disease-free. The following varieties have been found not to contract the disease in the field, and although some of them have become infected through grafting, they are resistant for all practical purposes: Ona, Penobscot, Cherokee, Chippewa, Katahdin, Kennebec, Pungo, Saco, Sebago, Houma, and Earlaine.

**Rhizoctonia,** or black scurf, is a fungal disease that first makes its presence known by dark brown cankers that "burn off" tender young sprouts. These cankers may also girdle the sprouts below their growing tips. Once growth is arrested in this manner, new sprouts appear below the dead areas, and these too may be killed; this process forms a rosette of branching sprouts. A later symptom is brown, sunken dead areas on mature stalks near the soil line. These cankers interfere with the distribution of nutrients in the plant, causing leaves to wilt and impairing development of tubers. In warm, moist conditions a grayish white collar of fungal growth may form on the stem just above the soil line. Hard, black bodies (black scurf) are evident on the tubers, especially after they have been washed. These structures, called sclerotia, are the resting bodies of the fungus. Rhizoctonia may also cause the tuber

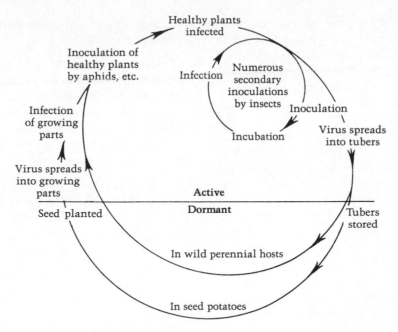

Life history of viruses causing potato mosaics and other viral diseases of potatoes

Reprinted, by permission, from Pyenson, *Fundamentals of Entomology and Plant Pathology*
(fig. 10.4), AVI Publishing Co., PO Box 831, Westport, CT 06881.

skin to be roughened in a crisscross pattern, a condition often referred to as russet scab. Both scurf and russeting reduce the value of tubers as table stock.

Plant clean certified seed that is as free as possible from sclerotia. If possible, grow potatoes once in every three to five years, rotating with corn and cereal grains, as these crops aren't susceptible to the same strain of fungus; this practice will not eradicate the fungus since it can live almost indefinitely in the soil, but the fungus population will be reduced. If conditions are wet and cold, either plant shallow or plant deep and cover with a thin layer of soil. The rationale behind this is that rapid initial growth of the sprouts is encouraged, giving them a good head start on the disease. Finally, avoid plant-

ing in heavy, poorly drained soils, and delay planting until warm weather, if possible.

**Ring rot,** or bacterial ring rot, is an extremely infectious disease caused by a bacterium that overwinters in slightly affected tubers. The disease cannot be detected in many of these tubers, and yet they contain sufficient bacteria

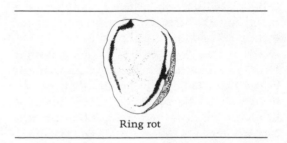

Ring rot

to contaminate the knife and planter, and these then inoculate healthy sets. Symptoms on plants do not become evident until late in the growing season, and some plants can be generally affected and show no visible signs aboveground; in other cases one or more stems in a hill may wilt and become stunted, while the remainder of the plant stays healthy. When affected plants are dug up you can usually find all gradations of the disease, from healthy to completely decayed tubers. The decay begins in the vascular ring, which is located about ⅛ inch beneath the skin of the tuber. If the tuber is squeezed by hand, a yellowish white ooze may appear in cheesy ribbons. The interior of the tuber eventually rots, leaving a shell of firm tissue.

You can prevent bacterial ring rot by the exclusive use of disease-free stock. This disease increases very rapidly, and experiments have shown that a crop having a mere trace of the disease may lose from 10 to 30 percent the next year. Because cutting and handling freshly dug seed potatoes can serve to spread the disease, the use of whole tubers is good practice. Store seed tubers in clean sacks, and containers and implements should be thoroughly disinfected if ring rot has previously been a problem.

**Root knot,** or nematode gall, is a condition in which galls are produced on tubers and roots by the parasitic action of nematodes. The galls vary in size and are more or less round, but they frequently run together to give the tuber a grotesque, knobby appearance. Nematodes are transmitted from field to field by running water, in soil clinging to implements, and on the hoofs of animals and the feet of people.

Never use potatoes from a field known to be infested for seed. A mere visual inspection of tubers may lead you to believe that they are healthy, but light infection can go unnoticed. In California, potatoes are planted in early spring, before the soil warms up and nematodes turn very active. For a number of control measures, see NEMATODE.

**Stem decay,** caused by bacteria and fungi, affects damaged, aboveground stems, especially in wet weather. The shoot above the decay wilts and dies. Since the disease organisms gain access through insect and machine injuries, infection is reduced as pest populations decline. Do not give plants more water than they need.

**Verticillium wilt** is evident late in the season, when the older, lower leaves become yellow and die. There may be some curling and rolling of the leaflets and instances of tipburning. Affected vines die prematurely, but the denuded stalk remains upright, looking like a pole with a flag of wilted upper leaves. The disease works its way up from the base until there is often only a small cluster of leaves at the top. The inside of the stem is invariably discolored a yellow or brown, from the base

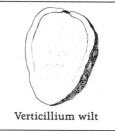

Verticillium wilt

well up into the top of the plant. The tubers often show a "pink eye" discoloration. A somewhat similar disease, fusarium wilt, produces a burning, bronzing, and slight yellowing of upper leaflets, and leaves eventually wilt and die. Stems of affected plants show a brownish flecking when cut longitudinally.

Wilt is best controlled by seed selection and an extended (four-year) crop rotation, as the disease carries over from one crop to another in seed potatoes and in the soil. The seed plot offers a good chance for roguing out affected seedlings before they can do much harm. To be safe, also take out and destroy the plants adjacent to the sick ones. Varieties resistant to wilt include Pontiac, Ona, Shoshoni, Katahdin, and Green Mountain.

Potatoes may be injured by a number of **viruses.** Symptoms include mosaic or leaf distortion and mottling, leaf roll, yellow and purple top, and "spindle tubers." In the field, viruses are usually carried about by insects, especially aphids and leafhoppers. Use clean seed, and keep these two pests from getting an early start in the garden. Early harvest of seed potatoes in Maine substantially reduces leaf roll, mosaic, and spindle-tuber infection.

# PRAYING MANTIS

There are about 20 species of praying mantis now established in the United States, some brought over from Europe and China, and all play a similar biological role as destroyers of other insects. Mantises are also cannibalistic. The female often devours the male after mating, and newly hatched youngsters may feed on their siblings.

The eggs are laid in large foamlike, straw-colored masses in fall. In May and June little mantises hatch out, resembling their parents except for the absence of wings. They grow slowly, reaching maturity sometime in August. In the northern states, mantises only live a single summer, the onset of cold weather marking the end of their lives. In the deep South they may enter a quiet interval, known as diapause, during extremely dry weather.

You should have little trouble introducing praying mantises to the garden, but there's no guarantee that they will stay put. Enough hatch

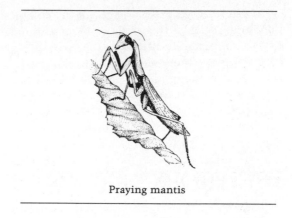

Praying mantis

from a single egg case, however (anywhere from 100 to 300 or more), that several cases stocked in strategic spots around the yard will provide enough for both you and the neighbors as well. Insects, like all animals, have a tendency to go where food is plentiful. If your garden develops a sudden outbreak of pests late in the season, the local mantises will usually be back to feed on them.

The first food of the newly hatched, wingless predators is probably aphids and other small, slow-moving prey. As they molt and grow, mantises can handle larger food, up to an occasional salamander, shrew, or even a small toad or frog. No matter what the victim, the praying mantis always begins its meal in the same way—by biting into the back of the neck, apparently to sever the main nerves and render the animal helpless.

You can gather egg cases from late fall into early spring. Look for them wherever plants grow in clumps a few feet high. Goldenrod clusters, hedgerows between fields, and roadside borders may all yield egg cases for your garden. When gathering your own cases, leave them attached to a section of twig. They can then either be stored in the refrigerator or, better yet, immediately placed in the garden area.

Place the egg cases off the ground so that they aren't soaked by spring rains or gnawed upon by hungry field mice. Lash the egg-bearing sticks to plant branches or to three-foot sticks set firmly in the ground. If you store the cases over winter, keep them in the refrigerator or outdoors. They're very winter-hardy, able to stand subfreezing temperatures.

# PREDATORY INSECTS

Predatory insects, mites, and molluscs are the meat eaters of the garden and field, as is suggested by the names of several: antlion, dragonfly, mealybug destroyer, tiger beetle, and pirate bug. They are generally larger than their prey and are equipped with sturdy mouthparts adapted to their feeding habits. Some predators are built to tear and chew the prey into manageable swallows; others have piercing mouthparts through which they suck the life out of victims. The larvae must find and devour several insects in order to sustain themselves, and are sometimes of equal or greater importance to the gardener than the adult stage.

Many helpful predators are active at night and are virtually unknown by growers, who tend to assume that the most visible insects do all the work. According to specialists, however, this is not so, and ladybugs and praying mantises have been somewhat overrated. Ladybugs sold commercially are collected in the Sierras when they are full of fat and in a dormant stage, so that when released in the garden they must fly off much of this energy before they're ready to feed and reproduce. This flight will likely take them right off your land. Growers who have released commercial ladybugs often mistake the naturally occurring helpers for the ones that were introduced. Praying mantises present a different problem—they will eat bugs, all right, and at a steady rate, but their appetites are somewhat indiscriminate and they may attack beneficial insects and even an occasional frog.

The point is that some beneficials are better than others. The level at which a predator or parasite will keep a pest population is determined in part by the specificity of the beneficial's hunger. When the population of a given pest balloons, the predacious insect with a broad variety of hosts will not respond quickly to the sudden glut of easy meals, and its population is not likely to keep pace with the pest's. Another requirement of a good natural enemy is searching ability: if the hunter blunders around the garden, it is not likely to be able to keep the host population at a low level, and will fail to check a sudden upsurge.

You can encourage beneficial insects to prey and parasitize around your plants by growing pollen-rich flowers and weeds. Goldenrod alone has been found to host more than 75 species of them. Other noteworthy flowering weeds are dandelion, wild carrot (Queen Anne's lace), lamb's-quarters, and evening primrose.

# PRIMROSE

## Insects

The feeding of **foliar nematodes** causes stunting of both crown and buds, and leaves are curled and distorted. Rogue diseased plants and change the surrounding soil.

You may also encounter APHID, flea beetle (see POTATO), fuller rose beetle (ROSE), MEALYBUG, SPIDER MITE, and WHITEFLY.

## Diseases

Several of the most common diseases of primrose can be prevented or reduced by picking and destroying sickly leaves and by a thorough cleanup in spring.

# PRIVET
## Insects

The **privet mite** causes privet leaves to yellow or fade. The mites feed on the undersides of leaves, sucking plant juices and causing foliage to drop in severe cases. The orange or red mites are nearly too small to be seen with the naked eye, being only $1/100$ inch long. In serious cases, control by spraying with lime sulfur. Care must be taken to cover the undersides of leaves.

The most serious insect pest of this shrub is the **privet thrips.** Infested foliage may be small and puckered. The leaves become yellowish or grayish, and seriously injured foliage may fall to the ground in late summer. As the plants produce new foliage the thrips abandon the older ruined leaves for newer ones at the top and side. Thus, each successive generation of the pest injures new foliage as it is produced. The winged adults are flat, elongate, and about $1/16$ inch long. They are yellow to blackish with

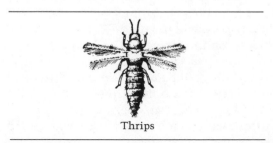

Thrips

gray markings. Nymphs are translucent, becoming light green to yellowish white as they mature. Eggs are typically deposited on the undersides of leaves, with a few on the uppersides. There are three generations a year in northern states. Adults overwinter in leaf litter under privet plants, under tree bark, and in moss.

Spray the bush thoroughly with a strong soap solution. The thrips will soon seek out a more amiable environment. It is also a good idea to remove any crumpled or unhealthy leaves to the compost heap. See THRIPS.

Many pests of privet shrubs are also common to LILAC, such as the lilac borer, lilac leafminer, and San Jose scale. See also APHID, MEALYBUG, and SPIDER MITE.

## Diseases

**Anthracnose** (also canker or twig blight) is a fungal disease that causes leaves to dry out and cling to the stem. The twigs may become blighted, and cankers dotted with pinkish fruiting bodies are formed at the base of the main stems. Often the bark and wood of diseased portions become brown and split. When the stems become completely girdled, the affected plants die.

Plant resistant varieties. Possible choices are Amur, Ibota, Regal, and California privet. European and Lodenese privet are susceptible. Once the disease hits, it is best to cut out and burn or compost all diseased branches.

**Galls** are probably the result of a fungus. Nodular galls measure up to six inches long and 1½ inches in diameter. It is best to destroy the entire stem on which they occur, pruning back until arriving at healthy wood. If you use plenty of properly composted humus around privets, the plant should have enough strength to produce new shoots and remain healthy. Once gall disease is allowed to go too far, plants will have to be removed from the garden.

# PROTOZOA

One protozoa, *Nosema locustae,* is currently on the market for microbial control of grasshoppers.

# PRUNE

See PLUM.

# PUMPKIN

When pumpkin fruit first appears, set protective layers of plastic, cloth, or paper under each to prevent insect attacks and damage from contact with the soil. Some growers place summer pumpkins between rows of corn to shade them from the direct rays of the sun; plants can also be set to good advantage near bushes or trees. Pumpkins and all of the winter squash varieties store best when harvested at full maturity before frost and when cured before being placed in storage. To cure the rind, clean it of any clinging debris and place the pumpkin in a warm, dry environment for three to seven days. This will dry and harden the rind, making it more resistant to decay.

Several kinds of **aphid,** such as melon, squash, and potato aphids, may damage pumpkin foliage; see APHID.

The **squash borer** may injure pumpkins, but as this crop is usually planted late it is better off than squash; see SQUASH.

Dark brown, hard-shelled **squash bugs** [color] are often mistaken for stink bugs, because

Squash bug adult and nymph

they put off a vile odor when stepped on. As the bugs suck sap from leaves, they apparently inject a toxic substance that causes wilting, especially at midday. Squirt water all along the stalk, since this causes them to crawl out into view so that they can be handpicked. Keep an eye out for their brick red eggs, laid between vines on the undersides of leaves; scrape them from the leaf into a container for disposal. See SQUASH.

Other pests include the melonworm (see MELON) and the striped cucumber beetle (CUCUMBER).

# PYRACANTHA

See FIRETHORN.

# PYRETHRUM

Pyrethrum is a well-known insecticide made from the pyrethrum flower, a member of the chrysanthemum family. The plants do not work when interplanted. While pyrethrum is safe for warm-blooded animals, as natural insecticides go, it's a heavy. Dozens of fruit and vegetable pests are controlled by it: all kinds of caterpillars, beetles, aphids, mites, leafhoppers, thrips, moths, and others. The esters (called pyrethrins) that give pyrethrum products their power are relatively nontoxic to bees and ladybug larvae. Nevertheless, to use pyrethrum is to grievously tamper with life in the garden, insect enemies and insect allies alike. And as mentioned before, a temporary setback for the good bugs, brought about by an injudicious use of pyrethrum, could wipe out many patient weeks of biological and cultural control. Adult ladybugs are killed at higher rates of application.

Although testing continues, pyrethrins appear to be readily metabolized by animals without ill effects, and rapidly break down into harmless substances in the environment. You can grow flowers containing the active principle in your garden. To make your own pyrethrum spray, just grind up a few flower heads of *Chrysanthemum cineraiifolium* and mix with water. A little soap can be added to improve the consistency.

Pyrethrum quickly loses power when stored in open containers, but retains effectiveness up to three years in closed containers. Watch out for pyrethrum formulations that are combined with synergists, things added to enhance their effectiveness. The most popular such additive, pipernonyl butoxide, may affect mammals' liver microsomal systems.

Insects are paralyzed by pyrethrum, sometimes after a short period of jumping around. If a bug receives less than a lethal dose, it will revive completely. For this reason, insects that still show any sign of life after an application should be stepped on or otherwise destroyed; or, if you are so inclined, the stunned pests can be removed from the garden area to continue their lives elsewhere.

Pyrethrum controls leafhoppers on potatoes (and acts as a growth stimulant on the crop as well); the sixspotted leafhopper, which is a vector of lettuce yellows; the cabbage looper (but is considerably less effective in controlling the imported cabbageworm); lygus bugs on peach; grape, flower, and citrus thrips; the grape leafhopper; and the cranberry fruitworm. Pyrethrum can be mixed with oil and dropped in corn ears to check the corn earworm.

Pyrethrins are more useful in controlling adult populations than larvae. In either case, the insecticide is most potent when applied as a spray. A pyrethrum-based aerosol is marketed by Raid as Tomato and Vegetable Fogger. Triple-Plus is a wettable powder or dust that also includes rotenone and ryania.

# QUASSIA

A botanical insecticide. See SPRAYS AND DUSTS.

# QUINCE

## Insects

Dormant-oil sprays give satisfactory control of **scale,** although beneficial insects may effect control if left undisturbed.

## Diseases

**Amylovora** may severely blight quince flowers. Since this bacterium is the same one that causes fire blight, seek out and destroy any diseased and neglected pear, quince, or apple trees that may harbor the disease. Prune back the branches of quince until healthy wood is uncovered, and break out any blighted fruit spurs. Use a dormant-oil spray.

# RABBIT

Rabbits plague almost every gardener at one time or another. The cottontail has adapted well to the suburbs, to the dismay of backyard gardeners.

Unlike groundhogs, rabbits are not able to climb, and a good chicken wire fence of three feet or more in height will discourage them.

Rabbits are considered game animals in most states and are therefore protected by law during certain seasons. Then, too, the discharge of firearms inside city and town limits may rule out shooting. If fences aren't doing the trick, try a live trap baited with apples or carrots. Or call the local game warden, who will either trap or shoot the rabbits or come up with another method of getting rid of them.

Rabbits are often kept from seedlings with open-ended coffee cans, but once the plants get taller than seven or eight inches they again become vulnerable. Until plants get to tall, you can keep the clear plastic lid on the top end. Sunlight passes through the lid and the water droplets that condense inside keep the young plants watered sufficiently. Other remedies: hang strips of aluminum foil on lines across planted areas; dust plants with talcum powder and an occasional dose of hot pepper; scatter blood meal about the garden; if liquid blood is available, add a spoonful to two gallons of warm water and spray on tree trunks or around garden plants; sprinkle powdered rock phosphate on seedling leaves; spray foliage within rabbits' reach with a repellent mixed from seven pounds of powdered rosin in one gallon of ethyl alcohol; or buy a commercial rabbit-and-deer repellent.

To prevent rabbit damage in winter and early spring, wrap fruit trees and ornamentals with cloth or heavy paper. A wire cylinder is more durable. The shields should be at least 18 inches high, depending on the depth of snow cover that you expect.

# RACCOON

Other than shooting, the best control for coons is trapping them alive. When the corn is sweet, however, it may be difficult to draw a coon into a trap, as there's a smorgasbord all around him, so it's advisable to begin your control program before the corn ripens. The raccoon is notorious for having a sweet tooth, and honey-soaked bread, marshmallows, and peanut butter should work well. You might also try fish. Some growers place a transistor radio (tuned to an all-night station, of course) in the corn patch with the volume turned all the way up. The only expense is the replacement of batteries.

# RADISH

## Insects

The only insects likely to cause much trouble in the radish patch are several species of **root maggot.** All concerned are yellowish white, legless, and measure from ¼ to ½ inch long. Seedcorn and cabbage maggots are particularly common. They tunnel into the roots, making them unsalable and less than appetizing. Because maggots are so fond of this crop, radishes are often planted as a trap crop to protect other vegetables. The maggot-infested roots are removed from the soil before they can complete their life cycles.

But if it's the radishes you're after, protect them by spading in generous amounts of hardwood ashes. Ideally, radishes should not be planted in soil that has hosted a member of the cabbage family for at least three years. In planning your garden for next year, try interplanting radishes with cucumbers, as radishes repel the cucumber beetle.

For other pests, see CABBAGE.

## Diseases

Radishes are little troubled by diseases. **Black root** appears as blackened areas where the secondary roots emerge. Rotate crops, and grow globe rather than long varieties if black root has been a problem.

The variety Red Prince is resistant to **fusarium wilt.**

# RASPBERRY

## Insects

The **raspberry cane borer** is one of the worst pests of raspberries, causing weak canes that often die before fruiting. The adult borer, a long-horned, black-and-yellow beetle, makes two rows of punctures about an inch apart on the cane, causing it to wilt. Between the two rows an egg is laid which later hatches into a small worm that burrows toward the base of the cane.

Cut the cane a few inches below the puncture marks. If the plant is vigorous and healthy, it will survive this pruning and put out new growth.

The **raspberry cane maggot,** a northern insect from the East to the West Coast, will cause the tips of the canes to wilt, often with a purplish discoloration. Sometimes the tips look as if they were cut, and there may be galls on the canes. The adult insect appears in late April, lays its eggs, and the maggot that does the damage reaches maturity in late June or early July. These white, $\frac{1}{3}$-inch-long maggots tunnel into the canes, eventually emerging as adult flies ready to deposit their eggs in canes and young shoots for another cycle.

Prune back and destroy infested canes in May to keep the cane maggot under control. Make the cut several inches below the infested part.

The **raspberry fruitworm** is a light brown beetle, $\frac{1}{8}$ to $\frac{1}{6}$ inch long, that feeds on buds, blossoms, and new leaves. Eggs laid on blossoms

and young fruit will hatch into grubs that bore into the fruit, causing it to dry up. Later they drop to the ground and overwinter in the soil. To prevent pupation, cultivate the soil thoroughly in late summer.

The **raspberry root borer,** also called the raspberry crown borer, does its work in the base of the canes and crown. The round, rust-colored eggs, about the size of a mustard seed, are laid on the undersides of leaves late in the growing season. The larvae hibernate and in spring begin to bore into the crown or canes, weakening the plant through partial or complete girdling. The adult moth has a black body with four yellow stripes and transparent wings. This pest prefers red raspberries to black. It is hard to spot once in the crown, but the cane will produce poorly and the plant may become stunted. The best control for this borer is to cut and destroy the affected canes at or below the soil line.

The **raspberry sawfly** is much like the blackberry sawfly, except for the coloring: larvae are pale green and spiny and adult flies are black with yellow and red markings. For symptoms and control, see BLACKBERRY.

**Spider mites** such as the twospotted mite are common pests of raspberries. They gather on the undersides of leaves and are so minute that they are scarcely visible until there are large numbers of them and much damage is done. Leaves will be pale or undersized. The use of chemical sprays has killed off many of the spider mite's natural predators, while the mites themselves quickly become resistant to new chemicals.

Spider mites are less often a problem on well-watered plants. Some gardeners use a hard spray of water to knock the mites off. You may need to use an angular or curved nozzle to spray the undersides of leaves. For further information, see SPIDER MITE.

The rednecked cane borer, rose scale, and tree crickets are pests of raspberries as well as blackberries; see BLACKBERRY. Raspberries will sometimes be troubled by APHID, European corn borer (CORN), and JAPANESE BEETLE.

# Diseases

Raspberries are more susceptible to viral diseases than any other berry in the United States. The chief diseases are leaf curl, mosaic, and streak. Other viruses may cause plants to decline in vigor and production, but have no other symptom.

Viruses are often spread from plant to plant by feeding aphids. Once infected, all parts of the plant will be diseased, usually beyond recovery. Since viruses are not curable except by a heat treatment that must be done in a laboratory, concentrate on prevention. Be sure to buy high-quality, certified virus-free plants, and choose disease-resistant varieties. Because aphids may be blown in the wind, you should plant berries 500 or even 1000 feet away from wild raspberries or old domestic plants that may be infected. Blackberry plants and other brambles sometimes harbor diseases that infect raspberries, so keep them separate.

There are other preventive measures that should be followed as a matter of course: check plants regularly for viral symptoms; keep the berry patch free of weeds; remove all old canes after harvest; and destroy the entire plant which shows evidence of disease.

Of the fungal and bacterial diseases, **anthracnose** is the most prevalent. The BLACKBERRY section describes the symptoms and control of the disease for black raspberries. On red raspberries, the lesions are very small and inconspicuous. During late summer the fungus produces an extensive grayish white growth on the shoots of the current season. This is known as the gray-bark phase of the disease. Pinpoint-size black dots, the fruiting bodies of the fungus, appear on these areas of the canes.

**Botrytis fruit rot** may be a danger to ripening bramble berries, especially during a wet spell of several days. Bruised or overripe fruit are most susceptible to this fungus, which may cause a grayish mold on the berries. The fungi causing this rot are airborne and the spores are easily spread.

Control by continuous picking to prevent accumulation of overripe berries on the plants. Handle the fruit gently to avoid squeezing or crushing it. Give the containers of picked fruit good ventilation to allow drying.

**Cane blight** may affect any bramble fruit plants, particularly weak plants or those with insect injury, broken stems, or injury from pruning. The fungus enters through the wounds and advances rapidly through the bark and cambium tissues. Spores on the infected areas are spread by splashing rain and air currents. Infection is not apparent until late in the season, when brownish purple discolorations appear on the cut or broken part of the cane and move down the cane or encircle it. Eventually, all lateral branches in that area will show discoloration and may wilt and die. Dead infected canes may keep on producing spores and remain a source of infection for two years or more.

Cane blight is more common on black raspberries than red. Vigorous plants are less susceptible. Control insects to help prevent the plant injuries that lead to cane blight. Remove and burn all dead canes, prunings, and stubs.

**Cane gall** affects black raspberries, and only rarely the red ones. It is much like crown gall (see BLACKBERRY) except that the galls are found on the fruiting canes rather than the crowns. The galls change from white to light brown; late in the season they disintegrate and the cane may split. Roots and new shoots are not infected. The bacterium that causes cane gall infects the plant through wounds. See CROWN GALL.

**Mild streak** affects the black raspberry, but not the red. This viral disease is hard to detect. Faint purplish streaks appear on the lower parts of new canes in summer, the leaves appear slightly hooked or twisted, and mottling may develop in lower leaves. The plants remain vigorous and produce well, although the fruits are seedy, crumbly, and of poor quality. Be sure planting stock is known to be free of mild streak, and see to it that no wild blackberries or wild raspberries are growing near cultivated plants.

A similar disease, severe streak, is less common but far more noticeable. The symptoms are generally the same as those of mild streak, except much more pronounced: leaves will be severely hooked, and vertical streaks are dark blue and very definite. Severe streak is controlled by rooting out all infected plants and removing any wild blackberries in the vicinity.

**Mosaic** is probably the most common and most damaging of the raspberry viral diseases. With some strains of mosaic there are no distinguishing symptoms, but the infected plants will be weaker and produce fewer canes and berries. Some strains of mosaic will be evident from short canes and yellowish, mottled foliage. Fruit from infected plants may be dry, seedy, flavorless, and crumbly. One symptom of mosaic diseases on red raspberries is large green blisters on the foliage in late spring, with yellowish tissue around them. The tips of infected black and purple raspberry canes may bend over, turn black, and die. Leaves become mottled and plants are stunted.

Dig out and burn all infected plants, including the roots, within three feet of the infected plant. Isolate new plantings well away from viral disease sources such as wild brambles or old infected plantings. Some varieties of red raspberries are mosaic-resistant; purple and black raspberries are susceptible. Remember that mosaic is not evident at planting time, so inspect your plants regularly.

Mosaic may be mistaken for other conditions. Late spring frosts, for example, sometimes produce a mottling on the leaves of older canes that resembles mosaic, but the young canes produced in early summer will be normal. Powdery mildew (see below) is sometimes mistaken for mosaic, as are the symptoms of a spider mite infestation (see above). Lloyd George, Pyne's Royal, Pyne's Imperial, Indian Summer, Newburgh, and Willamette are resistant to the aphids that transmit viruses.

**Orange rust** is a disease of black raspberries only; both wild and cultivated plants are susceptible. See BLACKBERRY for symptoms and control.

Although **powdery mildew** may occur on all bramble fruits, it is usually not a serious problem on any but red raspberries. The first symptom is a dwarfing and twisting of tip leaves in early summer, followed by a white powdery fungal growth on the undersides of leaves. Affected parts are yellowed and stunted. This fungal disease is spread by spores, which germinate best in warm temperatures when the humidity is very high. Latham is especially vulnerable.

Treatment is seldom necessary unless the mildew becomes severe, in which case you can spray or dust with sulfur. If plants are kept thinned out to let in sun and air, there will be less chance of powdery mildew developing.

There are two strains of **raspberry leaf curl virus**, one affecting black raspberries and the other red raspberries. The disease is first evident on the tips of canes, where the new leaves of red raspberries curl downward; those of black raspberries arch upward. The next season, all the leaves will be stunted and curled and the shoots will be abnormally short. The affected shoots will be pale and yellowish green when they first appear aboveground, and later will darken. In late summer the leaves turn from pale green to reddish brown. The fruit of affected canes is small, dry, seedy, and unpleasant-tasting. Infected plants never recover.

Destroy all diseased plants, and make new plantings from disease-free stock. Keep healthy raspberries away from diseased plants, and do not plant red raspberries near black raspberries, even when both are disease-free. Wild bramble fruit bushes should be eliminated from the vicinity of cultivated raspberries.

**Spur blight** is a fairly common disease of red raspberries, and less of a problem with black ones. It is seldom serious. Spur blight is first evident in late spring as discolored brown or purplish areas on the shoot at the spur (the point of leaf attachment). Usually buds near ground level are affected more than those higher up on the canes. The spots enlarge, and by late summer the bark in these areas splits and small, black, spore-producing growths appear. The following season, growth from buds in the affected areas will be weak, the leaves stunted, and fruit production severely reduced. Plants with spur blight are especially vulnerable to winter injury from low temperatures.

As this fungus is spread by splashing rain and nurtured by moisture on the plants, careful sanitation will help control it; eliminate weeds and thin plants to let air circulate and dry the foliage. Always remove and destroy old canes after harvest. To avoid spur blight, plant the fall-bearing variety Durham. Its canes may all be cut back to the ground to eliminate the source of this disease without sacrificing a crop.

**Verticillium wilt,** also known as bluestripe or bluestem, affects many kinds of plants, including raspberries (particularly black raspberries), potatoes, tomatoes, peppers, eggplants, stone fruits, and such weeds as horse nettle or nightshade. On black raspberries, the lower leaves develop an off-green or yellowish bronze tinge in June or early July. Leaves curl upward, become yellow, and then brown and fall off. The disease progresses up the plant. Canes show blue or purple streaks, beginning at the soil line and moving up the cane. Often the canes die before the fruit matures, and eventually the entire plant dies.

On the red raspberry, which is generally more resistant, verticillium wilt is evident later in summer, and the discolored leaves curl downward. The cane discoloration is often missing or hard to distinguish from the plant's normal red or bronze color.

The verticillium fungus lives in the soil and enters the plant through injured tissue, e.g., roots damaged by frost, transplanting, or nematodes. Once present in a planting it will be spread by soil cultivation or flowing water. Often the fungus will be borne in the soil if the crops mentioned above have been grown previously. Verticillium wilt is most likely in heavy, poorly drained soils, and is at its worst during cool seasons. For these reasons, do not plant raspberries where wilt-susceptible plants have grown. Avoid heavy soils, and be sure the drainage is good. Your planting stock should come from a nursery that practices good verticillium wilt control.

**Winterkill** hits the green, immature wood of raspberry canes. Fertilizer should be applied in late winter or very early spring so that growth will slow down and harden before frost. You can provide winter protection in severe climates by bending the canes to the ground and covering at least a third of their length with soil before the ground freezes. The bent canes will form a snow trap, and the drifts that cover the canes will protect them from fluctuating temperatures and drying winds. When snow melts in spring, the canes can be raised and tied to their supports, and winterkilled portions can be removed with the spring pruning.

## Wildlife

The most effective means of keeping **birds** from harvesting your raspberry crop is also the most troublesome: construct a tepee of lashed pipes, poles, or lumber, and then drape a net over the structure. Or nail up a frame of 2 × 3s and nail on wire screen. You'll want to include a door of some sort to allow two-legged animals to harvest the berries. You might consider planting a trap crop of elderberries or mulberries, which are less valuable to most humans but preferred by birds to cane berries.

See also BLUEBERRY.

# RED CEDAR

See JUNIPER.

# RED SPIDER MITE

See SPIDER MITE.

# REPELLENT PLANTING

When a pest takes over a garden, it doesn't attack everything. Some plants appeal to it little or not at all. A few actually repel bugs, because of a powerful smell perhaps, or for a reason not obvious to humans. Plants have been found to produce insect hormones that either cause developmental abnormalities or prevent metamorphosis into adults. Growers long have taken note of plants that particular pests avoid, and

have tried placing them among and around vulnerable crops.

Some bugs evolve sensory mechanisms to detect and avoid plants with defenses. Others develop metabolic mechanisms to cope with the poisons. For example, several tobacco pests rapidly excrete poisonous nicotine so that toxins do not build up within their bodies. Another pest, the tobacco wireworm, renders nicotine safe by transforming it into a relatively nontoxic alkaloid through the action of an enzyme. A third means of metabolic adaptation involves storing the poison as is within the insect's body. The handsome monarch butterfly stashes away the heart poisons manufactured by one of its important hosts, milkweed. Storing up toxic substances has an important side-benefit for the monarch, making it inedible to lepidopteran predators. The very distinctive wing coloration emphasizes the monarch's bad reputation—so effectively, in fact, that the entirely edible viceroy butterfly successfully mimics its coloration to avoid being eaten. Some plants repel a variety of pests, while others are especially noxious only to one or two: garlic offends most insects you'll come across, but the herb borage is noted for repelling only one, the tomato hornworm. The table on page 343 shows which plants bother which bugs. Don't stop with this list. Not all of these recommendations will be effective all of the time, and you may come up with your own repellent plantings.

Many of these plants can be put in the blender with water and a touch of soap to make a repellent spray. Leaves can be simply scattered around vulnerable plants, or even tied directly to the plants' stems.

**THE ONION FAMILY.** Chives, garlic, leeks, shallots, and allium—just reciting the list out loud is enough to taint your breath. And the aromatic power of the family is not lost on the insect world. *Garlic* is the mightiest of these, and is probably the most widely used living repellent. It helps out no matter where in the garden you plant it, and does so without taking up a lot of valuable space. (A fringe benefit you'll no doubt discover is that garden-grown garlic has a better-mannered taste than the cloves that come in boxes.) Garlic also does the job in the berry patch, grape arbor, and around fruit trees. It will not tamper with delicate fruit flavors, and the potato bugs, bean beetles, and stem borers will be kept from achieving pest status. Many gardeners see to it that a few onion plants mingle with their flowers each year too.

## PROTECTIVE HERBS.

A variety of herbs can protect your plants, although they might not occur to you as being unusually virulent. These successful interplanting combinations aren't the result of disciplined research, but were discovered by observant gardeners.

*Tansy,* also known as bitter or yellow buttons, is a tall, strong-growing and strong-smelling herb with jagged, fernlike foliage and flat clusters of orange yellow flowers that resemble the centers of daisies. Don't confuse this plant with look-alikes that have taken on similar names: silverweed or goose tansy grows in wet waste places; tansy ragwort is a low-growing invader of pastures and fields, and is reported to be a poison to livestock and a treatment for hemorrhoids; and one of the hedge mustards sometimes goes by the name of tansy mustard. But only *the* tansy, *Tanacetum vulgare,* has the fragrant power—it was used long ago as a substitute for pepper. You'll probably find the smell attractive, as did the old-time Maine loggers who used tansy boughs for bedding. Perhaps the boughs also served as a pest repellent as well as padding; insects can't cope with the plant, and keep a good distance.

Because of its size (in good soil, it may reach five or six feet), tansy is not suitable everywhere. Grow it with crops that can hold their own, such as in grape arbors or in berry patches to keep out Japanese beetles. Or clip the tansy occasionally, and sprinkle the tops about vulnerable plants. Since flowers may have pests too, tansy can be used as an ornamental perennial in the back of the border. If your vegetable garden is big enough, let tansy stand guard in several positions throughout the plot. To propagate it, sow seed or dig up a clump, then divide and reset. Herb gardens sell seed and young plants; you can occasionally get hold of a curly leafed variety, *T. vulgare* var. *crispum,* which is shorter and considered more ornamental. Every few years the woody center portion of a clump should be dug out and removed in order to let the new outer growth continue. Old parts can be chopped up and used as a pest-repellent mulch.

One tansy look-alike can help your garden —*yarrow,* an herb with a cool, clean smell a bit like lavender. It is also fernlike, but the yellow yarrow flowers appear much earlier in the season, usually around the Fourth of July. Although it's not as powerful as tansy, yarrow has a general rather than specific effect on insect varieties. You can buy seed and occasionally plants from herb gardeners.

*Borage* performs two good acts, repelling the tomato hornworm and attracting bees in numbers. Once you scatter borage seeds, the blue-and-pink-flowered annual will usually sow its own seeds. It grows to a height of two or three feet and is often seen growing wild. *Thyme* also plays this dual role. Fruit growers benefit from growing *mother-of-thyme,* an herb which forms an aromatic, spreading mat.

Just as *rue* was once used to ward off pestilence, it now serves to repel the Japanese beetle. Because rue has attractive blue green leaves, it complements flower gardens and ornamentals. You can start it from seed or root division.

You'll find several appealing allies in the *mint* family: spearmint; peppermint; catnip; apple, lemon, orange, and pineapple mints (more delicate in fragrance but effective just the same); pennyroyal; and hyssop. As a group, these plants are sturdy perennials with a knack for escaping the garden and reestablishing wherever the ground is rich and moist. Earthworms seem to be happy under a canopy of mint. Mint can be a valuable companion to members of the cabbage family, although it has been observed to attract the imported cabbageworm.

One member of the mint family takes the form of a ground-hugging vine. It is known by many names—ground ivy, gill-over-the-ground, creeping Charlie (or Jenny), and cat's-foot, to recite a few—and comes back year after year to produce pretty purple flowers and an aroma that protects nearby plants from insects. It spreads readily, sending down roots wherever it touches the ground, and prefers a partially shaded location.

*Coriander* and *anise* are annuals that have a reputation for repelling aphids; coriander also works against the carrot fly if planted between rows of carrots. A border of *opal basil*, teamed with interplanted marigolds, will help to minimize tomato hornworm damage. A thick growth of *prostrate rosemary* will act as a border for snails and slugs; apparently the sharp foliage doesn't feel good next to their slimy, soft skin. Nematodes find life unpleasant when *calendula* (pot marigold), *salvis* (scarlet sage), or *dahlias* are grown in neighboring soil. If these unseen underground pests are sapping the life of tomatoes, interplanting with *asparagus* can do the trick. And the asparagus beetles in your garden

will shy away from the herby *tomato* aroma. A host of other aromatics deserve mention as bug chasers, even though not enough is known about them to associate them with a certain pest. Here are some good ones to experiment with: *marjoram, oregano, lavender* and *santolina* (both effective against moths), *blessed thistle, chamomile* (used in teas also), *lovage, chervil, lemon balm,* and *bergamot.* Instead of arranging these herbs in a stiffly formal herb garden, scatter them informally among crops. In time, you'll come up with your own favorites.

Several members of the artemesia family work as repellent herbs, and with white or gray foliage they work in aesthetic plantings too. Probably the best known is *wormwood.* Actually, this name is often loosely applied to many artemisias, but it properly applies to *Artemisia absinthum,* a hardy perennial with woolly gray leaves and a strongly bitter odor. Try spraying a tea of wormwood on the ground in fall and spring to discourage slugs, and on fruit trees and other plants to repel aphids. *Southernwood* is a notable cousin. This shrublike plant grows to at least three feet high, displaying gray green, finely divided foliage. It does best in well-drained soils, and is propagated by root division. *Silky wormwood (A. frigida)* is anathema to snails. It prefers full sun and a rather dry location. Other members of the family include *tarragon, mugwort, silver mound, fringed wormwood,* and *dusty miller.* While more moderate in smell than wormwood and southernwood, they all give a measure of protection to any garden graced by their silvered foliage.

Protective plantings can keep animals away too. *Spurge* (also known as the mole plant) will discourage moles from trespassing. Seeds are commercially available, and the plant is decorative and self-sowing. *Castor beans* will work too, but the beans are highly poisonous to humans and should not be grown where children might happen by. As a safety precaution, the large seed pods can be clipped off before the poisonous seeds mature. This plant is an annual in colder areas, perennial in the South, and is grown quickly and easily from seed. The effectiveness of the castor bean plant lingers on after it is yanked. Plant a border of mint to repel mice and rats, and try a line of alliums if rabbits are giving you much competition.

## REPELLENT FLOWERS.

Many vegetable gardeners have found *marigolds* valuable in keeping Mexican bean beetles away from snap beans; growing miniature varieties makes picking easier. Also, these flowers have been found to exude a substance into the soil that keeps down the nematode population in the immediate area. Marigolds also will keep whiteflies off greenhouse tomatoes. The taller African marigold is just as effective, but don't expect bugs to shy away from the odorless varieties.

The *pyrethrum* flower is so potent that it is widely used as a commercial pesticide, but curiously it doesn't do much as a neighbor to ward off pests.

You can either buy dried pyrethrum flowers as powders or sprays or grow the plants from seed yourself; the same properties that make the commercial stuff an effective killer will work at this modest level too. This perennial does best in soil with good drainage.

*Feverfew* is a daisylike flower that can be planted as a border around roses or scattered throughout the garden. Feverfew grows to about 1½ feet tall, has yellow green, ferny foliage, and self-sows readily. You can propagate it by seed (in February or March), root division (March), or cuttings (set between October and May). If you like to experiment with the effect of herbs on your own body, herbals abound with uses for feverfew.

The *nasturtium* has become a popular flower in the garden for its ability to attract some insects and repel others. Plantings draw aphids from crops, and put the whitefly and squash bug to flight. Nasturtiums can also be brewed into a potent spray. If you're bothered by woolly aphids in apple trees, try both planting a ring of flowers around the trunk and spraying the foliage with nasturtium tea.

White *geraniums* serve as a trap plant for the Japanese beetle: the bugs eat it and die. Geraniums keep beetles from roses. *Nicandra* (or Peruvian ground cherry), a pretty plant with little pale blue flowers, is poisonous to many pests. It grows best in the shade, but produces more flowers in full sun. Set the seeds in fairly rich garden soil. Its blooming period (and its effectiveness) can be extended by keeping fading blossoms plucked.

A number of other domestic plants have insecticidal powers, and might appeal to the backyard experimenter. See the appendix for a list of these.

Keep in mind that repellent plants may not perform consistently. Many of the applications listed in the table below are recommended on the basis of informal observation in the garden and not by hard testing. In a test done by researchers at the University of Illinois, interplanting with herbs and pungent flowers did not seem to protect cabbage against the cabbage looper and imported cabbageworm. Interplanting tents at the Rodale Research Center in Pennsylvania showed that some repellent plants actually may *attract* imported cabbageworm moths to lay their eggs on crops, tansy and catnip in particular. The message here is to go at repellent planting in your garden or orchard with a spirit of experimentation.

## REPELLENT PLANTINGS

| Pest | Plants | Pest | Plants |
|---|---|---|---|
| Ants (and the aphids they carry) | Pennyroyal, spearmint, southernwood, and tansy. | Gopher | Castor bean. |
| Aphids | The above, as well as: garlic, chives, and other alliums; coriander; anise; nasturtium and petunia around fruit trees. | Japanese beetle | Garlic, larkspur (poisonous to humans), tansy, rue, and geranium (white works best). |
| | | Leafhopper | Petunia, geranium. |
| Asparagus beetle | Tomato. | Mexican bean beetle | Marigold, potato, rosemary, summer savory, and petunia. |
| Borer | Garlic, tansy, and onion. | | |
| Cabbage maggot | Planted in adjacent rows: mint, tomato, rosemary, hemp (illegal just about everywhere), and sage. | Mice | Mint. |
| | | Mites | Onion, garlic, and chives. |
| | | Mole | Spurge, castor beans, mole plant, and squill. |
| Cabbage moth | Mint, hyssop, rosemary, southernwood, thyme, sage, hemp (look out for nosey neighbors), wormwood, celery, catnip, and nasturtium. | Nematode | Marigold (African and French varieties), salvis (scarlet sage), dahlia, calendula (pot marigold), and asparagus. |
| Carrot fly | Rosemary, sage, wormwood, black salsify, various alliums, and coriander. | Plum curculio | Garlic. |
| | | Rabbit | Allium family. |
| Chinch bug | Soy beans. | Rose chafer | Geranium, petunia, and onion. |
| Colorado potato beetle | Green beans, horseradish, dead nettle, flax, catnip, coriander, nasturtium, and tansy. | Slug (snail) | Prostrate rosemary, wormwood. |
| | | Squash bug | Tansy, nasturtium, and catnip. |
| Cucumber beetle (spotted and striped) | Tansy, radish. | Striped pumpkin beetle | Nasturtium. |
| Cutworm | Tansy. | Tomato hornworm | Borage, marigold, and opal basil. |
| Eelworm (see Nematode) | Marigold (French and African). | Whitefly | Nasturtium, marigold, and nicandra (Peruvian ground cherry). |
| Flea beetle | Wormwood, mint, and catnip; interplant cole crops with tomatoes. | Wireworm | White mustard, buckwheat, and woad. |
| Fruit tree moth | Southernwood. | | |

# RESISTANT VARIETIES

You can avoid many plant diseases and pests by growing resistant varieties, those that resist the nibbling, sucking, and parasitism. You couldn't hope to find an easier, more effective means of pest control. Imagine a gardenful of vegetables that the bugs fly right past, or rows of crops that sterilize any insects that make the mistake of stopping to feed. While this vision isn't likely to become reality—not just yet, anyway—the growing list of varieties on the market is a step in the right direction. The following list includes many varieties that are readily available at this writing. (In time, this list is apt to become dated. Suppliers may drop varieties for a number of reasons, one of them being that some resistant strains don't have a very long useful life. While some have lasted several decades, other varieties lose their special power in only five years. Within a plant or insect species there are usually many slightly different forms, or biotypes. This means that, given local variations in climate and environment, plants and insects of the same name won't act the same.)

Resistance occurs in nature, without man's help. Plants differ in their reactions to threats from the environment and, as expressed by Darwin's law of survival of the fittest, some plant varieties disappear while others survive to perpetuate their kind. Many of these survivors make it because they have developed ways of getting by the bugs—ways collectively called resistance.

The study of insect-resistant varieties of plants is still in its infancy. Use of pesticides and other chemical means of controlling insects has kept research to a minimum, and while some resistant varieties have reached the market, many more are on the way. Developing the varieties is a slow business. It takes at least ten years to come up with a plant resistant to one pest, and developing a plant that will discourage two or more pests takes twice as long. Another setback is that resistance to one bug is often linked with a susceptibility to another. So, it seems unlikely that you will ever be able to go out and plant a totally bug-free crop. But resistant varieties do give you one more way in which to grow your food without poison.

Why is it that insects will devour one variety of a crop while another goes relatively unscathed? Resistance may result from any of several defensive strategies that a plant evolves. Researchers have put these strategies into three categories: nonpreference, antibiosis, and tolerance. Resistant varieties generally are protected by more than one of these.

**NONPREFERENCE.** Nonpreference plants are those which either turn an insect off or lack the stimuli that would turn an insect on.

The colors of certain varieties earn them some protection: red cabbage shows resistance to the cabbage looper and imported cabbageworm, and yellow varieties of green pea are more resistant to the pea aphid than blue green varieties.

The downy hairs on some nonpreference varieties offer a measure of resistance, either because the hairs affect the way in which light is reflected or because the fuzz just doesn't feel good to insects. Wheat growers in Michigan gain protection from the cereal leaf beetle by growing varieties with fuzzy leaves. Hairy varieties of legumes and cotton have fewer problems with leafhoppers.

In seeking shelter, many insects prefer plants that they can snuggle up in, especially plants with leaves that are close together. Sorghum and milo varieties with loose-fitting leaf sheathes are attractive to the chinch bug, and thrips are drawn to onion varieties having a narrow (in other words, cozy) angle of contact between leaves.

Some varieties are safer from attack because they don't put off the good smells that pests home in on. The aromatic oils or esters in certain potato varieties render them resistant to the Colorado potato beetle. Damage that does occur is found on the lower, older leaves, which evidently give off less smell. Plants that taste bad also have built-in protection, but to seedlings just one bite can be fatal. Odors also guide egg-laying bugs to suitable plants. If the cueing odor is absent or disguised in a variety, the females spend much of their time finding a host. And with such short life spans, many would-be mothers never get around to laying their eggs.

Wild potatoes release an alarm pheromone that wards off many pests.

## ANTIBIOSIS.
A second type of resistance, antibiosis, is shown by plants that injure or kill pests. The effects are sometimes difficult to detect, and sometimes dramatic. An insect that partakes of such a plant may vomit or go into a stupor, come to be stunted, have poor fertility, die young, or become malformed.

Antibiotic plants contain volatile oils, known as phytoalexins, that repel insects and fungi. Some plants seem to produce these oils when under attack, while others, such as garlic and marigolds, always have a good supply. While both susceptible and resistant varieties respond with defenses when attacked, resistant plants react quicker, and the development of the parasite is discouraged. Some plants even react by killing the tissue around the area of infestation, as well as finishing off the offending pathogen. Plants may encase repellents in vacuoles—little packages within cells—that rupture when bitten by an insect. Unfortunately, they would also rupture when bitten by you, leaving a bitter taste in the mouth. So, it's not likely that you'll find this method of resistance incorporated in many food plants.

Antibiosis has been found to figure in a number of insect-plant relationships: aphids on beans, cantaloupe, peas, Northern Spy apples and resulting hybrids, and Lloyd George red raspberries and resulting hybrids; the European corn borer, southern corn rootworm, and corn earworm on corn; the potato leafhopper on potatoes; the wheat stem sawfly on wheat and barley; and the chinch bug on sorghum. A solid-stem variety of wheat, Rescue, gives sawfly larvae another sort of problem—after hatching, the worms have trouble boring through the solid stems to reach feeding areas elsewhere in the plant.

## TOLERANCE.
In contrast to the other two types of resistance, tolerance is passive. Instead of avoiding, repelling, or otherwise influencing the life of insects, a tolerant plant survives and produces a crop simply by regenerating tissue fast enough to remain healthy. That is, it can take the chewing and still come through at harvest time. Such plants may lose up to 30 percent of their leaf mass without much lessening of yield. Unlike nonpreference and antibiosis, tolerance does not work by keeping down the number of bugs.

Tolerant plants are characterized by their ability to replace leaf area or a large part of their root systems. Such growth is stimulated by the output of growth hormone, and is dependent on the plant's ability to heal the wound and resist any diseases that may enter the affected area. And some tolerant plants survive bug troubles because of their structure. The tough, resilient stalks of some corn varieties can take the burrowing of European and southern corn borers without breaking.

As you decide on a resistant variety, keep in mind that any plant will be better able to stand up to insects and diseases if it is growing vigorously in excellent soil. The health and nutritional balance of soil is as important in warding off insects as the variety planted.

**DISEASE RESISTANCE.** Disease-resistant varieties are easier to come up with than those resistant to insects, for several reasons: diseases generally tend to stick to one species more than insects, while if an insect is turned off by one resistant crop, it can usually get by on another; diseases land on target more or less by chance, while insects can sniff out the target crop; and the make up of a disease organism is easier to comprehend.

## INSECT-RESISTANT VARIETIES

Both varieties and the bugs that eat them are very subject to change. A pea aphid in Quebec is not the same as a pea aphid in Kansas, and resistant varieties may not work in some areas. Most lists are arranged from most resistant to least resistant (most susceptible); those few that aren't arranged in such order, because their relative resistance hasn't been determined, are marked(*).

### ALFALFA

**Alfalfa aphid**
- *Resistant:* Cody, Lahontan, and Zia.
- *Susceptible:* Buffalo.

### BARLEY

**Greenbug**
- *Resistant:* Omugi, Dictoo, and Will.
- *Susceptible:* Rogers and Reno.

### BEAN

**Cutworm**
- *Resistant snap beans:* Wade, Idaho Refugee, Gold Crop, and Regal.
- *Resistant limas:* Black Valentine, Baby Fordhook, and Baby White.

**Mexican bean beetle**
- *Resistant:* Wade, Logan, and Black Valentine (lima).
- *Susceptible:* State, Bountiful, and Dwarf Horticultural.

### BROCCOLI

**Diamondback moth**
- *Moderately resistant:* Coastal, Italian Green Sprouting, and Atlantic.
- *Susceptible:* De Cicco.

**Harlequin bug**
- *Resistant:* Grande, Atlantic, and Coastal.
- *Moderately resistant:* Gem.

**Striped flea beetle**
- *Resistant:* De Cicco, Coastal, Italian Green Sprouting, and Atlantic.
- *Moderately resistant:* Gem.

### CABBAGE

**Cabbage looper and imported cabbageworm**
- *Resistant:* Mammoth Red Rock, Savoy Chieftain, and Savoy Perfection Drumhead.
- *Moderately resistant:* Special Red Rock, Penn State Ball Head, Early Flat Dutch, Badger Ball

Head, Wisconsin Hollander, Red Acre, Danish Ball Head, Charleston Wakefield, Premium Late Flat Dutch, Glory of Enkhuizen, Globe, All Seasons, Midseason Market, Bugner, Succession, Early Round Dutch, Stein's Early Flat Dutch, Badger Market, Large Late Flat Dutch, Jersey Wakefield, Marion Market, Wisconsin Ball Head, Large Charleston Wakefield, Early Glory, Green Acre, Round Dutch, Resistant Detroit, and Wisconsin All Season.
- *Susceptible:* Golden Acre, Elite, Copenhagen Market 86, and Stein's Flat Dutch.

**Diamondback moth**
- *Resistant:* Michihli Chinese and Mammoth Red Rock.
- *Moderately resistant:* Stein's Early Flat Dutch, Savoy Perfection Drumhead, Early Jersey Wakefield, and Ferry's Round Dutch.
- *Susceptible:* Copenhagen Market 86.

**Harlequin bug**
- *Resistant:* Copenhagen Market 86, Headstart, Savoy Perfection Drumhead, Stein's Flat Dutch, and Early Jersey Wakefield.
- *Susceptible:* Michihli Chinese.

**Mexican bean beetle**
- *Resistant:* Copenhagen Market 86 and Early Jersey Wakefield.
- *Susceptible:* Michihli Chinese.

**Striped flea beetle**
- *Resistant:* Stein's Early Flat Dutch, Mammoth Red Rock, Savoy Perfection Drumhead, Early Jersey Wakefield, Copenhagen Market 86, and Ferry's Round Dutch.
- *Moderately resistant to susceptible:* Michihli Chinese.
- *Susceptible (Canada):* North Star and Northern Belle.

## CANTALOUPE

**Mexican bean beetle.** Cantaloupe is generally resistant to this pest, but serious damage was done to Rocky Ford Earliest during an infestation.

**Spotted cucumber beetle**
- *Resistant (foliage):* Edisto 47, Edisto, and Harper Hybrid.

- *Susceptible (seedlings):* Edisto, Edisto 47, Harper Hybrid, and Honey Dew.
- *Susceptible (foliage):* Honey Dew.

## CAULIFLOWER

**Diamondback moth**
- *Moderately resistant:* Snowball A.

**Harlequin bug**
- *Resistant:* Early Snowball X and Snowball Y.

**Striped flea beetle**
- *Resistant:* Snowball A and Early Snowball X.

## COLLARD

**Diamondback moth**
- *Resistant:* Green Glaze.
- *Moderately resistant:* Morris Heading, Vates, and Georgia Southern.

**Harlequin bug**
- *Resistant:* Vates, Morris Improved Heading, and Green Glaze.
- *Moderately resistant:* Georgia LS and Georgia.

**Mexican bean beetle**
- *Resistant:* Georgia LS, Green Glaze, and Vates.

**Striped flea beetle**
- *Resistant:* Vates, Georgia, and Georgia LS.
- *Moderately resistant:* Morris Heading.
- *Susceptible:* Green Glaze.

## SWEET CORN

**Corn earworm.** Any corn with long, tight husks physically helps to prevent ear penetration by earworms.
- *Resistant:* Dixie 18 (field corn), Calumet, Country Gentleman, Staygold, Victory Golden, Golden Security, Silver Cross Bantam, Silvergent, Ioana, Aristogold, Seneca Scout, and Seneca Chief.
- *Susceptible:* Ioana, Aristogold Bantam Evergreen, Seneca Chief, Spancross, North Star, and Evertender.

**Fall armyworm.** Late sweet corn crops and second crops are especially vulnerable. Resistance depends on the planting time and tolerance of a variety. The varieties are arranged by survival rates, from best to worst.

*(continued on next page)*

## SWEET CORN—*Continued*

- *Resistant:* Golden Market, Long Chief, Golden Security, Evertender, Marcross, Golden Regent, Silver Cross Bantam, Calumet, Victory Golden, Golden Sensation, Spancross, Golden Cross Bantam, Aristogold Bantam Evergreen, Golden Beauty, Triplegold, Deep Gold, and Ioana.

**Sap beetles.**    As with the corn earworm, any corn with long, tight husks physically helps to discourage the sap beetle.
- *Resistant\*:* Country Gentleman, Deligold, Gold Pack, Golden Security, Harris Gold Cup, Tender Joy, Trucker's Favorite, Stowell's Evergreen, and Victory Golden.
- *Moderately resistant:* Atlas, Duet, Eastern Market, Gold Strike, Golden Grain, Golden Security, Marcross, Merit, Midway, Royal Crest, Silver Queen, Spring Gold, Stowell's Evergreen, Tendercrisp, Wintergreen, and Victory Golden.
- *Susceptible* (In many cases, sap beetles gain access to these varieties by way of entrances previously made by corn earworms): Aristogold Bantam Evergreen, Carmelcross, Corona, Deep Gold, Floriglade, Gold Mine, Golden Beauty, Golden Fancy, Ioana, Merit, Northern Belle, Seneca Chief, Seneca Explorer, Silvergent, Sixty Pak, Spancross, Spring Bounty, Titian, Vanguard, and White Silk Tendermost.

## CUCUMBER

**Mexican bean beetle.**    While not normally a serious pest of cucumbers, this beetle severely damaged these varieties in an outbreak: Arkansas Hybrid No. 4, Colorado, Crispy, Hokus, Marketer, NK804, Nappa 63, Piccadilly, Pico, Pixie, and Triumph.

### Pickleworm
- *Resistant:* Arkansas Hybrid No. 4, Cubit, Gemini, Nappa 61, Nappa 63, Pixie, Princess, Spartan Dawn, Stono, Ashley, Colorado, Hokus, Long Ashley, Model, Piccadilly, Packer, and Table Green.

### Spotted cucumber beetle
- *Resistant (seedlings):* Ashley, Chipper, Crispy, Explorer, Frontier, Gemini, Jet, Princess, Spartan Dawn, and White Wonder.

- *Resistant (foliage):* Ashley, Cherokee, Chipper, Gemini, High Mark II, Ohio MR 17, Poinsett, Stono (Stono is reported resistant to both striped and spotted cucumber beetles, and Fletcher and Niagara are moderately resistant to the two pests), and Southern Cross.
- *Moderately resistant (seedlings):* Cubit, High Mark II, Hokus, Nappa 63, Pixie, Poinsett, and SMR 58.
- *Moderately resistant (foliage):* Colorado, Crispy, Explorer, Frontier, Long Ashley, Nappa 61, Pixie, and Table Green.
- *Susceptible (seedlings):* Cherokee, Coolgreen, Model, Nappa 61, Packer, Pioneer, Southern Cross, and Table Green.
- *Susceptible (foliage):* Coolgreen, Cubit, Hokus, Jet, Model, Nappa 63, Packer, Pioneer, Spartan Dawn, and SMR 58.

## KALE

**Diamondback moth**
- *Resistant:* Vates (protected by antibiosis, as a result of compact cell structure within the leaf).
- *Susceptible:* Early Siberian and Dwarf Siberian (has loosely arranged cells that are apparently easy for the larvae to mine).

**Harlequin bug**
- *Resistant:* Vale.
- *Susceptible:* Dwarf Siberian.

**Mexican bean beetle**
- *Resistant:* Dwarf Siberian.

**Striped flea beetle**
- *Resistant:* Vates, Dwarf Siberian, Dwarf Green Curled Scotch, and Early Siberian.

## LETTUCE

**Lettuce root aphid**
- *Resistant:* Avoncrisp and Avondefiance.

## MUSKMELON

**Striped and spotted cucumber beetle**
- *Resistant:* Hearts of Gold.
- *Susceptible:* Smith Perfect and Crenshaw.

## MUSTARD

**Diamondback moth**
- *Resistant:* Southern Giant Curled.
- *Moderately resistant:* Florida Broadleaf.

**Harlequin bug**
- *Moderately resistant:* Old Fashion.
- *Susceptible:* Southern Giant Curled, Green Wave, and Florida Broadleaf.

**Mexican bean beetle**
- *Resistant:* Green Wave.
- *Susceptible:* Southern Giant Curled.

**Striped flea beetle**
- *Resistant:* Florida Broadleaf.
- *Moderately resistant:* Southern Giant Curled and Green Wave.

## POTATO

**Aphids**
- *Resistant:* British Queen, DeSota, Early Pinkeye, Houma, Irish Daisy, and LaSalle.
- *Tolerant:* Red Warba, Triumph, President, Peach Blow, and Early Rose.
- *Susceptible:* Katahdin, Irish Cobbler, Idaho Russet, Sebago, and Sequoia.

**Colorado potato beetle**
- *Resistant:* Sequoia and Katahdin.
- *Susceptible:* Fundy, Plymouth, and Catoosa.

**Potato leafhopper**
- *Resistant:* Delus.
- *Moderately resistant:* Sebago, Pungo, and Plymouth.
- *Susceptible:* Cobbler.

## PUMPKIN

**Serpentine leafminer**
- *Resistant:* Mammoth Chili and Small Sugar.
- *Susceptible:* King of the Mammoth and Green Striped Cushaw.

**Spotted cucumber beetle**
- *Resistant (foliage):* King of the Mammoth, Mammoth Chili, and Dickinson Field.
- *Susceptible (seedlings):* Green Striped Cushaw, King of the Mammoth, Mammoth Chili, and Small Sugar.

- *Susceptible (foliage):* Connecticut Field, Green Striped Cushaw, and Small Sugar.

## RADISH

**Cabbage webworm**
- *Resistant:* Cherry Belle.
- *Moderately resistant:* Globemaster.
- *Susceptible:* White Icicle, Red Devil, and Champion.

**Diamondback moth**
- *Resistant:* Cherry Belle, White Icicle, Globemaster, and Champion.

**Harlequin bug**
- *Resistant:* Red Devil, White Icicle, Globemaster, Cherry Belle, Champion, and Red Prince.
- *Moderately resistant:* Crimson Sweet.

**Mexican bean beetle**
- *Susceptible:* Sparkler, Champion, and White Icicle.

**Striped flea beetle**
- *Moderately resistant:* Champion and Sparkler.
- *Susceptible:* Globemaster, Cherry Belle, and White Icicle.

## RUTABAGA

**Diamondback moth**
- *Moderately resistant:* American Purple Top.

**Harlequin bug**
- *Susceptible:* American Purple Top.

**Striped flea beetle**
- *Resistant:* American Purple Top.

## SORGHUM

**Chinch bug**
- *Resistant:* Atlas.
- *Susceptible:* Milo.

**Corn leaf aphid**
- *Resistant:* Sudan.
- *Susceptible:* White Martin.

*(continued on next page)*

## SQUASH

**Mexican bean beetle.**   Although this beetle is not normally a serious pest of squash, White Bush Scallop was damaged severely in an outbreak.

**Pickleworm**
- *Resistant:* Summer Crookneck, Butternut 23, Buttercup, Boston Marrow, and Blue Hubbard.
- *Moderately resistant:* Early Prolific Straightneck, Early Yellow Summer Crookneck, and White Bush Scallop.
- *Susceptible:* Black Beauty, U Conn, Marine Black Zucchini, Seneca Zucchini, Cozella Hybrid, Long Cocozelle, Benning's Green Tint Scallop, Short Cocozelle, Zucchini, Caserta, Black Zucchini, and Cozini.

**Serpentine leafminer**
- *Resistant:* Butternut 23 and Cozella.
- *Moderately resistant:* Blue Hubbard, Zucchini, Benning's Green Tint Scallop, Summer Straightneck, Boston Marrow, Buttercup, and Pink Banana.
- *Susceptible:* Seneca Prolific, Green Hubbard, Seneca Zucchini, Summer Crookneck, Black Zucchini, Cozini, and Long Cozella.

**Spotted cucumber beetle.**   These beetles are attracted to the odor of the germinating seeds of some varieties, and often dig through the soil to eat seedlings even before they've grown to the surface. Mature varieties with flowers having a strong, sweet smell attract these beetles in greater numbers than do other varieties.
- *Resistant (seedlings):* Blue Hubbard, Green Hubbard, Long Cozella, Seneca Prolific, Summer Crookneck, and Summer Straightneck.
- *Resistant (foliage):* Black Zucchini, Benning's Green Tint Scallop, Blue Hubbard, Royal Acorn, and Early Golden Bush Scallop.
- *Moderately resistant (seedlings):* Boston Marrow, Buttercup, and Pink Banana.
- *Moderately resistant (foliage):* Green Hubbard, Pink Banana, Seneca Zucchini, Summer Crookneck, and Summer Straightneck.
- *Susceptible (seedlings):* Benning's Green Tint Scallop, Black Zucchini, Cozella, Cozini, Seneca Zucchini, and Zucchini.

- *Susceptible (foliage):* Boston Marrow, Buttercup, Cozella, Cozini, Long Cocozelle, Seneca Prolific, and Zucchini.

**Squash bug**
- *Resistant:* Butternut, Table Queen, Royal Acorn, Sweet Cheese, Early Golden Bush Scallop, Early Summer Crookneck, Early Prolific Straightneck, and Improved Green Hubbard.
- *Susceptible:* Striped Green Cushaw, Pink Banana, and Black Zucchini.

**Squash vine borer**
- *Resistant:* Butternut and Butternut 23.

**Striped cucumber beetle**
- *Resistant:* Early Prolific Straightneck, U Conn, Long Cocozelle, White Bush Scallop, Benning's Green Tint Scallop, Early Yellow Summer Crookneck, Cozella Hybrid, Marine Black Zucchini, Butternut 23, Short Cocozelle, Summer Crookneck, Zucchini, Royal Acorn, and Early Golden Bush Scallop.
- *Susceptible:* Black Zucchini, Cozini, Caserta, and Black Beauty.

## SWEET POTATO

**Southern potato wireworm**
- *Resistant:* Nugget and All Gold.
- *Moderately resistant:* Centennial, All Gold, Georgia Red, Porto Rico, and Gem.
- *Susceptible:* Nugget, Red Jewel, Georgia 41, Nemagold, and Jullian.

## TOMATO

**Twospotted mite**
- *Resistant:* Campbell 135.
- *Moderately resistant:* Campbell 146.
- *Susceptible:* Homestead 24.

## TURNIP

**Diamondback moth**
- *Resistant:* Seven Top and Purple Top White Globe.

**Harlequin bug**
- *Susceptible:* Amber Globe, Purple Top White Globe, and White Egg.

**Mexican bean beetle**
- *Susceptible:* Amber Globe and Purple Top White Globe.

**Striped flea beetle**
- *Moderately resistant:* Seven Top.
- *Susceptible:* Purple Top White Globe and Amber Globe.

### WATERMELON
**Spotted cucumber beetle**
- *Resistant (foliage):* Crimson Sweet and Sweet Princess.

- *Susceptible (seedlings):* Blue Ribbon, Charleston Gray, Crimson Sweet, Sugar Baby, and Sweet Princess.
- *Susceptible (foliage):* Charleston Gray, Blue Ribbon, and Sugar Baby.

### WHEAT
**Hessian fly**
- *Resistant:* Ottawa, Ponca, Pawnee, Big Club 43, Dual, and Rus.

## DISEASE-RESISTANT VARIETIES

| Crop | Disease | Variety |
| --- | --- | --- |
| APPLE | Alternaria cork rot | Delicious, Rome Beauty, Winesap, and Stayman Winesap. |
| | Apple blotch | Grimes Golden, Jonathan, Stayman Winesap, and Winesap. |
| | Bitter rot | Delicious, Rome Beauty, Stayman Winesap, Winesap, York Imperial, and Yellow Transparent. |
| | Black pox | Transparent, York Imperial, and Gano. |
| | Cedar-apple rust | Baldwin, Delicious, Rhode Island, Northwestern Greening, Franklin, Melrose, Red Astrachan, Stayman, Transparent, Golden Delicious, Winesap, Grimes Golden, and Duchess. |
| | Collar rot | Delicious, Winesap, and Wealthy. Moderately resistant varieties include Jonathan, Golden Delicious, McIntosh, and Rome Beauty. A susceptible variety can be grafted onto a resistant one, with good results: Grimes Golden on a Delicious trunk is standard practice. Also, Malling IX and VII rootstocks are resistant. |

*(continued on next page)*

| Crop | Disease | Variety |
|------|---------|---------|
| APPLE—Continued | Fire blight | Baldwin, Ben Davis, Delicious, Duchess, McIntosh, Northern Spy, Prima, Stayman, and Winter Banana. |
| | Mildew | Prima. |
| | Quince rust | Jonathan, Rome, Ben Davis, and Wealthy. |
| | Scab | Prima, Baldwin, Jonathan, Gravenstein, Dolgo crab, and Wealthy. |
| | Scar skin | Golden Delicious (tolerant). |
| ASPARAGUS | Asparagus rust | Supplanting the older Mary and Martha Washington are Waltham Washington, Seneca Washington, and California 500. |
| BEAN | Bacterial blight and wilt | Tendergreen (some types). |
| | Common mosaic | Roma II, Golden Wax Improved, Robust, Great Northern, U.S. No. 5, Refugee, Idaho Refugee, and Wisconsin Refugee. |
| | Powdery mildew | Contender. |
| | Rust | Kentucky Wonder, Tendergreen (some types), Harvester, and Cherokee Wax (yellow). |
| DRY BEAN | Bean halo blight | Many, including Pinto, Great Northern, Red Mexican, and Michelite. |
| LIMA BEAN | Downy mildew | Thaxter. |
| SOYBEAN | Bacterial blight | Flambeau and Hawkeye. |
| BEET | Boron deficiency | Detroit Dark Red. |
| BLACKBERRY | Orange rust | Eldorado, Orange Evergreen, Russell, Snyder, and Ebony King. |
| | Yellow rust, cane rust | Nanticoke, Austin Thornless, Boysen Brainerd, Burbank Thornless, and Jersey Black, as well as most European varieties. |
| BLUEBERRY | Blueberry canker | Weymouth, June, and Rancocas. |
| | Mildew | Stanley, Rancocas, Harding, and Katherine. |

| *Crop* | *Disease* | *Variety* |
|---|---|---|
| CABBAGE | Yellows and fusarium wilt | Many, including Jersey Queen, Marion Market, Wisconsin Golden Acre, Resistant Detroit, Harvester Queen, Hercules, Hybrid Blueboy, Stonehead, Charleston Wakefield, Globe, Wisconsin All Season, Wisconsin Hollander, some strains of Jersey Wakefield, Market Topper, Market Prize, Greenback, King Cole, Resistant Danish, Vanguard II, Savoy King, Red Danish, Red Ball, Red Head, and Ruby Perfection. |
| CANTALOUPE | Downy mildew | Texas Resistant No. 1 and Georgia 47 (also resistant to aphids), Dixie Jumbo, Edisto, and Planters Jumbo. |
| | Powdery mildew | Dixie Jumbo, Edisto, Magnum 45, Planters Jumbo, Samson Hybrid, and Sweet 'N' Early. |
| CAULIFLOWER | Yellows | Early Snowball. |
| CELERY | Fusarium wilt, yellows | Grow green petiole varieties or somewhat resistant Michigan Golden, Cornell 19, Tall Golden Plume, Golden Pascal, and Emerson Pascal. |
| SWEET CORN | Bacterial wilt | Many, including Golden Cross Bantam, Golden Beauty, F-M Cross, Carmelcross, Ioana, Marcross, Seneca Chief, N.K. 199, Iochief, and two white varieties, Silver Queen and Country Gent. |
| | Helminthosporium smut | Gold Cup and Silver Queen (white). Golden Cross Bantan and Country Gent (white). |
| CUCUMBER | Downy mildew | Burpee Hybrid and M&M Hybrid, Saticoy, Salty, Poinsett (also resistant to anthracnose and leaf spot), Gemini, Marketmore 76, Streamliner Hybrid, Pioneer, and Suyo Long. |
| | Mosaic | Pacer, Marketmore 70, Tablegreen 65, Challenger, Victory, Early Set, High Mark II, Slicemaster, Spartan Dawn, SMR, Gemini, and Salty. |
| | Scab | Maine No. 2, Pacer, Marketmore 70, Wisconsin SR 10 and SR 6, Highmoor, Victory, Slicemaster, Salty, Northern Pickling, and Spacemaster. |
| RED CURRANT | White pine blister beetle | Viking and Red Dutch. |

*(continued on next page)*

| Crop | Disease | Variety |
|------|---------|---------|
| DEWBERRY | Orange rust | Leucretia. |
| EGGPLANT | Phomopsis blight | Florida Market and Florida Beauty. |
| GRAPE | Anthracnose | Concord, Delaware, Moore Early, and Niagara. |
|  | Black rot | Beta, Campbell Early, Clinton, Delaware, Elvira, Lutie, Missouri Riesling, Moore, Early, Norton, Portland, Sheridan, and Worden. |
| DUTCH IRIS | Iris rust | Early Blue, Gold and Silver, Golden West, Imperator, Lemon Queen, and Texas Gold. |
| LETTUCE | Downy mildew | There are many strains of this disease, so consult your extension service. |
|  | "Multiple resistance" | Grand Rapids and Salad Bowl. |
|  | Tipburn | Slobolt, Summer Bibb, and Ruby. |
| MIMOSA | Mimosa wilt | Charlotte and Tryon. |
| MUSKMELON | Alternaria blight | Harper Hybrid. |
|  | Fusarium wilt | Gold Star, Harvest Queen, Harper Hybrid, Saticoy Hybrid, Samson, Chaca, and Supermarket. |
|  | Mosaic | Harper Hybrid. |
|  | Powdery mildew | Samson, Supermarket, and Chaca. |
| MUSTARD | Turnip anthracnose | Southern Giant Curled. |
| ONION | Pink root | Nebuka (Welsh onion), Beltsville Bunching, and Brown Beauty (in Arizona, Granex bears best). |
|  | Smudge | Early Yellow Glove, Downings Yellow Globe, and Southport Red Globe. |
|  | Smut | Evergreen (bunching). |
| PARSLEY | Septoria blight | Paramount. |
| PARSNIP | Root canker | Model and Tender and True. |

| Crop | Disease | Variety |
|------|---------|---------|
| PEA | Fusarium wilt | Wisconsin Early Sweet, Little Marvel, Thomas Laxton (all early varieties), and Frosty, Pride, Early Perfection, Sparkle, Wando, and Green Arrow (all medium to late varieties). |
| PEAR | Fire blight | Old Home, Orient, and Kieffer. |
| SWEET POTATO | Black rot | Allgold. |
| | Internal cork | Allgold, Centennial, and Nemagold. |
| | Root knot | Nemagold. |
| | Soft rot | Allgold. |
| | White rust | Goldrush. |
| | Wilt | Allgold, Centennial, Goldrush, and Nemagold. |
| TOMATO | Blossom drop | Summerset, Hotset, Summer Prolific, and Porter. |
| | Curly top | Owyhee and Payette. |
| | Early blight | Varieties that do not bear heavily are somewhat more resistant: Manalucie, Southland, Floradel, and Manahill. |
| | Fusarium wilt | Many; consult catalogs. |
| | Gray spot | Manahill, Manalucie, and a number of Hawaiian varieties. |
| | Root knot | Better Boy. |
| | Spotted wilt | Pearl Harbor. |
| | Verticillium wilt | Many; consult catalogs. |
| TURNIP | Club root | Immuna, Bruce, May, and Dale's Hybrid. |
| WATERMELON | Fusarium wilt | Charleston Gray, Crimson Sweet, Sweet Princess, Improved Kleckly Sweet, and Klondike. |
| | Melon anthracnose | Charleston Gray, Congo, Fairfax, Black Kleckly, Crimson Sweet, Sweet Princess, and two seedless varieties, Tri-X313 and Triple Sweet. |

*(continued on next page)*

# RHODODENDRON

## Insects

The rhododendron **lace bug** [color] is one of the most common pests attacking this plant. When adults and nymphs suck sap from the undersides of leaves, the top surfaces take on a mottled appearance much as if they had been sprinkled with white pepper. These pale areas show where chlorophyll has been destroyed as the result of feeding. The undersides are spotted

Rhododendron lace bug

with brown excrement that looks like dots of varnish. Injury is prominent when plants are grown in full sun, and the shrubs take on a yellow cast. Adult lace bugs are ⅛-inch-long flies, flattened and dark brown or black. Their lacy wings are folded over the back to lend a squarish appearance to the insect. The nymphs are dark, spiny, and move with a strange sideways motion.

If you see the characteristic mottling, draw the affected leaves between thumb and forefinger to squeeze the bugs to death.

The **pitted ambrosia beetle** is a stout black beetle, about ⅛ inch long. The small whitish larvae make horizontal galleries in the wood at the base stems, causing plants to wilt or break off at ground level. Rhododendrons growing in the shade and those that are well mulched are most susceptible to attack.

Cut out and burn affected stems. Pull the mulch away from plant bases and remove any excess mulch. It helps to select a sunny planting site at the beginning of the season.

As larvae, yellowish white **rhododendron borers** bore in stems and larger branches of rhododendron plants, causing them to wilt or break off. As they get larger, these larvae bore

Rhododendron borer

into the woody part of plants, pushing out fine sawdust as they go. Rhododendron borers overwinter as partly grown larvae in their stem burrows. Adults emerge in June, and eggs are laid on leaves, new twigs, or the rough bark of the main stem. The adult moths are small and black with three yellow stripes on the abdomen.

Prune and burn affected stems. Coat wounds with tree paint or paraffin. In June, crush any eggs that you can find. For other controls, see the suggestions for peachtree borers under PEACH.

The larva of the **rhododendron tip midge** is a whitish maggot that prevents the normal growth of plants by rolling young leaves. Remove damaged leaves as soon as possible.

Rhododendron tip midge and damage

# Diseases

Many rhododendron troubles that appear to be diseases are caused by environmental factors. So, before pruning or roguing, wait a while to test your conclusions—your plants might very well grow out of whatever ails them. Brown spots around the edges of leaves can be the result of summer drought. Scorching during winter months causes regularly spaced yellow or brown spots on either side of the midvein. Glazed ice on rolled leaves may act as a focusing agent for the sun's rays if the ice has not fallen by the time the sun comes out. Unlike the progressive effects of nutritional troubles, the spots caused by ice injury remain static, with sharply defined edges.

Azalea **bud-and-twig blight** is first known by the dwarfing of flower buds, followed by their browning and death in late summer or early fall. Fruiting bodies on buds resemble tiny pins stuck into the plant. Pick and burn infected plant parts in either late fall or early spring. Remove seed pods as soon as flowers have withered.

**Chlorosis** (or yellowing of leaves) can progress to the point that leaves become very pale and turn brown at the tips; veins remain green. This condition is usually caused by a deficiency of iron, which results from a number of factors: destroying feeding roots by tillage or cultivation; injuring roots by allowing soil to dry out; sandy soil with insufficient organic matter; an overabundance of hard coal ash in the soil; poor soil drainage; and alkaline soil, caused by too much lime or by setting plants near cement walls.

The permanent solution to chlorosis lies in improving the condition of the soil. Chlorosis is also tied with winter injury. When more water is lost from the leaf surface than the roots can absorb, a yellowing of the interveinal areas may result, leaving the tissue immediately adjacent to the veins a normal green. Root injuries (possibly caused by mice feeding underground) may also be responsible for chlorosis. Other possible causes include soil that is not acid enough, excess water from a building's downspout, or too much fertilizer, lime, or phosphate.

**Leaf curl** causes shoot leaves to become swollen and curled, and forms an irregular gall that is first pinkish and later becomes white when spore formation occurs. These swellings are known as pinkster apples, swamp apples, or honeysuckle apples. The disease may be reduced by handpicking and destroying all galls.

**Leaf spots** may be on the tips, along the margins, or in the middle of the leaf blade. They are usually silvery gray above and brown beneath, but sometimes they are brown above and show concentric rings caused by the advancing growth of the fungus. Spots caused by disease are easily distinguished from those caused by unfavorable environmental conditions—look for the tiny black fruiting pustules of the fungus embedded in the upper surface of the discolored tissues. They are easily seen with a magnifying glass. The environmental spots are caused by winter injury (see below) and ice that focuses sunlight, burning the leaves.

In small plantings, pick off and burn badly spotted leaf areas. Leaf spot is a secondary effect of the bush being in poor condition from some other cause.

**Rhododendron dieback** is a fungal disease that causes terminal buds or leaves to brown and hang limp in winter. Twigs are cankered, and leaves and flower clusters of shaded plants become brown and wilted. High humidity favors this disease.

Select sunny planting sites with good air circulation. Rhododendrons should not be planted near lilacs, which share the disease. Cut off browned or cankered branches well below the infected area. Remove severely affected plants from the garden.

Death of terminal twigs may result from a disease known as **twig canker.** If no borers are found in the wilting or dead twig, scrape the bark with a sharp knife to see if a small reddish brown canker lies just beneath the bark. To eradicate canker disease, remove all affected tips, making sure that the cut is about two inches below the canker area.

If entire branches die and no insects can be found, the bark should be scraped from the branch at the soil line and the inner wood inspected for reddish streaks. These streaks indicate **wilt.** Wilt is caused when the fungus enters the young roots and works its way up to the crown. At first, single branches yellow and wilt; later, the entire plant may become affected. The disease is most destructive in cool locations in the seedbed or nursery, and where the soil is not acid enough for good growth. The pathogens can enter the injured roots of newly transplanted seedlings or through growth cracks that often occur during the second year of the seedlings' growth.

Do not plant rhododendrons in soil known to have harbored this disease. In the nursery, rotate frames. Use light soils in drainage beds and avoid overwatering. Adjust soil acidity to a pH of 4 or 4.5.

**Winter injury** can effect browning on the margins and tips of leaves, caused when the loss of water from the leaf surface is greater than the ability of the roots to absorb water (also see chlorosis, above). Rhododendron leaves normally remain green in winter, and curl and **hang** down in very cold weather (see also ice injury, above). March winds may aggravate injury to plants in an exposed location. Discourage late growth and protect roots with a good mulch to minimize damage; if possible, drape cloth over plant tops when necessary.

Cold winter weather can kill flower buds also, especially of those rhododendrons that are not adapted to the local climate. Such plants are particularly susceptible because of late fall development which occurs in a prolonged growing season (sometimes created by a dry summer period followed by a rain). If flower bud scales become separated and even a flower or two appear, water can enter these buds during the winter to freeze and kill them.

# RHUBARB

## Insects

The **rhubarb curculio** has a reputation as the most troublesome of rhubarb pests. This is a large (¾-inch-long) snout beetle that looks as though dusted with yellow powder. If the powder is rubbed off, the insect can be seen to be grayish brown beneath. It bores into stalks, crowns, and roots and punctures the stems. Curculios are distributed from New England to Idaho, and south to Florida and Louisiana.

The beetle is easily seen and handpicking is an efficient control. It deposits its eggs in certain species of dock plants, particularly curly dock, and all wild dock should be dug out and destroyed in July while the eggs lay unhatched on the leaves.

Other pests include the European corn borer, which sometimes tunnels in the stalks (see CORN), stalk borer (TOMATO), and yellow woollybear caterpillar (see imported cabbageworm under CABBAGE).

## Diseases

**Anthracnose** of rhubarb (known also as stalk rot) is characterized by watery soft areas on stalks. Eventually, leaves wilt and die. Once this malady has struck, there is little you can do to control it, but you can get a good start on next year's crop by cleaning up rhubarb debris after harvest.

**Crown rot** causes rotting of the crown and main roots of rhubarb plants by water mold fungi. Plants usually have yellowed leaves that subsequently collapse and leave the plant wilted. Infected crowns are soft and water-soaked. Apparently healthy petioles collected from diseased plants frequently decay after packing. Dig up and burn affected plants.

**Leaf spot** is seen as small, circular, brown spots scattered over leaf surfaces; a fungus is responsible. While common in some areas, the disease doesn't do enough harm to require control.

**Verticillium wilt** is caused by a soil-borne fungus and is capable of infecting many other kinds of plants. Symptoms include wilting and pronounced yellowing, and marginal and inter-veinal dying of basal leaves. Early in the season, yellowing frequently occurs without wilting and is often confused with nutritional disorders. Heavy losses have occurred in some fields. The disease is usually widely distributed in coastal areas devoted to rhubarb production, particularly where other susceptible host plants, such as strawberries, tomatoes, and peppers, have been grown.

Plant rhubarb on land that has not been used for growing other susceptible crops. Do not use planting stock from fields showing symptoms of the disease.

# ROSE

For gardeners who take pride in growing first-quality roses without resorting to poisons, garden sanitation is an important precaution. Remove garden trash and mulches from rose beds each spring, and let the sun dry the soil for at least a month. This drying process kills many spores and fungi that overwinter in the mulch or soil. It is good practice to assume that leaves that wilt or fall during the growing season are affected by a disease or insect pest; remove these leaves to the compost pile as soon as possible. When selecting plants, look for varieties with inbred resistance to the more important diseases.

Roses are the focus of more poisonous sprays and dusts than any other flower, but garden sanitation, resistant varieties, and the following suggestions—along with a good dose of common sense—will get your plants safely through the season.

## Insects

You are likely to find tiny **aphids** [color], the green rose aphid and green and pink potato aphids being most common, clustered about the tips of young growing shoots. Roses are as popular with aphids as they are with humans, and you may see other species as well, in various hues and stages of hairiness.

The easiest control is to wipe the pests off by hand. Aphids prefer new growth, so you'll know where to look for them. This won't kill all of them, and because they reproduce prolifically, you can expect them to return in numbers before long. Other gardeners get along with just a forcible spray from the garden hose (this works against mildew, as well). But even these simple controls may be unnecessary if a good balance of insect life is at work in the garden. The larvae of ladybugs and lacewings will eat great numbers of aphids if given a chance. So before you influence the insect world with your thumb or a flood, let nature take its course for a time. Interplant chives with roses to keep aphids at a distance. As the chives develop, the clumps can be separated and spread to new spots.

Several species of **bee** damage roses occasionally. The female leafcutter bee cuts neat circles from the margins of leaves. She then stores the bits of leaf as food for her young, either in the excavated stems of dahlia or in tunnels in wood. The bees also stash away aphids and other insects, and thus rate as beneficials.

Carpenter bees (known also as leafcutter bees) deserve less compassion, however, as they bore out the pith of rose canes to cause wilting. Control by pruning canes below the infested section in spring, or whenever wilted canes are noticed. To keep the bees from entering cut canes, insert a flat-headed tack in the end or plug the hole with grafting wax, putty, or paraffin; you might also paint the end with shellac or tree-wound paint.

The **buffalo treehopper** [color] may sound like an improbable bit of Wild West folklore, but is actually a tiny (¼-inch-long) insect distinguished by a pair of short horns on its shoulders. Viewed from above, the treehopper appears triangular and is pointed at the rear. When females lay their eggs they cut curved slits in the bark of rose twigs, and these wounds make entrance for rose canker and other diseases.

To control this pest, remove weeds and grass from the immediate area, if possible. Alfalfa and sweet clover are especially favorable to treehoppers. Planting roses near other susceptible hosts may aggravate trouble; these include fruit trees and elm, cottonwood, and locust. The eggs within the twigs can be killed with a light dormant-oil spray applied before spring.

If you find holes in canes, especially while pruning in spring or fall, suspect the **carpenter ant.** Discard as much of the cane as has been hollowed out; tunnels sometimes go all the way down to the roots.

**Flower thrips** attack rosebuds and cause discoloration of the petals. Their feeding usually results in deformity or failure of the flowers to mature. Adult thrips are extremely active, slender, and brownish yellow in color, with feathery wings. The young are lemon yellow.

Flower thrips

Thrips breed in grass and weeds neighboring the garden. Because a new generation can come along every two weeks, especially if the weather is hot and dry, populations tend to build up quickly.

Light-colored and many-petaled flowers seem to be the most attractive to thrips. It is sometimes practicable to remove the outer petals as the rose is just opening and to crush the offending thrips by hand. The rose should then open normally. Cut off and burn seriously affected flowers. See also THRIPS.

The gray brown **fuller rose beetle** eats ragged areas from the margins of many outdoor plants, including azalea, begonia, camellia, canna, carnation, chrysanthemum, gardenia, geranium, and primrose. This nocturnal pest has a cream-colored strip on each side, and is about ⅓ inch long. The yellowish, brown-headed larvae work underground, feeding on roots and girdling stems to cause the yellowing or death of plants. This rose beetle is distributed in the South and in California, and is considered a serious greenhouse pest in the North, where they are most abundant in December.

The beetle hides on twigs or foliage during the day, and can be handpicked. You might try placing a stick barrier on stalks to keep this wingless pest from crawling to the flowers.

The hairy, lemon yellow **goldsmith beetle** is a large (one-inch-long) cousin of the June bug. The larvae resemble grubs and attack the roots of rose, chrysanthemum, canna, and other ornamentals. They are common in the eastern states and the Southwest. For control of the larvae, see the discussion of milky spore disease under JAPANESE BEETLE.

**Japanese beetles** need little introduction. Roses are just one of the hundreds of plants that these pests dine on. Many gardeners patrol the garden in early morning or evening to handpick

Japanese beetle

them; at these times of day, the beetles are less inclined to fly and can be dropped into a can of water topped with kerosene. See JAPANESE BEETLE for a number of controls that will handle this and most any other beetle found on roses.

**Leafrollers** are caterpillars that gorge themselves on foliage and then pupate within the protection of rolled leaves. There are several species of occasional importance. Now that DDT is no longer around to kill the parasites of these worms, rose growers should have a lot less trouble with leafrollers. If the sensation isn't too unpleasant, pinch the worms while they are in their rolled hideouts.

Twospotted **red spider mites** appear as tiny dots on the undersides of foliage. Leaves become discolored in a stippled pattern, and you may see tiny webs strung across the leaf surfaces or

Twospotted spider mite

on new growth. As the injury progresses, leaves brown, curl, and drop off. Mites overwinter on perennials and weeds, and become abundant in hot, dry weather.

Frequent hosings will wash mites off foliage. Clean up trash and weeds from the garden before the growing season. See SPIDER MITE.

A frequent and often unrecognized cause of poor growth is the **root-knot nematode.** Stunting, yellowing, and premature death may be attributable to other causes as well, so it is necessary to examine the roots for the characteristic swellings—knots of between ¼ and ⅛ inch in diameter, and ½ inch or more in length. In some infestations, the roots decay and fibrous roots are scarce. Roots often look like a string of beads. A magnifying glass will enable you to see the tiny worms within a bisected knot.

Examine the roots of barefooted roses before planting, and discard any having root knots. If roses are growing in pots, infestation can usually be detected by examining the roots on the surface of the ball of soil after the container is removed. Once serious damage has occurred, remove the plant and replace the soil before planting other susceptible plants. You can extend the useful life of slightly injured plants by enriching the surrounding soil with organic matter. Resistant varieties are available. See NEMATODE for other controls.

The **rose chafer** (known also as the rose bug or rose beetle) is a tannish, long-legged beetle that measures ¼ inch in length. It skeletonizes

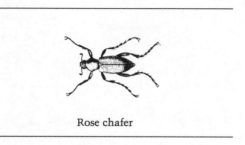

Rose chafer

foliage, and damages flowers (especially white ones) by feeding on the petals and soiling them with excrement. The creamy white, brown-headed larvae feed on the roots of grasses and weeds. Rose chafers are most common in the northeastern states, are troublesome as far west as Colorado, and breed most abundantly in sandy waste and lawns. Adults appear suddenly, damage plants for from four to six weeks, and then suddenly disappear.

Handpicking is the first line of defense against the chafer. An effective barrier can be made of cheesecloth or mosquito netting, either as a canopy or fence. The beetles can fly, but will seldom make it over a barrier. Suburban gardeners are rarely troubled much by this insect, but growers whose roses are near fallow fields shouldn't be surprised to see chafers

stalking about. Cultivated crops make better neighbors for roses. Rose chafers are poisonous to chickens, and may kill them. Grubs are vulnerable to milky spore disease; see JAPANESE BEETLE.

The **rose curculio** is a peculiar-looking beetle—it's colored red and carries a long, black snout. Adults eat holes in the buds of wild and garden roses, and the buds often fail to open as a result. White larvae feed on seeds and flowers. This pest is found throughout the United States, and is most common in northern, colder regions; North Dakota sees particularly severe infestations.

Handpicking works well enough in most cases. Collect and burn dried buds before larvae have a chance to complete their development and pupate in the soil. To control the larvae in the soil, see JAPANESE BEETLE.

**Rose galls** on stems are caused when any of a few species of tiny four-winged wasps lay their eggs. If a gall is cut open, larvae may be seen living in the plant cells. One variety, mossy rose gall, is fibrous mass of green to purple filaments on the stem. Another, rose root gall, is a conspicuous, rounded swelling at the base of the plant or belowground, reaching one or two inches in diameter. In either case, cutting open the growth will distinguish it from crown gall infection, which is caused by bacteria. Cut and burn infested galls as you find them.

The **rose leaf beetle** is a small, oval, metallic insect that appears late in spring to bore into buds and partially opened flowers. When abundant, the beetles also may eat shot holes in flowers at any stage of development. Larvae sometimes damage the roots of roses. The beetles can be handpicked or jarred from plants into a bucket of water topped with kerosene. Should an infestation get out of hand, you may have to fall back on a high-powered botanical such as pyrethrum.

A white, peppered effect on the top surfaces of leaves could be evidence that **rose leafhoppers** are feeding on the undersides. Both adults and nymphs suck sap, causing severely infested leaves to drop prematurely. These pests hop about nervously when disturbed. They are narrow in shape and colored yellow white. A second generation appears in fall, and is more damaging than the first.

Leafhoppers are too small and too energetic for the hands of more gardeners, so you had best resort to botanical sprays to repel them from valued plants.

The tiny yellow brown **rose midge** brings trouble when it lands on roses and deposits its yellow eggs on the tender growth near flower buds. These eggs hatch in just two days, and the larvae start feeding at the bases of flower beds or leaf stems, causing them to become distorted; the plant parts later turn brown and die. Twenty or 30 white maggots may be at work inside a single bud. Within a week, the larvae mature and build small white cocoons on the soil, and just one week later a new generation of midges flies out into the world.

In greenhouses, most injury occurs between May and November, and midges are seldom seen during the winter. They can be controlled by mixing tobacco dust with the greenhouse soil. Cut off and destroy all infested buds.

Snow white **rose scale** often thickly infests older canes. They feed on plant juices, resulting in weakened plant growth or even death. Remove and destroy infested canes. It helps to prune nearby berry canes of infested sections.

Rose scale

During the dormant season, spray with a white-oil emulsion.

Spiral mines that travel around canes are the work of **rose stem girdler** larvae. Stems swell up, split, and often die. Rugosa and hugonis roses are especially attractive to this pest. The metallic green, ¼-inch-long adult beetles appear in June and July, and place eggs under the bark of rose and raspberry stems. Clip off and burn the infested parts in winter or early spring.

Several types of **slug** feed on rose leaves. Both wild and cultivated roses are susceptible. Most damage occurs in the early spring on new foliage, but one species, the greenish white, bristly rose slug, continues to breed and eat all summer. See SNAIL AND SLUG.

Two species of **stem borer** larvae chew their way downward into the pith of stems until they reach the crowns, a journey that may take them one or two seasons. Branches are weakened and likely to die. Both species are yellowish white worms with brown heads, growing to ⅗ inch in length. While they spend most of this stage hidden from sight, keep an eye out for the adults—one, the rose stem sawfly, is a waspish insect with transparent wings, appearing in the garden in early summer; the other, the adult raspberry cane borer, is a slender beetle, striped black and yellow, and measuring about ½ inch.

Cut off rose canes a good six inches below the lowest point of injury, and either burn them or slit the stem and pluck out the borer (there is only one per cane). If possible, control borers on other hosts in the area, including blackberry, raspberry, and occasionally azalea.

One pest that shouldn't be handpicked, not without gloves at any rate, is the **stinging rose caterpillar,** a sluglike beast marked with red, white, and violet stripes. It measures up to ¾ inch long and bears seven pairs of large, spine-bearing tufts that can give a burning sensation for several hours. These caterpillars are usually found on the undersides of leaves. Since the overwintering cocoons are in plant refuse, a thorough fall cleanup will prove effective in preventing trouble.

Other pests of occasional importance are the celery leaftier (see CELERY), corn earworm (CHRYSANTHEMUM), EARWIG, fourlined plant bug (CHRYSANTHEMUM), harlequin bug (CABBAGE), potato leafhopper (POTATO), spotted cucumber beetle (CUCUMBER), SYMPHYLAN, and WHITEFLY.

## Diseases

Old rose varieties may not have the spectacular blossoms of modern hybrids, but they tend to be more resistant to disease. The four most common diseases damaging roses are blackspot, canker and other cane diseases, powdery mildew, and rust.

**Blackspot** may appear from the first flush of growth in spring to leaf drop in fall. Symptoms appear as coal black lesions on both the upper and lower surfaces of leaves. Blackspot can readily be distinguished from other leaf spots by the darker color and fringed or feathery margin surrounding the spot. Heavily diseased leaves tend to turn yellow and drop prematurely. When

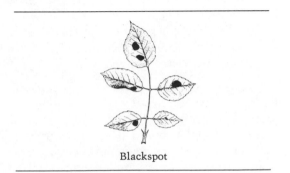

Blackspot

excessive premature defoliation occurs, the plant forms a new set of leaves, which causes a considerable drain on the food reserves in the roots. This results in a weakened plant with poorly matured wood. Such plants are especially subject to winter injury.

A few days after the spots first appear, little black pimples show up in the spots; this signals that the spores are about to be discharged, and that you had better act fast. Spores are carried by air currents, as well as on tools, insects, and the gardener's hands and clothing. Those that land on rose leaves will germinate and produce an infection, providing the environmental conditions are favorable. Free moisture on the leaves from rainfall, dew, or excessive sprinkling will favor the development of blackspot. With this moisture and favorable temperatures, the spots will appear in a week or ten days and then produce new spores to continue the disease cycle. Blackspot is indirectly responsible for a pale flower color; this is because diseased leaves and defoliated plants manufacture less of the sugars that help to intensify flower color.

All kinds of roses are susceptible to blackspot in some degree. The disease does its worst when highly susceptible plants are grown together, and it is best to buffer these varieties with roses that enjoy a tolerance to blackspot. Keep in mind that a tolerant variety isn't safe from all of the many strains of blackspot fungus, and therefore will do better in some areas than in others.

As virulent and persistent as blackspot may be, there are several preventive measures that, when used together, should get your roses through the season. Begin in fall by gathering and disposing of fallen leaves. In spring, after the danger of frost is past, remove the mulch and rake the ground thoroughly. Then, let the garden plot lie exposed to the sun's rays. When the first shoots appear, apply new mulch in generous amounts. Infected leaves and twigs should be removed from plants because they likely harbor overwintering fungus, but do so only if they are completely dry. Prune sickly canes freely. If worse comes to worse, you may have to rely on a dusting of finely ground sulfur to save your plants. The sulfur tends to burn leaves if applied in very hot weather.

Mulch keeps water drops and rain drops from splashing on leaves, and dry leaves won't support the fungus. Some gardeners push a hose nozzle right into the mulch to make sure the leaves stay dry. Never sprinkle roses from overhead, especially in the evening when foliage cannot dry before nightfall.

Although you're not likely to come across **brand canker,** is it a serious disease and deserves mention here. It is first known by dark red cankers with red or purplish margins that appear on stems. After a time the centers of the spots become light brown, and navy green spores may be seen. Cankers can girdle stems and kill them.

The disease develops in winter, when it is sheltered from the elements by the heavy protective covering that many growers put around the base of their plants. It is better to use a light winter mulch and to avoid piling dirt up around the canes. Or if canker has been a problem and winters are not too severe, omit the mulch entirely.

**Common canker** (stem canker) occurs in wounds on canes and in the cut ends of pruned canes, especially if the cut is not made close to a bud. Inspect canes in spring, and cut back damaged tissue. Prompt pruning is the best control. The danger of a canker infection is reduced by proper pruning—make clean cuts just above a bud. Such cuts heal quickly, while ragged cuts and those made too far from a bud heal slowly and are thus prime targets for infection. Disinfect shears and other cutting tools after using them on cankered plants.

Irregular or spherical galls on roots, crowns, or canes are symptoms of **crown gall.** While the organism causing the disease does not kill the tissue, it does stimulate abnormal growth.

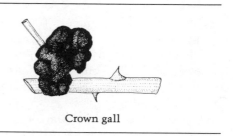

Crown gall

Control is largely a matter of prevention. Buy disease-free plants, and plant in uninfested soil. The causal bacteria can be kept out by avoiding wounds. Remove and destroy affected plant parts as soon as possible.

**Downy mildew** appears as irregular brown spots on the undersides of leaves of greenhouse roses. Plants grown outdoors are rarely affected. Keep the humidity in the greenhouse below 85 percent, and keep daytime temperatures high. Many of the newer rose varieties are resistant to mildew; check with the nursery or consult recent catalogs.

Three **physiological diseases** are worthy of note. These are not caused by disease pathogens, but result from problems with the plant's environment.

*Chlorosis* can be induced by an excess of lime in the soil; leaf blades of affected plants become uniformly yellow. Chlorosis may also be induced by a deficiency of iron (upper leaves are yellow with green veins), nitrogen (lower leaves are paled), or potassium (leaves turn gray and stems are weakened).

*Leaf scorch* can be traced to deficiencies of potash or, in the greenhouse, of boron or calcium.

Another greenhouse disease, *mercury toxicity*, occurs when paint containing mercury is used on sashes. The solutions for these physiological diseases lie in identifying them and eliminating the cause.

**Powdery mildew** [color] ranks as one of the most serious of rose diseases. It causes a white or grayish powdery coating on the surfaces of leaves, stems, buds, fruit, or other plant parts. Leaves and young shoots are most often affected. This visible coating consists of an intertwined network of threads and chains of egg-shaped summer spores that are easily detached and carried away by the wind. Such spores are produced throughout the growing season and new generations can occur every four days, providing that conditions are just right. In fall, black dormant resting bodies may be seen in the felty growth. These structures, along with infections in buds, serve to carry the fungus through the winter.

Unlike most fungi, spores of powdery mildew do not germinate readily in wet conditions. Frequent syringing and hosing therefore reduce the amount of mildew, but may encourage the spread of blackspot or rust at the same time. Powdery mildew is also somewhat special in that it feeds on the surface cells of plants, not inside them, and thus is one of the few rose diseases that can be cured. The plants are affected very gradually, and the powdery coating is therefore the first visible symptom. Other symptoms include: stunting and distortion of leaves, buds, growing tips, and fruit; distortion and stunting of leaves and buds; death of invaded surface tissues; a general decline in plant growth; yellowing of leaves; and premature leaf fall.

Rose varieties vary greatly in their susceptibility to powdery mildew. Hybrid teas, many red roses, and floribundas are usually considered susceptible, while shiny-leaved climbers, Welch multiflora, and Rugosa types get by without much damage. Resistant varieties among the hybrid teas include Aztec, Fragrant Cloud, Queen Charlotte, and Santa Fe; of the Floribunda, Fire King, Acapulco, and Spartan are resistant; and of the Grandiflora, Camelot, Montezuma, Mount Shasta, Pink Parfait, and Queen Elizabeth.

A common viral disease on the West Coast is **rose mosaic.** The symptoms include chlorotic spots that are most numerous at or near the midribs of leaflets. Ring, oak-leaf, and watermark patterns are also seen in some cases. Plants are often dwarfed, and buds are bleached. Although this disease is primarily of concern to West Coast growers, greenhouse roses and those shipped from the West are also vulnerable. Heat-treated rootstocks are free of disease. Any plant found to be affected should promptly be discarded.

Another viral disease, **rose streak**, affects roses grown in the East. Symptoms include brown rings on leaves, brown and yellowish vein banding, and necrotic lesions on canes above the buds. Streak is usually transmitted by grafting.

**Rust** [color] is an important fungal disease that produces bright orange pustules on the undersides of leaves and pale yellow spots on the tops. Pustules eventually become brick red, and wilting and defoliation follow. Pick off and destroy affected leaves as soon as they are noticed. Because leaves must be wet for four hours in order to become infected, careful watering and a loose mulch will go a long way toward preventing trouble. Rake up all leaves from the ground in fall or early spring. Periodic dustings of sulfur have been found effective, but you should not fall back on this unless preventive measures have failed.

# ROSE OF SHARON

See HIBISCUS.

# ROTENONE

This plant-derived insecticide has proven to be harmless to warm-blooded animals, although it will kill beneficial insects and fish. It occurs in several tropical plants, including derris (another name for rotenone), cube barbasco, and timbo. Rotenone is sold under brand names; when shopping for it, you should read labels to see if the product has been adulterated with synthetic toxins. Triple-Plus combines rotenone, ryania, and pyrethrum. Pure rotenone can be safely used on all crops and ornamentals.

It kills many types of insects, including certain external parasites of animals. However, it has little residual effect and the period of protection it offers is short—just three to seven days. Filter wettable powders through cheesecloth before adding to a sprayer.

A native weed, Devil's shoestring *(Tephrasis virginiana)*, contains rotenone. It is a perennial with yellowish white flowers, growing to about two feet, and is found in the eastern and southern states. The plant was once used by American Indians as a fish poison. The roots contain up to 5 percent rotenone and you might, with the aid of an illustrated field guide to wild plants, find some roots to grind into your own powder.

# RUTABAGA

See BROCCOLI, CABBAGE, and TURNIP.

# RYANIA

The ryania shrub, native to South America, is the source of a mildly alkaline insecticide that is considered quite safe to warm-blooded animals, including humans. Apparently, ryania works not by administering death to insects but by making them good and sick. It incapacitates the oriental fruit moth, European corn borer, corn earworm, cranberry fruitworm, codling moth, cotton bollworm, and imported cabbageworm. It is not persistent, meaning that it can be used close to harvest time.

As good as it sounds, ryania has not been in sufficient demand to warrant its being kept in stock by the sole source in the United States. The insecticide is available in Triple-Plus, a formulation including rotenone and pyrethrum and applied as either a dust or wettable powder. Another product, R-50, contains 50 percent ryania along with surfactants and clay.

# SABADILLA

The seeds of this South and Central American plant are ground into a powerful insecticidal dust. It is effective against a good number of pests, including grasshopper, European corn borer, codling moth larva, armyworm, webworm, silkworm, aphid, cabbage looper, imported cabbageworm, melonworm, squash bug, blister beetle, greenhouse leaftier, chinch bug, lygus bug, harlequin bug, and many household pests. The insecticidal effect is diminished soon after application. Sabadilla dust and seed can irritate mucous membranes and bring on sneezing fits. Honeybees are vulnerable to sabadilla.

# SCALE

Although inconspicuous individually, infestations of scale [color, oystershell scale] may encrust whole branches of trees. In great numbers, they cause host plants to become stunted and chlorotic by sucking sap. The honeydew they produce frequently supports the growth of dark sooty molds that can interfere with photosynthesis. Ants are attracted to honeydew,

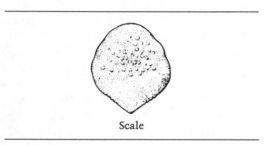

Scale

and carry scales from plant to plant as they do with aphids.

Scales are curiously simple organisms in the last stage of their life. After a brief mobile period, the female settles down permanently for a life of egg laying. Her legs drop off, almost all of her internal organs disappear, and she produces eggs—typically over a thousand—until there is little left of her.

The first nymphal stage is a pale, flat, six-legged creature with prominent eyespots. After a molt, the males become elongate and the females circular. In the third stage the female builds a curved shell within which the eggs are laid. Adult male scales have a single, puny set of wings, capable of propelling them only a

few feet at a time. They never eat, having no digestive system. They live to copulate, and usually die on top of the female after doing so. Fertilized females produce both males and females, while those that haven't been fertilized give birth to females alone.

Scales may be efficiently controlled by smothering them with a superior or supreme dormant-oil spray. Timing depends on the crop, but sprays should generally be applied before the buds break open. Scales can also be scraped from trees, and are vulnerable to insecticidal soap sprays and a spray of glue dissolved in water. See entries for each fruit for other control methods.

Scales have proven very good subjects for biological control. Two red scale parasites are commercially available, *Comperiella bifasciata* and *Aphytis melinus*. *C. bifasciata* was imported from the Far East early in the century and now offers economic control of yellow scale if there is not too much interference from ants. *Metaphycus helvolus* parasitizes black scale. In California, the chalcid *A. luteolus* keeps down soft scale. Growers can integrate the more traditional controls with this chalcid's help by spraying strips rather than a whole orchard at one shot; the beneficials can then retreat to unsprayed sections to continue their work. The damage of the purple scale to the citrus industry has been kept at very low levels by colonizations of the imported *A. lepidosaphes*. The vedalia ladybug helps control cottony cushion scale in the garden.

# SERVICEBERRY

## Insects

There are not too many insects that trouble the serviceberry. Among the more common ones are lesser peachtree borer (see PEACH), oyster-shell scale (LILAC), and shothole borer (peach-tree borer under PEACH).

## Diseases

**Rust** infects leaves, causing brownish orange spots of up to ¼ inch in diameter. In wet springs, light-colored spore horns are formed on the fruit and on leaf spots. Spores from these horns infect nearby red cedars, the alternate host of the rust fungus. The fact that the rust must infect the red cedar to complete its life cycle suggests that serviceberry not be planted within 500 yards of cedars or junipers.

**Witches' broom** (or black mildew) is a fungal disease that causes a stimulation of the growing point of a plant, giving rise to many new shoots with a broomlike appearance. There is usually a heavy coating of black mold on the undersides of leaves. The best control is to prune and burn affected leaves and twigs.

# SHREW

See MOLE.

# SNAIL AND SLUG

Slugs are not merely snails without houses, but make up another, closely related family under the phylum Mollusca. They have a mantle,

Snail

or internal shield, that corresponds to the snail's shell. When contracted, slugs draw their head and tentacles (eye stalks) below the mantle. Both slugs and snails have soft bodies and exude a slimy mucus. A silvery trail of mucus shows where these molluscs have traveled on their nocturnal forages.

These are pests of both garden and greenhouse. To keep them from the greenhouse, surround it with a border free from grass and weeds. Store pots and boxes away from weeds and plant debris.

Dry, cold weather and daylight send slugs and snails under boards, cabbage leaves, or pieces of potato and other garden debris. These shelters can be set out as traps in otherwise clean gardens. Dispose of the hiding creature by handpicking or sprinkling with a bit of salt. Such irritations cause slugs to produce great quantities of slime, and they destroy themselves through desiccation. (The snails you handpick can be served at that evening's dinner. They are a good food source, offering 12 to 16 percent protein and just 100 calories in 100 grams of meat.)

Snails will avoid protective borders of sand, lime, or ashes. A more elaborate defense is a metal barrier, several inches high and with the top two inches bent away from the crop at a 45-degree angle. A fence of window screen will also work.

These soft creatures are repelled by a mulch of wood shavings or oak leaves. They avoid tobacco stem meal and a soil drench of wormwood tea. Hellebore has long been used to keep slugs from grape vines. Quassia, another plant-based insecticide, is also potent. Commercial products include a herbal preparation, Fertosan Slug Destroyer, and Snailproof, made from incense cedar.

The best-known way to be rid of slugs and snails is to set out saucers or jar lids full of stale beer. The fermented liquid draws them from far and wide, and they drown in their drink. If it appears that they are taking a sip and then leaving, mix in a bit of pastry flour to make a sticky mixture.

Larvae of the lightning bug are able to climb up a snail's back and sink their sicklelike jaws into the victim's head. Garter snakes, grass snakes, rove and ground beetles, ducks, box turtles, and salamanders also have a penchant for escargot.

A natural enemy of the brown garden snail is the decollate snail, studied at Rincon-Vitova Insectaries and now sold by them. (As this is written, sales are restricted to Southern California.) Smaller brown snails are particularly vulnerable. Fortunately, when the predator switches to plant food, it seems to prefer fallen leaves and fruit.

# SNAPDRAGON

Several species of shield-shaped **stink bug** [color] spoil snapdragons by sucking sap. They are also found on columbine, sunflower, verbascum, lupine, and other flowering plants. If

Stink bug

you can't identify these pests by their shape, the odor they put off should clue you in. Stink bug populations will stay down if you carefully weed the garden area. The pests are vulnerable to insecticidal soap sprays.

Other pests include APHID, cabbage looper (see CABBAGE), corn earworm and fourlined plant bug (CHRYSANTHEMUM), NEMATODE, SNAIL AND SLUG, SPIDER MITE, and SYMPHYLAN.

# SOAP

Various nondetergent soaps have long been used in homemade remedies, and fish-oil soap was a common aid more than a century ago. Soap is mixed with water, and this may happen quicker if a tablespoon of alcohol is added to each quart of water base. These sprays don't remove bugs by washing them away (a forceful spray of plain water will accomplish this), but by the insecticidal power of fatty acids, which break up cellular membranes. (Our own hair glands produce a fatty acid that gives the scalp resistance to ringworm fungus.) Soft-bodied insects are most vulnerable—aphids, mealybugs, mites, and whiteflies. Beetles rank among the least susceptible. Potassium-based soaps don't make suds and aren't useful for cleaning, but they are more effective than most household soaps. A potassium soap is now marketed expressly for pest control: Safer's insecticidal soap is easier on plants than some household soaps, and kills many pests of vegetables and fruit.

Care must be taken with any soap spray to treat the undersides of leaves as well as the tops, especially if aphids, mealybugs, or mites are the problem. Soap is an ingredient in many homemade sprays because it thickens them and helps them adhere to plants. To control leafminers, it's best to treat plants with a soap spray about the time daffodils come into bloom, as the larvae are then most vulnerable. Soap sprays will cause various pest larvae to drop from tree foliage and bark.

Some soaps may seriously damage foliage. Test a spray on a small area if you are uncertain of its strength. Try diluting sprays that have a harsh effect.

In tests of soap sprays at the Rodale Research Center, a few household soaps controlled pests nearly as well as Safer's insecticidal soap, but had their disadvantages. Some proved harmful to foliage; Ivory Snow powder was less phytotoxic, but this powder takes time to prepare (in the tests, the powder was added to water a day before spraying, and the resulting gel was mixed in a blender just before use). Ivory liquid dish soap is easier to make into a spray, but it damaged plants at all concentrations tested.

In general, good results were had with a concentration of just over three tablespoons of soap per gallon of water. Higher concentrations were more injurious to plants, without giving better control.

Because plants differ in sensitivity to a spray, it is best to test a soap on a few leaves and watch for damage over the next three to five days. You can help sensitive plants by rinsing them with water a few hours after spraying with soap.

# SOIL POLLUTION

If your soil contains chemicals from past growing seasons, you can mix activated charcoal with the soil to help purge it. The charcoal can also be used as a root dip for transplants or can be poured as a water solution into the transplant hole. Only 1 or 2 parts activated charcoal need to be mixed in 100 parts water. However, because these methods are zonal in nature, they do not protect successive crops planted in the same area, nor do they always do a good job of protecting roots that grow out of the area of application. A more effective treatment is to mix the charcoal directly with the soil. Rates of application vary with the pesticide to be removed and the nature of the plant to be grown;

a general rate is 300 pounds per acre, and it is better to err on the excessive side. The cheapest source of activated charcoal is a chemical supply house. Drugstore prices would be substantially higher, but the convenience of smaller packages may make up for this in the case of very small applications.

# SOWBUG

The oval, round-topped sowbug is not an insect, but a crustacean and therefore a relative of the crayfish. The pillbug and sowbug are very similar, both in habits and controls, but the former performs the unique trick of rolling up into a ball to protect itself; the sowbug instead scurries for cover. The sowbug has 14 legs and 20 protective plates that run the width of the body.

Young tender plants may be nibbled upon, but sowbugs are rarely a demanding problem. Control them by keeping the garden clear of debris and old plant material. Tobacco dust and pyrethrum spray will hold the line if your problem is severe, but you should be able to get by with less powerful deterrents—an oak leaf mulch, wood ashes worked into the soil, or a weak solution of lime in water (stir two pounds lime into five gallons of water, let stand for 24 hours, and water the plants with the clear liquid). To keep sowbugs from taking over the compost pile, turn the compost frequently so that temperatures will stay high. You can sprinkle lime on the pile occasionally.

# SOYBEAN

## Insects

Soybeans are generally not susceptible to a large number of insect pests. They may be bothered by the **Japanese beetle,** but this once troublesome insect can be controlled by the use of milky spore disease. These beetles do not bother soybeans much early in the season, but by August they flock in ceaseless hordes and, if unchecked, will defoliate the plants very rapidly. See JAPANESE BEETLE.

The **velvetbean caterpillar** is frequently a pest in the southeastern states, where it attacks soybeans as well as velvetbeans, peanuts, kudzu, alfalfa, and horsebeans. The insect is a tropical species that does not survive the winter in the continental United States, except perhaps in the southern tip of Florida. The moths fly into this country sometime in June or July and may produce as many as three to four generations during a season. The insect does not usually become very abundant until late summer or early fall. However, a heavy infestation may completely strip the plants in a field within a few days. The small white, roundish eggs are laid singly on leaves and hatch within three to five days. The caterpillars feed for about three weeks. They are very active and will spring wriggling into the air when disturbed, at the same time spitting a brownish liquid. After completing feeding, the caterpillars enter the soil to pupate at a depth of ¼ inch to 2 inches. The adult moths emerge about ten days later.

Handpicking is an effective control because the caterpillars rarely become abundant, thanks in part to naturally occurring egg parasites and a fungus. Should the caterpillar population threaten to get out of hand, you can turn to the bacterium *Bacillus thuringiensis,* available commercially.

# Diseases

While commercial growers are confronted by many potentially significant diseases, the gardener who sows a few plants for the edible beans should have little or no trouble.

**Bacterial blight** is a disease that causes leaves to assume small angular brown or black spots. Cool, rainy weather favors this disease. Leaf tissue that is affected may drop out, giving the leaves a ragged appearance.

Reduce damage by crop rotation, deep plowing, avoiding overwatering, and planting seed from disease-free fields. Commercial varieties that have displayed some resistance include Hawkeye and Bethel.

**Bacterial pustule** is a warm-weather disease that usually appears in July. First symptoms are small, yellow green spots with reddish brown centers. The spots are most conspicuous on the upper surfaces of leaves. The central portion of the spots is slightly raised, developing into a small pustule, especially on the undersides of leaves. In the later stages the pustules rupture and dry. The disease-producing bacteria overwinter on diseased leaves.

Plant only from seed that is known to be free from bacterial pustule. Rotation of crops is a sound cultural practice. The variety CNS is highly resistant, and Ogden is somewhat resistant.

**Brown spot** is a fungal disease that is generally one of the first to appear on the leaves of young plants. When this disease strikes, the lower leaves become infected first, turning yellow and falling off. The fungus can be carried through the winter on diseased stems and leaves of plants. If you are growing your soybean crop in two successive years, be certain to plow the stubble deeply so that the stems and leaves are buried deep.

**Downy mildew** is a widespread fungal disease that causes small, pale green spots to appear on the upper surfaces of leaves. They soon enlarge, grow together, and become brown as the leaf dies. Grayish tufts of moldy growth appear on the lower surfaces of the leaves. Because this fungus overwinters on diseased foliage and seed, crop rotation and disease-free seed will prevent trouble. Although few varieties are resistant to downy mildew, Harosoy, Lindarin, Kent, Lincoln, and Clark are less susceptible than others.

**Weeds** left in fields until harvest time have been found to reduce yields as much as 17 percent. Proper seedbed preparation is the best method of control. By working the soil to start the weeds growing, then working the surface to kill those that have come up, you can be rid of them before planting time. Although the soil may have to be worked a number of times, it is necessary if a high-yielding crop is expected. Soybeans are particularly vulnerable to weeds because the crop grows very slowly during the early stages of life, allowing the weeds to outgrow them and offer severe competition for water, nutrients, and light.

# SPEARMINT

This is one plant that generally takes care of itself. Occasionally disease or insects (particularly the fourlined plant bug and garden flea hopper) may injure a few plants, but mint grows so prolifically that only commercial growers should have reason for concern.

# SPIDER

Maligned through the ages along with snakes and toads, the beneficial spider should instead be welcomed in the backyard garden. Spiders are exclusively insectivorous; indeed, they eat only live insects, since it is movement that attracts these eight-legged predators to their prey. Most spiders are not picky about the species of insect they eat, although wasps, hornets, ants, and the hard-shelled beetles are least preyed upon. Potential pests make up the mainstay of their diet.

If you will accept the spider's help in the garden, see to it that there is a variety of insect life for it to prey on. In other words, avoid the heavy-handed use of sprays and dusts.

Some species, such as wolf spiders and jumping spiders, run down their prey instead of snaring them in nets. Most spiders prefer dark, shaded locations where either moisture is available or the humidity is high. The new, naked spring garden offers a poor environment for spiders, but by the time the vegetables are tall enough to provide shade, they have moved in for the season.

Spider mites, other members of class Arachnida, are well known as pests. Not so well known is that some species are predators of plant-eating mites. See SPIDER MITE.

# SPIDER MITE

Mites aren't insects, but members of the anthropodan class Arachnida, along with ticks, scorpions, and spiders. This should suggest to you that they have four pairs of legs, unlike most other creatures in garden and orchard. The mites' relations also suggest their habit of spinning webs—not very large or dramatic ones, as they are used only for protection. The webs do not physically damage plants, but interfere with the looks of flowers. Mites are well known as pests, but some species are being used as beneficial predators. It's not easy telling the good from the not-so-good because mites are small and difficult to see without a hand lens.

Spider mite

Mites may make their presence known through any number of signs of poor health—slow growth, leaf drop, few fruit, discoloration, and so on. The only real test is to see the pests. A 5× lens will help you to spot mites on plants. Or you can hold a piece of paper beneath a branch you suspect to be infested and tap the plant to dislodge the mites. Look for bugs the size of salt grains.

Mite populations are very sensitive to changes in the habitat. Nutritional sprays of copper and zinc, for example, increase the numbers of the citrus red mite, while the citrus rust mite is favored by copper. On apple trees, populations of the European red mite and two-spotted spider mite increase if trees are fed nitrogen-rich fertilizer. This is bad news for orchardists, as mites can lower the chlorophyll content of leaves as much as 35 percent.

Probably at least two sprayings will be needed where mites have been a problem, but you may find that some orchard plots don't need any. Natural helpers such as ladybugs, lacewings, and predacious mites may do the job for you. Ladybugs are credited with controlling the citrus red mite in unsprayed California orchards. Other beneficials include damsel bugs and dance flies. Rotating oats, buckwheat, corn, or sorghum with wheat will break up the life cycle of the winter grain mite.

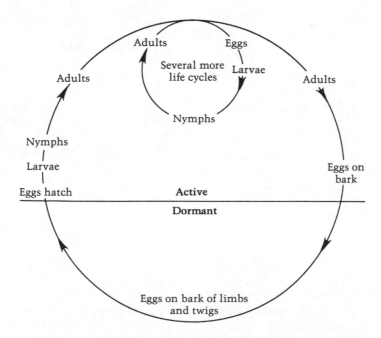

Life history of the European red mite

Reprinted, by permission, from Pyenson, *Fundamentals of Entomology and Plant Pathology* (fig. 5.5), AVI Publishing Co., PO Box 831, Westport, CT 06881.

BENEFICIAL MITES. Not all mites are pests, and some rate as important predators. In recent years a few species have come on the market. They are shipped as adults and should be scattered by hand throughout the crop area. Adults lay their eggs next to the pest mites on the undersides of leaves. You should release your beneficials near infested areas, not at random.

The mites occasionally feed on nectar or pollen, but will not injure plants. Their appetites are quite specific and they will not prey on aphids, scales, or whiteflies.

The predatory mites are susceptible to many pesticides (although resistant strains are now on the market). If chemicals must be used, release of the beneficials should be delayed two or three weeks; in addition, you can better their chances by rinsing treated foliage with water.

Until a few years ago, sales of predatory mites were almost entirely to commercial greenhouse growers. Now, many small-scale growers have discovered the mites. Some home gardeners have been disappointed—those who buy the mites only after an infestation has swept over their plants. They mistakenly hope to achieve control by dumping on the beneficials, as they might a pesticide. You have to catch the pests early. Otherwise, wash the plants, wait a few weeks, and then begin a program of introducing predators. To guard against a sudden infestation, frequently check the undersides of leaves with a 10× hand lens.

On a small scale, you can use purchased predators with success on your windowsill plants; but it's a matter of how much time and money you're willing to devote to a few plants. Minimum orders are fairly expensive, and you can't store for long those predators you don't use right away. How many predators do you need? One insectary's minimum order of 300 predators should guard 15 to 20 lightly infested plants. You can expect results in two to three weeks.

You can nurture your own naturally occurring populations of beneficial mites. Collecting them from the wild is a challenge, however. They aren't easy to track down, and you run the risk of introducing pest species along with the good. (The adult twospotted mites can be told by the spots; *Phytoseiulus persimilis* is red as an adult and lacks spots.) Finally, once you do establish the predators, you have to manage their population so that they don't eat every pest in sight and then starve.

Beneficial mites have been used outdoors, but mainly in commercial applications such as strawberries in coastal Southern California. In that state's San Joaquin Valley, grape growers bring in mites to speed up their shift away from chemical pesticides; the beneficial mites occur naturally and in time would rise up to control pest mites, but the transition is much easier by augmenting the natives with commercially raised mites.

Beneficial mites have been used extensively through Europe to control pest mites in greenhouses. Now, they are catching on in North America.

In western Canada, more than 70 percent of all commercial greenhouse cucumber growers have used one species, *P. persimilis*, as the primary control of the greenhouse spider mite, once their worst pest. This species is an excellent predator in mild, humid situations of up to 80°F. An "OP-resistant" strain is sold, raised from survivors in fields that were treated heavily with organophosphate chemicals. This strain, intended for greenhouse use, is valuable for integrated control: at the same time that chemicals are being used to control aphids or whiteflies or thrips, the OP-resistant mites can go about their business. *P. persimilis* is used to protect many vegetables, fruits, roses, and greenhouse crops. It is the principal predator of the common twospotted mite, and has been used to control that pest on many vegetables and fruits. Females lay an average of more than 50 eggs in a lifetime, and eat half that number of pest eggs. One of these mites can be counted on to eat an average of three dozen mite pest eggs over its short life span. The beneficials' eggs are orangish, oval, and larger than the twospotted mite's oval, clear or straw-colored eggs.

Effective at higher temperatures is *P. longipes. Amblyseius californicus* is recommended for control of Willamette mite on grapes, citrus red mite on citrus crops, and generally on those mites that do not form strongly webbed colonies; it is used in greenhouses, on house plants, and on outdoor ornamentals, remaining effective up to 85°F. *Metaseiulus occidentalis* can be put to work against the twospotted mite *Trichogramma mcdanieli* (in apples), Pacific mite, and tomato russet mite. It tolerates temperatures at 90°F. and more, and three strains are available. One is not very sensitive to photoperiods and won't go into diapause, staying active and continuing control after other strains have entered their resting stage. Another strain is resistant to organophosphates, carbaryl, and sulfur, and was originally gathered from treated vineyards in California's San Joaquin Valley.

You will get tree help from naturally occurring beneficial mites just by doing nothing. "Nothing" is the key here, because these mites are set back by pesticides and even fungicides. In California, three naturally occurring mite predators help to control a pair of prime strawberry pests, the twospotted and cyclamen mites. One beneficial mite preys on purple scale infestations, traveling to infestations on ladybugs.

Growers in the Midwest are grateful to the grasshopper mite, an important natural enemy that preys on grasshoppers. In its larval stage it is parasitic on grasshopper larvae and adults, and as a nymph and adult it becomes a predator of the eggs. When an adult grasshopper becomes heavily infested, it may be unable to fold its wings or fly properly.

# SPINACH

## Insects

Among the **aphids** that infest this crop, the yellow green spinach aphid (also known as the green peach aphid) is particularly voracious. See APHID for control measures.

The **beet webworm** sometimes injures spinach; see BEET.

The **spinach flea beetle** is one of several flea beetles that eat spinach. This is a ⅕-inch-long insect with greenish black wing covers and a yellow thorax. It hibernates as a beetle, and appears in the field in early spring to deposit clusters of orange eggs on the ground at the bases of plants. The short, dirty-gray to purplish larvae feed on the undersides of leaves. When disturbed, they drop to the ground and remain hidden until they perceive that danger is past. Pupation takes place underground, near the surface.

Control weeds in and around the garden area. Flea beetles avoid shaded areas, and you might plant spinach near houses or taller plants. Should infestation become a great problem, try garlic and, as a last resort, rotenone.

The **spinach leafminer** [color] is likely the most destructive pest of spinach; it also infests beet, chard, and a common weed, lamb's-quarters. It hatches from white eggs laid on the undersides of leaves by a small fly. The mines make leaves look blotchy. Plant growth may be stunted.

Some gardeners keep flies from laying eggs on spinach by erecting long structures over the

Spinach leafminer

rows with plastic on top and cheesecloth at the ends. If you aren't this ambitious, simply remove eggs by scraping them with a finger. You'll know if you're too late, because newly hatched miners make gray blisters when they burrow into the leaf. Using scissors, cut away affected leaf parts to keep the tiny green larvae from harming the plant further. Later in the life cycle, the fat worm drops to the ground and pupates in a cocoon. You can interrupt the cycle by cultivation. Rotation is effective because the miner feeds only on spinach, beet, and chard among garden plants. Keep down lamb's-quarters in the immediate area.

Spinach leafminer damage

## Diseases

Yellowed and curled leaves, stunted plants, and reduced yields are indicative of **blight** (or yellows), a disease caused by the cucumber mosaic virus and spread by aphids. Grow blight-resistant varieties, such as 56 and 612. Remove and destroy perennial weeds. See APHID for methods of controlling this vector.

Wet weather often brings **damping-off** with it. If you are growing spinach from seed, plant enough so that you can fill in the gaps with extra plants. See DAMPING-OFF.

A drawback to the preference of spinach for cool, moist days is susceptibility to **downy mildew** (chard blue mold), the most destructive disease of the crop. Affected plants show yellow leaf spots, with a fuzzy purple growth on the undersides. These spots may quickly run together, killing leaves or the entire plant. The causal fungus is soil- and seed-borne, and is favored by wet, cool weather; initial infection requires humidity above 85 percent and a mean temperature between 45° and 65°F. for a period of one week.

Prevent this disease by using clean seed, planting on fertile, well-drained soil, and using a three-year rotation. Do not crowd plants, and water on sunny mornings. Once the disease strikes, do not plant winter spinach, as it may carry over downy mildew from one growing season to the next. Resistant varieties include Dixie Market and Dixie Savoy (savoy); resistant hybrids include 7 (semisavoy), and 424 and 425 (smooth leaf). New varieties will likely be introduced as time goes on; check with your extension agent.

When **fusarium wilt** hits, the plant will generally turn yellow and the lower leaves lose their firmness and wilt. Look for a brown discoloration of the water channels. Virginia and Texas report the most severe damage. Although crop rotation reduces this disease, resistant strains offer the best answer.

## SPIREA

The foliage and flower buds of spirea are occasionally fed upon by the **obliquebanded leafroller.** This caterpillar conceals itself by rolling the leaf on which it is feeding. It also ties the terminal leaves together, thus marring the plant and interfering with its normal growth. The roller is yellowish to pale green and ¾ inch long when mature. There are two generations annually, one in spring and the other in late summer.

If the infestation is light you can pinch off and destroy affected parts. Should stronger measures be necessary, dust the plants with a mixture of equal parts tobacco dust and pyrethrum powder, or a spray of pyrethrum and rotenone. Both dust and spray should be made in two applications half an hour apart; the first drives the caterpillars from their hiding places, while the second one kills them.

# SPRAYS AND DUSTS

Sprays and dusts serve as your last line of defense in the organic grower's effort to maintain peace with pests and diseases.

The sprays and dusts discussed here work in any of a few ways. A water spray simply knocks pests and eggs from your plants, but even this method can be misused if you overlook beneficial insects among the pests; these predators and parasites could be set back more profoundly than the pests, resulting in an infestation. Other sprays and dusts are repellents or antifeedants, and work by chasing bugs or turning off their appetites. Insecticidal sprays and dusts are the organic growers big guns—plant-based or biological, for the most part, but killers nevertheless.

Natural controls can be abused. One overdose or mistimed application can upset the delicate ecology of a garden or orchard, undoing biological controls. Beneficials flee the area, leaving the door wide open for a sudden pest resurgence.

Use these aids only when other controls fail. Apply dusts and sprays locally, to take care of a specific problem, instead of blanketing the whole garden. Do your best to avoid disturbing beneficials on the prowl. Make certain you have correctly identified the enemy. This calls for close observation, and even entomologists occasionally misdiagnose problems.

HOMEMADE FORMULAS.    There is a little of the experimenter in every organic grower, and the result is a legacy of home-brewed alternatives to store-bought chemicals. A fundamental behind many preparations is that hot-tasting or stinky things make the best ingredients. Garlic, peppers, and the more powerful herbs are among those most often mentioned as bug chasers. Of course, there are powerful organic sprays and powders available commercially—ryania, rotenone, and pyrethrum, for instance—but those you make at home are tamer, cost next to nothing, and encourage a bit of thinking about insects and their worlds.

With most recipes you'll find a list of insects or diseases observed to be vulnerable to the spell of that particular blend of ingredients. This isn't to say that there aren't other vulnerable species, but it's true that most formulations are somewhat selective in their effect.

As you read over the many recipes that follow, keep in mind that most are the product of trial-and-error work by other growers, and as such shouldn't be looked upon as set formulas. Feel free to vary proportions and add or delete ingredients as you see fit. Keep an eye out for plants that aren't bothered by bugs, whether they are something you introduced or garden-side weeds: they might just be the stuff for a new recipe.

Most homemade sprays are derived from plants you can grow in your own garden, if they aren't already there. See REPELLENT PLANTING and the appendix for likely ingredients. Sprays are usually made by mincing repellent plants along with a little water in a blender. Before spraying, the resulting slurry is filtered and diluted. Sprays can be cooked on the stove, like a soup, but you stand to drive off or destroy the volatile, active ingredients. These ingredients can be coaxed out of a plant that produced them by mincing it and soaking it in mineral oil for a day or two, then adding it to water and using a little soap to render the oil miscible.

Pure soap is a common ingredient in many sprays. It helps keep sprays on the plant, and contributes its own proven insecticidal power. Soapy water alone has long been used as a spray. Naphtha soap seems to be recommended most often, but any nondetergent soap should do. Soap will dissolve better if you add a tablespoon of alcohol to each quart of water base. See SOAP.

Try pepper sprays against ants, mites, and various caterpillars, including cabbageworm and tomato hornworm. The sprays may also rescue plants from viruses, such as cucumber mosaic, ring spot, and tobacco etch. Dried pepper flakes can be spread about the bases of plants to ward off aphid-carrying ants.

Chamomile spray has been credited with preventing mildew and damping-off on seedlings. Wormwood sprays work against several soft-bodied pests such as slugs and aphids. Other popular sprays are based on cedar. An old-fashioned aid is Ced-o-flora plant spray, which contains hemlock oil. It can be sprayed to control mealybugs, spider mites, scales, and aphids.

Displaying a spark of intelligence we humans might envy, insects recognize that tobacco does them no good and they do their best to keep away from it. You can take advantage of their good sense by putting a handful of tobacco or tobacco wastes in water to make a spray. Let the meal steep for 24 hours, and then dilute the solution to the color of weak tea. Stems can be purchased from florists and seedsmen, and store-bought plug tobacco works fine. Or you can save a little time and effort by purchasing a tobacco extract and following the directions on the package. Commercial products may be in the form of nicotine sulfate. Tobacco sprays spread better if soap is added, but be sure

to rinse the plants with clear water after each application so that foliage is not burned. This is potent juice, and should be used well before beneficials are released in the garden. Tobacco sprays do funny things to roses, so unless the prospect of black roses piques your curiosity, use another spray. Tobacco smoke has been used also, blown over plants with a bellows. Nicotine has been found to inhibit growth in some plants and to cause early flowering in others. (And, nicotine you ingest from plants can make you sick.)

Nicotine is most effective against small, soft-bodied insects such as aphids; it also has demonstrated control over spider mites, leafrollers, fruittree borers, larvae of the cabbage butterfly, and ants.

Nicotine is not nearly as popular an insecticide as in the past (it was one of the three most important back in the 1880s). An obvious reason is the proliferation of chemical pesticides following World War II. Another factor is that we are smoking the plant scraps once considered a byproduct by the tobacco industry. Specifically, the stems and ribs that once went into insecticides are now processed into cigar wrappers.

Orchardists use oil sprays to smother mites, scale, and several other pests, as well as fungal and bacterial diseases. See FRUIT TREES.

High-powered botanical insecticides are made from pyrethrum, rotenone, ryania, and sabadilla. With their effectiveness comes the risk that beneficial insects will be eliminated along with the intended pest. See separate entries for each.

Of this group of lethal botanicals, quassia is perhaps the safest. It spares ladybugs and bees, while working against aphids, sawflies, and caterpillars. Quassia is purchased in the form of wood chips and shavings from a small Latin American tree. A close relative with similar properties is the familar ailanthus, or tree of heaven. To make a solution for spraying, quassia wood is simply steeped in water. Quassia has been found effective against aphids, sawfly larvae, and Colorado potato beetle larvae. Most adult insects go unharmed. A century-old recipe for combating aphids calls for quassia chips and larkspur seed boiled in water. The tea is syringed on plants.

False hellebore acts as a stomach poison for insects. It is especially useful against worm pests of the garden, sawflies, and chewing insects, such as beetles, caterpillars, grubs, cutworms, and grasshoppers. Although grown domestically, false hellebore is not listed with the grow-it-yourself remedies because of its potency, both to the bugs and you, and it should not be ingested. If you decide to try this botanical, keep its original use in mind—at one time the tips of arrows were treated with false hellebore to make a slight wound lethal.

A commercial insecticide called Hellebore is made from the plant's roots and sold as a dry powder. It is most often used as a dust to protect ripening fruits and vegetables. As a dust the powder is mixed with flour or hydrated lime; a spray is made by mixing one ounce of hellebore with two gallons of water.

You can grow your own false hellebore from seed sown in spring or early fall, or from root divisions in spring. The plant prefers light, sandy soil, and displays large yellow flowers.

Microbial sprays spread insect diseases about vulnerable plants. The living pathogen in mixed with water and likely a soap or other adjunct to keep the spray on plant surfaces. See MICROBIAL CONTROL. Many growers have found success with so-called bug juice sprays, made of ground up pests and water. The principle behind bug juice is not fully understood. See BUG JUICE.

SPRAYERS.   Some sprayers are simple and inexpensive; others are relatively elaborate. The smallest are pumped by a trigger and hold only a quart or so of liquid. A disadvantage is that your hand will tire before you've covered much foliage. And if you are trying to spray the undersides of leaves, you may find that the sprayer pumps air when tilted. More convenient are sprayers with the nozzle on the end of a flexible tube that runs from a storage tank holding a gallon or more. Because the tank stays on the ground, the pickup tube stays under the liquid. Lightweight plastic tanks seem to hold up better than those of galvanized steel. Stainless steel tanks last longest, but they are heavy and expensive. The spray is propelled either by a pump on the tank or by a long, cylindrical pump (called a trombone) between the hose and sprayer head. To handle a small orchard, you may want to invest in a backpack sprayer, a more expensive alternative that allows you to move a large quantity of spray in comfort.

# SPRUCE

## Insects

The best control for **aphids** [color] is a dormant-oil spray applied in early spring to kill overwintering adults. Use a miscible oil mixed 1 part in 25 parts of water. Oil removes the

Spruce aphid gall

bloom from blue spruces, however. A subsequent spray of nicotine sulfate, applied soon after the new growth begins in later spring, will control the young gall aphids before the pineapplelike galls are formed. When only a few galls occur, they may be cut off and destroyed before the insects emerge.

The **eastern spruce beetle** is a small bark beetle that emerges in June and July from small round holes resembling shot holes. Weak trees are particularly vulnerable, and the beetle prefers trees a foot or more in diameter. Its presence is made evident by an exuding gum mixed with sawdust on the trunk. The females lay their eggs in early summer and the larvae tunnel under the bark, hibernating in burrows in a partly grown condition and completing their development the following spring.

Infested trees should be cut and the bark removed before the middle of May to prevent the beetles from reaching maturity. See also the discussion of bark beetles under PINE.

The **spruce bud scale** is especially injurious to Norway spruce. The reddish brown adult scale, which is globular and ⅛ inch in diameter, closely resembles the new buds of trees. These insects pass the winter in an immature condition at the base of the terminal buds. Eggs hatch during early June and the young attack the new growth. A dormant-oil spray will control the scales as they overwinter.

Spruce trees are occasionally infested by the **spruce epizeuxis,** the larval form webbing together and feeding upon the needles. Larvae are brown and covered with wartlike protuberances, and resemble the spruce budworm (see FIR). The moths are a brownish gray, with a wingspread of less than an inch and narrow wavy bands marking the wings. They appear in the first half of summer. Control by removing and burning the dried needle masses in which larvae overwinter.

The **spruce mite** is reported to occur on spruce, pine, hemlock, and arborvitae. At times infestation may be extremely heavy. The mites are dull green to nearly black with a pale stripe on the back. The eggs are brownish and more or less flattened, and it is in this stage that winter is passed. The mite spins webs and causes a graying or browning of the needles.

Dormant-oil sprays usually give good control of the eggs if applied thoroughly. See ARBOR-VITAE and SPIDER MITE.

If dark brown moths are flying about your spruce trees, you may be in for a visit from the **spruce needleminer.** The moths lay their eggs early in summer and the resultant larvae bore into the bases of the needles. Often the heaviest infestation is on the lower branches, where many of the needles may be destroyed.

Control by washing the tree with a strong spray from the garden hose. Clean up and burn any trash that falls to the ground.

Other insect pests of spruce trees are gypsy moth (see TREES), spruce budworm (FIR), and white pine weevil (PINE).

## Diseases

**Cytospora canker** is a fungal disease that attacks the Norway and Colorado spruce, causing death first of the lower branches and then of branches higher on the tree. Cankers produced on affected branches are inconspicuous. They may be covered by quantities of resin which often drips to the lower branches. During rainy weather, threadlike, yellowish, gelatinous spore masses ooze from fruiting bodies on the cankers. These spores are spread by rain, wind, or pruning tools to other spruces where they enter through wounds.

Control involves pruning all diseased branches back to the trunk or nearest healthy lateral branch. Be careful to prune only when the trees are dry. It is beneficial to fertilize the trees to stimulate vigorous growth. Avoid nicking bark when mowing the lawn.

**Needle drop** may be a natural, seasonal phenomenon. See the discussion of diseases under TREES.

For **gray mold blight**, see FIR. For cankers, rusts, and witches' broom, see TREES.

# SQUASH

Many of the insects and diseases that plague squash also trouble pumpkins.

## Insects

Several species of **aphid** infest squash. Commercial growers protect summer squash by laying down metal foil. This reflects sky light and confuses the pests. See APHID.

Young plants may be chewed upon by **garden springtails.** See BEET.

**Pickleworms** [color] feed on the flowers and leaf buds of squash and tunnel within flowers, terminal buds, vines, and fruit. They are ¾ inch long, yellowish white with numerous spots when young, turning green and brown-headed. The worms are active in winters in Florida and Texas, and spread as far north as Connecticut, Illinois, Iowa, and Kansas late in the season.

You can often get by this pest by planting very early; late plantings may be destroyed. Some growers use squash as a trap crop to protect melons, but the squash vines should be removed and burned before the larvae are fully grown. An early fall cultivation will take care of underground pupae. Clean up the square patch thoroughly immediately following harvest. See RESISTANT VARIETIES.

**Squash bugs** [color] are dingy brownish black insects, about ⅝ inch long, that have a very disagreeable odor when crushed. The young bugs have this same odor. When newly hatched,

Squash bug adult and nymph

they have reddish heads and legs and green bodies. Later they become darker, the head and legs turning black and the bodies light to dark gray. In a more mature stage they are brownish black. The shiny gold-to-brownish eggs are found on the undersides of leaves. They are about 1/16 inch long and are laid in clusters along the center vein. The young bugs remain in clusters after hatching.

The squash bug feeds by inserting its needle-like mouthparts into the plant tissue and withdrawing the sap. It releases a toxin that causes wilting. Vine crops are easily killed by these bugs during the early part of the growing season. Older plants often have one or more runners damaged. The leaves on these damaged runners wilt, become crisp and dark brown at the edges, and later die.

The squash bug can be repelled from squash and other susceptible plants by growing radishes, marigolds, tansy, or nasturtiums nearby. Garden sanitation is important, as it has been discovered that the squash bug likes moist, protected areas and often hibernates under piles of boards or in garden trash. Remove vines from the garden at season's end. The bugs may hide in deep, loose mulch such as hay and straw; use sawdust, compost, or plastic sheeting instead. Try to plant the new crop as far away as possible from the crop of the previous year. Pick eggs and insects off plants. Trellised plants are less susceptible. Sabadilla is an effecive insecticide. Tachinid flies are a natural enemy. See RESISTANT VARIETIES and TIMED PLANTING.

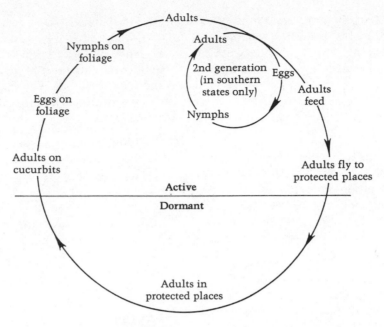

Life history of the squash bug

Reprinted, by permission, from Pyenson, *Fundamentals of Entomology and Plant Pathology* (fig. 3.1), AVI Publishing Co., PO Box 831, Westport, CT 06881.

The larval form of the **squash vine borer** [color] is a one-inch-long worm, colored white with a brown head and small brown legs. It makes its presence known by wilted runners and vines, and deposits of yellow, sawdustlike frass near the bases of stems. When mature, the larva leaves the plant and descends into the soil, where it spins a cocoon and remains until

Squash vine borer adult and larva

the following spring. There are two generations in the South, and usually but one in the northern states.

Keep a look out for a sudden wilt of plant parts, as you can control borers by slitting affected stems and then killing the worm within. A delayed planting of summer squash may miss the feeding larvae. Earlier varieties can work too, as the borer doesn't lay its eggs until July, by which time the plants are large enough to withstand attacks. See TIMED PLANTING. Baby blue and butternut squash are both somewhat resistant; winter squash varieties such as Hubbard are most severely injured. Encourage the vines to root at the joints from which the leaves grow by covering the joints with damp soil or decaying mulch. Clean up and dispose of vines after harvest.

You can keep the adults (orange-and-black, wasplike moths) from laying eggs on squash plants by wrapping strips of nylon stocking around stems. The stockings give as the plant grows. You should also pile up soil in a mound as high as the blossoms; add to the mounds as the plants grow taller. Trichogramma wasps will parasitize this pest.

Other pests of note include beet leafhopper (see BEET), corn earworm (CORN), MEXICAN BEAN BEETLE, onion thrips (ONION), spotted and striped cucumber beetles (CUCUMBER and RESISTANT VARIETIES), WHITEFLY, and yellow woollybear caterpillar (see imported cabbageworm under CABBAGE).

Tall cover crops can in some cases protect an undergrown crop from insect vectors of disease. Cucurbits have successfully been interplanted with maize.

## Diseases

**Bacterial wilt** is a disease spread by striped or spotted cucumber beetles. The bacteria are carried in the digestive tract, and as the beetles feed their droppings spread the pathogen in the feeding wounds. Usually a single leaf gradually wilts and dies, followed by a vine and then the whole plant. Plants are often stunted.

If a plant is suspected of bacterial wilt, cut across a stem and squeeze it to press out some of the plant juice. Touch the juice to your finger

Bacterial wilt

and slowly move your finger away. Juice from a diseased plant will be sticky and stringy. Once a plant has bacterial wilt, there is no cure. Control the cucumber beetles (see CUCUMBER) and remove and destroy wilted plants early in the season. Acorn and butternut squashes are resistant.

**Mosaic** virus is most common on straight-neck and crookneck summer squash. Plants develop yellow spots on the leaves and sometimes on the fruit. The plants become stunted and the yields are reduced. There are two species of virus: cucumber mosaic is carried by aphids; squash mosaic is spread on seeds and by the cucumber beetle, and is most common in the Southwest.

There is no cure for mosaic once the plants have been affected. Do not allow perennial weeds such as milkweed, wild cucumber, and catnip to grow near squash or pumpkin fields, as insects feed on the weeds and carry the virus to healthy crops. See APHID and cucumber beetle under CUCUMBER for methods of controlling these vectors.

# STRAWBERRY

## Insects

Two types of **aphid** [color] do harm to strawberry plants: the leaf aphid and the strawberry root aphid. Leaf aphids transfer viral diseases from old or infected plants, which should be rogued out. The strawberry root aphids, which suck juice from the plant, are distributed belowground by the cornfield ant. Frequent soil cultivation will help control the ant and thus the aphid.

Both the cyclamen **mite** (also called the strawberry crown mite) and the spider mite may cause trouble. The cyclamen mite, invisible to the naked eye, feeds on buds and leaves to cause distorted or blotched blooms, curled-up purplish foliage, and no fruit. It is a cool-weather mite, seldom active in the heat of summer.

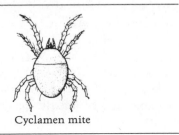

Cyclamen mite

The cyclamen mite is usually kept under control by its natural predators, unless these have been destroyed by insecticides. Buy healthy certified strawberry plants, and avoid touching clean plants after handling those that may be infested. For control of the spider mite, see SPIDER MITE.

**Nematodes** are nearly invisible worms that live in the soil. There are several varieties, notably the root-knot nematode in the North, the sting nematode in the South, and the root-lesion nematode in both areas. Nematodes cause swellings or galls on the roots and weaken the plants, causing a much lower fruit yield. Flourishing in sandy soil rather than in clay, they are rarely troublesome in fertile soil rich in organic matter.

Marigolds tend to repel nematodes, and frequent but shallow cultivation will prevent damage to the roots. Keep plenty of organic matter in your soil and nematodes should be no problem; see NEMATODE.

In the Pacific Coast area, the **omnivorous leaftier** may cause problems. Since the larvae overwinter in the crevices of rough-barked trees, the best preventive is not to plant strawberries near such trees.

**Root weevils** produce larvae that feed on the roots and crowns of strawberry plants. The adult strawberry weevil is a ¼-inch-long, reddish brown, black-snouted beetle that almost completely severs the unopened buds, leaving them hanging. You may first notice the weevil's presence by small holes in the petals of the blossoms. Plants are stunted, leaves bunched together, and roots and crown eaten away. The female weevil leaves eggs inside the unopened bud before crawling into the stem, girdling it and nearly severing it. The weevils emerge from the soil in early June and lay eggs throughout summer. A new beetle develops inside the bud, emerging in July after the fruit has been harvested. Larvae feed on roots until cold weather, when they move down deeper into the soil.

Strawberry weevil

Since there is only one generation a year, control the root weevil by destroying any stems hanging by a thread and those which have already fallen, for they are the ones containing eggs. Often trash, wild brambles, hedgerows, and similar foliage are hibernating places for the adult weevils, and if it is practical, such shelter should be eliminated. This weevil has recently made entomological news for its susceptibility to the caterpillar nematode, now commercially available. See NEMATODE.

Garden **slugs**, which are often present in mulch, may become a problem for strawberries.

While slugs usually feed on dead vegetation and help to break down organic matter, they sometimes develop a taste for strawberries. The molluscs do most of their damage on the bottom of berries that grow close to the soil or damp mulch, as they dislike dry surfaces.

If slugs are causing much damage to your berry plants, control them by placing large pieces of shingle between the fruit clusters. The slugs will crawl under them, and in the morning you can uncover the pests and sprinkle salt on them, whereupon they will melt before your eyes. See SNAIL AND SLUG.

The **strawberry crown borer** is a grub that bores into the main root of the plant, weakening or even killing it. The adult is a brown beetle, $\frac{1}{5}$ inch long, with reddish patches on the wings.

Strawberry crown borer adult

The adults feed on stems and leaves, but it is the grubs that do the chief damage. Crown borers pass the winter under leaves or just under the surface of the soil. In late March they become active and lay their eggs in the crowns. Since eggs are laid from March to August, the grubs can do their damage all summer.

Injury to strawberries can be avoided by taking advantage of the two weak points in the insect's habits. First, the adult beetles cannot fly. Second, they do not begin to lay eggs until the last of March. If new beds are some distance from old beds, preferably 300 yards away, adult crown borers will not migrate to spread trouble. To avoid setting infested plants, dig them in February or early March before the eggs have been laid. Since crown borers are harbored by wild strawberry plants and cinquefoil, be certain to destroy any that are in the neighborhood before starting a strawberry bed. The only remedy for serious infestations is to plow the plants under after harvest and set another bed on a new site. To be completely sure that the new bed will be free from infestation, use disease-free plants and situate the new bed as far away from the other bed as possible—at least 300 yards. Strawberries can repel the onslaughts of the crown borer if they are strong and vigorous, so see to it that the soil is bolstered with compost.

The **strawberry crown mite,** a $\frac{1}{4}$-inch-long, reddish larva, burrows into the crowns of plants to cause stunting and a general weakening of the plant. Eggs are laid about the crown. Control through sanitary measures and crop rotation.

The mark of the **strawberry leafroller** is the folded, webbed leaves within which the larvae feed. In spring, the larvae pupate and become inconspicuous reddish brown moths with a wingspread of $\frac{1}{2}$ inch. Eggs are laid on the undersides of leaves, and as soon as the larvae hatch they feed on the leaves, folding them together for protection. Plants are weakened, leaves turn brown, and fruit is deformed. The rolled leaves protect these pests from most insecticides, which only serve to kill the leafroller's natural predators. The full-grown larvae are $\frac{1}{2}$ inch long and colored yellowish green or greenish brown. There are two generations, the second of which feeds from August to cold weather; the larvae then pass the winter in their folded leaves.

Natural parasites usually prevent serious infestations of this pest. Rotenone has been found effective when applied to the undersides of leaves, but hold off with this powerful botanical as long as possible.

The **strawberry whitefly** occasionally puts in an appearance. Eggs are laid on the undersides of leaves, and the nymphs remain to suck the sap. The adult whiteflies are about $\frac{1}{16}$ inch long and covered with a grayish powder. Infestations are usually in patches, not in a whole field. If control is necessary, spray the undersides of leaves with rotenone.

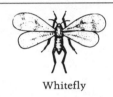

Whitefly

**White grubs** and **wireworms** seldom cause serious damage, and are usually not a problem if you avoid planting berries on land that was recently sodded.

## Diseases

Several precautions of strawberry culture, if carefully followed, will prevent many ailments. When making a new bed, plant the crop on soil that has not been used for strawberries for many years, and as far as possible from any old, infected beds. Plant in well-drained soil that is enriched with plenty of manure, compost, or other humus-rich organic matter.

Prevent crowding of runners; if necessary, thin them out several times through the summer. Pull all weak or diseased plants promptly. Your plants should go into the winter thinned and pruned and ready for spring. Mulches will protect plants from winter injury. Coarse hay, dried alfalfa, and sweet clover are excellent choices: they don't pack down, are free of weed seeds, and allow the berry plants to be uncovered gradually. Use clean tools to lessen the chance of spreading infection and avoid working with plants when they are wet.

You will have fewer problems if you buy inspected, disease-free stock from reputable nurseries. Because of the viral control program in the United States, most plants are now certified as virtually virus-free. This does not, of course, guarantee 100 percent protection, but it is decidedly safer to buy registered plants than unregistered.

Buy varieties that grow well in your area and those resistant to diseases prevalent in your area. A healthy, vigorous plant is much less vulnerable to disease. Rotate your berry patch regularly to avoid insect or disease buildup; a one- or two-year rotation is best. Plow under old fields soon after harvest and control wild strawberries nearby, as these are likely sources of virus.

**Black root rot** is a disease caused by a number of different fungi that are common in most soils. Healthy roots are white or clear while young, and with age they become brownish or darker on the surface and have a yellowish living core. The roots of affected plants are black throughout, with few if any healthy laterals. Plants have reduced vigor and may die. Feeder roots are quickly killed, and the darkened patches on the main root give way to the blackening that gives this disease its name. Winter injury, poor drainage, and acid or alkaline soil all favor attack by the root rot fungi. Plants affected with other organisms, such as red stele, verticillium wilt, or nematodes, are usually also attacked by root rot fungi. Black root rot fungi are present in most soils and can be introduced with contaminated soil or plants.

Be sure your plants have clear white roots, and plant them in soil that is well drained to a depth of at least two feet. Soil should be neither too acid nor too alkaline—a pH of 5.5 to 6.5 is best for strawberries. The use of mulch will help prevent injury to roots.

**Fruit rot** [color], of which there are several types, causes trouble on ripening or picked fruit. Gray mold, caused by the botrytis fungus, affects blossoms and green and ripe fruits. Infection often starts in frost-injured blossoms or in berries near the ground. Any injury to blossom or fruit will allow the fungus to enter, spreading a light brown, soft rot. Gray mold disease is favored by rainy weather and moist conditions, and thinning plants will help you to keep infection to a minimum.

Plan ahead by setting out plants in narrow rows. Do not fertilize plants heavily in spring, as this causes thick foliage, which in turn keeps berries moist and susceptible to rotting. A mulch of straw or pine needles will help prevent soil-borne fungus from infecting fruit.

Other fruit rots include hard rot, which attacks berries that touch the soil and causes a hard brown area on them, and leather rot, a discoloration and leathery texture on berries that touch the soil, either directly or through splashing. These minor rots are also prevented by thinning plants and laying down a mulch to keep berries off the soil.

Although foliage diseases are widespread, they seldom cause severe losses. **Leaf blight** leads to large, red to brown spots surrounded by a purplish margin, and is most destructive to weak or slow-growing plants. Some varieties are more susceptible, as are older plants. **Leaf scorch** causes small, dark purplish spots on leaves and sometimes on fruit stalks. Usually it occurs during wet spring months, or occasionally in autumn. In severe cases, plants may die. See RESISTANT VARIETIES. **Leaf spot** [color] is similar in appearance to leaf scorch, but the purple spots have graying or white centers.

This fungal disease attacks early in the season, especially where growth is very succulent or where heavy nitrogenous fertilizer has been used. Leaves become resistant as they mature. The leaf spot fungus sometimes attacks fruit and causes black seed, affecting the berry with one or two black spots.

Control for all the above foliage diseases is the same. Choose resistant varieties, and plant in well-drained soil in an unshaded area. Keep weeds down and avoid overgrowth. Rotate or renew plantings often.

**Leaf variegation**, also known as June or spring yellows, is a condition in some varieties of strawberry that is due to a hereditary factor. As the names imply, a mottling of the leaves is the chief symptom. Yellowing is severe in cool spring weather, but later disappears. Blakemore, Premier, and the everbearing varieties seem especially susceptible to this trait. The condition is not infectious. The only method of control is to plant stock that is free of the inherent weakness of leaf variegation.

Another of the foliar diseases is **powdery mildew**. It causes an upward curling of leaf margins, and a white powdery fungus may be visible on the upper leaf surfaces. It usually develops in spring or fall under overly moist conditions.

Choose resistant varieties, and plant in well-drained soil in an unshaded area. Keep weeds down, avoid overgrowth, and rotate or renew plantings often.

One of the strawberry plant's greatest enemies is the root disease **red stele,** a destructive fungal infection that causes plants to become severely stunted and to wilt, often just before the fruit becomes ripe. The plants have a dull bluish green color, and the small white feeder roots are lost, leaving long, rattail-like main roots that are often discolored at the tip. Some or all of the roots will show a reddish color when split, showing either near the tip or extending through the entire root. The red color is rarely seen during summer, but in spring and late autumn as a rule. The rotted roots prevent water from reaching the plant, and severely affected plants may die. Others may appear to recover in warm summer weather, but will develop symptoms again when cool, wet fall weather arrives and again the following spring. Because the fungus develops at cool temperatures with plenty of moisture, it is especially dangerous in spring. Red stele is extremely infectious. It can survive many years in the soil, and may be spread by soil adhering to cultivating and farm equipment, by water running off from affected fields, or by transplanting diseased plants.

If you know of red stele in the neighborhood, you might seriously question planting any strawberries. Control of red stele is difficult, but fortunately there are resistant varieties. Any plant with the symptoms should be dug up for inspection of the roots. Correct faulty drainage in the strawberry bed. Inspect plants carefully before putting them in the ground; if you spot the rattail or branchless root, there may be a case of red stele. Cut into the suspicious-looking roots to check for the reddish center. See RESISTANT VARIETIES.

**Rhizopus rot** or **leak** attacks berries after harvest, causing juice to run from the berries. It will destroy the fruit very rapidly, causing a whitish mold and brown color. It is spread through bruised or wounded berries and can also be spread by direct contact. This rot is most active at warm temperatures. Control by handling fruit carefully and keeping it at a cool 40° to 45°F.

**Verticillium wilt** is another fungal disease that is most active in cool, humid weather. Symptoms often appear just as the plant is about to ripen. Leaves wilt and dry at the margins and between the veins. New growth is retarded, giving the plant a stunted appearance. New roots growing from the crown are shortened and usually have blackened tips. Black streaks may appear on leaf stalks and runners. Severely affected plants sometimes suddenly collapse and die, while slightly affected ones often recover and produce normally the following year. (In the West, however, affected plants do not usually recover.) The disease may spare the rooted runners from a diseased mother plant.

The verticillium wilt fungus can survive many years in the soil. There are no strawberry varieties completely resistant to it, but some are better off than others. Varities that are somewhat resistant are Blakemore, Catskill, Guardian, Marshall, Redchief, Salinas, Robinson, Siletz, Sunrise, Surecrop, and Vermillion. Varieties most susceptible to verticillium wilt are Daybreak, Dixieland, Earlidawn, Jerseybelle, Klondike, Lassen, Molalla, Northwest, Raritan, Shasta, and Vesper.

Avoid rotating strawberry plantings with crops such as tomatoes, peppers, eggplant, or potatoes, as they may have left the fungus in the soil. Do not plant strawberries where cotton, okra, melons, mint, apricots, almonds, pecans, cherries, avocados, roses, or cane fruit grew anytime in the previous ten years.

Any one of several **viral diseases** may affect the yield of your strawberry plants. Some of these diseases do not show any clear-cut symptoms, but are responsible for a less vigorous or even stunted plant with fewer runners. Two viral diseases, aster yellows and leaf roll, will cause leaves to be twisted, rolled, or cupped downward. Multiplier disease results in spindly plants that have many crowns, few short runners, if any, and leaves a third to half of normal size.

For control, plant virus-free stock. Follow the precautionary measures described in the introduction to this section.

**Weeds,** if not controlled by cultivation, will compete with strawberries for soil nutrients, and you'll end up with less of a crop. Hoeing or hand weeding is necessary in small gardens. When a cultivator is used in large beds, it should be operated in the same direction each time, for going in different directions will often pull out runners that have been dragged into position at the edge of the row.

## Wildlife

**Birds** are fond of strawberries, and you may need to set up wire cages or plastic netting stretched over a frame.

**Mice** sometimes take up their winter quarters in a mulched strawberry bed, and may do damage by rooting around among the stems. In order to avoid this, thin out the plants in early fall before the mice begin to make their winter nests, and do not spread the winter mulch too early. By the time the top of the ground has a frosty crust, mice will have made their winter homes elsewhere and the winter protection may be safely spread on the strawberries.

# STRESS

Plants have several requirements for health: light, nutrients, water, and friendly temperatures. If any one is stinted, then a plant will be stressed, and therefore more susceptible to diseases, pest damage, and competition from weeds.

Plants need *light* in order to conduct photosynthesis, the process by which the sun's energy combines carbon dioxide from the air with water and nutrients from the soil to produce sugars. These sugars are used in a few ways: as fuel for the plant's metabolism, as food stored in the form of starch, and as the stuff from which fruit and other plant parts are made.

A lack of *nutrients* may make itself known by symptoms that look like those caused by pathogens—bacteria, fungi, and viruses. But appearances are very deceiving, and plants are slow to manifest the effects of a deficiency. Visual symptoms of nutrient deficiency may appear in any or all organs, including leaves, stems, roots, flowers, fruits, and seeds. A general note to keep in mind is that a deficiency of one element implies excesses of other elements; for example, too much potassium can block a plant's uptake of magnesium. See entries on individual nutrients and COMPOST.

Wouldn't a bug choose to pick on healthy plants, just as we instinctively look over the garden to select the most attractive vegetables for a salad? Apparently not. Bugs and people have dissimilar appetites. Similarly, it is the weak, underfed, rough-coated calves—and not the suckling, fat, smooth-coated ones—that are eaten up with lice. A weak, sickly hen in the flock will always carry most of the lice. Trees weakened by drought, leaky gas mains, or loss of roots due to excavation are more heavily attacked by borers than nearby healthy trees of the same kind. Chinch bugs tend to collect and breed more heavily on corn or wheat up on an eroded slope rather than down at the foot of the slope where eroded soil minerals and organic matter pile up to enrich the soil.

Judging by such evidence, bugs and diseases act as censors in nature. And it's all because insects thrive and reproduce better on unbalanced or inferior diets, as we think of them. While we are interested in proteins and minerals in our food, the bugs go for plants having excess carbohydrates. Other plant constituents figure in insect feeding habits, too. So the best way to make plants unattractive to bugs is to be sure that plants get a balance of nutrients, and this can only be brought about with a healthy soil, enriched naturally.

Relatively little of the *water* plants take from the soil is retained. Most is used to carry nutrients and to supply the photosynthesis process; it then evaporates from the leaves. Evaporation does more than draw water up through the plant: it also cools leaves in hot, dry weather, just as perspiration cools skin. Leaves may be discolored if they give up water faster than the roots can draw it in.

Any plant has a range of *temperatures* in which it does best, and this range may change with the plant's life stages—sprouting, vegetative growth, blossoming, and fruiting. Higher temperatures generally favor growth, not only of the plant but also of pests, disease organisms, and weeds.

Stress may be caused by pollution. See AIR POLLUTION and WATER POLLUTION.

# SULFUR

Sulfur has been used as a fungicide for thousands of years. Although it is a naturally occurring element, and important to plant nutrition, that doesn't mean the vineyard, garden, or orchard can handle it in concentrated doses. Some organic growers don't use sulfur, reasoning that as a fungicide it can disturb the soil microorganisms that work for the overall health of the plants. Entomologists point out that sulfur can have a negative effect on beneficial insects.

Many people cannot grow grapes without resorting to sulfur because they have planted grapes in areas of slow-moving air and high humidity concentrations—perfect for the development of mildew. Grapevines shouldn't be troubled by mildew if grown in sunny locations on slopes where the air moves, if properly pruned during the season for maximum air circulation and light penetration, and if cultured organically. In less than these optimum conditions, mildew may be a persistent problem, and growers usually have to choose between using sulfur or uprooting the vines.

Sulfur is used by fruit growers to avoid both fungal diseases and mites. Biological control research has shown that sulfur is lethal to many beneficials that, left to themselves, can keep down a number of important pests. So in the orchard too, sulfur is a last resort. Another thing to keep in mind is that sulfur and oil sprays should not be used within a month of each other.

You shouldn't need to use sulfur in the garden if plants are properly tended. This is just as well, as many cucurbits and roses may be severely damaged by this fungicide.

Dusting sulfur is ground very fine. Wettable sulfur is a powder combined with a wetting agent that helps the sulfur mix better with water to make a liquid for spraying. In greenhouses troubled by powdery mildew, a mixture of one pound wettable sulfur and one pint water can be painted on heating pipes to release a sulfur vapor. The applications should be repeated twice a week. For small greenhouses, you can make this simple vaporizer: Place flowers of sulfur in the bottom of a tin can, through the side of which is inserted a 60-watt light bulb. Lime sulfur is a reddish brown liquid used as a dormant spray in concentrations of one part to eight or ten parts water. Lime sulfur is caustic and must not be applied to summer foliage.

# SWEET POTATO

## Insects

**Root-knot nematodes** are small soil pests that attack the roots of many plants, including sweet potatoes. These nematodes produce small galls or swellings on the fine feeder roots, but they can also enter the storage roots and feed in the tissues beneath the skin without causing the common galls found on root crops. Symptoms include decayed areas under the skin, surface blemishes and pitting, deformed roots, poor color, and sometimes severe surface cracking. The vines of infested plants are usually stunted and yellowish, the leaves may show brown dead spots, and the plants may be killed in severe cases.

Whenever possible, plant sweet potatoes in soil that is free of this pest. Nematodes are found not only in the soil but also on the roots of plants, and it is important to obtain nematode-free plants for transplanting into the field. The sources of infestation for plants are infested seed stock and plant-bed soils. You can use vine cuttings in clean soil as a sure way of getting clean seed.

Yellow Jersey, Heartogold, and Nemagold have been found resistant to the common root-knot nematode. The nematodes enter the roots of Nemagold but fail to develop and usually die. See NEMATODE.

The **sweet potato weevil** is about ¼ inch long and resembles a large ant. The head, snout, and wing covers are dark blue, and the prothorax and legs are reddish orange. It has well-developed wings and is capable of limited flight. The eggs are yellowish white while the larvae are white, legless, and about ⅜ inch long. The pupae are white and somewhat shorter.

The adult places its eggs in small cavities that it punctures either into the stem of the plant near the ground or directly into the sweet potato. The eggs hatch in about a week, and the grubs then feed in the vine or potato for about two or three weeks. The pupa is formed within the stem. The adult may live for several months, the time varying with the weather and the

Sweet potato weevil

conditions under which the potatoes are stored. In a year, six to eight generations may be produced. The adult weevils damage sweet potato plants by feeding on leaves, vines, and roots and by pitting the potatoes with feeding and egg deposition cavities. However, it is the larvae that cause the gravest damage by their feeding and tunneling through vines and potatoes.

In areas of commercial production where weevil infestation is light, nonplanting zones can be established. The weevil is wiped out by depriving it of food for about a year; no sweet potatoes are grown, bedded, or stored within a zone extending at least half a mile from the point of infestation. The procedure has resulted in eradication of the weevil from both single farms and whole communities. Potatoes still in the ground when infestation is discovered should be removed from the premises at harvest time and disposed of in such a way as to prevent infestation of other properties. Before the potatoes are plowed out, vines should be cut off at the surface of the ground and burned when dry. Destroy all potato roots, crowns, small sweet potatoes, and scraps in the field by cultivating and by grazing livestock in the field after harvest. The old potato fields should be plowed at least twice during the winter in order to expose any roots or potatoes missed. After the waiting period is up, replant in the zones that have been out of production.

In areas generally infested with weevils and in places where the extent of commercial production does not warrant the establishment of nonplanting zones, effective control can be maintained by the recommended cultural and sanitary practices. Use state-certified seed sweet potatoes and be certain to destroy plants and tubers in seedbeds as soon as you have produced enough plants. Rotate crops and try to plant the new crop as far away as possible from the crop of the previous year. See RESISTANT VARIETIES.

# Diseases

**Black rot** is a fungal disease that causes potatoes to rot both in the field and in storage. It can also attack the underground stem of the growing plant, causing dark decayed spots and sometimes resulting in death. On the potatoes themselves this rot is characterized by dark, slightly sunken corky areas on the surface. Spots are usually circular, with the diseased areas sharply defined from the healthy areas. As decay develops the spots penetrate into the flesh of the potato. Decayed tissue often takes on a greenish tinge when cut open. A bitter taste is associated with this rot. All potatoes showing visible signs are culls, and black rot can sometimes cause serious cullage in the fields and after storage.

Plants can become infected in the seedbed from diseased seed roots or from infected soil. The disease is carried on seed potatoes in both active and spore form and can live in the fields or plant-bed soil for a year or longer. Sweet potatoes that are washed before packing may be infected by spores in the wash water. Black rot that develops in transit may make the whole lot unsalable when it reaches the market. Varieties enjoying considerable resistance are available; Yellow Jersey, on the other hand, is especially susceptible.

Discard seed potatoes with black rot, practice a three-to-four-year crop rotation, and cure quickly with high temperatures and humidity. One easy way to control black rot is to cut the sprouts above the seedbed soil line instead of pulling them. Since black rot occurs only on the underground stem, only the healthy part is planted. Sprout cuttings live and yield as well as field sprouts.

**Dry rot** causes sweet potatoes to mummify in storage. Cool storage discourages this fungus, as the ideal range of temperatures for the pathogen is 75° to 90°F.

**Scurf** is a fungal disease that lives on the outermost layers of cells on the potatoes and causes superficial round, brownish black spots. Individual infections are relatively small, but they may be numerous enough on a root to cause almost complete discoloration of the surface. While eating quality is not harmed, the poor appearance of the diseased potatoes greatly reduces market acceptability. Although this disease is not caused by moisture, moist conditions favor its growth. It develops on potatoes in the field and in storage when the spores are carried in from the field. The disease or its spores can live over for a season in hotbed sites or in the field as well as on the potatoes.

Use healthy sprouts only. They should be bedded in sand that has not previously hosted sweet potatoes.

**Soft rot** is a storage disease characterized by rapid, soft, watery rot and a whiskery growth. Nancy Hall and Southern Queen have some resistance to this disease. Sweet potatoes should be cured for 10 to 14 days at a humid 80° to 85°F., followed by storage at 55°F.

**Stem rot** or **wilt** is caused by a fungus of the fusarium group and lives within the roots and stem of the plant. Sweet potato plants may become infected from infected seed roots or infected soil in the field or seedbed. Aside from the direct infection of plants produced on infected seed roots, the principal entrance of the fungus appears to be through wounds incurred during pulling or transplanting. Considerable spread may take place when the plants are dipped or soaked in basins or sumps while being held before setting in the field. Infection in the field may occur when the stem or roots are injured by insects, tillage tools, or wind.

Young plants infected in the seedbed often die after transplanting. Surviving plants first show a few bright yellow leaves around the crown, then dwarfed leaves and crown, and later badly rotted stems. Although they may survive through the season, the potatoes are usually small and yields are poor.

The stem rot fungus lives in the potatoes through the storage season without causing external symptoms. Spores of the disease picked up in the field during harvest may adhere to the roots during storage.

Southern Queen, Yellow Stransburg, and Triumph are quite resistant; among highly susceptible varieties are Big Stem Jersey, Little Stem Jersey, Maryland Golden, and Nancy Hall. Use plants with clean white roots and remove and destroy diseased plants. Do not plant sweet potatoes in the same soil every year.

# SWISS CHARD

See CHARD.

# SYCAMORE

## Insects

The sycamore **lace bug** [color] is very common on the undersides of leaves. The upper surfaces will show a white-peppered spotting which indicates where the bug has sucked sap. The adult bugs hibernate under the edges of the bark and emerge to lay their eggs soon after the

Sycamore lace bug

leaves unfold. A spray of nicotine sulfate and soap twice in spring will control the lace bug on sycamores.

Other insects that may trouble sycamores are APHID, terrapin scale (see MAPLE), and white-marked tussock moth (HORSE CHESTNUT).

## Diseases

For description and control of **anthracnose,** see TREES.

# SYMPHYLAN

Also known as the garden symphilid, the many-legged symphylan [color] is often confused by gardeners with the centipede and millipede. Each is in a different class and has distinguishing habits. The centipede is a beneficial predator, feeding nocturnally on soil insects and millipedes. The millipede is anything but beneficial, and is discussed in a separate entry.

Symphylans have bodies broken up into 14 segments, and move about on 12 pairs of legs. Because of their superficial resemblance to centipedes, they are sometimes called garden centipedes. They feed on young plant roots and are capable of downing their own weight's worth in a day. Affected plants tend to wilt in strong sunlight and are prone to fungal and bacterial root rots. Lettuce leaves may be eaten full of holes if they rest on infested ground. On tomatoes, stunted plants show bluish stems, yellowed lower foliage, and dark green upper foliage.

To check garden soil for symphylans, dunk the root ball of a plant in a bucket filled with water. The pests should rise to the top. Or you can dig a shovelful of soil and sift through it. If you find an average of ten or more symphylans in each of a sample of several shovelfuls, then your garden is in need of help.

On a large scale, fields can be flooded for a time to wipe out the symphylan population. A safe, nonchemical control for the gardener has not been devised, but you might try drenching the soil with garlic or tobacco teas. It's best to move the compost pile some distance from the garden if symphylans have been troublesome. You can reduce their numbers by thoroughly mincing the soil with a power tiller.

# TACHINID

As larvae, tachinid flies [color] parasitize many insect pests. One species, *Tachineaphagus zealandicus,* is commercially available. See FLY.

# TANGERINE

See CITRUS FRUIT.

# TENT CATERPILLAR

The larvae of several moths and butterflies are collectively referred to as tent caterpillars. The name is applied especially to *Malacosoma americana,* known as the eastern tent caterpillar

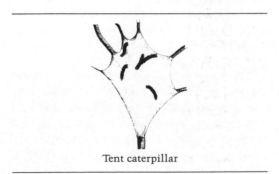

Tent caterpillar

[color] and sometimes as the apple tent caterpillar. Tent caterpillars multiply rapidly and can defoliate many deciduous trees and shrubs over a wide area in a short time. Wild cherry trees and apple trees are most often attacked; peach, pear, plum, rose, hawthorn, and various shade and forest trees are occasionally infested. In spring the pests' unsightly nests, or tents, are conspicuous on susceptible trees by the roadside or in neglected orchards. These caterpillars are abundant and troublesome for several years in a row. They often eat all the leaves on a tree, weakening but seldom killing it. Once the caterpillars mature in early summer they cause no further feeding damage.

LIFE CYCLE.   One generation of the eastern tent caterpillar develops in a year. Larvae are present in late spring, cocoons and moths in early summer, and eggs for the remainder of the time.

The larvae, or caterpillars, hatch in spring from egg masses about the time the first leaves are opening. The young caterpillars keep together and spin threads of silken web. After feeding for about two days, they begin to weave their tent in a nearby tree crotch, sometimes joining with caterpillars from other egg masses. As the caterpillars grow, they enlarge the tent until it consists of several layers. In good weather they leave the tent several times a day in search of food, stringing silk after them, while bad weather sends them between the layers of the tent. When they finish feeding on a tree, the caterpillars leave the nest and search for food. Upon reaching maturity they spin cocoons on tree bark, fences, brush, weeds, or buildings, or among dead leaves and debris on the ground. When full grown, about six weeks after hatching, the caterpillar is almost two inches long and sparsely hairy. It is black, has white and blue markings, and shows a white stripe along the middle of its back.

The pupal cocoon is about one inch long and white or yellowish white. In early summer, reddish brown moths emerge, and the females deposit masses of eggs in bands around twigs. The eggs are covered with a foamy secretion that dries to a firm brown covering that looks like an enlargement of the twig. An egg mass usually contains about 200 eggs.

## CONTROL METHODS.

You can control the eastern tent caterpillar by hand if you have only a few infested trees. Since many insects are concentrated in a few groups, they can be easily destroyed. Take action when you first see the nests before the larvae start to feed. Tear the nests out by hand or with a brush or pole, and either crush any surviving caterpillars on the ground or burn them with a torch made from oily rags tied to a pole. If you elect the latter method, be careful that the tree is not singed and that the fire does not get out of hand. In winter, you can destroy the egg masses by cutting off the infested twigs and burning them. Remove wild cherry trees growing in the vicinity of orchards, if possible.

The star insect enemy of a western species, the Rocky Mountain tent caterpillar, is a digger wasp labeled *Podalonia occidentalis.* This insect is somewhat less than an inch in length and is entirely black with the exception of the abdomen, which is gleaming red orange with a black tip.

Full-grown caterpillars are attacked by an important parasite that you might mistake for the common housefly. *Sarcophaga aldrichi,* however, is twice the size; its light gray thorax bears three longitudinal black stripes and the abdomen is divided into alternating light and dark squares. These flies insert their living maggots beneath the skin of caterpillars. The pests usually live long enough to spin their cocoons and transform into pupae before succumbing.

A commercially available pathogen, *Bacillus thuringiensis,* is a very effective control. Baltimore orioles have been known to clean up entire infestations, and the birds' hanging sacklike nests are a sign that your trees are insured against tent caterpillar trouble.

## SIMILAR PESTS.

The forest tent caterpillar *(Malacosoma disstria)* and the fall webworm *(Hyphantria cunea)* are sometimes mistaken for the eastern tent caterpillar. The forest tent caterpillar may be distinguished from the eastern tent caterpillar: it has a row of creamy white spots, instead of a stripe, along its back; it is found more often on forest trees than on fruit trees; and it does not form webs.

The fall webworm makes a nest resembling that of the eastern tent caterpillar, but it is located at the tip of a branch instead of at the crotch. In addition, the fall webworm is smaller and hairier, and is present from midsummer to autumn. It feeds on many kinds of trees, including most of those attacked by the eastern tent caterpillar.

These insects may be controlled with the same sprays used on the eastern tent caterpillar. For more information, consult your county agricultural agent, state agricultural experiment station, or state agricultural college.

# THRIPS

Thrips are small, slender insects with feathery wings. They damage plants by sucking their juices, stinging them, and rasping at fruit and leaves to cause scars. Leaves may turn pale and silvery, then die. The pests are attracted

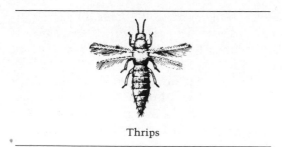

Thrips

especially to white blossoms. Damaged rosebuds turn brown and may open to show distorted petals. Thrips may transmit the virus that causes spotted wilt on tomato.

Infested flowers and buds should be promptly removed to avoid infesting plantings. Plants under water stress are particularly susceptible to trouble; see to it that plants are properly watered or irrigated in hot seasons, especially in dry areas. On a large scale, you can get around thrips by not growing susceptible crops in consecutive years; a three-year crop rotation (e.g., red clover or potatoes, oats, and then flax) will help.

Because many thrips have a wide range of hosts that includes many weeds, they can often survive in the garden area even if no susceptible crops are grown. So, control of weeds and volunteer plants may be well worth your while for this pest alone, especially if tomato spotted wilt virus has been a problem. The area of clean culture should extend for some distance; thrips are weak fliers, but they can beat the air hard enough so that the wind will blow them a good ways. In still air the smaller species can manage a speed of 10 centimeters a second, while larger thrips can hit 50 centimeters a second. But the wind usually blows them wherever it will. Thrips stick to the ground when stiffened by the cold, and won't take off if the temperature is below 64° to 69°F.

Aluminum mulches are effective in keeping thrips off low-growing crops. The pests will avoid rose blossoms if you place foil-wrapped boards around the plant base so that they extend a foot or two beyond the canopy. As is true of aphids, thrips apparently lose track of what's up and what's down.

Green lacewings, commercially available predators, prey on thrips. The pests are vulnerable to soap, oil-and-water, and tobacco sprays, and to diatomaceous earth when applied to blossoms and soil. Rotenone works too, and you might try making a repellent spray of common field larkspur. Tobacco and sulfur dusts are proven stand-bys. Some thrips species are beneficial predators of mites, other thrips, and small insects; good thrips are usually distinguished by bands or mottling on the wings. See also gladiolus thrips under GLADIOLUS.

# TIMED PLANTING

Insects appear at about the same time every year, their exact emergence dependent on temperature, moisture, and availability of food. After feeding for a while, they will either hibernate, migrate to other areas, or lay eggs and reemerge later to resume feeding. By planning to harvest before, or planting after, the heaviest feeding stages, you may protect your crops.

The following chart, based on data from entomologists around the United States, shows planting dates which you can use to avoid insect damage. Insect emergence times will vary within the regions shown. By keeping records of the dates you first spy insect pests in your garden, you can create your own very precise emergence time chart within a few years. Such knowledge of insect life cycles will further enable you to plant around their feasts, and correctly time releases of beneficial insects. You may also want to consult your county extension agent for more-precise local data.

Use the map of the United States to find your region. Say a reader in Tennessee has problems with striped cucumber beetles every year. She finds that she's in region 5, and that, according to entomologists in her area, the beetle emerges in early May. Rather than set out her

tender young cukes at the same time, she may decide to wait until mid-June, after the beetles have emerged, found her garden lacking in cucurbits, and moved on.

Or say that our Tennessee reader has a problem with codling moths damaging her apples. The chart shows that the moth has a first generation in early May and another in July. If she sprays with rotenone, she will time her sprays at these periods.

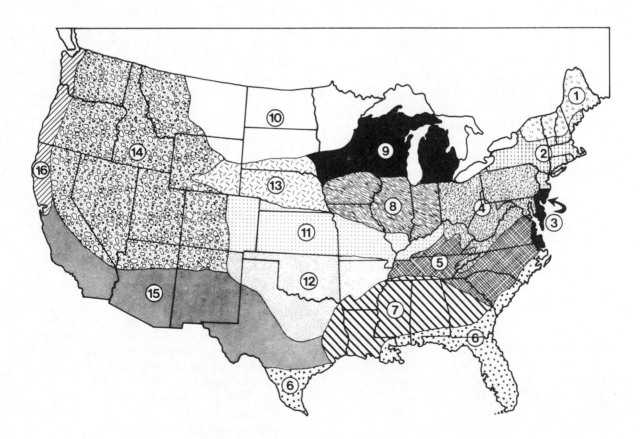

The 16 zones represent areas where emergence times of various insects are approximately the same. In zone 14, the Rocky Mountains, emergence times depend more on elevation than latitude.

## INSECT EMERGENCE TIMES

| Region | Bean Aphid | Green Peach Aphid | Pea Aphid | Woolly Apple Aphid |
|---|---|---|---|---|
| 1 | No problem. | June 1. | End of May. | Late June to early July. |
| 2 | No problem. | Late April. | April. | Does not apply. |
| 3 | No problem. | Eggs hatch late April. | April. | Migrates in June. |
| 4 | June 1. | May through June. | April. | Does not apply. |
| 5 | May through June. | Late April, early May. | May. | April. |
| 6 | Year-round. Worst from Oct. to May. | Year-round. Worst Oct. to May. | On peas in fall, winter, spring. Worst Nov. to Feb. | Does not apply. |
| 7 | Mid- to late-March. Active year-round. | Mid- to late-March. Active year-round. | Mid- to late-March. | Mid-April. |
| 8 | When warm weather arrives. | At peach bloom. | Active May 1, most abundant during June. | May and June. |
| 9 | June. | Early June. | April. Most destructive in May. | June. |
| 10 | Not much problem. | May and June on fruit. | Builds up May-June, again in fall. | Does not apply. |
| 11 | No problem. | Early spring and fall. | Active in April. | On roots, early April. Aboveground, mid-June. |
| 12 | No problem. | Spring and fall. | March and April. | On elm in spring. Migrates to apple early summer. |
| 13 | No problem. | May. | May. | Mid-May. |
| 14 | Active June through Aug. | Active all year; peaks early July. | Active mid-May to June; mid-Sept. to Oct. | Late April. Active through season. |
| 15 | Spring and fall. | Spring and fall. | Does not apply. | Spring. Most active in late June. |
| 16 | Active mid-May through July. | Mid-April to early June. | May through June. | Early to mid-June. |

(continued on next page)

| Region | Asparagus Beetle | Colorado Potato Beetle | Striped Flea Beetle | Japanese Beetle |
|---|---|---|---|---|
| 1 | Mid-June. | First week of June. | June 1. | Mid-July. |
| 2 | May. | Late May. | Mid-May to early June. | Late June. Active through Oct. |
| 3 | Late April to early May. | Early May. | Late April to early May. | End of June. |
| 4 | Early May. | Early May. | Early May. | End of June. |
| 5 | Late April to early May. | May. | April. Active through summer. | June. |
| 6 | Does not apply. | March. Active through May. | Spring, summer, and fall. | Does not apply. |
| 7 | Does not apply. | Does not apply. | March and April. | Does not apply. |
| 8 | May. | April to May. | Late May. | Most abundant in July. |
| 9 | Early May. | Mid-May to late June. | May through June. | Does not apply. |
| 10 | May and June. | June. | Does not apply. | Does not apply. |
| 11 | Late April through May. | Late April through May. | April to May. | Does not apply. |
| 12 | Early spring. | Does not apply. | When seedlings come up. Active all season. | Does not apply. |
| 13 | Mid-June. | Mid- to late-June. | June. | Does not apply. |
| 14 | Active early May through June. | Mid-June. | Active all summer. | Does not apply. |
| 15 | Does not apply. | Does not apply. | Does not apply. | Does not apply. |
| 16 | April through June. | Late May to early June. | Does not apply. | Does not apply. |

| Region | Corn Earworm | Eastern Tent Caterpillar | Forest Tent Caterpillar | Fall Armyworm |
|---|---|---|---|---|
| 1 | Late July. | Eggs laid May 1. | Eggs laid May 1. | Does not apply. |
| 2 | August. | Late April to early May. | Does not apply. | Does not apply. |
| 3 | June, peaks in Aug. | Eggs hatch in April. | Does not apply. | Late July to early Aug. Peaks late Aug. |
| 4 | July to Aug. A problem after Sept. 1. | April. | April. | July through Aug. |
| 5 | Mid-June. | April. | Moths emerge in June. Active through fall. | Late July. |
| 6 | North: Summer, early fall. South: Early spring, late fall. | Eggs hatch in March. | Does not apply. | North: June through July. South: Feb. through April. |
| 7 | 1st generation—June. 2nd—mid-July, most destructive. | March. | March. | Mid-June. |
| 8 | Spring and early summer. | Eggs hatch March. Larvae grown in 4 to 6 weeks. | Eggs hatch in May. | Fall in north, mid-summer in south. |
| 9 | End of July. Active through Aug. | Does not apply. | Does not apply. | Does not apply. |
| 10 | Mid-July through Aug. | May. | May. | Does not apply. |
| 11 | Migrates late June. Causes damage through Sept. | Eggs hatch April 15. | Does not apply. | July and Aug. |
| 12 | Plant so corn silks between June 25 and July 1. | When first buds come out. | Does not apply. | July 4. Next 3 generations appear monthly. |
| 13 | Causes most damage in July. | Mid- to late-April. | Mid-May to June 1. | Aug. and Sept. |
| 14 | Early July. | Does not apply. | Mid-April. | Active July through summer. |
| 15 | Does not apply. | Does not apply. | Does not apply. | Does not apply. |
| 16 | June and July. | Does not apply. | Late spring to early summer. | Does not apply. |

*(continued on next page)*

| Region | Gypsy Moth | Tomato Hornworm | Apple Maggot | Plum Curculio |
|---|---|---|---|---|
| 1 | Eggs laid May 10. | Does not apply. | June 30 to July 1. | Mid-May. |
| 2 | 1st generation— May. | Late July to early Aug. | End of June. Active into Oct. | Mid-May. Active through early July. |
| 3 | Eggs hatch late April, early May. | Does not apply. | July 4. | May. |
| 4 | Eggs hatch May 1. | June. | July 4. | May. |
| 5 | Does not apply. | Late June. | May and June. | April 1 to mid-April. |
| 6 | Does not apply. | Apr. to Sept. | Does not apply. | Early March. |
| 7 | Does not apply. | Mid-June. | Does not apply. | Does not apply. |
| 8 | Late April to May. Full-grown larvae by late June. | May or June— eggs hatch a week later. | Late June to early July. | Active when apples and peaches bloom. |
| 9 | Does not apply. | Late June to early July. | Mid-Aug. Active until first frost. | Does not apply. |
| 10 | Does not apply. | Does not apply. | First half of July. Active through Aug. | Does not apply. |
| 11 | Does not apply. | July and Aug. | Does not apply. | Does not apply. |
| 12 | Does not apply. | June. | Does not apply. | Does not apply. |
| 13 | Does not apply. | Late June. | Mid-June to Sept. 1. | Does not apply. |
| 14 | Does not apply. | July and Aug. | Does not apply. | Does not apply. |
| 15 | Does not apply. | Does not apply. | Does not apply. | Does not apply. |
| 16 | Does not apply. | Late May through June. | Does not apply. | Does not apply. |

| Region | Mexican Bean Beetle | Northern Corn Rootworm | Spotted Cucumber Beetle | Striped Cucumber Beetle |
|---|---|---|---|---|
| 1 | June 1 to mid-June. | Does not apply. | Mid-June. | Mid-June. |
| 2 | Mid-June. | Does not apply. | Mid-June. | Mid-June. |
| 3 | May. | Does not apply. | May to June. | Mid-May. |
| 4 | June 1. | Eggs hatch mid-June. | June. | End of May. |
| 5 | May. | Does not apply. | Early May. | Early May. |
| 6 | Does not apply. | Does not apply. | A continuous pest. | Does not apply. |
| 7 | Does not apply. | Does not apply. | Does not apply. | Does not apply. |
| 8 | Late March in south. June in north. | Larvae in spring. Adults in July. | First 70°F. day. | May. |
| 9 | Does not apply. | Larvae in late June. Adults in late July. | Mid-May. Active through season. | Does not apply. |
| 10 | Does not apply. | Eggs hatch June. Adults emerge late July. | June. | Does not apply. |
| 11 | Does not apply. | Does not apply. | Early May. | Does not apply. |
| 12 | Does not apply. | Does not apply. | Does not apply. | Does not apply. |
| 13 | Does not apply. | Larvae in June. | May. | Does not apply. |
| 14 | Adults in May. Larvae feed May to July. | Does not apply. | Does not apply. | Does not apply. |
| 15 | Does not apply. | Does not apply. | Does not apply. | Does not apply. |
| 16 | Does not apply. | Does not apply. | Does not apply. | Does not apply. |

*(continued on next page)*

| Region | Squash Bug | Squash Vine Borer | Cabbage Looper | Codling Moth |
|--------|-----------|-------------------|----------------|--------------|
| 1 | Does not apply. | Does not apply. | Mid-July. | June 1. Ten days later, fruit will drop. |
| 2 | June. | Mid-June. Most active in July. | Late June to early July. | 1st generation—third week of May. 2nd—Aug. |
| 3 | Late May to June. | Adults—May. Larvae—second week of June. | July. Active through Sept. | 1st generation—late May. Larvae—early June. |
| 4 | June. | Mid-June. | 1st generation—June. | 1st generation—May 20. |
| 5 | About May 1. | Late June. | Late April. | 1st generation—early May. 2nd—late July. |
| 6 | March. Active through Aug. | Aug. through Oct. | North: Early spring to late fall. Year-round in south. | Does not apply. |
| 7 | June. | Late June to July 1 on garden crops. | Early March. | 1st generation—at petal fall. |
| 8 | June. | Adults lay eggs when vines run. | Moths lay eggs in spring. | Early May. 2nd generation—June to Sept. |
| 9 | Does not apply. | Does not apply. | Early Aug. | 1st generation—mid-June. 2nd—late July. |
| 10 | Does not apply. | Late June to early July. | After mid-June. | 1st generation—mid-May. 2nd—July. |
| 11 | When squash is fruiting. Active through fall. | Early to mid-July. | When cabbage comes up. | 1st generation—May 2nd—June to July. |
| 12 | Late May. | Late May through June. | Active year-round. Most destructive in fall. | 1st generation—at petal fall. 2nd—June and July. |
| 13 | Late spring. | July. | May. | 1st generation—May. 2nd—June. |

| Region | Squash Bug | Squash Vine Borer | Cabbage Looper | Codling Moth |
|---|---|---|---|---|
| **14** | Aug. to Sept. | Does not apply. | June. Active until frost. | 1st generation—mid-May. 2nd—Aug. through Sept. |
| **15** | Does not apply. | Does not apply. | Sept. and Oct. | 1st generation—May 1. 2nd—25 days later. |
| **16** | April and May. | Does not apply. | April and May. | 1st generation—early April. 2nd—late June. |

# TOBACCO

For uses as an insecticidal spray, dust, and fumigant, see SPRAYS AND DUSTS.

# TOMATO

In planning the tomato patch, look into recent varieties; the old standards are gradually being replaced by disease-resistant and better-yielding varieties.

Soil should be worked well to encourage plants to send down a good root system. Plants discouraged early on by compacted or poorly drained soil will be handicapped throughout the season and vulnerable to diseases and pests. Stay out of the tomato patch when the soil is wet in order to avoid compacting the soil.

Nutrient deficiencies may manifest themselves in a variety of symptoms—stunting, poor fruit development, or chlorosis spots and discoloration on leaves. By the time you notice that the plant is hungry for something, it's often too late to set things right. Nitrogen can be added to good effect, in the form of such supplements as blood meal, cottonseed meal, and manure. While animal manures are excellent fertilizer, an overdose often results in excessive vine growth and poor fruit set. Add lime, if necessary, to bring the pH up to between 6 and 6.8. See the subentry "deficiencies" in the following discussion of diseases of tomato.

Early dangers for the young plants are damping-off disease, root-knot nematodes, sun-robbing weeds, and leaf-eating insects. Many growers start their plants indoors. Seedlings will have a better chance of resisting early problems if transplanted into the garden, but also have less developed root systems and therefore would not fare as well in compacted and otherwise unfavorable soils. Seedlings should be thinned so that they are two to three inches apart by the time they reach a height of one inch; transplant the extras to flats or peat pots. If plants are left unthinned, they will become spindly and more subject to disease. Tomato plants grown in protected places tend to become somewhat weak and spindly; give them a few days' exposure to the outdoor environment while they are still in the containers to help them survive transplanting. Transplant when conditions are at their best—soon after a rain, when cloudy, or in late afternoon. Ward off cutworms with a cardboard band or tin can, inserted into the soil around each plant.

A mulch of straw, leaves, dried lawn clippings, or plastic will serve to keep down weeds. To conserve space and keep fruit off the ground, stake the plants. You can encourage plants to produce larger fruit by removing side branches as they emerge. Leave two or three main stems to each plant.

Large-scale plantings should be rotated so that tomatoes are grown once every three years. It is not a good idea to grow tomatoes in rotations with potatoes, eggplant, okra, or peppers, as these crops are susceptible to many of the diseases that hit tomatoes. Keep out of the field when the plants are wet, to avoid spreading such diseases as gray leaf spot, early blight, late blight, and septoria leaf spot. If you are a smoker, be aware that tobacco often carries a virus that can easily pass to tomatoes; wash your hands thoroughly before working with the plants. One researcher found that 40 percent of cigarette tobacco is infected with tobacco mosaic virus, among the most serious diseases affecting tomatoes. So, here's another good reason for dropping the habit.

While you're setting out the tomatoes, sow marigolds, borage, or opal basil around and through the rows to repel the tomato hornworm through the season. Interplant asparagus with your tomatoes, and the roots of the perennial will discourage nematodes that parasitize tomato roots. Nematodes not only damage plants by feeding, but also are carriers of disease—fusarium wilt, verticillium wilt, root rot, tomato ring spot, tomato black ring, and bacterial wilts, among others.

Once plants are well under way and growing rapidly, they are less susceptible to leaf damage. But as fruit appears, most of the plants' energy will go to growing it, and pests may seriously slow vegetative growth, stunting plants and reducing the harvest.

## Insects

**Aphids** of several species may trouble tomatoes, but they are usually kept in check by a variety of natural enemies, including the larvae of the syrphid fly, ladybugs, lacewing larvae, and parasitic wasps. Infestations may cause blossom drop, necrotic spots on leaves, leaf roll, and stunted growth. Mold may grow on the honeydew exuded by the aphids. Aphids transmit certain mosaic diseases to tomatoes. See APHID.

Both larval and adult **Colorado potato beetles** feed on tomato foliage and may strip plants if they occur in numbers. The oval, hard-shelled adult is about ⅜ inch long, with alternating black and yellow stripes on the wing covers. The full-grown young are sluggish, soft-bodied, humpbacked grubs, colored red with two rows of black spots on each side of the body. See POTATO for controls.

Several species of **cutworm** [color] have earned reputations as primary tomato pests. Although they vary considerably in habits and appearance, they are usually gray to brown or black in color and measure from 1 to 1½ inches long when fully grown. Cutworms are ordinarily most troublesome in spring, doing their damage

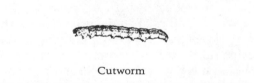

Cutworm

in the hotbed or in the field and garden soon after transplanting. They may cut the plants off near ground level, feed on the foliage or fruit, or destroy the roots. Most obviously, they cut off the young plants.

Clean cultivation of the crops that precede tomatoes will help reduce cutworm damage. In late summer or early fall, plow under sod or weedy ground on which tomatoes are to be grown the following year; keep the area clean to discourage the cutworm from laying its eggs. In a small garden, cutworms can be collected by hand. Protect young tomatoes by pressing tin, tar paper, or cardboard cylinders into the soil around plants; or wrap stems with four-inch collars of paper at transplanting time. Cutworms are vulnerable to *Bacillus thuringiensis*, a microbial control.

The **European corn borer** occasionally moves into tomato rows if early corn is planted nearby. The borers migrate from the dying corn to find a new source of food. See CORN.

Many species of **flea beetle** attack plants, especially young transplants. Leaves look as though shot full of holes. These insects are about $\frac{1}{16}$ inch long, and jump about when disturbed. A garlic spray should repel them. For other control measures, see potato flea beetle under POTATO.

The **greenhouse whitefly** commonly infests tomato under glass and is often carried into the field, where it may persist on the plants. The tiny mothlike adult has a mealy appearance due to small particles of wax that it secretes. It

Whitefly

lays groups of eggs on the undersides of leaves, which hatch into small oval larvae that suck sap. Spraying a solution of nicotine sulfate and soap on the undersides of leaves is a sure means of control, either in the greenhouse or out. See WHITEFLY.

The **stalk borer** is a slender insect that grows up to 1¼ inches long. When young it is creamy white with a dark purple band around the body and several brown or purple stripes running lengthwise. The full-grown borer is creamy white to light purple and is without stripes. The stalk borer eats a tunnel in the stem of the plant, causing it to wither and die. This tunnel has an opening up to ¼ inch in diameter at its lowest end, and is a telltale sign of trouble within. To locate the borer, simply split the stems lengthwise above the opening to the tunnel. The plant may be saved by puncturing the insect and binding up the split stem. Give the plant liberal doses of water for several days.

**Stink bugs** [color] cause white lumps under the skin of tomatoes. These are formed by the punctures the bugs make when feeding. The saliva breaks down the tissue of the fruit. True to their name, stink bugs put off an odor if handled. They move into the garden from weedy areas. Keep down weeds if the bugs have been a problem in past years. Spray with insecticidal soap.

The **tobacco hornworm** [color] is nearly identical to the tomato species, and can be quickly distinguished by the color of the horn —red on the former worm, and green and black on the latter.

Because they're so easy to see, hornworms are most often controlled in small plantings by handpicking. This task becomes even easier if dill is interplanted as a trap crop, because the fat pests are impossible to miss on the spindly herb stalks. If you spot a worm carrying about white pupae, or cocoon-spinning grubs, let it be—the pest is the victim of a parasitic braconid wasp and should be left to nurture a new brood of beneficials. Some gardeners have success with sprinkling hot pepper on plants. The moth can be plucked out of the hornworm life cycle by drawing it to a blacklight trap.

Hornworms are particularly vulnerable to biological controls. A spray of the commercially available *Bacillus thuringiensis* bacterium is effective; see *BACILLUS THURINGIENSIS* for more information. Trichogramma wasps parasitize eggs, and are available through the mails as eggs. Larvae are parasitized by the wasp *Hyposoter exiguae*. Southern tobacco growers have taken full advantage of a naturally occurring wasp predator, *Polistes exclamans*, by setting up inexpensive shelters or nesting boxes to save the wasps from other predators; locally, the hornworm lost its pest status. The males and young queens arrive in late summer and die with the approach of winter, leaving new young queens to find shelter in cracks and crevices and under the bark of trees—or in the shelters provided by growers. *Bacillus thuringiensis* is not harmful to this wasp. If all else fails, use rotenone to kill young caterpillars. Some gardeners plant four-o'clocks as a trap crop.

The **tomato fruitworm** [color], also known as the corn earworm and bollworm, comes in green, brown, or pink, has light stripes along its sides and back, and reaches a length of 1¾ inches. It eats holes in fruit and buds.

In the southern states and California, the fruitworm is an annual pest. In the extreme South the moths may emerge from their pupal cells as early as January, although most of them appear later in spring. Soon after she emerges, the female moth begins to lay single eggs, no larger than a pinhead, on the leaves. The new larva crawls over the leaves, feeding sparingly until it finds the fruit. Arriving at a tomato, it

Tomato fruitworm

cuts a hole and burrows inside, usually at the stem end. A worm may feed until fully grown upon a single tomato, or it may move from one tomato to another, injuring several as it completes its growth. The mature worm leaves the fruit and enters the soil, where it transforms into the pupal, or resting, stage. There may be two or more broods in a single season.

In home gardens having a few plants, the fruitworms may be picked by hand. Garlic and onion sprays should reduce the number of feeding pests. Rotenone has been used successfully as both a dust and a spray. Disc tomatoes after harvest to kill pupae.

Various trichogramma wasps parasitize eggs. *Trichogramma pretiosum* is important to California tomato growers, accounting for up to 40 percent of the eggs of several pests. This species is commercially available. *Hyposoter exiguae* is a parasite of fruitworm larvae.

The tachinid fly *Lydella stabulans grisescens* was imported to parasitize the fruitworm, and is now established in the Northeast. Tomatoes have been protected from the fruitworm moth and other winged pests by blacklight traps. Fruitworm larvae are susceptible to both *Bacillus thuringiensis* and a nuclear polyhedrosis virus. Although this virus is not commercially available, you can mix up your own viral spray; see the controls for corn earworm under CORN.

The **tomato hornworm** is one of the most important and certainly the largest of tomato insect pests. This green worm, distinguished by a series of white diagonal bars along each side and a prominent horn at the back end, reaches a length of four inches—big enough to make handpicking an ordeal for the squeamish. The greenish eggs are deposited singly on the undersides of leaves and hatch in three to eight days. Larvae feed on leaves and sometimes fruit, reaching full size after three or four weeks of steady eating. Winters are spent as hard-shelled pupae, three or four inches below the ground. The moths that appear in May or June are variously known as sphinx, hummingbird, or hawk moths; have a wingspan of four or five inches; and fly on long, narrow gray wings. They fly at twilight and pause in the air to sip nectar from deep-throated flowers in the manner of a hummingbird. While northern growers have but one generation to contend with, there may be two or more in the South. See TIMED PLANTING.

Tomato hornworm adult and larva

Leaves are given a greasy appearance, then are turned dry-looking and bronzed by **tomato russet mites.** Lower leaves are affected first, and this is the symptom to look for—the mites themselves are easily overlooked because they are so small. Younger plants are not often troubled seriously. The pests can be viewed under a good hand lens. They are pale and conical in shape. Hot, dry weather encourages them.

Predatory mites help keep these pests under control. *Phytoseiulus persimilis* and *Metaseiulus occidentalis* can be purchased. Use sulfur, as a dust or spray, as a final measure.

## Diseases

The following discussion of tomato diseases is long not only because the plant is vulnerable to so many ailments, but also because tomato is the most popular garden crop and therefore merits a lot of attention.

The list of diseases seems less daunting if classified by cause. Some are brought on by pathogens—fungi, bacteria, and viruses. Fungal organisms living in the soil are the agents behind root rot and damping-off, fruit rots (anthracnose and buckeye rot), blights, fusarium and verticillium wilts, gray mold, and gray leaf spot and septoria leaf spot. Seeds and seedlings may carry bacteria that cause canker and bacterial spot. Viruses are usually transmitted by insects—aphids, leafhoppers, cucumber beetles, and thrips—and cause mosaics, curly top, and spotted wilt, all primarily diseases of foliage.

Nonpathogenic diseases may be traced to nutrient deficiencies and environmental factors —both the weather and the microclimate created by watering, plant density, and so on.

Note that seed catalog descriptions of tomatoes may give the initials VF or VFN along with variety names; these indicate resistance to verticillium and fusarium wilts, or both wilts and nematodes.

**Anthracnose** [color] is a fungal disease that primarily affects ripe fruit and causes a slight sunken spot rot. Because symptoms are not prominent until the fruit ripens, many growers do not become concerned about this disease until it's too late to prevent infection. The earliest symptoms are circular sunken spots in the skin, which look as though they had been made with a match head or eraser. On ripe fruit these spots enlarge rapidly and the central portion may appear to be dark because of black fungal structures just under the skin. Soon the entire fruit rots as other organisms enter and break down the tissues. Sunken, circular, dime-size spots with black specks in them are typical of this disease. When August temperatures are high and heavy rains or dews occur, be on the lookout for anthracnose. The fungus overwinters both in diseased vines and on and in the seeds; plowing under is not effective. Long rotations and hot-water seed treatment (122°F. for 25 minutes) facilitate control. Place plants in well-drained soil.

**Bacterial canker** is a destructive tomato disease caused by a seed-borne bacterium. In recent years, losses from this disease have been slight, principally because of the increased use of certified seed by southern plant growers. However, growers who save their own seed and ignore control measures can still expect to suffer losses from canker if weather conditions are favorable. The disease is occasionally found in the greenhouse as well as the field, and attacks plants at any stage of growth. Seedlings may be rapidly destroyed or may produce plants that remain stunted and valueless. Occasionally there will be no evidence of the disease until some time after transplanting. On older plants the first symptom is wilting of the margins on leaflets of the lower leaves. This wilting usually appears first on only one side of the leaf. As the margins become dry, the leaves curl upward, turn brown, wither, and die. The petioles remain attached to the stem, a characteristic that distinguishes this disease from fusarium and verticillium wilts. The plant itself often shows a one-sided development of the disease, causing it to lie over in a characteristic fashion. On larger plants a single shoot may be killed early, while the remainder of the plant appears normal. Eventually, however, the entire plant is affected as the dying continues up the stem and much of the foliage is destroyed. As the decay progresses, the pith becomes yellow and mealy in appearance and cavities form within the stem. Later, the destruction of the tissues extends to the outer surface of the stem, forming the open cankers that give the disease its name. Diseased plants may die early, but they often survive until harvest.

Often the fruit is infected on the surface, where the symptoms appear first as small, raised, snowy white dots, 1/8 to 1/4 inch in diameter. The centers of those spots later break open and become brown and roughened. The white color persists as a halo about the margin and produces a bird's-eye typical of the disease. These spots are caused by rain washing the bacteria from open cankers to the fruit and do not extend deeply into the flesh. When the plants are severely diseased, however, the fruit may also be internally affected. If this occurs when the fruit is small, it is stunted and deformed. Internally infected fruits show no external evidence of the disease, but small cavities may be found in portions of the fruit.

Never save seed from fields known to have bacterial canker. If you buy seed, it is worthwhile to buy certified seed; such seed is usually canker free. If it is necessary to use seed that is of unknown origin, it can be put in a loosely woven bag and soaked for 25 minutes in water at 122°F. This treatment will eliminate most internal and external disease organisms that are seed-borne. It is very important that the water does not get any hotter than the recommended temperature, as the seed is easily damaged by heat. The bacteria are also sensitive to acid conditions, and a cloth bag of seeds can be soaked for 24 hours in a mixture of one ounce pure acetic acid in one gallon of water. This amount of acid and water will treat about one pound of seed. Use only 3/4 ounce of acid if seeds are already dried. Diseased seed can also be rendered safe if it is fermented in the crushed pulp of the fruit for 96 hours. Keep the temperature at 70°F. and stir twice a day. In addition, it may be helpful to either sterilize the seedbed soil or replace it if it is not known to be free from disease.

If you have soil that is known to be infected with canker, you should grow another crop for at least three years before planting tomatoes again. In small gardens where three-year rotations are not practical, it's important to pull up and destroy tomato vines after harvest. All tools used in either the greenhouse or field should be disinfected in one part household bleach to ten parts water.

Rainy seasons favor the growth and spread of **bacterial spot,** a disease that first shows up as small, dark, greasy-looking spots on leaflets. Soon there may be blossom drop and small, water-soaked spots, slightly raised and rapidly enlarging from $\frac{1}{8}$ to $\frac{1}{4}$ inch on young green fruit. These spots have whitish halos at first, which later disappear, and the centers become rough, light brown, and sunken. Especially in the South, buy only certified seedlings if possible. Pasteurize the soil in which you will start seedlings, and avoid areas of the garden that have previously carried the disease. Don't water plants with a sprinkler.

**Bacterial wilt** is most common on tomatoes in the southern states, but is occasionally found in other tomato-growing areas. Symptoms include a rather rapid wilting and death of the entire plant without any yellowing or spotting of the leaves. If a stem of a wilted plant is cut across near the ground line, the pith shows a darkened, water-soaked appearance, and a slimy gray exudate appears when the stem is pressed. In later stages of the disease there is also a brown decay of the pith, and cavities may be formed in the stem. Bacterial wilt does not cause spotting of the fruit.

The causal bacteria live in the soil and infect the plant through the roots or stem. Pathogens are most common in low, moist soils and are most active at temperatures above 75°F. The bacteria occur both in newly cleared land and land that has previously grown susceptible crops. Fields frequently are first infected as a result of planting infected seedlings. The bacteria may also be carried by drainage water from an adjacent field.

If the disease occurs, tomatoes and susceptible crops such as tobacco, potato, eggplant, and pepper should not be grown on or near infested soil for a period of four or five years. Instead, try grains, legumes, or corn. If a few wilted plants are noticed in the field, they should be removed at once and destroyed to prevent further spread of the disease. Resistant varieites include Venus and Saturn.

**Blossom drop** is a sign of a stressed plant, and stress can have any number of causes: temperature, soil moisture, nutrient imbalance, and disease. Stressed plants save themselves by not setting fruit. Particularly problematic are low soil moisture concurrent with hot, dry winds, or a sudden change from hot to cool, wet weather. Some varieties—those with smaller fruit—are less vulnerable, and include Red Cherry and Tiny Tim. Early-bearing varieties will produce their good-size tomatoes before the heat of midsummer; try New Yorker or Burpee's Big Early Hybrid. Large, beefsteak-type tomatoes are especially vulnerable.

You can irrigate in hot, dry weather. Certain varieties, such as Walter, Floramerica, Porter, and Hot-set hold their blossoms better in hot weather. If plants are grown close together, blossoms will tend to be shaded. See that you don't add a surplus of nitrogen to the soil, especially early in the plant's life. If you suspect too much nitrogen is a problem, you may be able to save blossoms by pinching back some terminal growth.

Blossom drop is aggravated by verticillium and fusarium wilts. Resistant strains and rotation are the easiest answers. Nitrogen isn't the only nutrient that can cause problems. Deficiencies in potassium or phosphorus will increase the chances of blossom drop. You can check for nutrient imbalances with a soil test before the growing season, in time to add missing components of good soil.

Blossom drop ceases to be much of a problem as the season progresses and becomes temperate. But if tomatoes are still producing blossoms as the season nears its close, pollination is less likely to happen and again the symptom may be blossom drop. Although the reason is not clear, fruit set can sometimes be increased by shaking the plant or hitting the top of the stake in the middle of a warm sunny day.

Gardeners often experience disappointment when they reach for that first tomato and find that the bottom half is rotten. This trouble is known as **blossom-end rot** [color], a nonparasitic disease of tomatoes that causes damage in both the field and greenhouse. Like other nonparasitic diseases, it is not limited to any particular region and occurs wherever conditions are favorable. Control depends entirely upon avoiding these conditions.

The first evidence of injury is a brown discoloration of the tissues near the bottom end of the fruit. These spots enlarge and darken until they cover a third to half of the fruit. As

Blossom-end rot

the spots increase in size, the tissues become shrunken and affected surfaces become sunken, black, and leathery. Soft rotting of the fruit may occur if secondary fungi or bacteria invade the spots.

Blossom-end rot is most apt to occur when the plants have been grown under favorable conditions during the early part of the season and are then subjected to a long period of drought at the time the fruit is in an early stage of development. Breakdown and shrinkage apparently occur when cells at the blossom end fail to receive sufficient water to support their growth. Blossom-end rot also occurs after periods of considerable rainfall, possibly because the soil becomes so nearly saturated with moisture that many small rootlets are killed by lack of aeration or destroyed by certain fungi that are particularly active in moist soils. The damaged root system cannot supply fruit with sufficient moisture.

In the greenhouse this disease can largely be prevented by maintaining a uniform supply of moisture. In dry weather, field plants should only be cultivated as necessary to avoid drying out the soil. Take care not to hoe or cultivate closer than one foot from the plants so that the roots aren't pruned. Water evenly and only as needed. Overdoses of nitrogenous fertilizer are suspected to aggravate this disease by blocking the plant's uptake of calcium. Check to see if the soil has sufficient calcium by testing in fall or very early in spring. If you find a deficiency, add finely ground limestone before setting out plants. To raise the pH of the soil by one unit, use about ½ pound of limestone to each 10 square feet.

Other factors hindering a plant's uptake of calcium and water might include parasitic fungi, nematodes, and a lack of aeration resulting from soil compaction. Generally the disease is more serious on staked plants than on prostrate or bushy ones. Tomatoes planted unusually early in cold soils are likely to develop this rot on their first fruit. Certain varieties are less vulnerable to blossom-end rot than others; check with your state extension agent.

Mulch the soil around the plants with sawdust, peat moss, grass clippings, or even tin foil to prevent the loss of moisture and cut down on the amount of blossom-end rot.

**Botrytis fruit rot** (gray mold) may be an important disease of greenhouse tomatoes in periods of cool temperatures and high humidity. Tremendous numbers of fungal spores are loosed whenever infected plants are jarred, and travel in air currents to infect other plants. The disease first shows as small water-soaked spots that enlarge rapidly, sometimes involving the entire fruit. Diseased areas are covered by a gray powdery growth. Control by raising the temperature and reducing the humidity in the greenhouse.

**Buckeye rot** is a fungal disease that affects green fruit which comes in contact with wet soil and splashed water. This disease causes large, leathery, brown lesions on the fruit that tend to become soft; look for distinct rings. Affected fruit breaks down quite easily and fails to ripen. Mulch plants to prevent water from splashing. Water with care. Don't overstimulate growth with strong fertilizers, as this seems to predispose tomatoes to buckeye rot.

**Cloudy spot** describes the marks left by the feeding of stink bugs. White, spongy cells can be seen beneath the site of injury. A garlic spray should keep these pests from puncturing fruit. Keep down weeds in the vicinity of the tomato patch; these may host stink bugs, which fly in to feed on green fruit.

A **cracking** of the surface at the stem end of the fruit occurs during periods of rapid growth brought on by abundant moisture and high temperatures. Cracks may radiate from the stem or extend longitudinally around the shoulders of the fruit. They vary in depth, but often extend deep into the flesh. Where they develop slowly, the surface tissue heals and becomes fairly firm, but it is likely to rupture in handling. These cracks not only blemish the fruit but also are points of invasion by early blight and other rots of the fruit.

Little can be done to control this trouble except to provide an even supply of moisture, and thereby prevent alternate periods of slow and rapid growth. Since the tendency to crack increases with ripeness, loss can be reduced in part by picking the fruit before they reach full color. Use such resistant varieties as Campbell's 17 and 29 and H1350, H1409, H1370, and H6201. Paste tomatoes Roma, Chico, and Parker are all markedly resistant to cracking. Homestead 24, Manalucie, Marion, Manapal, Indian River, Floradel, Floralou, and Pearson Improved are listed as relatively free from trouble.

**Curly top,** or western yellow blight, is caused by the same virus as curly top of sugar beets. It appears to be carried by a single insect species, the beet leafhopper, a migratory insect that has breeding grounds in weedy abandoned lands and sagebrush areas west of the Rocky Mountains. In those sections where great numbers of the leafhoppers appear each year, it is almost impossible to grow profitable crops of tomatoes, beans, spinach, and other susceptible plants. Although the tomato is susceptible to curly top infection at any stage of growth, the susceptibility decreases markedly as the plants grow older. Infected seedlings soon show a yellowing of the foliage accompanied by curling and twisting of the leaves. Such plants ordinarily die within a short time.

On well-established plants in the field, the first symptoms consist of a pronounced upward rolling and twisting of the leaflets that exposes their undersurfaces. The foliage becomes stiff and leathery and the entire plant assumes a dull, yellowed appearance. The branches and stems are abnormally erect, and the petioles of the leaves curl downward. There often is some purpling of the veins of the leaflets, and the entire plant is usually very stunted. Many of the roots and rootlets are killed. Very few, if any, fruit are produced after infection, and those that are already set turn red prematurely. All standard varieties of tomato are severely injured by curly top, but early varieties appear to be killed more quickly than those that set fruit later in the season. Owyhee and Payette show resistance.

Leafhoppers become carriers of the curly top virus by feeding on infected wild or cultivated host plants. See the discussion of curly top under BEET.

Tomatoes are more likely to be injured if grown near sugar beets, Russian thistle, or other favorable hosts. Local experience should suggest a time for planting to avoid the period of leafhopper migration. Because leafhoppers prefer to feed on plants in the open, shading plants with muslin offers considerable control, and it helps to set plants as close as six inches apart. As a rule, the infection is reduced whenever plants are grown in partly shaded areas. Because of the extra labor and material involved, these methods are suited to home gardens rather than commercial plantings.

**Damping-off** describes the wilt that attacks tomato seedlings at the soil line and kills them. Usually the roots are killed and the affected plants are water-soaked and appear shriveled. The disease typically occurs in small patches scattered through the seedbeds. Seedlings grow out of their susceptibility after about two weeks; by this time the stem is hard and sturdy enough to stave off trouble.

Pasteurize soil, and provide warm, well-drained soil seedbeds. See DAMPING-OFF.

A variety of nutrient **deficiencies** may be signaled by plant symptoms. Fortunately, manure and composted kitchen scraps can supply a wide range of nutrients. Blossom-end rot results from too little calcium; see the above discussion of that disease. Nitrogen deficiency leads to very slow growth of plants, followed by a progressive paling of the leaves that starts at the tip and at the top of the plant. Leaves are small and thin, and you may notice some purple veins. The stems are yellow and rigid. Flower buds turn yellow and drop off. Apply compost immediately and supplement it with blood meal, cottonseed meal, hoof and horn meal, or other material rich in nitrogen.

If, on the other hand, your plants grow an abundance of bright, light green leaves, and there are few blossoms and fruit, suspect an excess of nitrogen. The cure is to cut down on fertilization. You can also increase the supply of bone meal and granite dust to bring phosphorus and potassium up to be in correct ratio to nitrogen content. This trouble is most likely to appear during the seedling period, but it can also affect plants in the garden or field if the soil is overnourished with nitrogen.

Phosphorus deficiency contributes to slow growth, and plants are tinged purple. Leaves are unnaturally small and seem to be rather fibrous. Phosphorus deficiency will also delay the setting of fruit. Cool temperatures may interfere with the uptake of the mineral. A good supply of bone meal, applied from the time you put the tomato plants in the ground, will help prevent this problem. Once plants are set, you can still side dress with bone meal.

Calcium deficiency can cause blossom-end rot, as mentioned above, and also retard growth and cause stems to get thick and woody. Upper leaves appear yellow (as contrasted to the yellowing of leaves of the lower plant, as in nitrogen, phosphorus, or potassium deficiency). Plants seem weak and flabby, fruit is prone to blossom-end rot, and terminal buds will probably die. Use any good grade of lime, preferably dolomitic limestone, to counteract this deficiency.

Limestone can also remedy magnesium deficiency. Tomatoes lacking this nutrient may have brittle young leaves that curl up and turn yellow. The yellowing appears in the leaves rather than the stems, in the areas farthest away from the veins; but look for a green border on the margins of yellowed leaves. Use dolomitic limestone or put a handful of Epsom salts in the hole before you set the plant.

Boron deficiency causes blackened areas at the growing tip of the stunted stem. The terminal shoots curl, then turn yellow and die. In severe cases, fruit may darken and dry out in certain areas, and the whole plant will have an abnormally bushy appearance. A teaspoon of borax will quickly save the plant.

Iron deficiency shows up as yellowed leaves on which veins remain green. Plants in alkaline soil may not be able to absorb sufficient iron. Check pH, and if necessary, add an acidic supplement such as leaf mold. Apply manure, sludge if you can get it, or dried blood to correct the trouble.

Tomatoes deficient in copper show stunted shoot growth and very poor root development, which will make the whole plant undersized and sickly. The foliage turns bluish green and the leaves curl upward. Few or no flowers form, and leaves get flabby. Use plenty of manure as a deterrent.

Use manure also for zinc deficiency, which is identified by small, narrow leaves that are colored yellow between the veins and mottled with dead areas. Stems may crack.

Manganese deficiency shows up as very slow growth and yellowed areas between veins on leaves. There are few blossoms and no fruit. Manure is a good cure.

**Early blight** is one of the most common tomato diseases in home gardens. It is primarily a foliage blight but also may cause fruit rot around the stem end of the fruit in late fall. Early blight is characterized by irregular brown spots with concentric rings in a target pattern on the lower leaves. These spots soon enlarge to ¼ to ½ inch in diameter, run together, and cause

the leaf to turn brown and usually to drop off the plant. Robbing the plant of its leaves serves to decrease the size of fruit and expose it to sunscald. If cool, humid weather conditions linger after a rain, this blight will move up the plant until all leaves on the lower half have fallen. Spotting and girdling of the stems also may occur.

The early blight fungus overwinters in old tomato debris and on weeds such as Jimson, horse nettle, ground cherry, and nightshade. Seeds or transplants also may carry the fungus. A three-year rotation should help prevent trouble; do not alternate tomatoes with members of the potato family. Infected plant waste should be turned under immediately following harvest; or, in small gardens, simply gather the old plants and burn them. Plant resistant varieties, such as Manalucie, Southland, Floradel, Floramerics, and Manahil. If you're growing tomatoes from seed, be sure that they come from healthy plants.

**Fusarium wilt** is one of the most prevalent and damaging diseases of tomato. The fungus causes trouble by growing in the food and water channels of the plant. It is found in the greenhouse as well as the field and can live for long periods in the soil. This organism generally does not cause damage unless the soil and air temperatures are rather high during much of the season. Susceptible varieties are either killed or damaged so badly that they produce little fruit. Once fields become infected they remain so and may afflict crops years later if conditions favor the disease.

Many growers know this wilt disease as yellows, because of the characteristic color of infected plants. The first symptom is a slight yellowing of a single leaf or a slight wilting and drooping of the lower leaves. Only a single branch on one side of the plant may be affected. A distinct brown discoloration of the xylem can be seen in a cross section of a stem close to the base of the plant. This discoloration often extends upward for some distance and is particularly evident in the petioles of wilted leaves at the point where they join the stem.

The fusarium wilt fungus enters through the roots and passes upward into the stem, where toxic substances that cause the wilting and death are produced. In severely affected plants the fungus may pass into the fruit and seed. However, infection from internally infected seed is rare, as fruit usually decays and drops.

To control fusarium wilt, grow seedlings in clean soil; wilt-free fields may become contaminated by the use of infected transplants. Seedlings should not be located on old garden soil or on land where wilt is known to have occurred. If you intend to purchase the plants, be certain that the grower's fields are free from wilt. Tomatoes should not be grown on the same land more than once in four years if possible; a slight wilt infestation soon can be greatly increased by growing tomatoes repeatedly in the same place. Pull up and destroy affected plants at the end of the season. Certified seed is less likely to carry the wilt fungus than uncertified seed. If you use home-saved seed, be sure that it comes from plants completely free from wilt. A number of varieties show resistance to wilt. Susceptibility varies with time and location, so consult with the local agricultural agent. Regardless of variety, nematode injury renders tomatoes especially vulnerable to wilt; see NEMATODE and RESISTANT VARIETIES.

**Ghost spot** is caused by the same fungus responsible for gray mold. Look for small whitish rings, measuring ⅛ to ¼ inch in diameter, on young green tomatoes. The spots do not extend far into the fruit, and so do not greatly affect marketability. High temperatures and low humidity discourage the disease. Avoid planting where gray mold has been, especially in damp locations. In the greenhouse, lower the humidity, raise the temperature, and improve ventilation.

Gardeners in the southeastern states are apt to be visited by **gray leaf spot** fungus on tomatoes, peppers, eggplant, ground cherries, and other related plants. It can be distinguished from other fungal diseases that thrive in warm, moist weather by the appearance of dark brown spots on the undersurfaces of older leaves. These leaves eventually yellow, wither, and drop. Serious infestations will kill all but a few tip leaves, and few fruit will appear.

Because this fungus is carried over from season to season on the remains of diseased plants in the soil, use a three- or four-year rotation. See RESISTANT VARIETIES.

**Gray mold** first appears as a heavy gray growth on dead leaves or stems at the base of the plant. The gray spots then travel to growing leaves, tan markings appear on stems, and gray or yellowish soft spots show up on fruit. If this mold appears in the greenhouse or hot bed in spring, raise the temperature and lower the humidity to discourage growth of the fungus.

**Late blight** is a fungal disease that causes severe defoliation and destructive rot of fruit. The first symptoms of the disease are irregular, greenish black, water-soaked patches on the older leaves. These spots enlarge rapidly and in moist weather may show a white, downy growth of the fungus on their surfaces. In moderately warm, wet weather, the spread of infection is so rapid that plants look as though they had been frosted. Fruit may be attacked at any stage in its development. Affected areas on the fruit are usually large, dark colored, firm, have a rough surface. The disease is favored by rainy, foggy weather with temperatures of 40° to 60°F. at night and 70° to 80°F. in the day. It is checked when the weather turns hot and dry. With favorable weather conditions, late blight may be spread to tomatoes from diseased potato vines, be brought into the area on infected transplants, or blow in from other regions. This disease is particularly important east of the Mississippi and along the Pacific Coast.

In the greenhouse, late blight can be controlled by lowering the relative humidity in the house. Although this will quickly check the enlargement of lesions, it will not kill the fungus, and the growth of the fungus will be resumed as soon as cool humid conditions recur. The best way to reduce relative humidity is to raise the temperature inside the house and provide ample ventilation. Keeping night temperatures to 65°F. or above usually holds the relative humidity sufficiently low to avoid further development of the disease. In the field, control begins with destruction of infected plants and tubers; it is believed that blight cannot live on in soil alone. In the South, be sure to get certified plants from nurseries that grow tomatoes far from potatoes. If you buy seedlings in the North, check they are not from a greenhouse that suffers from blight. See RESISTANT VARIETIES.

**Leaf mold** is principally a greenhouse trouble, although on occasion it may develop in the garden or field. All aboveground parts are affected, particularly the leaves. The first symptoms usually appear on the surfaces of the older leaves in the form of diffuse whitish spots that rapidly enlarge and become yellow. Under humid conditions, the lower surfaces of these patches becomes covered with a velvety, olive brown coating of the causal fungus. When conditions favor the development of the disease in the greenhouse, much of the foliage is killed and the crop is greatly reduced. Although fruit infection is rare, the fruit stems and blossoms may also be affected.

The fungal spores are produced in great numbers on the undersides of the leaves and are spread by air currents, watering, and contact with the plants. The spores are tough and may survive around the greenhouse for several months after the plants have been removed. Occasionally the spores may also be carried on the seed.

The disease can be controlled to a considerable degree if adequate ventilation is provided during fall and spring months. During the warmer spring and fall months, the ventilators should be handled so as to provide maximum ventilation and sufficient heat at temperatures of at least 60° to 65°F. If heat is provided at night, the relative humidity will be kept down. The disease will usually not be severe if the relative humidity is kept below 90 percent. Resistant varieties are available, but the fungus mutates into new forms and older varieties may lose their resistance. See RESISTANT VARIETIES.

**Leaf roll** is a nonparasitic disease that generally follows periods of wet weather and is likely to occur on plants in poorly drained soils. It has also been observed after close cultivation and extremely close pruning. Overwatering and deep cultivation may also cause this disease.

The rolling generally starts on the lower leaves and proceeds upward until, in some instances, almost all the leaves are affected. In severe cases the rolled leaves are somewhat thickened and tend to rattle when the plant is shaken. Ordinarily, affected plants bear a normal or nearly normal crop of fruit, but plants may lose a third to half their leaves with a corresponding loss in fruit quality and yield. This defoliation is most severe on staked plants.

The conditions that cause the plants to lose their leaves are not fully understood. One theory is that heavy rains that whip and riddle older rolled leaves may be partly responsible. Or it could be that aphids cause the trouble. Perhaps the best explanation is that loss of leaves is caused by an accumulation of carbohydrates in the plant. Heavy pruning and the removal of suckers is thought to restrict plant growth and increase carbohydrates in the lower leaves.

This condition is not particularly damaging, and plants recover with more-favorable weather. To lessen damage from leaf roll, choose well-drained areas for tomatoes, particularly if you are growing them in the garden. Do not cultivate deeply near the plants, and avoid extremely close pruning.

Misshapen fruit—tomatoes that are puffy or caved in—is sometimes caused by cool nighttime temperatures around the time blooms are fertilized.

**Mosaic diseases** are common tomato troubles. They may be transmitted from nearby weeds and flowers: ground cherry, horse nettle, pokeweed, plantain, Jimsonweed, nightshade, catnip and other mints, Jerusalem cherry, milkweed, motherwort, white cockle, burdock, wild cucumber, flowering spurge phlox, marigold, petunia, zinnia, and hollyhock, gladiolus, and geranium, among others. Remove weed hosts within 150 feet and consider growing these flowers elsewhere. Don't grow tomatoes next to potatoes or cucumbers.

*Tobacco mosaic* is the most common viral disease of tomatoes. The symptoms are so varied that it is often difficult to see any difference between healthy and infected plants. Young plants first display malformation of the leaflets. On older plants very little malformation is present, and the mosaic pattern of mottling is the distinguishing symptom. Leaflets may tend to be long, stringy, and fernlike, or they may only show a strong tendency to be pointed. Generally, the young leaflets point straight up and the plants exhibit a slightly grayish appearance. Mosaic-infected plants wilt much more severely in a bright period following a cloudy period. In older plants the main symptom of the disease is mottling of the leaves. When looking for diseased plants, work with the light to your back so that the plants are shaded, as the

deformation and mottling will be much easier to see. Fruit may be mottled, ripening unevenly.

Tomato plants affected with cucumber mosaic are stunted, yellowed, and bushy. Although leaves may show a mottling suggestive of common mosaic, the most pronounced symptom is a shoestring appearance of the leaves. Sometimes the leaves are so distorted that very little remains of the leaf blade but the midrib. Severely affected plants produce but few fruit, and these are usually smaller than the fruit from healthy plants.

The virus overwinters in certain perennial weeds, such as pokeweed, catnip, and milkweed. At least five species of aphid as well as striped and spotted cucumber beetles are capable of transmitting this disease. However, cucumber mosaic is not transmitted in tomato seed, is not easily transmitted by rubbing and handling the plants, and does not persist in the soil or on the hands of those working with affected plants. This low infectivity means that you needn't remove adjoining plants when roguing diseased tomatoes.

Mosaic viruses are easily spread by handling or, by tools and stakes—anything that comes in contact with plants. A strain of tobacco mosaic can be passed on to tomatoes by smokers or tobacco chewers. Should you have either habit, wash your hands thoroughly before entering the tomato patch or greenhouse.

The disease can live in the soil on the decaying plant matter, and tomatoes should be grown elsewhere until the old plants are decomposed completely. Seed can be sterilized by soaking it in 1 part concentrated hydrochloric acid in 19 parts water (a 5 percent solution) for four to eight hours. Rinse and dry seed. In Europe plants are inoculated with a mild strain of mosaic virus. Seedings may benefit from occasional spraying with milk. Constantly inspect and rogue diseased plants. Remove one plant on either side of the diseased one, as it is almost impossible to remove a diseased plant and not contaminate healthy adjacent ones. When pruning, pinch off blossoms without touching any other part of the plant; do not use a knife.

In the greenhouse, do not grow tomato plants in seedbeds used for flower production. Do not plant cucumbers and melons next to a greenhouse in which a fall or spring crop of tomatoes is to be grown; otherwise, virus-bearing aphids and cucumber beetles may move into the greenhouse as the cucumbers and melons mature. Aphid control is important in preventing the spread of mosaic diseases once a few plants in the garden are infected (see APHID).

**Psyllid yellows** is a disease caused by the feeding of the tomato (or potato) psyllid. The damage is principally due to a toxic substance that is released into the plant as the insect feeds. Look for a thickening of the older leaves and an upward rolling at their bases as they turn yellow and develop purplish veins and margins. Younger leaves may curl and plants are sometimes dwarfed. The stems and petioles seem unusually slender, and plants take on a spindly appearance. If the disease hits young plants, they will produce little or no fruit; if the disease happens later, tomatoes that are already set will likely be yellowish red, quite soft, and of poor quality.

Psyllids cause trouble in the nymph stage; they are flat and look something like fringed scales, and change in color from yellow to orange to green. To control them, clear the area of weed hosts, including Chinese lantern and ground cherry. A garlic spray may keep psyllids from feeding.

**Root knot** is caused by tiny parasitic worms known as nematodes; see NEMATODE. Better Boy is a resistant variety.

**Septoria leaf spot,** or septoria blight, is one of the most damaging of tomato diseases and often causes severe losses in the Atlantic and Central states. It also occurs frequently as far south as Arkansas, Tennessee, and North Carolina. The disease is most severe during rainy seasons and in fields where plants are crowded and are bearing a heavy fruit load.

As a rule, the disease is slow in getting started and is not much in evidence before early or middle July and not before the plants have begun to set fruit. The first infection is ordinarily found on the older leaves near the ground; look for small, water-soaked spots scattered thickly over the leaf. These spots soon become roughly circular and have gray centers surrounded by darker margins. Later, the centers show tiny dark specks in which the spores of the fungus are produced. The spots are more numerous and smaller than those of early blight, usually measuring $1/16$ to $1/8$ inch in diameter. If they are numerous, the leaflet usually dies and drops from the plant. When conditions favor infection, there is a progressive loss of foliage until only a few leaves are left at the top of the stem so that the fruit is exposed to sunscald. The fruit is rarely affected, but there may be spotting of the stem and blossoms.

It has been found that this leaf spot fungus will not live over on the remains of plants buried deeply in the soil. If all the vines are covered, deep cultivation or plowing in fall or spring will do much to prevent infection. The plants in small gardens may be collected in fall and burned. Clean cultivation, control of weeds, and rotations will help to free growing areas of the fungus. Stay out of the tomato patch when plants are wet, as the fungus is easily transmitted by brushing against them.

**Soil rot fungus** is the same pathogen responsible for damping-off of seedlings. The first symptom is a slightly sunken brown spot on the fruit, outlined with concentric markings. The spot enlarges and may break open, unlike the somewhat similar condition known as buckeye rot. Infections occur when plants touch the ground or are splashed by rainwater, and a good mulch should help. Avoid setting tomatoes in poorly drained soil, and use varieties that can be staked.

**Spotted wilt** first appears on young plants as numerous small, dark, circular dead spots on younger leaves. Spots may turn bronzy before becoming dark and withered. The tips of stems are darkly streaked and often wither. If young plants are able to survive, the new growth is very dwarfed and the leaflets appear distorted. On older plants the growing tips are somewhat damaged and the foliage has a yellow tinge to it. The fruit has many spots with concentric, circular markings.

Control involves eliminating weed hosts and keeping the tomato patch isolated from such vulnerable vegetables as lettuce, celery, spinach, peppers, and potatoes. Infected seedlings can be removed to stop an infestation, and can be replaced with healthy plants, as the disease is not soil-borne. Varieties developed in Hawaii and considered resistant to spotted wilt virus include Pearl Harbor, Anahu, and Kalohi.

## Environmental problems

**Cat-facing**—a peculiar puckering—is caused by poor fertilization, often the result of cool weather. The warmer the site of your patch, the better the chances for adequate pollination. High temperatures also can discourage pollination; Porter is a variety resistant to heat.

To protect plants against **frost,** use bottomless milk jugs, cloches, or plastic tunnels. The soil in raised beds will be warmer than soil in a standard garden.

**Sunscald** can occur whenever green tomatoes are exposed to the sun, but is most frequent in hot, dry weather. This injury is common on plants that have suffered a premature loss of foliage from leaf spot diseases, such as early blight or septoria leaf spot, and is a major cause of loss from these diseases. Fruit of plants suffering from verticillium or fusarium wilt also is likely to suffer from sunscald as a result of the loss of the lower foliage. Varieties resistant to graywall, leaf mold, nailhead spot, and verticillium wilt are listed under RESISTANT VARIETIES.

# TRAP CROPPING

Growers use certain plants not only to repel potential pests, but also to attract them—away from valued crops. These so-called trap crops can be set either around plantings or between rows. Because pests congregate on them, handpicking is made easier. Trap crops also serve as breeding grounds for parasites and predators. Here is another nontechnical means of plant protection that has evolved from years of observation and experimentation by growers.

On a large scale, cotton growers plant strips of alfalfa in their fields to attract and concentrate lygus bugs which would otherwise raise havoc. At the garden level, a number of happy combinations have been discovered. Dill and borage attract the tomato hornworm from tomato plants (and the fat larvae are easy to spot on herbs, facilitating handpicking). Japanese beetles can be lured from valued crops with plantings of white or pastel zinnias, white roses, odorless marigolds, soybeans, and knotweed. Mustard, planted early as a trap crop, saves cabbage from the ravages of the gaudy harlequin bug. On a larger scale, soybeans are protected from the Mexican bean beetle when sown with a trap crop of green beans.

Once pests flock to the trap crops, you can release purchased or collected beneficials on the plants to clean up. Thus these plants can serve as little insectaries. With plenty of food at hand, your beneficials will be off to a good start. Rose geraniums serve this purpose well in sustaining populations of whiteflies and the parasite *Encarsia formosa*.

# TRAPS

The most obvious way to be rid of insect pests is to chase them from the garden with repellents or to nail them on the spot with insecticides. But you can also render pests harmless by luring them into traps. Experienced gardeners and orchardists also use traps to monitor the appearances and populations of pests. Most traps work by appealing to an insect's need for food, shelter, or sex.

One of the most popular traps is nothing more than a shallow dish or jar lid containing a bit of stale beer, and set out in the garden to draw snails and slugs. If you find these pests are taking a sip and then leaving, mix a bit of flour with the beer to make a sticky mixture. As simple as this beer trap may be, it incorporates the two things that make a good trap: an effective lure (fermented malt, in this case) and a means of detaining the pest (a pool of liquid made more treacherous with flour). Simpler still is a board set out in the garden to lure pests that seek shelter. The board is regularly lifted and the insects squashed.

Traps can be made sticky with commercial preparations sold for that purpose. Tanglefoot, Stickem, and others are widely available. A simple whitefly trap for greenhouse use employs a yellow card treated with a sticky compound; whiteflies are drawn to yellow. (In fact many insects are, because yellow is a component of the green in foliage.) Such traps are most effective early in the growing season, before pest populations have had a chance to balloon. Insects can also be trapped in shallow containers of water; either the water or the container should be colored yellow. In the orchard, fruit growers hang sticky cards and bright plastic fruit. A few scattered traps will serve only for monitoring; place traps in each tree and enough pests may be snared to reduce their populations significantly. Pheromones are used to lure specific pests.

Light traps draw a wide variety of night-flying lepidopterans, including adults of the tomato and tobacco hornworm, codling moth, oriental fruit moth, corn earworm, European corn borer, cabbage looper, cotton leafworm, pink bollworm, European chafer, fall army-worm, bagworm, spotted cucumber beetle (a carrier of bacterial wilt), cutworm, and hundreds more. Species differ in color preference. Even regular light bulbs will draw some pests, including the adult European corn borer. Pink or green fluorescent lamps work against striped and spotted cucumber beetles, and an argon lamp appeals to the pink bollworm moth. Red, orange, and yellow lights are ignored or even avoided by almost all insects.

Light traps are easy to use. All you do is turn them on at dusk and off in the morning, and you can buy models with electric eyes to spare you even this little bit of effort. If they're to have an effect on summer larva populations, traps must be in operation early in spring, as soon as the adult moths and flies first appear. Be sure to inspect a trap often to see what kinds of bugs you're catching. For one thing, these inspections will tell you just when each pest appears in the garden and when it departs. This can help in timing plantings to avoid egg-laying periods. Also important, frequent inspections of the trap will let you know when to turn it off. If the pests are few in comparison with innocuous insects, the trap is doing more harm than good. Chances are that many of these bugs would have otherwise ended up as meals for insect predators and parasites, fly-catching birds, toads, and fish.

Some light traps incorporate an electric grid that makes insects disappear with a pop and a flash. Such devices are often set up outside soft-ice-cream stands to handle the flying clouds that hover about the fluorescent lights. While there's no denying the effectiveness of electrocutors in keeping the number of bugs down, your aims as a gardener aren't the same as those of ice-cream sellers. They want to kill bugs, any bugs, but you are only after those that may damage crops. With an electrocutor, there's no way of knowing if you're killing pests or just harmless bugs that happened by. All you hear is that little pop.

See also BORDERS AND BARRIERS.

# TREES

For the sake of convenience, many of the troubles that trees have in common are dealt with in this entry. You'll find these troubles arranged as follows:

General care, page 422
Insects, page 425
Diseases, page 427
Environmental problems, page 431

## General care

A tree is unique among all living organisms. It lives longer and grows taller and larger than any other organism of land or sea. Yet a tree, like a human being, can be hurt and can die from its wound. When a tree is wounded, a chemical reaction takes place to protect the wound and compartmentalize it. But often the wound provides an entry for bacteria, fungi, and viruses.

A tree wound might be caused by a bird, animal, or insect; by fire or storm damage; by heavy mechanical equipment; by a nail; or by a branch breaking off to leave a poorly healed stub. Any part of the tree is susceptible—roots, trunk, branches, and bark. The better you are at detecting tree troubles, the better you can prevent and control decay, the major cause of damage to trees. First, take care to avoid wounding trees.

Second, help the tree help itself after wounding. Remove the obviously injured bark and wood as well as dead and dying branches. Shape the wound like an ellipse if possible. Fertilize and water properly, and remove less valuable trees or shrubs that may be crowding the injured tree.

Third, if the tree becomes too unsightly or too dangerous because of the threat of falling limbs, remove it. A professional arborist can help make the decision as to when this is necessary.

Fourth, when planting new trees, plan ahead to avoid likely accidents to trees. For example, trees should be planted away from driveways, curbs, and walkways.

PLANTING.   Transplanting is a precarious time for trees, and extra care must be taken to help them become established.

The best time of year for planting, usually early spring or late fall, varies with species and location. Check with your local nursery for this information. Planting holes should be wide and deep enough to accommodate bare-rooted trees without any cramping of the roots. For balled rootstock, make the holes at least a foot wider than the diameter of the ball. Untie and roll back burlap wraps from the trunk after planting. Plastic burlap and plastic bag wraps will not decompose in the soil and should be completely removed. Fill the hole with topsoil and cover to the depth that the tree was growing in the nursery—that is, the depth at which it grows naturally, but no deeper. Add water immediately after planting and continue to water periodically for two seasons. Evergreens require more water than deciduous trees and shrubs. Water heavily near the roots about once a week for several hours unless there is enough rain to saturate the soil. Begin to water seven to ten days after the last rain, and continue through the growing season and well into fall. The soil should be saturated but not waterlogged.

When trees three feet or over are planted, additional precautions may be advisable. To help support the trees until the roots become firmly established, hose-wrapped guy wires can be connected from the tree to supporting poles. These supports should be left in place for one to two years, when they must be removed to prevent them from girdling the trunk. Damage from sunscald and minor wounds can be prevented by wrapping bare trunks with burlap, creped kraft paper, or a couple of thicknesses of aluminum foil.

Prevent animal damage with wire mesh supported by stakes around the tree, several inches from the trunk. It should be a heavy enough gauge to repel larger animals, but a fine enough mesh to keep small rodents from getting to the bark. The aluminum mesh sold to protect gutters from leaves seems to work well. Anything tied around the tree trunk or around a branch should be loosened each year and removed as soon as possible.

## MULCHING AND FERTILIZING.
Tree leaves are a natural mulch, and the leaves that fall from healthy trees and shrubs should be put back into the soil below. In time, the layer of leaves weathers down to rich, black, humusy soil. This method of mulching provides both aeration for absorbing moisture and necessary minerals for the root system. Trees thrive on it. If the leaves are thought to carry insect eggs or disease spores, however, it is advisable to rake

them and add them to the compost heap. If the compost is properly made so that it is heated to 130° to 140°F., pests on the leaves will be killed. Then in spring, when it is needed for plant growth, return the fully rotted compost to your trees.

Although mulch is important in preserving soil moisture, there are times when it is wise to rake off all the old mulch and let the sun shine on the soil for a while. This is especially true where trees or shrubs are prone to mildew or fungal diseases fostered by dampness. If you garden in a damp climate, give the soil a sun bath each spring. Remove all the old mulch and let the sun kill off the disease spores in the top layer. After a few weeks of baking, mulch again with fresh material.

To make sure your trees get all the nutrients they need, apply a mixture of balanced minerals and fertilizers. Try a mixture of equal quantities (by weight) of: 1) cottonseed meal or dried blood; 2) phosphate rock or bone meal; 3) wood ashes, granite dust, or greensand; and 4) dolomitic limestone. For small trees set in open soil, these nutrients may be spread and filled into the earth. But the most practical way to fertilize large trees or those whose roots are covered with sod is to punch 18-inch holes in the soil with a pipe or crowbar, slanting them in toward the trunk. These holes should be 2 inches apart, covering the entire area beneath the crown out to the drip line. Fill each hole with six to eight ounces of the fertilizer mix and then cap it with peat moss or topsoil.

## PRUNING.
Pruning of trees and shrubs adds to their health and good looks, helping nature to grow the plant in the most attractive and productive form. By ensuring adequate sunlight and air circulation, and by removing weak or crowded growth, you also strengthen the plant's resistance to infection and insects. Dark, overly

damp growth often makes a favorable breeding environment for trouble.

Three types of pruning contribute to the appearance and health of trees and shrubs. *Basic* pruning gives plants a structure that prevents weakness, intemperate growth, and deformity. This includes pruning to balance top and roots after planting, training the young plant, and general control of its size and shape. If a tree grows where machinery must pass, prune to allow plenty of room, as bruised or broken-off branches are unsightly and unhealthy.

*Maintenance* includes regular pruning to maintain an optimal balance between old and new wood, foliage, and the flowering and fruiting parts of the plant. Maintenance pruning also means prompt removal of all dead, diseased, or injured parts that would allow rot and disease organisms to enter the plant. Suckers and watersprouts are other energy stealers that must be pruned out.

*Renewal* is a type of pruning that many gardeners hesitate to try, though it is often simple. This pruning aims at the rejuvenation of old trees or shrubs, which in most cases is accomplished by thinning out the older growths and heading back the larger of the young branches to force development of strong, healthy new growth. Such an operation should be done gradually over several years, not all at once.

Most pruning can be done any time, although maples and birches "bleed" heavily in spring and should be trimmed in summer. Those evergreens that are pruned to induce bushiness or extra foliage, such as white pines, Norway spruce, and Colorado blue spruce, should be pruned back to a lateral bud in spring. Summer is a good time to observe which branches are diseased or dying; these can be pruned in late fall.

Cut out all broken, dead, or diseased wood to prevent the spread of decay. Live stems are often removed to improve air circulation, to admit more light, or to push the remaining part of the crown to more foliage. Restrict your cutting when possible to small branches, as heavy pruning can lead to shock from loss of food manufacture and may bring on sunscald.

Start at the top and work your way down, looking for misshapen, diseased or obstructing branches as you go. Weak growths and crotches should also be removed to prevent splitting. If you must remove a larger branch, make a preliminary undercut so you don't strip the bark when the branch snaps off. Coat all wounds and cuts with a wound dressing.

Don't neglect wounds that expose the wood beneath the bark. Such openings are invitations to bacteria, fungi, and insects that normally leave healthy trees alone. The tree itself will put forth a normal callus growth to close the opening; you can help by carefully trimming and paring off loose, rough, or protruding bits of torn wood. When you cut a branch, don't leave a stub that will interfere with the normal growth of a protective callus. Ornamental shade trees such as dogwood, Japanese weeping cherry, and flowering crab apple easily become malformed from improper care of wounds.

If the wound is large, shape it into an ellipse by careful paring and trimming to encourage and promote healing. Cut the bark down to solid, sound wood, removing all bruised tissues as you work the incision into this shape. Allow it to dry and then apply the wound dressing over the entire surface. Repeated applications may be necessary once a year until the callus has covered the wound.

**RESTORATION OF OLD TREES.**   If you acquire land with old trees on it, they are likely to need some attention. The first step is a survey of the old trees to estimate what kind and how much repair they will need. A professional arborist, either public or private, should be called in for advice. Often the most picturesque old trees are imperfect ones that take their unique character from a gerontic deformity, scar, or irregular growth pattern, and it would be folly to try to change the very quirks that make them interesting.

Safety should be your first concern. While you may be able to spot a heavy branch hanging perilously over the house, walk, or driveway, less obvious are defects such as a faulty crotch, cavity, or imbalanced tree that could become dangerous in high winds or snowstorms. Once the old trees have been made safe, the next step is to renew their vigor. You will need to dig a few test holes to determine the soil strata and the levels at which the trees' roots run. If the root systems are predominantly shallow, spiking the turf as described above may be the best way for fertilizer to reach them. If they are deeply rooted, the food insertions will have to be close together, shallower, and far beyond the branch spread. Root systems of trees reach out much further than you might think—as far as twice the reach of the crowns.

Sometimes a steadily declining old tree will suddenly burst into exaggerated bloom or fruit bearing, and then die in a year or two. This is a natural phenomenon for some trees, but a program of feeding, pruning, and root doctoring is not likely to cause such quick, dramatic results; instead, leaf color, annual growth, and revived vigor will be gradual.

## Insects

If you were to prepare a list of the insects that are found on trees and shrubs throughout the United States, it would run to several pages. Most of these insects are harmless, many are helpful, and all but a few of the remainder are kept under control by their natural enemies— birds, diseases, and insect predators and parasites.

Most insect attacks are not fatal to trees, and it is not necessary to resort to chemical insecticides. Try water sprays, organic fertilizers, and pruning the infested branches. If the pest persists, find out if a biological control agent can be of help; see BIOLOGICAL CONTROL. As a last resort, turn to an organic pesticide, such as an oil or botanical spray; see SPRAYS AND DUSTS.

Tree Tanglefoot is a commercial compound that is applied as a band around the tree trunk to stop cankerworms, gypsy moth larvae, ants, and other pests that travel to and from the tree top. Don't apply great gobs of it, or repeat applications on the same spot. You can protect trees with tender bark by applying the compound on a paper band.

Recent research has found that trees apparently are able to warn one another of danger. Once warned of an impending insect invasion, a tree can generate antifeedants that will keep the pests at bay.

For ease in identifying problems, the following insects have been grouped according to the way in which they affect plants.

**SUCKERS OR SAP-FEEDERS.** These insects feed by piercing the tissues and sucking out the juices, causing slowing of growth and death of twigs or limbs.

**Aphids** [color] are small, soft-bodied sucking insects that attack most trees and shrubs. On otherwise healthy trees, they are rarely more than a nuisance. They excrete a honeydew which attracts ants and flies, provides a substrata on which mold grows, blocks light from leaves, and causes spots on automobiles and outdoor furniture. On injured trees or trees subject to a stressful environment, aphids may cause severe injury. When plants are starved or become thirsty, aphids may multiply to the point that ladybugs and other beneficials can no longer control them. If an aphid-ridden plant is treated for nutritional deficiency and thoroughly watered, the aphid population should decline. See APHID.

**Lace bugs** [color], whose wings show a lacy pattern under magnification, are found on sycamore, basswood, hawthorn, cherry, white oak, and many broadleaved evergreens. These flat, small insects, about ⅛ inch long, suck juices from the leaves, causing whitish or yellowish spots. Their molasseslike drops of excrement can be seen on the undersides of leaves. Lace bugs seldom cause serious injury to deciduous trees, but leaves may drop from evergreens, causing dieback of twigs and small branches. If necessary, control by spraying or dusting the foliage with nicotine or pyrethrum.

Most **scale** insects are so small that they are not easily seen in their early states of development. Some types of scale secrete a honeydew on which an unsightly black covering of sooty fungus soon develops on foliage, twigs, and branches. The fungus blocks the light needed by chlorophyll, and growth may be stunted and branches killed.

Scale can be controlled by a dormant-oil spray. Many trees and shrubs will not tolerate petroleum oils, including various species of beech and maple, hickory, mountain ash, red and black oak, walnut, butternut, and yew. Those plants relatively tolerant to oils include apple, boxelder, dogwood, elm, linden, pin oak, burr oak, white oak, and post oak%

## CHEWERS OR FOLIAGE-FEEDERS.
This group of tree pests includes various caterpillars, sawfly larvae, beetles and their grubs, and leafminers (which split the leaf apart and mine out the tissues). These insects devour foliage, and some, such as the gypsy moth, can deforest whole areas of woodlands.

There are two generations of **cankerworms** (or inchworms), one in spring and one in fall. Full-grown cankerworms are about an inch long, pale yellow, green, brown, or black, with several light-colored stripes running lengthwise down the body. See APPLE for their life cycles and controls.

Luckily, the **gypsy moth**'s natural enemies, including birds, tachinid flies, and certain species of beetle and wasp, will help even a heavily defoliated area recover naturally. Several species are *not* attractive to this pest: ash, balsam fir, butternut, black walnut, catalpa, red cedar, dogwood, holly, locust, sycamore, and tulip poplar. See GYPSY MOTH.

**Tent caterpillars** like to pupate in woodpiles and under old baskets and boxes, so clean up these areas. You can remove the egg bands of tent caterpillars by hand or scrape them off with a dull knife. This egg band is a ringlike mass that has the appearance of brownish black, glossy caviar. Toward spring, the gloss dulls and the bands are harder to spot. Once they've hatched, you can prune out each individual nest and destroy all the worms. In cleaning out the egg masses, be sure you don't destroy praying mantis cocoons, which are silvery or tan, papery, spindle-shaped, and spun around the branches of shrubs and low trees.

Large numbers of tent caterpillar moths and moths of many other species may be captured in blacklight traps. Burning tent caterpillars is not recommended because of both the fire hazard and the fact that many caterpillars fall to the ground unburned and escape. See also TENT CATERPILLAR.

## GNAWERS OR WOOD-FEEDERS.
This third group of tree-damaging insects includes bark beetles and borers that eat small holes in trunks and limbs, forcing out the sap; carpenter

ants that honeycomb the interior wood of trees; and the oak twig pruning beetle, whose presence is made known by numerous twigs of oak lying on the ground with the pith eaten out and grubs inside.

The **carpenterworm** often makes large galleries in the trunks and larger branches of locust, ash, oak, and maple. Large unsightly scars are evident wherever this insect occurs. Fortunately, the carpenterworm is usually not in sufficient abundance to cause serious injury. The females deposit their eggs in the vicinity of wounds. The young larvae first feed on the inner bark, then burrow in the wood. The mature larvae are about 2½ inches long and pinkish with brown heads. The adult moths are gray with brown or black markings, have a wingspread of about three inches, and are thought to live about three years.

Remove and destroy infested branches. Avoid damaging or wounding the tree when working around it. All wounds should be dressed to keep the moths from laying eggs in the tree. A light trap may be effective against the adult moths. It is best to use it in early summer when the moths emerge.

Carpenterworms can be killed by squirting a liquid into their holes—not a chemical insecticide, but a soup of live caterpillar nematodes. They seek out the enemy, enter its body, and eat. These beneficials can be ordered by mail. See NEMATODE.

**Flatheaded borers** are among the worst enemies of deciduous trees and shrubs. Damage is especially severe to young, newly transplanted trees. The larvae burrow underneath the bark, girdling the tree. Adult borers are beetles with a metallic sheen that emerge in early summer and lay eggs in bark wounds. When the borers hatch, they eat through the bark to the cambium layer, and their holes diminish the value of the lumber when the trees are cut; hence, they are a serious problem on commercial plantations. See flatheaded appletree borer under APPLE.

A group of insects called **twig girdlers** gnaws through the bark of twigs or branches, breaking the central part of the stem. You may see branches hanging from the trees in late summer, fall, or winter. Or the larvae may cut the twig from within, hollowing out the center and then cutting the edge off cleanly and leaving the bark irregularly broken. Whatever you can do to encourage bug-eating birds, especially fall tree-dwellers, will keep these girdlers and borers from doing too much damage. But in making your diagnosis, remember that playful squirrels can also be the cause of fallen twigs.

## Diseases

Take care not to injure trees; promptly dress wounds. Once a disease begins, prune and burn the affected branches to halt its spread. Pruning tools should be disinfected afterward to prevent spreading the disease-causing fungus or bacterium. If only the leaves are affected, rake and burn them when they fall.

DECAY.   Before decay begins in a tree's wood, a series of events take place. The first is an injury, whether from fire, birds, animals, breaking branches, lightning, jarring from mechanical equipment, or another cause. The tree reacts to its injury: chemical changes take place in the wood and the wood discolors, either darkening or bleaching lighter. A growth ring forms, acting as a barrier to seal off the injury. Antidecay fungi and bacteria may aid the tree.

In northern hardwood trees, the discoloration and decay take a definite pattern, forming a column in the tree related to the location of the injury. This column spreads up and down inside the tree, but seldom spreads outward. Because the tree will grow healthy new wood around the defect, the column of discoloration and decay is no larger in diameter than the tree was at the time it was injured. If there have been repeated injuries, as is often the case with forest trees, several columns may be present in various stages of development at the same time and place. These multiple columns can sometimes be seen on the ends of logs, where they take on a concentric pattern or a cloudlike pattern.

The defects that form in a tree under cankers produce a somewhat different pattern. A canker (a dead area in the bark) tends to form a localized defect rather than a column of discoloration and decay. When a canker develops about a wound, the defect usually does not spread around the entire stem, but lies immediately beneath the canker.

Of northern hardwoods, red maple and yellow birch are highly susceptible to this decay process, and sugar maple ranks as the most resistant.

**Anthracnose** is a fungal disease most evident as spotted, blotched, distorted leaves that turn brown and often fall prematurely. On twigs and branches, the effect of anthracnose is usually disfigurement. Buds and small twigs may be blighted early in the season, appearing to be frost-injured. Cankers (dead areas in the bark) often form at the juncture of buds and twigs. Anthracnose fungi overwinter in leaves on the ground or are carried over to the next year in infected buds and twigs. The spores develop under cool, moist conditions. Trees may be weakened over successive seasons of trouble, and are most sensitive to drought.

Gather and destroy the diseased leaves when they fall, or compost them under several inches of soil. Prune out infected twigs and branches. Consider fertilizing in fall or spring.

Watering may be indicated in dry spells. A Bordeaux spray is effective. When planting new trees or shrubs, consider using anthracnose-resistant species. London plane is more resistant than American sycamore, and black oaks are much more resistant than white oaks.

Leaves of ash, beech, dogwood, elm, horse chestnut, linden, maple, and oak are often affected by **leaf scorch,** a noninfectious disease that causes a browning between veins or along margins of the leaves. Scorch usually develops during July or August and is especially severe following periods of drying winds and high temperatures, when the roots are unable to supply enough water to the tree to keep up with the large amount of water lost through the leaves. The condition may also result from, or be made worse by, shallow soils, roots that girdle the crown of the tree, a diseased root system, drought, and other diseases that weaken the tree. Fir, pine, and spruce often show leaf scorch as a brown discoloration of the needle tips. The more severe the scorch, the farther down the needle the browning is found. Scorch on needles may result either from hot, dry weather or from high winds during cold weather.

Trees should be fertilized to prevent new growth from becoming easily susceptible. Watering is very helpful, particularly to those trees that have been recently planted. Apply water to the ground around the trees, not to limbs or foliage.

The group of **leaf spot** diseases is characterized by yellow, brown, or black dead blotches in leaves. Heavily infected leaves may turn yellow or brown and fall prematurely. The leaf spots of deciduous trees and shrubs are caused by fungi that overwinter in dead leaves on the ground. Under the warmer, moist conditions of spring, the spores germinate and the fungus grows into the leaf. Some leaf spot fungi produce summer spores that splash about in rainy weather and intensify the disease. On conifers and broad-leaved evergreens, the fungi pass the year on the host plant. All leaf spot diseases caused by fungi are favored by cool, moist weather, especially early in the growing season as new leaves are developing.

To prevent leaf spot, gather and compost fallen leaves in autumn. If only a few leaves are infected, they may be removed by hand. A strong stream of water often helps remove dead foliage from dense evergreen shrubs such as boxwood or arborvitae. For evergreens, spacing the plants well and keeping down weeds and grass under the lower branches provides better ventilation and reduces the moist conditions that favor infection.

**Mistletoe** is a parasite of trees found in many areas of the country, particularly the East and West Coasts, and all across the South. The pale, waxy berries are carried by birds and dropped into bark crevices, where they germinate. Their rootlets attach themselves onto the host tree, and the parasite takes its nourishment from the tree's sap veins; it cannot live in soil. Where mistletoe fastens on, grotesque swellings ensue, and the tree may eventually die. See MISTLETOE.

**Nectria canker** attacks birch, elm, linden, black walnut, and other hardwoods in the Northeast. Water-soaked areas, darker in color than the adjacent healthy bark, are formed on the trunk and large limbs. The edge of the diseased areas cracks, and callus tissue that forms under the cracked bark becomes infected and dies. The annual repetition of this process forms concentric circles of dead callus tissue. When a canker completely girdles the trunk or branch, the portion above the canker dies.

Where the cankers are not too large, it is important to cut out the diseased areas and treat the wounds with a dressing.

**Needle rust** [color] is a fungus that attacks needles of two- and three-needled pines. Red pine is very susceptible. The disease develops in spring as small cream-colored, baglike pustules on needles. The pustules rupture and orange spores are blown to infect goldenrod and asters. The rust overwinters and can live indefinitely in the crowns of these alternate hosts. During summer and autumn, spores from goldenrod and asters infect needles of the pines. This disease may cause needle drop and stunt young pines, but it seldom causes much damage on older trees.

Control by destroying goldenrod and asters near valuable pine plantings or nurseries.

Fungi of the **powdery mildew** group infect the leaves of various trees, producing powdery patches of white or gray on leaf surfaces. Tiny black fruiting bodies of the powdery mildew fungus are often found on the white patches. Trees commonly affected with powdery mildew in the Northeast are catalpa, dogwood, horse chestnut, linden, magnolia, and sycamore.

In most cases mildew is more unsightly than harmful. Sanitary measures, as described above, are usually sufficient to control the trouble.

**Rust** fungi found on junipers in the Northeast cause reddish brown round galls up to 1½ inches in diameter on the twigs, or slight swellings on the branches or trunks. During rainy weather in the spring, sticky orange spore masses or horns protrude from the galls or swellings. The spores are carried by wind to leaves or fruits of nearby alternate host plants, such as apple, pear, ornamental crab, hawthorn, shadbush, quince, mountain ash, or chokeberry. These alternate hosts are needed for completion of the life cycle of rust fungi, since spores produced on junipers are unable to reinfect junipers. If the alternate hosts are removed, rust fungi cannot survive. Handpicking galls in spring before spore horns appear will give good control of rust.

**Shoestring root rot** (also known as mushroom or armillaria root rot) infects many trees, including birch, black locust, chestnut, maple, mountain ash, sycamore, poplar, oak, pine, spruce, larch, and yew. Affected trees show a decline in vigor of all or part of the top of the tree. Foliage becomes scant, withers, turns yellow, and drops prematurely. Fan-shaped white fungal growths are found between bark and wood close to and below the ground line. Rootlike dark brown or black "shoestrings" of the causal fungus grow beneath bark and in soil near affected roots. These strands are rather brittle, and they fuse or grow together where they cross. When one of them is opened, the inside appears to be a mass of white compressed cotton. The strands may cause infection of roots of nearby trees, especially if these trees are in poor vigor. In late fall, clusters of honey-colored mushrooms are often found growing around the base of affected trees. Prevent trouble by avoiding injuries to the roots of healthy trees.

**Sooty mold** fungi grow as saprophytes on honeydew secretions of such insects as aphids and scales. The heavy sooty growth covers needles of various evergreens and leaves of elm, linden, magnolia, maple, and tulip poplar. Although the heavy coating of mold on leaves is unsightly, it does not often interfere seriously with food manufacture in the leaf.

Control the insects responsible for secreting the honeydew on which the mold exists. Aphids are usually at the root of the trouble. Ladybugs may keep their numbers down; you can help by applying a sticky band around the trunk to keep aphid-carrying ants on the ground.

**Spanish moss** is not a true parasite, but an air plant, as are lichens and orchids. The hanging, grayish strips of this moss are common in the South. Though it does not suck the tree's juices like mistletoe, this plant may smother the tree to death if it runs rampant.

**Verticillium wilt** is a fungal disease that attacks many trees, causing a sudden wilting and yellowing of foliage that is followed by premature defoliation. One limb or the entire tree may be affected. Some trees wilt and die suddenly, while others gradually fade over a period of years. Discolored streaks are present in the outer rings of the wood of infected branches, the color depending on the tree species.

Prune all dead branches and fertilize affected trees to stimulate vigorous growth. Remove badly infected trees, together with as many roots as possible; the disease is thought to be transmitted through the soil. Do not replant ailanthus or other wilt-susceptible trees in the same location; these include black locust, catalpa, elm, Kentucky coffee tree, linden, maple, redbud, smoke tree, tulip poplar, and yellowwood. Infected trees can be replaced with resistant species, including most conifers and certain broadleaved trees: beech, birch, boxwood, dogwood, fruit trees, holly, linden, locust, mulberry, oak, pecan, serviceberry, sweet gum, sycamore, walnut, and willow.

A powdery mildew fungus and gall mites are usually associated with broomlike growths on branches known as **witches' broom.** Several hundred galls may be found on a single tree, causing an unsightly appearance in winter. Affected branches are weakened and break easily during wind storms, and the broken wood is exposed to wood-decaying fungi. Affected buds are larger and more open and hairy than normal. Mites may be found inside the buds along with small black fruiting bodies of the mildew. Threadlike strands of mildew are found on the outside of the bud. Branches that develop from these buds are dwarfed and clustered, giving the witches' broom effect. There is no practical control for this disease. If the brooms are unsightly, prune them off.

# Environmental problems

A well-fed, vigorous tree, planted in the proper climate and surroundings, is not often seriously troubled by pests or diseases. However, trees never exist in isolation, and a great variety of environmental factors that you take for granted can make all the difference between health and sickness.

**Construction** of buildings, roads, and ditches is one of the main sources of environmental damage to trees. The traffic of heavy equipment may damage tree roots through the effects of soil compaction. Diseases may enter the wounds caused by construction equipment. Trees are hurt when the ground is covered with soil fill or by asphalt or concrete. Even the addition of a few inches of soil will change the amount of water and oxygen available to the roots. Feeding roots are very sensitive to changes in the water level, and if they cannot reach their normal water supply, trouble will result. Carbon dioxide and other gases may build up in the filled soil to produce a toxic effect. Growth slows down, foliage may become discolored, and the trees decline and often die within a few years.

The symptoms caused by these factors are often described as **decline.** Another general term is dieback. Older trees tend to be more susceptible. Common signs of distress include small and discolored leaves, premature fall coloration and loss of leaves, reduced twig growth, dying twigs in the upper crown, and the decline of parts or all of the tree. Trees may become worse over a couple of years and die, or they may continue to live indefinitely. The list of possible causes is long, and isolating the one responsible for a particular case of decline can be difficult.

Construct a small well around the trunk of an established tree to prevent smothering of trunk tissues but not of the roots. If possible, avoid adding excessive soil or impervious materials within the drip line (the entire area under the branches) of the tree.

If soluble salts from **dog urine** enter the soil, tree roots may be killed. A metal collar on the trunk protects only the bark, not the roots, and the entire planting area may have to be screened off from dogs.

Trees will show the effects of **drought** the following summer, especially if they were not watered during the dry period. If leaves are scorched or drop off from lack of water, the tree will not be able to store its normal supply of food. Keep your trees well watered in dry times.

**Electric currents** from service wires may make rough wounds where wires come in contact with trees. Entire limbs or trees may die. Wet weather helps to cause the short-circuiting of current that causes high temperatures to reach lethal levels in tender tree tissue. Be sure that wires close to the trees are thoroughly insulated. Wires may also cause mechanical injury by chafing.

**Flooding** and changes in normal drainage patterns, especially during the growing season, may kill tree roots by depriving them of oxygen. Low-lying areas subject to flooding are poor sites for some species, including eastern white pine, hemlock, paper birch, red cedar, red pine, white spruce, and sugar maple, all of which will decline and eventually die if flooded. Trees that tolerate occasional flooding and can be planted in damp sites are ash, black gum, cottonwood, elm, overcup oak, red maple, river birch, silver maple, sweetgum, sycamore, white cedar, and willow.

**Frost** and unseasonably cold weather may kill parts of trees in autumn and new growth in spring. To harden your trees against such damage, do not stimulate them to grow late in the season, as new growth cannot stand up to the winter cold. Plant native trees or those known to be hardy. Valuable and small exotic trees that may be especially frost-susceptible should be protected with shelters in the winter. Frost cracks appear as longitudinal splits in the tree trunk that tend to form successive layers of callus until a large so-called frost rib protrudes from the trunk. These cracks are often a result of ice formation and sudden changes in winter temperature. They occur particularly in early winter, after autumn rainfall has caused a spurt of late growth of immature bark and sapwood. You can help prevent frost cracks by making sure your trees have proper drainage so that excess moisture doesn't collect about the base. Dead bark can be cut away from the edges, and a disinfectant and a wound dressing applied. See winter injury, below.

**Lightning** strikes trees and roots to cause a wilting of foliage and possibly trunk abrasions. The lightning may burn a complete ring around the cambium layer of the lower trunk, in which case the whole tree will die. Sometimes a tree whose cambium has been destroyed from top to bottom will remain green through the rest of the season and even put out new leaves the following spring, only to die later that year. The tree can be helped to recover from its wound by fertilizing and watering. Lightning rods may help prevent trouble; they should be professionally installed.

You may observe **mosses** and grayish green growth lichens on the trunks of trees and bases of shrubs, especially those growing in damp areas. Lichens, made up of certain fungi and algae, take the form of crusts, leaves, or cushions. Mosses usually have leafy, erect, or creeping stems that develop from a heavily branched green structure; the plants develop in dense clusters or cushions. Mosses and lichen are no cause for alarm as they seldom affect the growth of the tree or shrub, and it is not necessary to remove them.

**Natural gas** from a leaking or broken gas main can cause yellowing foliage, slow growth, or even the death of the tree. You may see blue or brown streaks in the wood of roots and trunk. Breaks in the main some distance down the street may miss other trees in the neighborhood and nevertheless hit yours because of some quality of the subsoil layers. The utility company will respond immediately to a call for detection of gas leakage, but here's how to make the test yourself. In the lawn area between the tree and the suspected gas leakage, dig a hole large enough to accommodate a half-bushel basket. In this hole, place a potted tomato plant. Then cover the hole with boards, some paper or burlap, and finally a small amount of the soil. After 24 to 36 hours, remove the plant and examine it. If all the branches and leaves bend down sharply and firmly from the main stem (not just presenting the wilted appearance of a plant needing water), gas is present in the soil.

Once the gas leak is repaired, you can help your tree back to health by digging a trench on the side of the leak or all around the tree and then aerating the soil thoroughly with compressed air. Next, water heavily. This process should also be followed when replacing dead trees with new trees.

An evergreen tree may be just that, but all coniferous trees do go through a seasonal **needle drop.** Although the sight of foliage turning yellow or brown and then falling to the ground may be alarming, chances are good that this is a natural phenomenon, brought on by the changing seasons and perhaps encouraged by dry weather. Still, it is wise to keep an eye open for diseases or insects that also can cause needle drop.

Needles stay on the tree for a particular length of time before turning color and dropping, and that period varies from two to several years, depending on the species. Once you become familiar with a tree's seasonal cycle, you won't be apt to misdiagnose the cause. Another indication that all is well is the simultaneous yellowing of many trees; and on the individual tree, yellowing occurs throughout the older needles of the interior. Look at the affected needles themselves to see if they have marked blemishes or webs of mites or needleminers. The new growth at branch ends should be green and unblemished. Needleminers and mites attack spruce. Aphids concentrate on white pine, and new growth will turn yellow.

Unless needle drop is the work of a bug, there is little to worry about. You can help guard evergreens against the drying effects of winter weather by giving them plenty of water before the ground freezes.

If **pesticides** or herbicides have been used nearby, your trees may suffer. A neighboring building may have been treated for termites, for example. Many of the materials used by exterminators for this purpose are extremely toxic to plants. If poisonous gas is employed, the vapors may find their way to your property. Some trees show such poisoning by a browning of the midveins of the leaves; on others, the whole leaf turns brown. The injury will show in one month if the tree is in leaf, but not until the following spring if the leaves have fallen.

It is a mistake to assume that **pollution** only damages trees in and around large cities —air pollutants occur also in rural areas. There are two main classes of gaseous air pollutants, point sources and oxidants. Point source pol-

lutants from a point source include sulfur dioxide ($SO_2$) and fluorides. $SO_2$ comes mostly from burning coal and oil for the generation of electricity. However, $SO_2$ can also be produced by smelting sulfur-containing ores or manufacturing sulfur products. Fluorides are produced by reduction of aluminum ore, manufacture of phosphate fertilizer, and stone-processing operations. Other less common pollutants of this class are ethylene, hydrogen chloride, and ammonia.

Oxidants are formed in the atmosphere from chemical reactions powered by sunlight. Ozone and PAN (peroxy acetyl nitrate) are produced mostly from industrial and auto emissions. Low concentrations of ozone also occur naturally in the atmosphere. The oxidants are common components of smog.

Factory pollution, in the form of gases, dusts, or smoke, may cause discoloration, mottling, or browning of foliage, especially on evergreens. Some gases contain arsenic and forms of sulfur that are poisonous to leaves, while dusts may simply smother the foliage. In general, hardwoods are the most resistant to noxious fumes.

Domestic poisons can affect your trees. Strong detergents or chemicals poured into nearby sewage may kill a tree whose roots are in the drainage field. Oil from a nearby oil burner intake pipe may soak the root area. Soaps, oils, and salts that leak from nearby auto service stations, laundries, tennis courts, or refuse dumps may all suffocate or poison tree roots.

Roadside trees are sometimes menaced as much by salt as by weedkillers. See below.

**Root girdling** may stunt trees and eventually kill them. Actually, the trees are killing themselves, with roots that constrict the trunk. Such trees turn earlier than nearby, healthy ones. They are more apt to suffer insect problems, as well. Leaves tend to be smaller. Roots should grow outward from the trunk, but may encircle the trunk if the sapling was kept too long in too small a container. Girdling may also be caused by an overly small planting hole.

---

## REACTION OF TREES TO SALT

| *Sensitive* | *Tolerant* |
|---|---|
| **DECIDUOUS TREES** ||
| American beech | Austrian pine |
| American hornbeam | big-tooth aspen |
| black cherry | black locust |
| crabapple | black walnut |
| eastern redbud | European horse |
| hackberry | chestnut |
| hawthorn | honey locust |
| littleleaf linden | mountain ash |
| pin oak | Norway maple |
| red oak | poplar |
| shagbark hickory | Russian olive |
| speckled alder | silver maple |
| tulip tree | tree-of-heaven |
| white oak | white ash |
| **CONIFEROUS TREES** ||
| American arborvitae | Austrian pine |
| balsam fir | Colorado spruce |
| Canada hemlock | eastern red cedar |
| dawn redwood | European larch |
| eastern white pine | jack pine |
| red pine | |
| scotch pine | |
| white spruce | |
| yew | |
| **SHRUBS** ||
| American elder | alpine currant |
| common box | buckthorn |
| dogwood | fragrant sumac |
| European euonymus | Pfitzer juniper |
| flowering quince | Siberian peashrub |
| hazelnut | staghorn sumac |
| Indian currant | winged euonymus |
| coralberry | |
| Japanese barberry | |
| spirea | |

You can rescue a tree that is strangling itself by chiseling away a section of the guilty root or roots. Treat the severed ends. The top should be pruned to compensate for the reduced root system.

When streets and highways are treated with **salt** during winter snow and ice storms, adjacent plants are attacked both on foliage and through their roots. Sodium chloride (common table salt) and calcium chloride, another deicer, are both toxic to plants. When salty slush is sprayed on trees by passing cars, evergreens show yellowed or browned needles and, in early spring, twig dieback. On deciduous trees, symptoms are bud death, twig dieback, and the formation of tangled twigs known as witches' broom. If trees are harmed by salt in the soil, symptoms may develop gradually, and are often most obvious toward the end of summer and in hot, dry spells.

Diagnosis is not straightforward, because salt-induced decline looks much like the injury caused by a number of other environmental stresses. Some state extensions will test foliage to determine if road salt is the cause.

Roadside plantings can be isolated from salt in the ground by digging a drainage ditch to carry away salt runoff. Salty soil will benefit from a rinse with plenty of watering in spring. A simple barrier of plywood or burlap will keep spray from hitting plants.

If you are planning to grow roadside plants, consider the salt-tolerant and salt-sensitive plants on the list printed here, adapted from a pamphlet published by the University of Wisconsin-Extension.

**Shade** and lack of sunlight often result in spindly growth and the death of the lower limbs or inner twigs. Light-loving trees may do very poorly under the heavy shade of larger trees. When transplanting, it is important to keep trees, especially evergreens, well spaced. Plant shade-tolerant trees in areas shaded by larger trees, and prune out the inner limbs of such heavy shade trees as the Norway maple.

Poor foliage and twig growth, lichen on the bark, and general weakening of the tree are occasionally due to **starvation,** which in turn may be caused by planting in poor soil, exhaustion of soil food, lack of water, extensive paving, packing of surface soil, or a combination of these. The remedy is to fertilize and water properly, to leave space between pavement and trees, and to aerate or cultivate the soil in the root area.

High-intensity sodium **street lights** in your neighborhood may do indirect damage to trees by causing them to grow faster and keeping them growing longer into autumn than normal. This increases trees' vulnerability to air pollution and to frost damage, particularly in the case of young trees. London plane trees are particularly susceptible. Combat such damage by planting dormant trees in fall, by choosing more resisant trees such as ginkgo, or by getting rid of such lights, if possible.

**Sunscald** affects the bark tissues, and you will see dry, cracking, or curling bark on limbs and trunks that had been smooth. When bark accustomed to shade or protection is suddenly exposed to drying and heat from sun or wind, as in the case of thinning out or planting young trees in open areas, sunscald is apt to result. Young trees can be protected on the south exposure by setting up a vertical board or by winding the trunk with burlap or wrapping tape.

**Sunscorch,** in times of high temperature or drought, will cause injured or diseased roots and keep parts or all of the tree from receiving adequate moisture. The visible symptoms of sunscorch will be the yellowing, browning, and withering of leaves on one side or the whole

tree. The condition starts on leaf tips and edges. Avoid disturbing the roots, if possible, but if they already have been disturbed, prune the trees to balance the reduced root system. Keep your trees well watered in dry spells, and conserve soil moisture with mulch.

The process of **transplanting** may cause somewhat of a shock to trees, and you should expect a recovery period before they again grow well and flourish. Proper pruning at transplanting time helps minimize the shock.

In addition to frost damage, your trees may suffer **winter injury** or drying. The usual symptoms in late winter and spring are the browning and withering of foliage and twigs. This is a common condition on evergreens. It is caused by dry winds that remove moisture while the roots and soil water are still frozen. Sun reflection from buildings and pavement may also cause drying.

You can help prevent such winter injury by mulching about the base in early fall and watering thoroughly. Burlap screens and wax emulsions also give some protection.

# TRICHOGRAMMA

Several species of this tiny wasp parasite are marketed for use in controlling some two hundred insect pests. Wasps are sent through the mails as eggs. Three species are available

Trichogramma wasp

currently: *Trichogramma platerni,* for avocado groves and orchards; *T. pretiosum,* for vegetables and field crops; and *T. minutum,* for ornamentals, orchard crops, and grapes.

The wasp lays its egg in the host egg, which then ceases development. The wasp larva feeds on the egg and emerges as an adult. Thus the pest never gets a chance to damage a crop.

The eggs sometimes are mailed on cards— eggs so tiny that the cards have the surface texture of sandpaper. The cards are set out at the first sight of pest moths. The moth populations should be monitored, perhaps with pheromone traps, to determine if repeated releases are necessary. One company will ship trichogramma eggs weekly, biweekly, or monthly, as requested on the order form.

# TULIP

## Insects

Tulips avoid the summer rush of insects, being spring bloomers. This is not to say that tulips are entirely untroubled, however. A number of **aphids,** the grayish tulip bulb and green tulip leaf aphids in particular, impair growth by sucking plant juices. The tulip bulb aphid is found in most northern and western states, and causes damage by robbing the juices from exposed leaves on growing plants to cause severe distortion, stunting, and sometimes death of the plant. They build up dense colonies on stored bulbs, especially those with cracked outer scales, and may also pick on iris bulbs. The tulip leaf aphid is somewhat smaller, and occurs in clusters on leaves and shoots, causing leaves and flowers to fail to open. It winters between crops on dormant bulbs. For controls, see APHID.

**Millipedes** frequently attack bulbs in tulip beds, especialy in old beds in which the bulbs are not reset each year. The holes they eat into the bulbs are often followed by decay, and good numbers of bulbs may be destroyed. When abundant, these pests also injure bulbs and roots of other plants, strawberries in particular. Dig up tulip bulbs after they have flowered, and keep them cool and dry until fall planting. See MILLIPEDE.

Other tulip pests include SPIDER MITE and WIREWORM.

## Diseases

**Breaking** is a viral disease that causes variegated flowers with stripes or streaks of color. Leaves are distinctly mottled, and the size and production of bulbils (aerial bulbs) are reduced. In normal flowers the color is uniform, except at the base of the flower.

Aphids can transmit this disease, and some gardeners will want to separate parrot and bicolored tulips from solid-color plants; white flowers usually aren't changed. Other growers actually enjoy the exotic color variations wrought by this virus. At any rate, foliage and plant health is not frequently damaged.

**Tulip fire** (botrytis blight) is a common disease that is especially damaging in rainy springs. Spotting and collapse of stems, leaves, and flowers are usually accompanied by the brownish gray mold of the fruiting stage. Tiny black sclerotia carry the fungus to the soil to infest plants the following spring. The disease may be spread by windblown spores from infected plants.

Once plants are infected, they should be pulled and burned. The disease is most serious when tulips are grown in the same area year after year. Tulip fire will have a hard time catching up with the flowers if the tulip bed is moved each year. Stems should be removed from bulbs right after they are dug up, and any diseased bulbs must be stored during the summer in a cool, dry place if they are to be of any value. Remove husks from bulbs just before they are planted, and any bulbs that look sick at this time should be discarded. Once plants are up, remove and destroy infected leaves, buds, and blossoms, and burn plant refuse at season's end.

Tulip fire

# TULIP TREE

See POPLAR.

# TURNIP

Turnips have no pests that are unique to them, but share many with cole crops such as broccoli, cabbage, horseradish, and kale. Insects you're likely to run into on turnip include aphid, cabbage looper, cabbage maggot (tunnel into the underground stems and roots of early plants to cause wilt and stunting), harlequin bug, and yellow woollybear caterpillar; see CABBAGE. Diseases, too, are very similar to those affecting CABBAGE. Also see RESISTANT VARIETIES.

# VERBENA

## Insects

The **verbena budworm** bores within the new shoots of verbena, causing them to wilt. This pest is a greenish yellow worm with a black head, growing to a little less than ½ inch when mature. The adult is a purplish brown moth with a wingspread of ½ inch. Ordinary infestations can be controlled by handpicking or clipping and then burning the infested tips. Growers outside the eastern states should have little trouble.

The small, **yellow verbena leafminer** feeds between leaf surfaces, making mines that appear as blisters or blotches. This pest is found throughout the United States, wherever verbena is grown. Handpick infested leaves.

The **yellow woollybear caterpillar** is indeed woolly and yellow. Verbena is one of many plants that feed this common garden pest. Because of its sloth and size—it grows to a length of two inches—handpicking is the recommended control.

## Diseases

**Powdery mildew** is apt to seriously injure greenhouse verbena. It appears as white moldy patches on leaves and young shoots late in the season. See ZINNIA.

# VIBURNUM

The snowball **aphid** causes young leaves to become severely curled, deformed, and discolored with a sooty mold that develops on aphid honeydew. The aphid varies in color from gray to dark green. It attacks new growth early in the season, and in June moves on to other plants. For control, see APHID.

# VIOLET

Deep red **aphids** occasionally infest violet leaves. See APHID.

The **violet gall midge** is a small two-winged fly that lays its white eggs in the curled margins of unfolded new leaves. The larvae that hatch remain in the curled margins, causing further curling, distortion, and twisting of leaves. There are likely several generations in greenhouses each year. Clean up leaves as they drop.

If leaves are skeletonized or eaten through overnight, suspect the larvae of the **violet sawfly.** These ½-inch-long worms are either colored olive green or blue black, and are marked with whitish tubercules. They hide under the lower leaves or in the soil during the day. The adult is a black, four-winged fly that measures ⅝ inch in length. It lays its eggs in blisterlike incisions in leaves. There is only one brood in the North, and overlapping generations in the South.

Nocturnal handpicking raids may do for the energetic, but other gardeners will have to rely on daily inspections of lower leaves and botanical sprays on foliage.

Keep an eye out for the celery leaftier (see CELERY), cyclamen mite (AZALEA, DELPHINIUM), flea beetle (POTATO), MEALYBUG, red-banded leafroller (ROSE), SLUG AND SNAIL, SPIDER MITE, and yellow woollybear caterpillar (see controls for imported cabbageworm under CABBAGE).

# VIRUSES

Viruses are termed obligate parasites, which means they cannot live outside of living cells. It is this characteristic that makes viral pathogens so difficult to produce for large-scale applications—production must be done with live, infected insects.

Because no way has been found to mass-produce viruses, commercial production is expensive. Dead insects and the usual culture media like agar won't work. One way of getting around this problem has been to spray pest-infested trees with virus, collect the larvae before they spin cocoons, and then trap the emerging adults for release in infested stands. Viruses have the advantage of being passed on from infected adults to eggs. Also, they have a great potential for biological control. It has been estimated that nearly half of California's most damaging pests are vulnerable to viral diseases.

Until several years ago, no viruses had been approved for interstate sale by the Environmental Protection Agency. Today, nuclear polyhedrosis virus (NPV) is sold for control of many pests. You can find NPV-infected insects and spread the disease yourself. See NUCLEAR POLYHEDROSIS VIRUS. A granulosis virus (GV) may soon be marketed as a control for codling moth in orchards. Elcar is a viral insecticide registered for use against the corn earworm (also known as the cotton bollworm and tomato fruitworm) on many crops.

# VOLE

See MOLE.

# WALNUT

The **navel orangeworm** is parasitized by *Goniozus legeneri,* also known as the summer navel orangeworm parasite. See ORANGE.

The worst insect pest of black walnuts is the **walnut caterpillar** [color]. If you see a black walnut tree partially stripped of leaves during midsummer, it likely is the work of this worm.

The caterpillars are black with white hairs. They lift head and tail when disturbed. There is one generation in the North, and two in the South. Like many insects, their population goes in cycles; they may be bad for a year or two and nearly disappear for several seasons. Their habit of congregating at the bases of branches each night makes them easy to eradicate on small trees. Just rub them out during late evening with a rolled-up burlap bag. If you don't mind climbing a ladder, you can get to pests on larger trees.

The **walnut husk fly** can be a serious pest on walnut trees. It overwinters in the soil under the trees in small hard, brown cases. Adults emerge in late summer, usually August, and spend two or three weeks on the foliage before they mate and begin to lay eggs. Adult flies are about the size of a housefly and colored brown with a yellow semicircle on the back. The female

Walnut husk fly

penetrates the husk of the nut to lay her eggs. The hatching larvae feed on the husk and then tunnel to the outside of the nut and drop to the ground.

The chief injury to the walnut is caused by the feeding larvae, which release a dark liquid stain over the shells and sometimes the kernels. Otherwise, the nut is not harmed. Worms may be destroyed by dropping the infested nuts into a pail of water; once drowned, the maggots can easily be removed along with the husk.

## Diseases

For **anthracnose,** see TREES.

**Black line** of walnuts is a disorder of grafted walnut trees. It rarely develops in trees less than 10 years of age and is most common in 15- to 25-year-old trees. The first symptoms are weak growth, sparse foliage, and yellowing and premature dropping of leaves from certain branches. The death of the affected branches follows. Finally the entire treetop dies. This disease is caused by the failure of the newly formed wood (xylem) of the walnut top to unite with the wood of the black walnut rootstock at the graft union. A dark brown or black layer of corklike tissue forms between the xylem of the rootstock and the scion. At first this layer is very narrow, but later it may become as wide as ¼ inch.

After black line becomes well established, decay of the bark frequently occurs beneath the graft union. Affected trees usually die within four to six years after symptoms are first noted, although some trees live longer. The exact cause of black line is not known, but trees lacking in vigor appear to be especially susceptible to the disease.

Once infected, a tree cannot be cured. The only known preventive measure is to plant Persian walnut varieties that have been grafted on Persian walnut rootstocks. The more vigorous types of Persian seedlings, such as Manregian, should be used for the rootstocks. Late-maturing varieties will be less vulnerable.

**Crown gall** is caused by bacteria that enter the tree through wounds. See CROWN GALL.

# WASP

Wasps are recognized as important parasites of pest insects. They also serve as predators; the potter and papernest wasps, for example, sting and capture caterpillars, then take them to the nest as food for wasp larvae. These beneficial wasps are attracted by blooming flowers, as are the adult wasps mentioned below.

BRACONID WASPS [color].    Among the most important of aphid parasites, these tiny wasps characteristically leave a round exit hole in the backs of victims. Braconids also parasitize the larvae of moths and butterflies (gypsy moth, satin moth, browntail moth, codling moth, oriental fruit moth, strawberry leafroller, sugarcane borer, tent caterpillars, and cutworms), and the larvae of many beetles.

If you've seen a fat tomato hornworm covered with egglike cocoons, then you have likely been the beneficiary of a braconid. These cocoons [color] are usually the pupae of wasps that have gone through their larval stage within the worm's body. In one instance more than 500 braconid larvae were counted in the body of a hornworm that showed no signs of injury. But in time, the feeding of these tiny parasites will take its toll, and they will go on to lay their eggs in other larvae. In handpicking these worms from your plants, leave any that are parasitized.

Another internal parasite of the braconid family specializes on the cabbageworm. When the larvae are finished feeding within the host, they exit the worm's body and form silky cocoons on a leaf. Although the affected cabbageworm does not die right away, it will not be able to pupate.

Satin moth larvae are the victims of an imported braconid, *Apanteles solitarius.* First introduced to New England in 1927, the wasps there now hit at least 60 percent of the overwintering larvae of this pest of shade trees. The braconid parasites of the browntail moth have knocked this moth from pest status in New England and eastern Canada.

This family of wasps has also had a tremendous effect on gypsy moth populations, and includes both parasites of both pest eggs and larvae. A braconid parasite of codling moth larvae, *Ascogaster quadridentata,* would be more effective if it weren't for secondary parasites and broad-spectrum insecticides. In Nova Scotia —where the plant-derived ryania has been used instead of DDT and parathion, and where there are no secondary parasites—codling moths don't cause as much trouble to apple growers.

The most serious pest of peaches, the oriental fruit moth, is parasitized by a good number of braconid wasps, some species abundant and others quite rare. The numbers of all beneficials become rarer still in orchards that have been thoroughly sprayed for control of the oriental fruit moth. One wasp, *Macrocentrus ancylivorus,* has shown great promise in controlling the oriental fruit moth. Growers in the region for which this parasite is known to be effective —roughly Massachusetts to Michigan and southward to eastern Missouri, Arkansas, and northern Georgia—can prevent nearly half the damage caused by the pest by making liberations when the second or first and second generations are present, without spraying.

CHALCID WASPS [color]. Mealybugs, aphids, scale, and larvae of beetles, moths, and butterflies are parasitized by these small ($\frac{1}{32}$-inch-long) wasps. Some chalcids are a metallic black, while others are golden. They are highly rated as control agents, and have been used to attack pest eggs, larvae, and pupae of dozens of pests worldwide. The golden chalcid, *Aphytis melinus,* is sold for use against scale. The egg parasites of the trichogramma family are also commercially available (see TRICHOGRAMMA).

The success of chalcid wasps is dependent on favorable temperature and moisture, and is seriously affected by sulfur fungicides and miticides, road dust, and chlorinated hydrocarbon insecticides. DDT had been a major threat to these beneficials.

ICHNEUMON WASPS [color]. These wasps are highly important as parasites of moth and butterfly larvae. Typically they are slender, have a long abdomen, and may bear a formidable-looking ovipositor. But this organ is used only for placing eggs, and not for stinging people.

Adult wasps feed on pollen and nectar, and often from puncture wounds made in host larvae; fly maggots often enter these wounds to infest larvae of the gypsy moth. Wasps deposit eggs either within or near hosts.

Several ichneumon species were successfully used to control the spruce sawfly in eastern Canada and New England. Other species have been used to keep down populations of the European pine sawfly in New Jersey, the larch sawfly in Canada, and the European corn borer in the United States.

# WATERMELON

## Insects

The melon **aphid** generally makes its appearance in the field late in summer and soon becomes abundant on the undersides of leaves. Leaves curl, the vine becomes too stunted to produce a crop, and plants may wilt and die. The small, dark green insect may attract other insects such as wasps or flies to the melons with the honeydew it produces. Eliminate as many weeds as possible from the growing area, especially live-forever in northern states, to deprive these pests of overwintering sites. See APHID.

**Cucumber beetles** attack vines, especially those of young plants. As with cantaloupes, much of the damage can be prevented by covering the vines with plastic hotcaps or netting until they have developed half a dozen leaves. See CUCUMBER.

The **melon fly** female lays eggs in the stems of the vines and in young fruit. After the eggs hatch, the developing larvae feed on the surrounding plant tissue and usually kill the vine or spoil the fruit. The melon fly rests on castor bean, cocklebur, and Jimson weed in preference to crop plants. Under dry conditions or in areas where weeds have dried, and in areas where there are no broadleaved weeds, a corn border will help attract flies away from the melons. Infested fruit should be removed and destroyed, and nearby tomato and cucurbit crops should be plowed under as soon as harvesting is completed. If you haven't too many plants, you can protect fruit by stapling paper bags around them. These bags may be left on until the fruit is ready for harvest.

## Diseases

**Anthracnose** is a common disease of cucumbers, muskmelons, and watermelons. Leaves, fruits, and stems of infected plants show water-soaked areas, followed by the death of the plant. Crop rotation helps somewhat. Resistant varieties, such as Charleston Gray, Crimson Sweet, Sweet Princess, Improved Kleckly Sweet, and Klondike, will get you by this disease. For other prevention methods, see CUCUMBER.

**Fusarium wilt** is probably the most serious disease of watermelon. It is caused by a fungus that lives in the soil and penetrates the roots. Symptoms are a brown discoloration of the stems, followed by a wilting of the branches and death of the plant. Planting wilt-resistant seed and long rotations with other vegetables are the best methods of controlling this disease. Resistant varieties include Charleston Gray, Congo, Fairfax, Black Kleckly, Crimson Sweet, Sweet Princess, and two seedless varieties, Tri-X313 and Triple Sweet.

**Mosaic** is a viral disease transmitted by aphids that causes mottling of the leaves, stunting of the plants, and misshapen fruits. Good control of aphids and eradication of other cucurbit plants from the vicinity may help to check the spread of this disease. See also APHID and CUCUMBER.

For other diseases of watermelons, see CUCUMBER and CANTALOUPE.

# WEEDS

Weeds are as much a part of gardening as bugs. And like bugs, weeds aren't all bad. Perceptive observers of the garden have credited weeds with loosening and enrichening the soil, conserving moisture, and supplying beneficial insects with shelter and an alternate food source.

Just how many wild plants you can tolerate among the vegetables is in part a matter of personal preference: some gardeners aren't comfortable unless every last weed has been yanked and consigned to a healthy compost pile with its seed-killing temperatures; others are more casual.

Begin pulling weeds early in the season, soon after sowing, while they're still easy to pull and before they can compete with your seedlings. Weed regularly so that the plants don't have time to produce seeds. These aggressive plants excel at producing seed, quickly and in profusion, and buried weed seed may wait a century for the right opportunity to sprout. Don't give the garden up to weeds come fall.

Weeding is less troublesome if you plant intensively—in beds, for instance, rather in rows that permit weeds to pop up. Corn grown close together will reduce the light available to opportunistic wild plants; those weeds that do emerge will be puny and easy to pull.

Tilling may take care of annual weeds, but perennials can overwinter as roots, even as fragments of roots. These include nettles and thistles, dock, quackgrass, and milkweed.

Many gardeners spare time and their backs by mulching to keep down weeds. Use any material that excludes light; clear plastic won't work for that reason. Try hay in the garden, ground bark around flowers, and stones around the bases of trees.

Ground covers can smother weeds. Pachysandra, myrtle, and vinca are good choices. Bigger plants that will crowd out weeds include ferns of many kinds, Siberian iris, day lily, lily-of-the-valley, violet, feverfew, and catnip, all of them perennials. Artemisias or crown vetch will do, but they are inclined to take over if given half a chance.

Annuals that can be used as ground covers include marigolds, nasturtiums, snapdragons, petunias, some of the herbs, and even lettuce.

One weed can be used to control other weeds: purslane. The low, thick-growing plants are pleasant to walk on and they make it possible to go into the garden after a rain without compacting the earth. Purslane also helps keep the earth cool in hot weather. And it is good to eat—in fact, the early settlers brought it over as a garden plant.

In other parts of the vegetable garden, the vegetables themselves can control weeds and protect the soil. Squash grown among corn serves both ends. Volunteer tomatoes do, too. Even the tomatoes you set out, if you refrain from staking them and let them spread, will discourage weeds. Of course tomatoes like a mulch, and some clean hay under the plant will keep the fruit clean and unattacked by pests, especially slugs who can't stand the prickliness of hay on their soft bodies.

Natural enemies of weeds—insects and pathogens—have been used to control leafy spurge, curly dock, and a musk thistle.


**WEEDY LAWNS.** Trouble here may indicate low soil fertility, poor soil structure, low clipping, or a seed mixture high in weed seed.

Feeding programs that furnish lawn grasses with necessary organic plant food elements throughout the growing season tend to discourage weeds by enabling the grass to compete more successfully. Fertilize cool season grasses in the fall and spring and as needed during the summer months. Withhold spring fertilization of warm-season zoysia, buffalo, and

Bermuda until May 1; stop fertilizing them about August 31.

Mowing is an effective method of keeping weeds in check because it allows the grass to crowd out the weeds. The best mowing height is usually from two to three inches. Although a lower mowing clearance will cut back the weeds decisively, it will also hamper the grass from growing. However, if you have an area of a lawn that is almost entirely weeds, you might want to take advantage of the physical makeup of some perennial weeds. Normally they store a reserve food supply in their underground parts; when the plants begin to blossom, their reserve food supply has been nearly exhausted and new seeds have not yet been produced, that's the time to lower your mower and cut back the weeds to severely weaken them.

In most lawns, however, a higher mower setting is needed. Mowing at a height of 2½ to 3 inches shades the soil and protects the bluegrass roots from the damaging effects of summer heat. High mowing of common Kentucky bluegrass is an excellent deterrent to the germination and growth of many annual weed species. Some of the newer lawn grasses perform best when mowed at 2 inches or lower. Remember that the mowing technique you use on your lawn is one of the most critical maintenance procedures. Don't be misled by thinking that if you cut your grass tall that you will have to mow it more often. Regardless of the mowing height, the rule of thumb is to cut the lawn frequently enough so that you do not have to remove more than one-third of the green leaf area of the plant at each mowing. (This means low-cut turf will have to be mowed more often.) Increasing the mowing height from 1½ to 2½ inches will greatly decrease the number of broadleaved weeds.

Undercut and cut around small patches of undesirable grass with a sharp spade. Lift the patch and use it as a pattern to cut a replacement piece of the same thickness from an inconspicuous place elsewhere in the lawn. Make certain the replacement sod is firmly tamped into place, and water it well until it becomes established.

Most store-bought grass seed contains a certain percentage of weed seed; check the label. Most often cheap seed turns out to be the most expensive you can buy because it is high in noxious weeds.

## WEEDS AND BIOLOGICAL CONTROL.

The balance between plant eaters and beneficial insects can be maintained with the help of weeds. Adults of predatory and parasitic species often sustain themselves on nectar and pollen of flowering weeds. By allowing wild plants to grow between corn rows, Florida gardeners and farmers can reduce the number of fall army-worms. Johnsongrass and sudangrass have been found to discourage the Willamette mite in California vineyards. The grape leafhopper is kept in check by an egg parasite that thrives in borders of wild blackberry. Soybeans have been protected from the velvetbean caterpillar in Georgian fields by growing a cover crop of sicklepod.

You can encourage beneficials to move to your crops by mowing nearby weeds when the predators and parasites are flourishing.

# WHITEFLY

The various species of whitefly [color] are among the most universal pests of both green-house and garden. The adults are about ¹⁄₁₆ inch

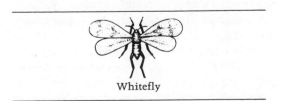

Whitefly

long and fly on white powdery wings. They feed and lay their eggs on the undersides of young leaves. Eggs are pale yellow at first, turn gray in five days to a week, and then hatch into tiny white "crawlers," or nymphs, that move around on the leaves for a few days before settling down in one place to feed. The nymphs develop fully

*Encarsia formosa*

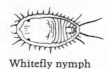

Whitefly nymph

in two weeks at normal greenhouse temperatures. The pupae are slightly larger and thicker than the nymphs. Adults appear in about ten days, and the entire life cycle takes about a month.

In cooler regions the insects can survive winter only indoors or in greenhouses. In spring the adults leave through doors and ventilators or on vegetable transplants, and are soon established on garden or field crops and weeds. The critical time in the annual cycle is fall, when whiteflies enter the greenhouse. They may fly in, or be taken inside on ornamental plants such as geraniums or chrysanthemums. A few weeds left in a supposedly empty greenhouse may support the whiteflies through the winter. Adults cannot survive for more than a week without food plants.

## BIOLOGICAL CONTROL.

An effective biological control for homes, gardens, and greenhouses is a small parasite, *Encarsia formosa*. It occurs naturally in Canada, the United States, and England.

The adult parasite is about 1/40 inch long, and all (except one or two in 1000) are females who produce without mating. The female searches for whitefly nymphs and pupae on the

leaves, and lays an egg in each whitefly nymph. The egg hatches into a larva inside the whitefly, causing the pest to turn black and appear like specks of black pepper on the leaf undersurface. Inside each black speck is a parasite adult which emerges by cutting a round hole in the top of the whitefly. Inside the home or greenhouse, it takes about 20 days for the parasite to complete its life cycle. Temperatures above 70°F. are best.

It's rare to get 100 percent control—usually about 80 to 90 percent is more like it—but the surviving whiteflies are not numerous enough to harm your plants. Actually, you need some whiteflies to keep the parasites alive, since they eat nothing else. They are absolutely harmless to plants, and do not bite, sting, or otherwise bother people. In fact, you have to look closely even to see them.

*Encarsia* is not greatly attracted to yellow sticky traps, so long as there are plenty of whiteflies about to be preyed upon. To ensure that you'll have whiteflies on hand to sustain the parasite, keep temperatures below 70°F. and reduce the number of traps. You can help *Encarsia* to overwinter. In fall, collect leaves from greenhouse tomato plants bearing pupae of both pest and parasite, and place the leaves near a rose geranium plant. This plant will slow the activity of both insects, and you should have some of the tiny wasps to release in spring.

Another species of the pest, the citrus whitefly, is vulnerable to fungal diseases. You can take advantage of one naturally occurring disease, known as red aschersonia, by making an inoculative spray. The first step is to find infected whiteflies. They become swollen and secrete more than the normal amount of honey-dew. After the pest dies, fungal strands extend through the body to make a fringe and red spore-bearing pustules develop. To cultivate the disease, sterilized slices of sweet potato are inoculated with the fungus and placed in pint jars for 30 to 40 days. Water is added to the jar, the jar is shaken, and the contents are filtered and sprayed. In moist periods of June and July, just one pint of the preparation proved sufficient for controlling citrus whiteflies in an acre of orchard. A second fungus, known as yellow aschersonia, causes similar effects as the red variety, except for the color of the pustules. A spray can be prepared by the same method.

Whiteflies are prey for lacewings and lady-bugs. Populations of these beneficials can be supplemented with commercially available reinforcements.

**CONVENTIONAL CONTROLS.** Modest numbers of whiteflies can be controlled with sticky yellow traps. These are sold to manage or monitor many pests, and you can make them by placing a commercial goo on bright yellow cards. Once traps are placed throughout the green-house, at plant canopy height, you can shake the plants to stir the whiteflies so that they'll notice the yellow lures.

Whiteflies are put off by anything with nicotine in it, and tobacco dust and tobacco tea have long been used to keep them from plants. The tea can be thickened with a bit of soap or white flour to make it stick better to the undersides of leaves. Sprays made from ryania should also do the trick. Oil sprays suffocate adults, nymphs, and eggs. Safer's insecticidal soap works against the pest, and has low toxicity to *Encarsia* if used as directed.

Unusually vulnerable plants may be suf-fering from deficiencies of magnesium and phosphorus.

# WILLOW

## Insects

The **leopard moth** lays its eggs in bark openings, and the larvae bore into the twigs and limbs of the tree, causing them to wilt. The borers are yellow or pinkish with brown or black hairs. When infestation is heavy, many of the limbs wilt, hang down, and even die. Sawdust may protrude from the holes in the bark.

Control procedures involve cutting and pruning all infested branches. If this is done at the first sign of trouble, the larvae may be caught inside the pruned branches. On lightly infested or large branches, probe into the holes with a wire to kill any borers that may be present. Heavily infested trees may have to be destroyed. Keep trees in such a vigorous state that the borers will have to go elsewhere.

The **willowshoot sawfly** is a ½-inch-long, wasplike insect that appears in early spring to lay eggs in the new shoots of willow, girdling the stem below the eggs in the process. The larvae bore in the shoots, sometimes tunneling them for two feet. They reach maturity in November. Cut and burn the wilted shoots.

Many of the pests of poplar also trouble willows; see POPLAR. Other pests include APHID, gypsy moth (see TREES), and oystershell scale (LILAC).

## Diseases

Two types of fungus cause a **blight** that blackens and shrivels willow leaves and pro-duces black lesions on branches. Repeated infections will cause the tree to die. Control the disease by pruning and burning infected wood in early spring.

The small black sunken **cankers** that appear on twigs, branches, and trunk may kill the tree within a few years. No control is known, but vigorous trees apparently resist this infection better than those weakened by blight (see above) or winter injury (below).

**Crown gall** is caused by a bacterium that invades roots, trunk, or branches of the willow, stimulating cell growth to cause tumorlike swellings that grow to the size of a pea or larger. Galls on the trunk sometimes grow to one or two feet in diameter. Growth of the tree is sometimes retarded. Leaves turn yellow and branches or roots may die.

Infected young trees and nursery stock should be removed and burned. Do not replant the area with willow, poplar, chestnut, syca-more, maple, walnut, or fruit trees. Older trees may survive an attack of this disease without being greatly injured. Make every effort to avoid wounding the stems and roots of healthy trees, since infection occurs through such openings in the bark.

Various stages of **rust** appear as orange and red blisters on the undersides of leaves. The alternate hosts of these different rusts are balsam fir, larch, currants, and gooseberries. Control is usually unnecessary, but in nurseries the rust-infected parts of the trees should be promptly pruned and destroyed.

**Tar spot** is a fungal infection that causes thick black raised spots to appear on leaves. Raking and burning fallen leaves gives suffi-cient control, as the fungus overwinters on them.

**Winter injury** describes the effects of alter-nate freezing and thawing of the bark on the south or southwest side of the tree during late winter. Symptoms include long vertical cracks, which may later be invaded by disease organisms.

Lean a wide board against the tree on the exposed side to prevent the sun's rays from warming the bark; heat on the bark is a danger when contrasted with freezing temperatures at night. Painting the trunk white will accomplish the same result.

# WIREWORM

Wireworms [color] are the larvae of a family of beetles commonly called click beetles or skip-jacks. The larvae are slender, jointed, unusually hard-shelled worms. They range in color from light to dark brown, and grow up to a length of

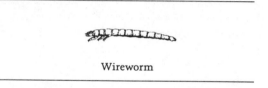

Wireworm

1½ inches. These chewing insects feed entirely underground, attacking germinating seeds and the roots, underground stems, and tubers of growing plants. Potatoes, beets, beans, cabbage, carrots, corn, lettuce, onions, and turnips are among the crops subject to injury. Damage is most likely to occur on poorly drained soil and on land that has recently been grass sod. Because they operate out of sight, wireworms are not suspected of much of the trouble they cause.

Millipedes are often confused with this pest, but they have many pairs of slender legs located on the segments from front to back, while wireworms have but three pairs of legs posi-tioned well forward. Millipedes, or thousand leggers as they are often called, characterist-ically curl up into a loose spiral position when disturbed; wireworms do not.

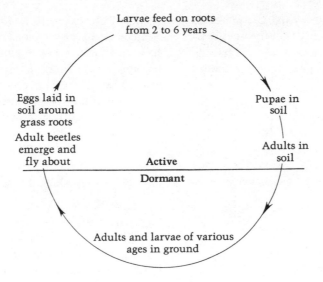

Life history of the wireworm

Reprinted, by permission, from Pyenson, *Fundamentals of Entomology and Plant Pathology*
(fig. 3.27), AVI Publishing Co., PO Box 831, Westport, CT 06881.

The wheat wireworm is somewhat typical of the several important species. It takes usually three years to complete the cycle from egg to adult. Eggs are laid in June or July and hatch into tiny worms that eat their way through two growing seasons. In August of the third summer, they enter their pupal cells and transform into beetles, which emerge from the soil the following year. Click beetles flip up into the air with a clicking sound when placed on their backs. They vary in color from light tan to dark brown, and are incapable of sustained flight.

If sod land must be used for the garden, thoroughly plow or stir it once a week for four to six weeks in the preceding fall. This aerates the soil, exposes the insects to weather and natural enemies, and crushes many others. Hardpan can be broken up by trenching—that is, digging a deep trench and throwing the soil over the surface. The trench is then filled by digging a second, parallel trench.

On a large scale, it is best to select crops known not to be favored by wireworms. This is especially important when the crops chosen are cultivated annuals, because cultivation of the soil at regular intervals makes conditions unfavorable to the egg-laying adults. Just as in the garden, cultivation exposes all stages of the pest to weather and enemies. Thorough cultivation is effective in controlling most species of wireworm, although fields hosting witchgrass are likely to continue to support wireworm infestations. It is best to cultivate when temperatures are sufficiently low to slow up the exposed wireworms, as they are then easy prey to natural enemies.

You can keep wireworms to a minimum by growing annual green manure crops. Clover is a good choice, but hay mixtures including timothy or red top will support a pest population. Oats, wheat, barley, and rye are seldom seriously injured by wireworms. In the Northwest, farmers may avoid damage by growing alfalfa for at least three years, then one year of potatoes, and finally one or two years of a suitable vegetable crop such as corn, beans, peas, or sugar beets.

Potatoes are attractive to wireworms and can be used as a simple trap. Cut a potato in half, cut out the eyes to prevent it from growing, and run a stick through the middle. Bury the spud about one inch under so that the stick stands vertically as a handle, and pull the traps out after a day or two. Some potatoes have yielded as many as 15 to 20 pests.

# WOODCHUCK

See GROUNDHOG.

# YELLOW POPLAR

See POPLAR.

# YEW

## Insects

The **black vine weevil** (or taxus weevil) feeds on a variety of plants, but does the most damage to yew. The adult weevils feed on the foliage and buds of the plant during the night, hiding by day either in dense foliage or in debris on the ground beneath. Injury by the adult is negligible; the larvae cause the chief damage by feeding on the roots. Leaves of root-damaged plants turn yellow and eventually wither. Severely injured plants become stunted or die. The adult is an oblong brownish black insect measuring 1/3 inch long. It does not fly, but walks from place to place. The weevil deposits its tiny white eggs in the soil during summer

months. The larva is white with a yellowish head, and is approximately 1/2 inch long when fully developed. It overwinters in the soil and pupates the following spring, after a short period of feeding.

The adults emerge from the pupae in June, and it is best to anticipate them by placing bands of something sticky, such as the commercially available Tanglefoot or Stikem, around the trunks. The weevils will be trapped as they try to walk to the foliage above.

Several sucking insects, including **mealybugs** and **scales,** occasionally attack the yew. See MEALYBUG. Scale insects are brown and inconspicuous; they attach themselves to the foliage and remain there the rest of their lives. Control them with a dormant-oil spray in late winter or early spring.

## Diseases

Most diseases of yew trees stem from unfavorable environmental factors.

Too much water around yew roots may cause **wet feet.** Affected plants grow poorly or even die. Such yews are most often found in poorly drained soils, at corners of buildings near rainspouts, or in areas that have been overwatered. When these conditions are corrected, the plant will regain its healthy condition.

A general **yellowing** of the plant may be the result of placing yews with acid-loving plants such as laurel, rhododendron, or azalea. Yews thrive in a sweet or neutral soil with a pH of 6 to 6.5, whereas health-type plants grow better in an acid soil (pH of 4.5 to 5.5), and it is best to keep these two types of plants separate. If this is impossible, ground limestone may be added around the base of yews; apply three pounds of limestone per 100 square feet of soil area around the yews, and repeat every third year.

# ZINNIA

## Insects

If zinnias wilt without an apparent cause, look to the **stalk borer.** The young borer is creamy white with a dark purple band around its body and several brown or purple stripes running the length of its body. The full-grown larva is creamy white to light purple and has no stripes. When young borers hatch in the spring from eggs on grass and weeds, they first feed on the leaves of the nearest plants and then gradually work their way to larger, stemmed plants.

Remove and destroy nearby weeds, leaving a clean area around the zinnia bed. Infested plants can sometimes be saved by slitting stems, destroying the borer within, then binding the stems and keeping plants watered. If it is necessary to pull a badly damaged plant, pinch the stem to kill the offending borer. If affected stems are cut off below the boring, the plant may develop side shoots that will later flower.

Other bugs to look out for are APHID, black blister beetle (see ASTER), cyclamen mite (AZALEA, DELPHINIUM), European corn borer (CORN), flea beetle (POTATO), flower thrips (PEONY, ROSE), fourlined plant bug (CHRYSANTHEMUM), JAPANESE BEETLE, MEALYBUG, redbanded leafroller (ROSE), SPIDER MITE, spotted cucumber beetle (CUCUMBER), and WHITEFLY.

## Diseases

**Bacterial leaf spot** is known by brown spots of irregular shape, surrounded by a yellow border. Do not sprinkle water on foliage. Seed can be soaked for two minutes in a solution of one part household bleach to nine parts water; do this just before sowing the seed.

**Fungal leaf spot** is characterized by irregular black spots on the foliage. It seldom causes much damage, but is sometimes a problem following rainy periods. Zinnias grown in confined areas such as cold frames, greenhouses, and cloth enclosures are more commonly affected than those grown in open fields or gardens. If you have had trouble with leaf spot, do not grow zinnias in the same soil year after year. Remove any infected flower heads promptly.

**Mildew** is generally the most serious obstacle to zinnia growers. The first noticeable symptom is a powdery white growth on leaves. Under favorable weather conditions, the mildew grows rapidly and may soon cover all leaves, stems, and buds. The plants often live until frost, producing dwarfed, distorted leaves and flowers. The disease is most severe in late summer during cool, damp, cloudy weather, and plants grown in shady and confined areas are most susceptible.

# ZUCCHINI

See SQUASH.

---

## Sources

As more and more gardeners and orchardists abandon chemical pesticides, an increasing number of plant protection products have hit the market: botanical powders and spray bases, insect predators and parasites, insect diseases, nets, exploding scares, and so on. Your local garden supply center may stock some, and many of those products you can't find are available through the mail. A handy list of suppliers is available from Readers Service, *Organic Gardening*, 33 E. Minor Street, Emmaus, PA 18049.

# Appendix

## INSECTICIDAL PLANTS

Of the many plants used locally against insect pests, either as poisons or repellents, relatively few have been studied closely. Interest in doing so waned with the boom of chemical insecticides. But as the limitations and dangers of petrochemicals become known, botanical sources are getting a second look.

| | |
|---|---|
| American plum *(Prunus americana)* | Leaves and flowers toxic to insects. |
| American wisteria *(Wisteria frutescens)* | Acetone extract of seeds somewhat toxic to codling moth larvae. |
| Balsamroot *(Balsomorhiza sagittata)* | Powder of stems and leaves somewhat toxic to pea aphids. |
| Bear hops *(Humulus lupulus)* | Powdered leaves toxic to southern armyworms and melonworms. |
| Black Indian hemp *(Apocynum cannabinum)* | Extract of twigs and stems controls codling moth larvae. |
| Black walnut *(Juglans nigra)* | Odor of leaves repellent to insects. |
| Buffalo gourd *(Cucurbita foetidissima)* | Powdered root somewhat toxic to cucumber beetles; mix a little water with powder for hand soap. |
| California buckeye *(Aesculus californica)* | Flours made with meat hulls of nuts toxic to larvae and adults of Mexican bean beetle; parts of plant toxic to humans. |
| Canadian fleabane *(Erigeron canadensis)* | Can be ground up to make melonworm repellent. |
| Chinaberry *(Melia azedarach)* | Shade tree, repellent to grasshoppers and locusts. A repellent tea can be made with live or dried leaves. Powdered fruit slightly toxic to European corn borer larvae. |
| Chinese wingnut *(Pterocarya stenoptera)* | Powdered leaves fairly toxic to Mexican bean beetle larvae. An ornamental tree. |
| Chinese wisteria *(Wisteria sinensis)* | Acetone extract of seeds somewhat toxic to codling moth larvae. |
| Common oleander *(Nerium oleander)* | Effective against codling moth. |
| Cucumber, cantaloupe, and pumpkin | Acetone extract of seeds (and aqueous extract of pumpkin seeds) toxic to mosquito larvae and might be lethal to other pests. |
| Dwarf or red buckeye *(Aesculus pavia)* | Flowers attract and kill Japanese beetles. |
| False indigo *(Amorpha fruticosa)* | Acetone extract of flowers repellent to chinch bugs and striped cucumber beetles. Powdered mature pods with seeds moderately toxic to Mexican bean beetle larvae. A sugar derivative, amorpha, effective as dust against chinch bugs, cotton aphids, squash bugs, tarnished plant bugs, potato leafhoppers, blister beetles, and spotted cucumber beetles. The fruit is more insecticidal than roots. |
| Fishberry *(Anamirta cocculus)* | Berry used as insecticide. |
| Japaca, yellow oleander *(Thevetia peruviana)* | All parts, except leaves and fruit pulp, used to make cold-water extraction effective against a number of pests, especially aphids. |
| Larkspur *(Delphinium* sp.) | Powdered roots toxic to bean leafrollers, cross-striped cabbage-worms, cabbage loopers and melonworms. |

| | |
|---|---|
| Manroot, wild cucumber *(Echinocystis fabacea)* | Powdered root toxic to European corn borer larvae. |
| Mescal or coral bean *(Sophora secundiflora)* | Powdered seeds of this flowering shrub toxic to armyworms. |
| Nutmeg *(Myristica fragrans)* | Oil has toxic properties. |
| Osage orange *(Maclura pomifera)* | Roots, wood, and bark repel insects. |
| Pawpaw *(Asimina triloba)* | Powdered aerial portion has some effect on mealyworms. |
| Peach *(Prunus persica)* | Leaves and flowers toxic to insects. |
| Pine (various species) | Pine tar oil improved performance of standard codling moth baits. |
| Prairie zinnia *(Zinnia grandiflora)* | Slightly toxic to celery leaftiers. |
| Rayless chamomile *(Matricaria matricarioides)* | Powdered heads fairly toxic to diamondback moths. |
| Scentless false chamomile, mayweed *(Matricaria indora* or *M. chamimile)* | Flower heads as effective as commercial pyrethrum in controlling face flies; may well have applications in the garden. |
| Sesame *(Sesamum indicum)* | An effective synergist for pyrethrins. |
| Soap plant, soap root *(Chlorogalum pomeridianum)* | Grows from California north to Oregon. Powdered bulbs toxic to armyworms and melonworms. |
| Sour sop *(Annona muricata)* | Powdered seeds toxic to armyworms and pea aphids. |
| Spanish dagger *(Yucca shidigera)* | Powdered leaves toxic to melonworms, bean leafrollers, and celery leaftiers. |
| Spindle tree *(Euonymus europaeus)* | Fruit has paralyzing action on aphids. |
| Sugar apple | Seeds and roots extracted with ether, converted to resinous substance to make contact poison for aphids. Toxic and repellent to diamondback moth larvae. Hot-pressed and heat-extracted oils of seeds highly toxic contact poisonous to several pests. |
| Sweet flag, calamus *(Acorus calamus)* | Alkaloid root works as contact poison to insects, even though it is edible to humans. Grows commonly in swamps and along brooks. |
| Thundergod vine *(Tripterygium wilfordii)* | Used as garden insecticide in China. Powdered bark is mixed with liquid soap. |
| Tung-oil tree *(Aleurites fordii)* | Tung-oil soap somewhat toxic to sugarcane woolly aphids. Oil sprayed to control boll weevils. |
| Turkey mullein *(Eremocarpus setigerus)* | Used by American Indians as fish poison; toxic to cross-striped cabbageworms. |
| Wood fern, shield fern *(Dryopteris felix-mas)* | Powdered rhizome toxic to armyworms. |
| Wormseed, Jerusalem tea *(Chenopodium ambrosioides)* | Some parts toxic as extracts or dusts on several species of leaf-eating larvae. |
| Yellow azalea *(Rhododendron molle)* | Dried and pulverized flowers work as contact and stomach poison. Powdered flowers can be sprayed to control certain species of lepidopterous larvae. Roots and leaves not insecticidal. |

# INDEX

The word *color* in brackets following an entry indicates that the item is illustrated in the color plate section of the book. Bold-faced page numbers indicate the main pages where the insect or disease is discussed.

## A

Acerola, **5**
Agri-Strep, for fire blight on pears, 302
Agrocin, 84, 173
Air pollution, **5**
  trees and, 433
Alfalfa
  BT protects, 35
  resistant varieties of, 346
Alfalfa aphid
  control of, 8
  resistant varieties for, 346
Alfalfa caterpillar, BT controls, 35
Algae, lawns and, 211
Allelopathy, **5**
Allium
  aphids and, 6
  rabbits and, 342
Almond, BT protects, 35
Alternaria fruit rot, cherries and, 91
Alternaria leaf spot, carnations and, 78–79
Aluminum foil
  to control mosaic virus, 124
  as a mulch, 6, 147
*Amblyseius californicus*, 103, 173, 376
Ambrosia beetle
  mango and, 222
  pitted, 356
  rhododendron and, 356
Ambush bugs, [*color*], 63–64
Ammonia, as a bait for cherry fruit flies, 90
Amylovora, quince and, 333
Anise
  aphids and, 6
  as a repellent plant, 341
Anthracnose, [*color*]
  ash, 25
  avocado and, 32
  beans and, 44
  blackberries and, 56
  cabbage and, 73
  citrus fruit and, 105–6
  cucumber and, 123
  currants and, 126
  mango and, 222

  mint and, 230
  orchids and, 283
  peppers and, 307
  privet and, 331
  raspberries and, 336
  rhubarb and, 359
  sycamore and, 394
  tomatoes and, 409–10
  trees and, 428
  turnip, 73
  watermelon and, 441
Anthracnose spot, citrus fruit and, 106
Anthracnose stain, citrus fruit and, 106
Antibiosis resistance, 345
Antlion, 3
Ants
  aphids and, 6–7
  carpenter, 360
  citrus fruit and, 102
  control of, 7
  hills of, 202
  peonies and, 305
  repellent plants for, 343
  roses and, 360
*Apanteles ornigis*, 58
*Aphelinus mali*, 7, 8
Aphid mummies, [*color*], 8
Aphids, [*color*], 3, **5–8**
  alfalfa, 8, 346
  allium and, 6
  anise and, 6
  ants and, 6–7
  apples and, 9
  apricots and, 21
  asparagus, 26
  asters and, 28
  beans and, 38, 234, 399
  begonia and, 50
  birch and, 52
  cabbage, [*color*], 66
  campanula and, 76
  carnations and, 78
  carrots and, 80
  celery and, 85
  chard and, 88